Recent Developments in Elastomers: Science, Characterization and Implementation

Recent Developments in Elastomers: Science, Characterization and Implementation

Edited by **Stephen Louis**

NY RESEARCH
PRESS

New York

Published by NY Research Press,
23 West, 55th Street, Suite 816,
New York, NY 10019, USA
www.nyresearchpress.com

Recent Developments in Elastomers:
Science, Characterization and Implementation
Edited by Stephen Louis

International Standard Book Number: 978-1-63238-388-4 (Hardback)

Printed in China.

Contents

Preface

Every book is a source of knowledge and this one is no exception. The idea that led to the conceptualization of this book was the fact that the world is advancing rapidly; which makes it crucial to document the progress in every field. I am aware that a lot of data is already available, yet, there is a lot more to learn. Hence, I accepted the responsibility of editing this book and contributing my knowledge to the community.

An elastomer is a natural or synthetic polymer having elastic properties. This book offers a thorough insight of recent advances in the field of elastomers and their composites through contributions by internationally recognized specialists. The book describes latest synthesis, modelling and experimental methods in elastomers, and also discusses new means of crosslinking, liquid crystal elastomers, nanocomposites, smart elastomers, elastomers in microelectronics and microfluidics, elastomers in cement concrete and mortar, experimental testing and modelling. It also represents how enhancements in materials, processes and characterization techniques can boost performance in the area of engineering. It offers a distinct opportunity to find the latest research on elastomer advances from laboratories around the world. This book accosts to industrial and academic researchers in the fields of physical, chemical, biological sciences and engineering.

While editing this book, I had multiple visions for it. Then I finally narrowed down to make every chapter a sole standing text explaining a particular topic, so that they can be used independently. However, the umbrella subject sinews them into a common theme. This makes the book a unique platform of knowledge.

I would like to give the major credit of this book to the experts from every corner of the world, who took the time to share their expertise with us. Also, I owe the completion of this book to the never-ending support of my family, who supported me throughout the project.

Editor

Section 1

New Approach to Elastomer Crosslinking

Chapter 1

Aspects Regarding Radiation Crosslinking of Elastomers

Elena Manaila, Maria Daniela Stelescu and Gabriela Craciun

Additional information is available at the end of the chapter

1. Introduction

Elastomers or rubbers are materials used in a large range of industrial and household applications. A common physical-chemical treatment is curing (crosslinking), imparting the rubber mechanical and thermal stability. Elastomers show low thermal conductivity values, and therefore, require complex and high cost heating methods; thus, the ionizing (accelerated electrons) method shows high interest for the grafting and crosslinking processes. In addition to the lack of environmental impact, reliability, flexibility and low costs render the radiation technologies especially attractive.

Vulcanisation is the process by which rubber is changed from essentially a plastic material to either an elastic or a hard material (Stelescu et al., 2010). In this process, an elastomer is transformed from a 'plastic', 'formable' material into an 'elastic' material by the formation of a three-dimensional network with different types of junctions. The word vulcanisation derives from *Vulcan*, the Roman God of fire. Not accidentally, it also means volcano, a hot place where quite some sulphur species can be found. The term vulcanisation was therefore originally exclusively applied to the crosslinking reaction achieved by sulphur at high temperatures.

Nowadays this term is also applied to refer to other crosslinking processes, such as peroxide cure. There are several possibilities for the crosslinking of rubber. The already mentioned sulphur vulcanisation was the first to be discovered and still is today's most common cure system. The sulphur vulcanisation process requires the presence of carbon-carbon unsaturation in the polymer and it leads to a three-dimensional rubber network in which the polymer chains are linked to each other by sulphur bridges. As a result, sulphur cured articles have good tensile and tear strength, good dynamic properties, but poor high temperature properties like ageing, for instance (Alvarez Grima, 2007; Dluzneski, 2001).

Other vulcanisation systems, i.e. peroxides, ultraviolet light, electron beam, microwave, resins, etc. were later discovered and gained more importance with the progressive development of synthetic rubbers.

The use of organic peroxides as crosslinking agents for rubber was first reported by Ostromislenski in 1915 (Alvarez Grima, 2007) and at present it is the only vulcanization method that can compete with accelerated sulphur cure, with respect to vulcanization rate. Interest in the industrial use of peroxides as curing agents increased with the introduction of a number of fully saturated elastomers, such as ethylene-propylene rubber (EPM), fluoro elastomers (FKM), etc., which cannot be cured by sulphur vulcanization (Alvarez Grima, 2007). Peroxide vulcanisation leads to a rubber network in which the polymer chains are linked to each other by very stable covalent carbon-carbon bonds. Peroxide cured vulcanisates have therefore good high temperature properties, like heat ageing and compression set, compared to sulphur cured articles (Alvarez Grima, 2007; Dluzneski, 2001)

In addition, peroxide cure allows vulcanisation of both unsaturated and saturated polymers, which is not possible with sulphur vulcanisation. On the other hand, some mechanical properties of peroxide cured articles are inferior to those achieved by sulphur cure, i.e. tensile strength, dynamic properties, and therefore entail a limitation on the use of this vulcanization system. Another serious problem with peroxides is the lack of sufficient scorch time. Scorch time: the time elapsed until vulcanisation starts, is of major importance in order to control the vulcanisation reaction. The term scorch safety refers to a certain scorch time which is enough to provide good processing of the material before it starts vulcanising; this is of great importance especially in industry. Upon reaching the typical dissociation temperature of a peroxide, the crosslinking reaction immediately proceeds at full speed, leaving the processor little or no time for shaping of the rubber article. The addition of certain scorch-retarders is sometimes of help, but this usually goes at the expense of crosslink density obtained: a waste of money (Alvarez Grima, 2007).

2. Electron beam and microwaves vulcanization

Radiation curing has historically been used as an alternative to peroxides in applications where the curatives themselves or sideproducts of vulcanization are viewed as impurities in the final product. Peroxide cure progresses through a series of radical intermediates, each of which can undergo side reactions which may not necessarily contribute to crosslink density. Radiation cure, on the other hand, has been promoted as a cleaner and more homogeneous cure process. Electron beam irradiation has been used in the wire and cable industry for longer than 30 years and applied to a wide range of commodity and specialty elastomers. A survey of the types of elastomers susceptible to radiation curing is available, as are review articles describing the electron-beam curing of commercially significant grades (Bhowmick & Vijayabaskar, 2006; Henning, 2008). Variables such as radiation dosage and the effect of polymer microstructure and chemical additives on the efficiency of electron beam cure have been studied. Some researchers (Zaharescu et al., 2000; Chowdhury & Banerji, 2005) studied radiation-induced crosslinking in thermoplastic elastomers based on ethylene-propylene rubber (EPDM) and polyethylene (PE) or polypropylene (PP) plastics.

Elastomer crosslinking by means of electron beam (EB) is done without heating and in the absence of vulcanization agents. The reaction mechanism is similar to crosslinking with peroxides, but in this case, reaction initiation is due to the action of EB and in the presence of the polyfunctional monomers. Ionizing radiation produces an excitation of polymer molecules. The energies associated with the excitation are dependent on the irradiation dosage of electrons. The interaction results in formation of free radicals formed by dissociation of molecules in the excited state or by interaction of molecular ions. The free radicals or molecular ions can react by connecting the polymer chains directly or initiating grafting reactions.

EB vulcanization has demonstrated extremely positive results compared to the conventional curing system such as: no polymer degradation due to high temperature as EB crosslinking occurs at room temperature, no oxidative degeneration in polymers as observed in classical crosslinking, direct crosslinking by C-C linkage by EB, extremely strong bonds, high degree of crosslinking, extremely short curing cycles, zero blooming effects; extremely high tensile strength; extremely high resistance to compression set; extremely high resistance to oils, grease, lubricants; highly improved accelerated ageing properties, very high productivity, perfect for thin products, lower material waste (MGM Rubber Company-Research and Development, 2007). However, the radiation crosslinking of rubbers was not used in larger technical applications because of the high cost of irradiation to bring about vulcanization, but could become an industrial process when the radiation dose decreased with the use of some sensitizers.

Modification of thermoplastic and rubbery materials by EB and microwave (MW) radiations is a potential method for the development of new materials like polymers and composites. Interaction of each of the two physical systems, EB and MW, with a substance has revealed their ability for changing physical and chemical proprieties of the treated substance.

Therefore, the physical and chemical effects of the EB and MW radiation have drawn high interest, resulting in a large industrial application range of the new materials with improved characteristics. A comparative study of the effects of separate interaction of EB and MW with the substance, effects on which the modern processes for the production of new materials with special characteristics have been developed, has revealed the following: (a) production of materials by EB (ionizing) radiation is based on the coulomb interaction of accelerated electrons with the atoms of the substances they are penetrating. From this interaction, secondary electrons, excited molecules, free ions and radicals result, which are precursors of some chemical reactions resulting in new materials; (b) production of materials by MW (non-ionizing) radiation is based on thermal effects (dielectric heating) of the interaction of the electromagnetic waves with the substance. MW energy absorbed within the substance is dependent on the molecule characteristics, that is on the permittivity and magnetic permeability resulting in a high selectivity level for the chemical reactions induced by MW interaction with the constituents of the material they are penetrating; (c) by separate/ combined EB and MW irradiation, materials with single characteristics are obtained with high yields, inaccessible by the traditional curing methods.

Radiation (EB-ionizing radiation and MWs-non-ionizing radiation) are used successfully instead of the traditional methods and, in addition, provides new possibilities like as crosslinking in thick polymer articles, in complex sections or production of recycled rubbers of high quality.

The high-molecular compounds (elastomers) are made up of a large number of elementary molecules – monomers. These molecules can be of the same class or of different classes, resulting thus homopolymers and copolymers, respectively. The radiation acts on the high-molecular compounds as follows: (1) at a relatively low radiation dose (radiation dose = the energy absorbed per unit of mass of the irradiated material; it is measured in J/Kg; unit of measure for the absorbed dose is a gray: 1Gy = 1 J/kg), only „pure polymerization" occurs. With the linear increase in the radiation dose, the linear polymer chains increase progressively by addition of a molecule from a blend with low molecular weight to a radical at the end of an increasing chain; (2) with the dose increase during the pure polymerization, a new process named „grafting" takes place. By this process side chains are made in the polymer, resulting in modified surface properties in some materials, like as: biocompatibility, moisture-absorbent and moisture-repellent, and mechanical, chemical and thermal characteristics. This process is very promising for the biomaterial synthesis. (3) with the further increase in radiation dose, the „crosslinking" process takes place, when links are formed between polymer chains resulting in two- and three-dimensional structures. During this process insignificant chemical changes occur but significant structural changes. Simultaneously degradation occurs. Crosslinking and degradation (through chain scission) are two competing processes that always co-exist under radiation. Both processes have many applications in the field of new materials with improved characteristics, like as elastomer crosslinking. (4) the combined effects of the two processes are used in many industry processes, being known as „radiation curing". Three radiation induced processes are used in „radiation curing": monomer polymerization, polymer chain crosslinking and chemical bonding between polymer and substrate molecules.

Use of ionizing and non-ionizing radiation to obtain materials with improved characteristics is based on their advantages like as: *(I) electrons beam (EB):* (I.1.) accelerate the polymerization/vulcanization process, thus reducing the preparation time with tens to hundreds times; (I.2.) increase the conversion of raw material into the finished material up to a 100% yield and, therefore, the percentage of the residual monomer is nearly 0%, which is very important with noxious monomers; (I.3.) reduce the energy required in crosslinking and grafting with 5 –10 times as compared with the traditional method; *(II) microwaves (MW):* (II.1.) promote a narrow distribution of the molecular masses, as the crosslinking and grafting occurs simultaneously in the bulk due to the rapid energy transfer, volumetric and selective properties of MW dielectric heating , (II.2.) promote fast crosslinking and grafting processes, by tens to hundreds times faster than in classical procedure because the electromagnetic energy is directly transferred from the MW to the atoms or molecules of the irradiated material

3. Polyfunctional monomers (co-agents)

Reported papers suggest that appropriate polyfunctional monomers (co-agents) in polymer matrix (Vijayabaskar & Bhowmick, 2005; Yasin et al., 2005) could be used to obtain desired rubber physical properties at lower irradiation doses (Hafezi et al., 2006; Stelescu et.al., 2011; Stelescu et al., 2012). Co-agents are multi-functional organic molecules which are highly reactive towards free radicals (Alvarez Grima, 2007). They are used as reactive additives to boost the vulcanization efficiency (Endstra, 1990). The most used coagents are molecules with maleimide groups, (meth)acrylate groups, or allylic groups, (Dikland et al., 1993) but polymeric materials with a high vinyl content, i.e. 1,2-polybutadiene, can also act as co-agents.

The co-agents can be divided into two groups: Type I and Type II co-agents.

Type I: Addition and hydrogen abstraction reactions: these co-agents consist of rather polar molecules with a low molecular weight and activated double bonds. Their main characteristic is that they are highly reactive towards radicals, so scorch takes place very fast, which sometimes can be a disadvantage. By using this kind of coagents not only the rate of cure is increased but also the crosslink density or state of cure. A disadvantage that may be present when using this type of co-agents is that, due to polarity, the compatibility of these co-agents with the polymer matrix is limited. Some examples of Type I co-agents are: acrylates, methacrylates, bismaleimides and zinc salts.

Type II: Addition reactions: these co-agents are, in general, less polar molecules, which form more stable free radicals, so scorch does not take place as fast as with the previous type of co-agents. The use of these co-agents leads to an increase in crosslink density of the vulcanisate but, unlike Type I, they are not capable of increasing the cure rate. Due to their low polarity, these co-agents have a good compatibility with many elastomers. Some examples are: high-vinyl 1,2-polybutadiene, divinylbenzene, allyl esters of cyanurates, isocyanurates and sulphur.

This review gives an overview about our research (Stelescu & Manaila, 2007; Zuga et al., 2007; Zuga, Miu et al., 2007; Manaila et al., 2008; Stelescu et al., 2010; Manaila, Martin, Stelescu et al., 2009; Stelescu et al., 2009; Manaila, Martin, Craciun et al., 2009; Stelescu et al., 2008; Manaila, Stelescu, Ighigeanu et al., 2011; Manaila, Stelescu et al., 2011) on elastomer crosslinking by irradiation with accelerated electrons, a much more ecologic method that does not need to add crosslinking agents into the blend.

In addition, the main advantages that the new technique brings are: (a) almost 100% efficiency in converting raw material into finished materials; (b) reduction from tens to hundreds of times of the production length; (c) achieving unique properties of materials which cannot be obtained by conventional methods; (d) perfect adaptability to the highest demands on the environment because no reaction by-products are released into the atmosphere; (e) adaptability to any degree of automation (allows technological lines with speeds of over 500 m/min) and strict control of technological processes; (f) in many cases, reduction of energy consumption by 20 to 30 times compared to "classic heating" or conventional processes; (g) the process is simple and can be controlled by only one single parameter, i.e. absorbed dose, the quantity that varies with the application; (h) radiation crosslinking is technically and economically feasible as a pure physical process, i.e. without the addition of sensitizers, there are no residues of alien substance needed for the chemical processes, or of their decomposition products.

At the interaction of ionizing radiation with (co)polymers, breaking of covalent bonds occurs, as well as the emergence of free radicals (transitional chemical species) on the main chain (if the lateral groups break) or in the main chain (if it breaks itself). The final effect is either crosslinking of macromolecular assembly or cutting the main chain of macromolecules and decreasing average molecular weight. In fact, the two effects, crosslinking and degradation, coexist and we need to point out the predominance of one of them.

In our study we used five polyfunctional monomers: TAC (triallylcyanurate) and TAIC (triallylisocyanurate) of type II, and TMPT (trimethylopropane trimethacrylate), EDMA (ethylene glycol dimethacrylate) and ZDA (zinc diacrylate) of type I.

Table 1 presents the chemical structure, type, functionality, and characteristics of polyfunctional monomers (co-agents) PFMs used.

PFMs	**Type**	Chemical structure	**Characteristics**
Triallylcyanurate Luvomaxx TAC DL 70 (TAC)	II	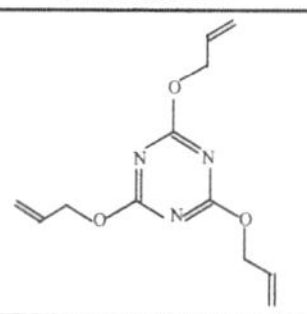	Melting point: 26–28 ^{0}C; Boiling point: 149–152 ^{0}C; Density: 1.34 g/cm^{3}; 26% percentage of ash, 30% active synthetic silica.
Triallylisocyanurate Luvomaxx TAIC DL 70C (TAIC)	II	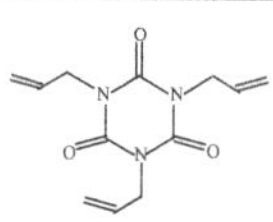	Melting point: 26–28 ^{0}C; Boiling point: 119–120 ^{0}C; Density: 1.34 g/cm^{3}; 30% active synthetic silica.
Trimethylolpropanetrimethacrylate Luvomaxx TMPT DL 75 (TMPT)	I	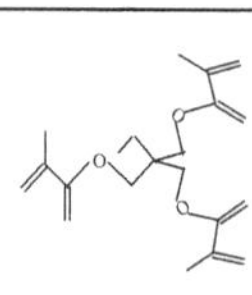	Melting point: -25 ^{0}C; Boiling point: >200 ^{0}C; Density: 1.36 g/cm^{3}; 22% percentage of ash, 75 ± 3% active ingredient.
Ethylene glycol dimethacrylate Luvomaxx EDMA DL 75 (EDMA)	I	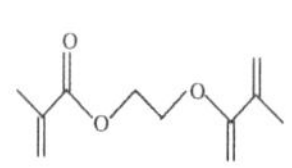	Melting point: -40 ^{0}C; Boiling point: 85 ^{0}C; Density: 1.25 g/cm^{3}; 23% percentage of ash, 75 ± 3% active ingredient.
Zinc-diacrylate ZDA GR 75 (ZDA)	I	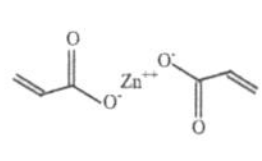	Melting point: 240-244 ^{0}C; Boiling point: 141 ^{0}C; Density: 1.23 g/cm^{3}; 75 ± 3% active ingredient.

Table 1. Characteristics of polyfunctional monomers (co-agents) PFMs used

4. Experimental installations, sample irradiation and laboratory tests

The experimental installation consists mainly of the following units: an accelerated electron beam source, a microwave source of 2.45 GHz and a multimode rectangular cavity used as reaction chamber. As accelerated electron beam source is used the electron accelerator ILU-6M and electron linear accelerator ALIN-10. Two experimental installations for the separate EB irradiation and combined irradiation with EB and MW are carried out: *experimental installation A* and *experimental installation B*.

Each experimental installations consists mainly of the following units: an electron beam source, a microwave source of 2.45 GHz, a multimode rectangular cavity in which are injected both, EB and MW. Experimental installation A uses as electron beam source the electron linear accelerator ALIN-10 of 6.23 MeV and 70 Gy s^{-1} maximum dose rate (built in the NILPR Institute, Bucharest, Romania). Experimental installation B uses as electron beam source the industrial electron accelerator ILU-6M of 1.8 MeV and 10.8 kW maximum output power, built in Russia, Institute of Nuclear Physics-Novosibirsk. It is placed at Electrical Project and Research Institute from Bucharest, Romania.

The ALIN-10 electron accelerator was built in Romania, National Institute for Lasers, Plasma and Radiation Physics, Electron Accelerator Laboratory-Bucharest, is of the travelling-wave type, driven by 2-MW peak power tunable EEV M5125-type magnetrons operating in S-band The optimum values of the EB peak current I_{EB} and EB energy E_{EB} to produce maximum output power P_{EB} for a fixed pulse duration τ_{EB} and repetition frequency f_{EB} are as follows: E_{EB} = 6.23 MeV; I_{EB} = 75 mA; P_{EB} = 164 W (f_{EB} = 100 Hz, τ_{EB} = 3.5 μs). The main characteristics of the ALIN-10 electron accelerator are presented in Table 2.

Characteristic	*Value*
EA energy	6.23 MeV
EA power in impulse	75 mA
EA impulse duration	τ= 3.5 μs
EA impulse repeat frequency	f_r = 50-100 Hz
Mean power	13 μA at 50 Hz
Maximum mean power	82 W

Table 2. The main characteristics of electron accelerator ALIN-10

The ILU-6M accelerated electron beam source is placed at Electrical Project and Research Institute from Bucharest, Romania. The ILU-6M is a resonator-type accelerator, operating at 115±5 MHz. This accelerator generates electron beam pulses of 0.375 ms duration, up to 0.32 A current peak intensity and up to 6 mA mean current intensity. The cross-sectional size of the scanned EB at the ILU-6M vacuum window exit is 1100 mm x 65 mm. The EB effects are related to the absorbed dose (D), expressed in Gray or J kg^{-1}. The single pass dose with conveyor under the ILU-6M scanner is adjustable from 12.5 kGy to 50 kGy. The main characteristics of the ILU-6M electron accelerator are presented in Table 3.

Characteristic	*Value*
EA energy	1.8 MeV
EA power in impulse	0–1 A
EA impulse duration	τ= 500 μs
EA impulse repeat frequency	2, 3, 5 10, 15, 25, 50 Hz
Mean power	0–6 mA
Maximum mean power	10.8 kW
Electron scattering method	Electromagnetic scanning device
Useful section of EA field at scanning device output	100 cm x 6.5 cm
Conveyor belt for samples to be irradiated with the following characteristics:	Moving speed, V_{belt}=1.56-12.8 cm/s; Dimensions: length = 125 mm; width = 290 mm; thickness = 160 mm; Distance between terminals: 1282 mm Distance from the scanning device output Window: H = 100–500 mm; Dimensions of sample holder: 500 mm x 300 mm (with possibility of extension up to 1500 mm x 600 mm)

Table 3. The main characteristics of cavity electron accelerator ILU-6M

The Experimental installation A uses the multimode rectangular cavity of a mechanical and electrical proper modified MW oven (MEM-MWO) of 2.45 GHz and 710 W output power. The conventional operation of the 2.45 GHz oven magnetron was properly modified in order to permit the use of an electronic regulator that ensures variable magnetron output power (Martin et al., 2001). Microwaves are generated as 10 ms pulses at 50 Hz repetition rate. For MW irradiation was used the SAR values which prevents during MW+EB irradiation the rise of the samples final temperature above 60-70°C.

The rubber samples were obtained from raw rubber mixtures, as compressed sheets of 2 mm in the polyethylene foils to minimize oxidation. For radiation treatments the sheets were cut in rectangular shape of two different sizes: *type 1* of 0.1 m x 0.03 m^2 and *type 2* of 0.15 x 0.15 m^2. The treatment was performed with layers of 10 sandwiched *type 1* sheets irradiated using ALIN-10 accelerator and three sandwiched *type* 2 sheets using the ILU-6M accelerator.

The EB effects are related to the absorbed dose (D) expressed in Gray or J kg^{-1} and absorbed dose rate (D*) expressed in Gy s^{-1} or J kg^{-1} s^{-1}. The MW effects are related to SAR (Specific Absorption Rate, expressed in W. kg^{-1}) which is equivalent to D* and SA (Specific Absorption, expressed in J. kg^{-1}) which is equivalent to D.

According to the Technical Report Series No. 277 (Andreo et al., 1997), the absorbed dose is the major parameter in the accelerated electron radiation. The vulcanizing and grafting process performances are provided by the severe control of this parameter. The relation defining the absorbed dose is:

$$D = d\varepsilon / dm \quad (1)$$

where:

dε is the mean energy given up by the ionizing radiation to the mass amounts **dm** of the substance interacting with this ionizing radiation.

dm is emphasized to be very low but not so low that the mean energy **dε** given up by the radiation would undergo a significant fluctuation.

Absorbed dose is measured in **J/kg**. The SI unit measure for the absorbed dose is the **gray** (Gy): **1 Gy** = 1 Joule/kg. The **rad** unit is also used, with the following relation between the **Gy** and **rad**: **1 Gy = 100 rad.** A relevant example: a material irradiated by 2 Mrad (20 kGy) means that AE (accelerated electron beam) have deposited 2.10^8 ergs or about 10^{19} eV per gram substance.

After irradiation the samples are analyzed by mechanical testing. Tensile strength and tear strength tests were carried out with a Schoppler strength tester with testing speed 460 mm/min, using dumb-bell shaped specimens according to ISO 37/1997, respectively angular test pieces (type II) in according to ISO 34-1/2000. Hardness was measured by using a hardener tester according to ISO 7619/2001. Elasticity was evaluated with a test machine of type Schob. The cure characteristics of the compounds were measured at 160°C using an oscillating disk rheometer (Monsanto), according to the SR ISO 3417/1997.

5. Electron beam and microwaves processing of elastomers/rubbers with polyfunctional monomers

5.1. Aspects regarding crosslinking of a natural rubber (NR) blend

Natural rubber is a heavily researched material. The outstanding strength of natural rubber has maintained its position as the preferred material in many engineering applications. It has a long fatigue life and high strength even without reinforcing fillers. Other than for thin sections it can be used to approximately 100 °C, and sometimes above. It can maintain flexibility down to -60 °C if compounded for the purpose. It has good creep and stress relaxation resistance and is low cost. Its chief disadvantage is its poor oil resistance and its lack of resistance to oxygen and ozone, although these latter disadvantages can be ameliorated by chemical protection. An natural rubber structure is illustrated in Figure 1.

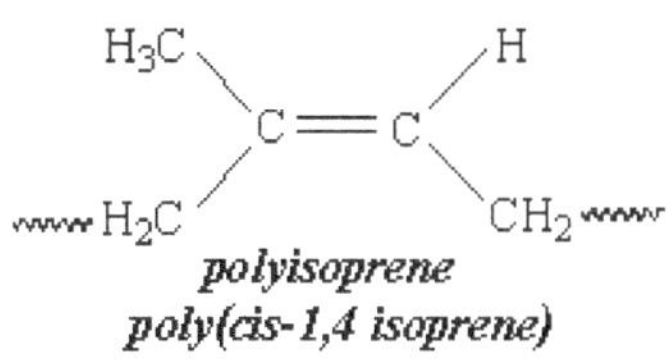

Figure 1. Chemical structure of natural rubber (NR)

The vulcanisation of natural rubber (NR) by sulphur in presence of organic accelerator is a complicated process. The mechanism of vulcanisation and its acceleration depends on the structure of the rubber, type and concentration of accelerators and activators (zinc oxide and fatty acid) and on the thermodynamics of each particular reaction. The chemistry of vulcanisation is complex and the resulting crosslinks may be mono-, di-, tri- or higher poly-sulphides, with a proportion which is among others largely determined by the vulcanisation system, the cure time and the temperature.(Stelescu et al., 2010).

This section presents the influence of the vulcanization method – of the crosslinking mechanism – on the characteristics of a natural rubber blend (Stelescu et al., 2010) .

The following raw materials have been used: Crep natural rubber, Ultrasil VN3 precipitated silica (50 phr), zinc oxide quality I (5 phr), stearic acid (0.5 phr), polyethylene glycol coupling agent (3 phr), Irganox 1010 antioxidant (1 phr). For crosslinking blends, the following were used: Perkadox 40 benzoyl peroxide (8 phr) – for peroxide vulcanization; sulphur (1.5 phr) and vulcanization accelerators (tetramethyl thiuram disulfide TMTD (1 phr) and mercaptobenzothiazol MBT (0.5 phr) – for sulphur vulcanization; accelerated electrons (EB) and polyfunctional monomer – trimethylopropane trimethacrylate TMPT DL 75 C (6 phr) for electron beam vulcanization.

Blends have been made by means of blending technique, on a laboratory roll with electric heating at 65-70°C, total blend time of 12′. Plates for physical-mechanical determinations have been made by means of hydraulic press. Blends crosslinked with *peroxide* or with *sulphur and accelerators* have been reticulated at 160°C and vulcanization time was chosen depending on curves obtained on the Monsando rheometer (rheologic characteristics) of blends in order to obtain the following blend samples: subvulcanized (T_{50}), vulcanized (T_{90}) and supravulcanized (T_{140}). Thus, the time needed to obtain T_{50} and T_{90} blends was determined from rheograms and corresponds to T_{50} and T_{90} in Table 1, and the time needed to obtain T_{140} supravulcanized blends was 22′ for both types of blends. Plates for irradiation were modeled by pressing at low temperatures of maximum 100°C for 3′.

The EB vulcanization rubber processing is performed with the accelerator ILU-6M of 1.8 MeV and 10.8 kW. For EB treatment the rubber sheets were cut in rectangular shape of 0.15 x 0.15 m^2. The layers of three sandwiched sheets were irradiated by repeatedly passing on a conveyor under the ILU-6M scanner. Samples were irradiated with 5, 10, 15 and 20 Mrad respectively (1 Mrad = 10 kGy).

Rheologic characteristics of blends crosslinked with peroxide (symbol NRI-P) and with sulphur (symbol NPI-S) respectively, obtained by means of Monsanto rheometer (Table 4), show that the minimum moment and the maximum moment have high values because the charge and the other ingredients introduced in the blend have led to an increase of blend viscosity, and as a result, the blend opposes a high resistance force to the rotation of the oscillating disk of Monsanto rheometer. Physical-mechanical properties of obtained blends are presented in Table 5, Table 6 and Table 7.

Blend name	*Moment$_{min.}$* *(dN/m)*	*Moment$_{max.}$* *(dN/m)*	*T_{90}*	*T_{50}*
NRI-P	22	48.9	12′30″	3′45″
NRI-S	31	59.2	12′30″	1′45″

Table 4. Rheologic characteristics of blends

Characteristics / Symbol blend	*NRI-P*			*NRI-S*		
	T_{50}	*T_{90}*	*T_{140}*	*T_{50}*	*T_{90}*	*T_{140}*
Hardness, ºShA	80	83	85	75	74	74
Elasticity, %,	39	38	41	36	34	32
Tensile stress at 100%elongation, N/mm^2	3.3	4.6	5.26	2.4	1.82	1.9
Tensile strength, N/mm2	12.7	10.3	6.6	20.6	17.6	16.8
Elongation at break, %	346	246	153	600	640	646
Permanent set, %	42.5	32.9	36.8	81.7	62	62
Tear strength, N/mm	20	14.6	7.3	46	42	43

Table 5. Physical-mechanical properties of NR rubber blends vulcanized with peroxide (NRI-P), and sulphur and accelerators (NRI-S)

Characteristics / Symbol blend/ Irradiation dose	*NRI*			
	5Mrad	*10Mrad*	*15Mrad*	*20Mrad*
Hardness, ºShA	50	55	56	66
Elasticity, %	28	28	30	32
Tensile stress at 100%elongation, N/mm^2	0.73	1	1.3	1.7
Tensile strength, N/mm^2	4	5.3	8	9.3
Elongation at break, %	500	520	533	487
Permanent set, %	148	110	80	61
Tear strength, N/mm	14.5	21.5	45.5	50.5

Table 6. Physical-mechanical properties of NR rubber blends vulcanized with EA (NRI)

Characteristics / Symbol blend/ Irradiation dose	*NRI-TMPT*			
	5Mrad	*10Mrad*	*15Mrad*	*20Mrad*
Hardness, ºShA	50	76	81	81
Elasticity, %	30	31	34	36
Tensile stress at 100%elongation, N/mm^2	0.82	3.1	4	5.2
Tensile strength, N/mm^2	3.6	13.8	15.9	14.6
Elongation at break, %	527	427	407	347
Permanent set, %	157	37	30	23
Tear strength, N/mm	18	67.5	72.5	72.5

Table 7. Physical-mechanical properties of NR rubber blends vulcanized with EA in the presence of TMPT (NRI-TMPT)

Comparing physical-mechanical properties of subvulcanized (T_{50}), vulcanized (T_{90}) and supravulcanized (T_{140}) samples from each type of NRI-S and NRI-P blend (table 5), it is noticed that: (a) in the case of blends crosslinked with peroxides, as the vulcanization time increases, hardness, elasticity and tensile stress at 100% elongation increase and tensile strength, elongation at break, permanent set and tear strength decrease; (b) for blends crosslinked with sulphur and accelerators, as the vulcanization time increases, a slight decrease of hardness, elasticity, tensile stress at 100% elongation, tensile strength, permanent set and tear strength, and a slight increase of elongation at break take place; (c) hardness, elasticity and tensile stress at 100% elongation of blends with ingredients vulcanized with sulphur are lower than those vulcanized with peroxide, instead, tensile strength, tear strength, elongation at break and permanent set have higher values; (d) characteristics of blends crosslinked with peroxide decrease significantly when increasing vulcanization time compared to samples obtained through other vulcanization methods, indicating degradation of polymer chain in the presence of peroxide; (e) physical-mechanical properties of subvulcanized, vulcanized and supravulcanized samples of NR blend depend on the crosslinking technique and mechanism.

Comparing blends cross-linked with peroxide to NRI and NRI-TMPT type (table 6 and table 7), irradiated with EA (in both cases crosslinking is done by radicalic mechanism), it can be noticed that: (a) the optimal dose needed can be 20 Mrad and 10-15 Mrad respectively in the case of blends containing 6 phr TMPT; (b) upon increasing the irradiation dose, hardness, elasticity, tensile stress at 100% elongation, tensile strength and tear strength increase and permanent set decreases; (c) blends crosslinked with EA in the presence of TMPT exhibit superior values of tear strength, have good values of permanent set – similar to those obtained through other methods – indicating an efficient crosslinking, and tensile strength, elongation at break, hardness, elasticity and tensile stress at 100% elongation have values ranging between those obtained in vulcanization with peroxide and those made by vulcanization with sulphur and accelerators; (d) natural rubber blends containing 6 phr TMPT irradiated at 15 Mrad and 20 Mrad respectively, exhibit superior physical-mechanical properties compared to control blend NRI-P tip T_{90} (obtained by vulcanization with peroxide). Based on results obtained and existing literature studies, reaction mechanisms are suggested for crosslinking natural rubber using the crosslinking systems presented above.

Vulcanization with sulphur and accelerators of NR (Figure 2) is done in general by ionic mechanism and leads to the formation of sulphur bridges between ($C-S_x-C$) macromolecules or cyclic combination of sulphur. At high temperatures, desulphuration takes place, determining the formation of shorter sulphur bridges. As a consequence of thermal instability of sulphurs, NR vulcanized with sulphur can be devulcanized in the presence of disulphur diaryls or amines, at temperatures of over 300ºC (Stelescu et al., 2010).

Vulcanization with peroxides (Figure 3) is done by radicalic mechanism when bonds form between C-C macromolecules; crosslinking is initiated by thermal decomposition of peroxide, which is considered the determining stage of crosslinking speed. Then free radicals formed extract hydrogen atoms from natural rubber chains in order to form macroradicals (Stelescu et al., 2010; Hofmann, 1967). Macroradicals recombine forming crosslinked structure.

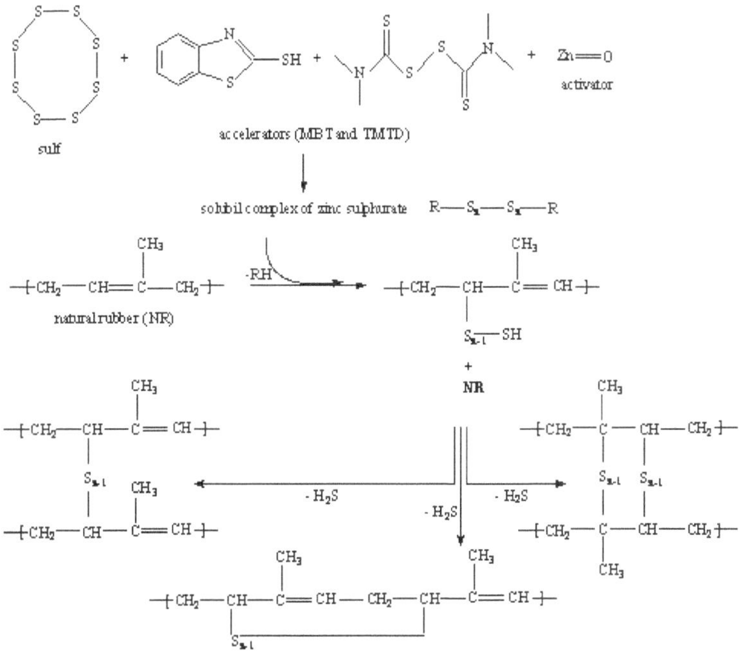

Figure 2. Reaction scheme – Mechanism of crosslinking natural rubber with sulphur and accelerators (Stelescu et al., 2010).

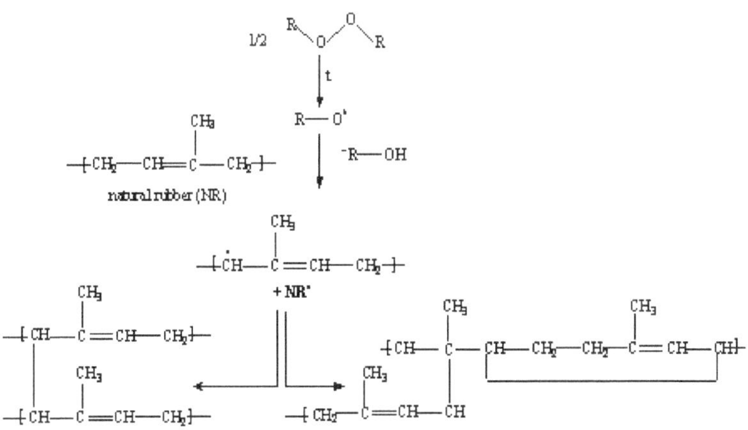

Figure 3. Reaction scheme – Mechanism of crosslinking natural rubber with peroxides (Stelescu et al., 2010) .

$$NR\text{-}H \xrightarrow{EB\ irradiation} NR\text{-}H^{\bullet}\ \textit{(excited state)}$$

$$NR\text{-}H^{\bullet} \longrightarrow NR\text{-}H^{+}\ \textit{(positive ion of polymer)} + 1e^{-}\ \textit{(electron)}$$

$$NR\text{-}H^{\bullet} \longrightarrow NR^{\bullet}\ \textit{(free radical of polymer)} + H^{\bullet}\ \textit{(hydrogen radical)}$$

$$NR\text{-}H^{+} + NR\text{-}H \longrightarrow NR^{\bullet} + NR\text{-}H_2^{+}\ \textit{(radical ion of polymer)}$$

$$NR^{\bullet} + NR\text{-}H_2^{+} + 1e^{-} \longrightarrow NR\text{-}NR\ \textit{(crosslinked polymer)}$$

$$NR_1\text{-}H^{\bullet} + NR_2\text{-}H \longrightarrow NR_1\text{-}H + NR_2\text{-}H^{\bullet}\ \textit{(energy transfer)}$$

$$H^{\bullet} + NR\text{-}H \longrightarrow NR^{\bullet} + H_2$$

Figure 4. Reaction scheme – Mechanism of crosslinking natural rubber with EB

Elastomer crosslinking by means of EB (Figure 4) is done without heating and in the absence of vulcanization agents. The chemistry of the process is based on macroradical formation from elastomer chains, which recombine, causing structuring. Isomerisations, double bond migrations, cyclizations, destructions etc. take place simultaneously with vulcanization. The reaction mechanism is similar to that presented in crosslinking with peroxides, but in this case, reaction initiation is due to the action of EB (Boye, 2008).

5.2. The influence of trimethylol-propane trimethacrylate (TMPT) co-agent on the mechanical properties of the natural rubber (NR) cross-linked by EB irradiation

In the following experiment (Stelescu & Manaila, 2007) the natural rubber blends were grafted and crosslinked by means of the accelerated electrons in the presence of a multifunctional monomer - trimethylol-propane trimethacrylate (TMPT), and the influences of the trimethylol-propane trimethacrylate (TMPT) percentage and accelerated electron irradiation dose on the physical-mechanical characteristics of the natural rubber (NR) blends were investigated.

The used materials were: natural rubber Crep 1X (NR), trimethylol-propane trimethacrylate TMPT DL 75 (TMPT), Ultrasil VN3 precipitated silica (50 phr), zinc oxide quality I (5 phr), stearic acid (0.5 phr), polyethylene glycol coupling agent (3 phr), Irganox 1010 antioxidant (1 phr) and benzoyl peroxide Perkadox 14-40B-GR (8 phr) as curing agent for the control blend.Natural rubber blends containing 0, 3, 6 and 9 phr of TMPT were prepared by blending on a laboratory roller mill. From these samples in a shape of plates were obtained by means of hydraulic press at 150 MPa and 100°C. The resulted plates were treated by irradiation, using the accelerator ALIN-10. The accelerated electrons dose rate was established to 2.4 kGy/min in order to accumulate 5 Mrad, 10 Mrad, 15 Mrad and 20 Mrad respectively (1 Mrad = 10 kGy). The control blend was obtained with benzoyl peroxide as curing agent. The blend was prepared on a laboratory roller mill and the control sample curing was accomplished on hydraulic press at 160°C. The best curing time (12'30") was determined by means of a Monsanto Rheometer.

Figures 5-10 illustrate the changes in physical-mechanical characteristics depending on the TMPT percentage in the blend and irradiation dose. Hardness (figure 5) increases as the irradiation dose and TMPT percentage increase; the hardness values for the blends with 6 and 9 phr of TMPT each irradiated at 15 and 20 Mrad are similar to those of the control blend (83°ShA). Tensile stress at 100 % elongation (figure 6) for the blends with 6 and 9 phr TMPT at an irradiation dose of 20 Mrad shows a higher value than tensile stress at 100 % elongation for the control sample (4,6 N/mm²). Tensile strength (figure 7) for the samples from blends with 6 and 9 phr TMPT, irradiated by means of accelerated electrons shows higher values than for the control sample (10,3 N/mm²), even at an irradiation dose of 10 Mrad. Values of the elongation at break (figure 8) for the samples irradiated with accelerated electrons are higher than those for the control sample (246 %). This aspect decreases as the irradiation dose and TMPT percentage increase. Permanent set (figure 9) decreases as the irradiation dose and TMPT percentage increase; for the blends with 6 and 9 phr of TMPT irradiated with 15 and 20 Mrad, the permanent set values are lower than those for the control sample (14,6 %), revealing a high crosslinking level. Tear strength (figure 10) for the samples from the blends irradiated with accelerated electrons shows higher values than those for the control sample (14,6 N/mm).

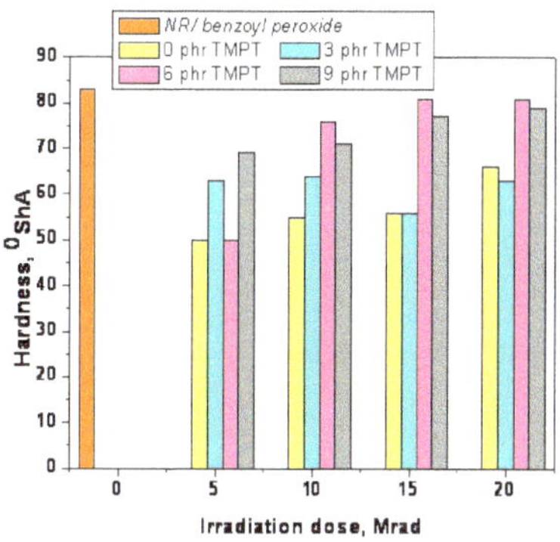

Figure 5. Changes in hardness versus the irrad iation dose and TMPT percentage

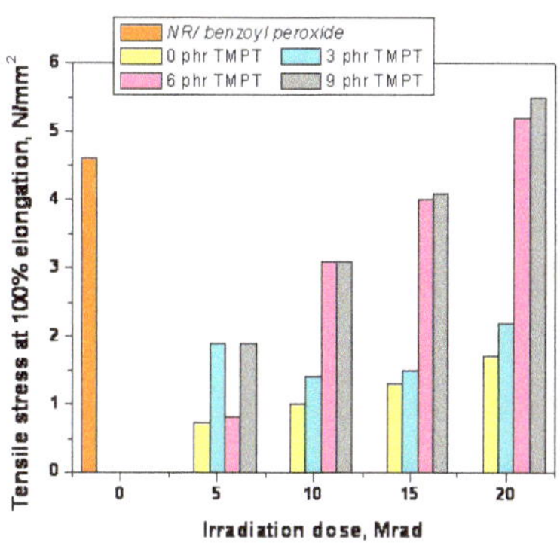

Figure 6. Changes in tensile stress at 100% elongation versus the irradiation dose and TMPT percentage

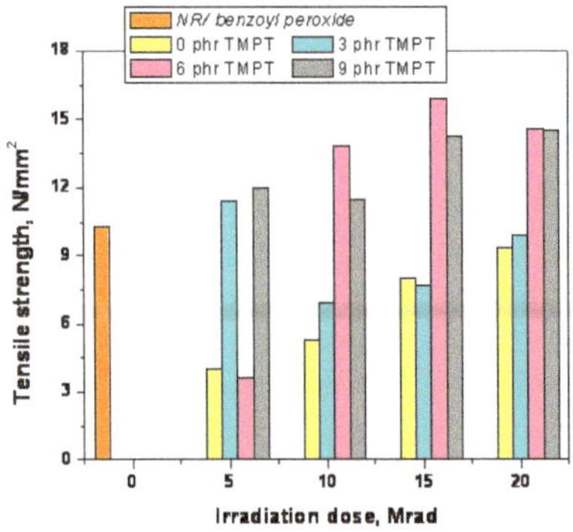

Figure 7. Changes in tensile strength versus the irradiation dose and TMPT percentage

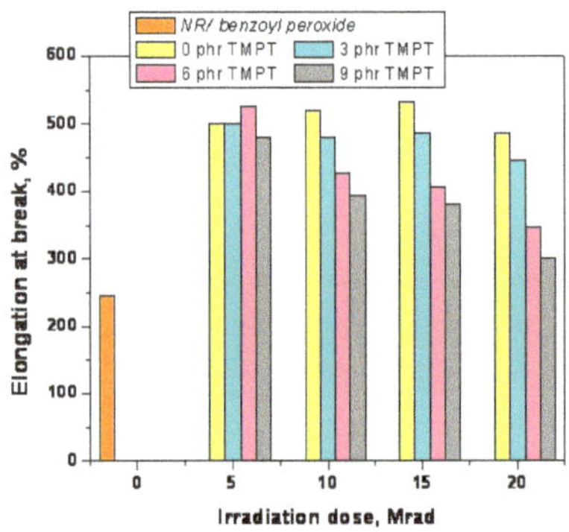

Figure 8. Changes in elongation at break versus the irradiation dose and TMPT percentage

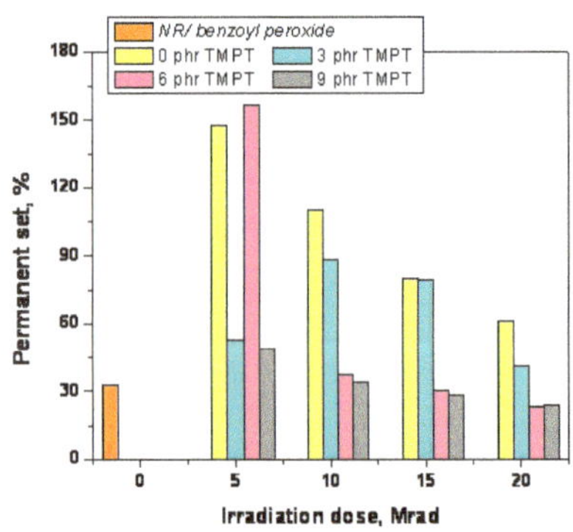

Figure 9. Changes in permanent set versus the irradiation dose and TMPT percentage

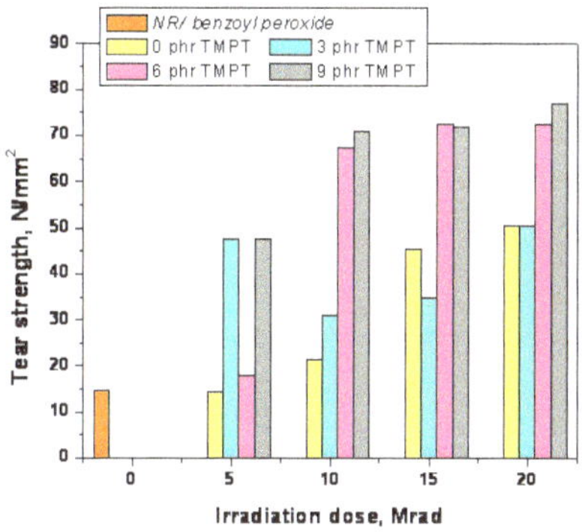

Figure 10. Changes in tear strength versus the irradiation dose and TMPT percentage

As a conclusion, the natural rubber blends containing 6 and 9 phr of TMPT irradiated each of them at 20 Mrad have shown higher physical-mechanical characteristics than those for the control sample (obtained by curing with peroxide at 160°C). This proves the advantages of a new processing technique for the NR blends resulting in 10 times shorter curing time, removing the curing agents, and lack of wastes.

5.3. Characteristics of natural rubber (NR) blends cross-linked by electron beam and microwave irradiation

Crosslinking of elastomers or rubbers by electron beam (EB) and microwave (MW) radiations could be a new method for improve their mechanical properties. Interaction of each of the two physical systems, EB and MW, with a substance has revealed their ability for changing physical and chemical proprieties of the treated substance. Therefore, the physical and chemical effects of the EB and MW radiation have drawn high interest, resulting in a large industrial application range of the new materials with improved characteristics. In the following we present the effect of the polyfunctional monomers (PFMs) on the mechanical properties of the NR (natural) rubber crosslinked by electron beam (EB) and microwave processing (Manaila, Stelescu et al., 2011).

The following materials were used in the study: natural rubber Crep 1X; triallylcyanurate TAC DL 70 (3 phr); triallylisocyanurate TAIC DL 70C (3 phr); trimethylopropane-trimethacrylate TMPT DL 75 (3 phr) and zinc-diacrylate ZDA GR 75 (3 phr). Blends were prepared on an electrically heated laboratory roller mill. For preparation of NR with polyfunctional monomers, the blend constituents were added in the following sequence and amounts: 100 phr NR and 3 phr polyfunctional monomers (TAC, TAIC, TMPT and ZDA respectively). Process variables: temperature 25-50°C, friction 1:1.1, and total blending time 5 min. Plates required for physico-mechanical tests were obtained by pressing in a hydraulic press at 110 ±5°C and 150 MPa. Dibenzoyl peroxide vulcanized samples were prepared similarly to the experimental ones with the following specifications: 8 phr of dibenzoyl peroxide as vulcanizing agent was added and the blend vulcanization was achieved in a hydraulic press at 160°C; the vulcanization time was measured by means of Monsanto Rheometer. The resulted plates were treated by irradiation, using the accelerator ILU - 6M. For EB and EB+MW treatments the rubber samples were cut as compressed sheets of 2 mm thick in the polyethylene foils to minimize oxidation. The layers of three sandwiched sheets were irradiated by repeatedly passing on a conveyor under the ILU-6M scanner.

The mechanical properties of samples are summarized in Tables 8-11.

Mechanical characteristics	*BP vulcanization*	*EB vulcanization*		*EB+MW vulcanization*
	NR - P	**5 Mrad**	**10 Mrad**	**5 Mrad + 55′**
Elasticity, %	50	44	44	42
Tensile stress at 100%elongation, N/mm²	-	0.3	0.4	0.19
Tensile strength, N/mm²	0.9	0.8	0.9	4.6
Elongation at break, %	87	487	300	935
Permanent set, %	5	15	8	19
Tear strength, N/mm	2	9.5	14.5	11

Table 8. Physical-mechanical characteristics of blends: NR+TAC

Mechanical characteristics	*BP vulcanization*	*EB vulcanization*		*EB+MW vulcanization*
	NR - P	**5 Mrad**	**10 Mrad**	**5 Mrad + 55′**
Elasticity, %	68	42	42	50
Tensile stress at 100%elongation, N/mm²	-	0.4	0.4	0.21
Tensile strength, N/mm²	0.96	0.7	2.2	4.8
Elongation at break, %	87	387	647	850
Permanent set, %	9	11	13	15
Tear strength, N/mm	2.5	7	13	13

Table 9. Physical-mechanical characteristics of blends: NR+TAIC

Mechanical characteristics	*BP vulcanization*	*EB vulcanization*		*EB+MW vulcanization*
	NR - P	**5 Mrad**	**10 Mrad**	**5 Mrad + 55′**
Elasticity, %	68	48	53	48
Tensile stress at 100%elongation, N/mm²	-	0.4	0.4	0.29
Tensile strength, N/mm²	0.82	3.6	8.3	8.7
Elongation at break, %	60	673	727	785
Permanent set, %	3	9	7	14
Tear strength, N/mm	1.39	13.5	21	13

Table 10. Physical-mechanical characteristics of blends: NR+TMPT

Mechanical characteristics	*BP vulcanization*	*EB vulcanization*		*EB+MW vulcanization*
	NR - P	**5 Mrad**	**10 Mrad**	**5 Mrad + 55′**
Elasticity, %	66	46	50	50
Tensile stress at 100%elongation, N/mm²	-	0.39	0.51	0.26
Tensile strength, N/mm²	1.3	0.99	3	3.7
Elongation at break, %	140	340	607	675
Permanent set, %	2	10	6	11
Tear strength, N/mm	4.5	7.5	18	18

Table 11. Physical-mechanical characteristics of blends: NR+ZDA

Analyzing the mechanical characteristics obtained from the EB cross-linked samples compared with those of cross-linked with peroxide mixtures in the presence of the same types of PFMs can be observed: (1) significant improvements in tensile strength (between 130% for NR+TAIC and NR+ZDA at 10 Mrad, and 912% for NR+TMPT at 5 Mrad; (2) elongation at break increases, for all polyfunctional monomers type, the smallest increase being 142% for NR+ZDA at 5 Mrad and highest 1112% for NR+TMPT at 10 Mrad; (3) tear strenght increases with radiation dose, (between 180% for NR+TAIC at 5 Mrad, and 1410% for NR+TMPT at 10 Mrad; (4) for all samples was noticed a decrease in elasticity.

Relatively low permanent set values indicate a good return to its original shape after applying a force, so an efficient curing for all samples. In conclusion, even at a dose of 5 Mrad was achieved an efficient crosslinking of NR. The polyfunctional monomer influence on these parameters for the samples vulcanized with EB is the following: TMPT > ZDA > TAC > TAIC.

Comparing the mechanical characteristics obtained from the EA+MW cross-linked samples with those of cross-linked with peroxide mixtures in the presence of the same types of MP can be observed a significant improvement in tensile strength (up to 411% for NR+TAC, 400% for NR+TAIC, 960% for NR+TMPT and 184% for NR+ZDA), in elongation at break (up to 974% for NR+TAC, 877% for NR+TAIC, 1208% for NR+TMPT and 382% for NR+ZDA) and tear strenght (up to 450% for NR+TAC, 420% for NR+TAIC, 835% for TMPT and 300% for NR+ZDA). The order of influence of polyfunctional monomers on the studied parameters for EA+MW is identical to that in the case of irradiation with EA (TMPT > ZDA > TAC > TAIC).

Improved characteristics of hardened mixtures with EA and EA + MW from the hardened peroxide is due on the one hand the advantages of accelerated electrons (the process is very fast, and due to high penetration power of radiation there is an effective and uniform curing), and on the other hand, the advantages of microwave (promote a narrow distribution of the molecular masses, as the crosslinking occurs simultaneously in the bulk of material because of the microwave interaction with all material under irradiation; promote fast crosslinking processes) (MGM Rubber Company-Research and Development, 2007; Martin, 2002). Also, due to reduced processing time, it is removed the degradation due to thermal degradation at high temperature elastomer maintenance of about 160°C for 10-30 min.

Analyzing the characteristics influence of polyfunctional monomers type, can be seen that the best results were obtained by using TMPT followed by ZDA (coagents curing Type I). Type I polyfunctional monomers are highly reactive and increase the rate and state of cure. Type II (TAC and TAIC) polyfunctional monomers increase the state of cure only. Also, the influence of polyfunctional monomers to increasing the mechanical characteristics is determined by the reactivity, the number of reactive groups (functionality): TMPT has functionality three and ZDA has functionality two (Boye, 2008; Henning, 2008).

5.4. Characteristics of materials based on chlorinated polyethylene (CPE) obtained by cross-linked by electron beam in the presence of triallylcyanurate (TAC).

Chlorinated polyethylene (CPE) is a synthetic elastomer produced by the means of controlled chlorination of polyethylene and has been commercially produced since the late 1960s (Stelescu et al., 2011; Stelescu et al., 2008; U.S. Patent, 1969). A generalized chemical structure for CPE is shown in Figure 11.

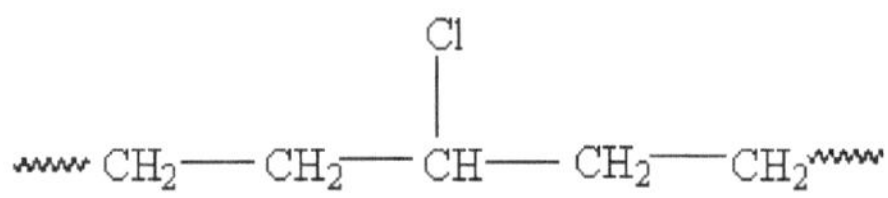

chlorinated polyethylene

Figure 11. Chemical structure of chlorinated polyethylene (CPE).

Chlorinated Polyethylene Elastomers (CPE) are produced from HDPE that is randomly chlorinated in an aqueous slurry. Polymers are differentiated by chlorine content, molecular weight and crystallinity. Chlorine contents generally range from 25 to 42%. Advantages in using CPE, include very good resistance to ozone, oxidation, abrasion and flex cracking. CPE also has good resistance to alcohols, alkalis and acids. Limitations for CPE include moderate resistance to aromatic oxygenated solvents (Hallstar, 2009) Cure agents for CPE compounds are typically based on (1) peroxide cure systems with coagents; (2) thiadiazole-based chemistries; or (3) irradiation crosslinking techniques (Stelescu et al., 2011; Stelescu et al., 2008). The choice of cure system depends upon a number of factors such as compound cost, processing equipment and curing equipment. Peroxide cures are preferred when extra scorch safety, shelf life, bin stability, low permanent set and high-temperature performance are desired. Irradiation-curable compounds are usually formulated in a manner similar to the peroxide-curable compounds, except that no peroxide is necessary (Stelescu et al., 2011; Flynn et al., 1985). About 70% of CPE is used in wire and cable applications with trimellitates the plasticizer of choice. The remaining applications, each at 15%, are hydraulic hose and molded/extruded automotive parts (Hallstar, 2009).

In our experiments (Stelescu et al., 2008), we chose an efficient coagent for crosslinking of CPE by electron beam irradiation: triallyl cyanurate (TAC), to increase the rate and degree of cure. Materials used in the study: CPE TX10 chlorinated polyethylene (35 % chlorine), barium sulphate, titanium dioxide, calcium silicate, chlorinated paraffin, Uvinul 5050 antioxidant and TAC DL 70 trialylcyanided polyfunctional monomer; in the control blends the vulcanizing agent di(tert-buthylperoxiisopropyl) benzene (Perkadox 14-40B-GR) was used. Blends were prepared by mixing on a laboratory roller mill electrically heated up to 100-110°C for 10'. Ingredients were added in the following sequence: CPE, titanium dioxide after CPE was adhered to the roller, calcium silicate and the antioxidant, barium sulphate and chlorinated paraffin, and finally the polyfunctional TAC monomer. To the control blends 8 phr Perkadox was added after the blend preparation. Blends were homogenized on the laboratory roller mill, and from these plates were obtained by pressing in an hydraulic press. The resulting plates were packed in a polyethylene film and subjected to 5, 10 15 and 20 Mrad irradiation in the ILU-6M electron bem accelerator.

The crosslinking and grafting of CPE blends by accelerated electron radiation was proved by comparing physical-mechanical characteristics of the irradiated blends with those of the control blends with the same composition but crosslinked by classical method with peroxides. Physical-mechanical characteristics for the blends of CPE TX10 chlorinated polyethylene containing barium sulphate, titanium dioxide, calcium silicate, chlorinated paraffin, Uvinul 5050 and TAC, irradiated with different doses of accelerated electrons are shown in Figures 12-17. Hardness has shown higher values than that of the non-irradiated blend (63°ShA), and a slight decrease with the increase in the TAC amount (Figure 12); increasing further the EB dose above 5 Mrad has resulted in no improvement of this characteristic. The control blends have shown an increase of the hardness from 69 °ShA up to 76 °ShA with the increase in the TAC amount. The resulting values for these are comparable with the values obtained by irradiation.

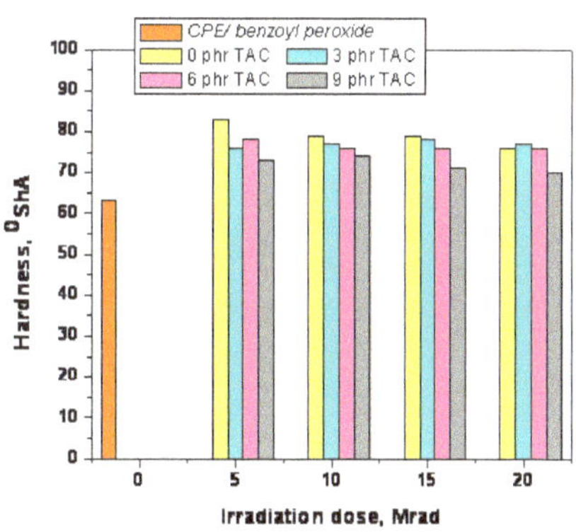

Figure 12. Changes in hardness versus the irradiation dose and TAC percentage.

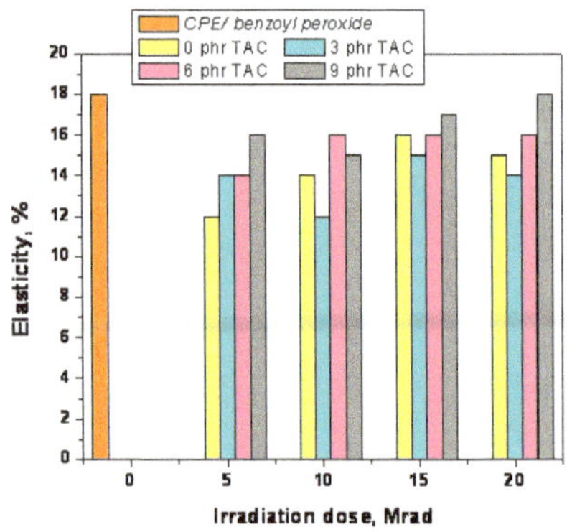

Figure 13. Changes elasticity versus the irradiation dose and TMPT percentage

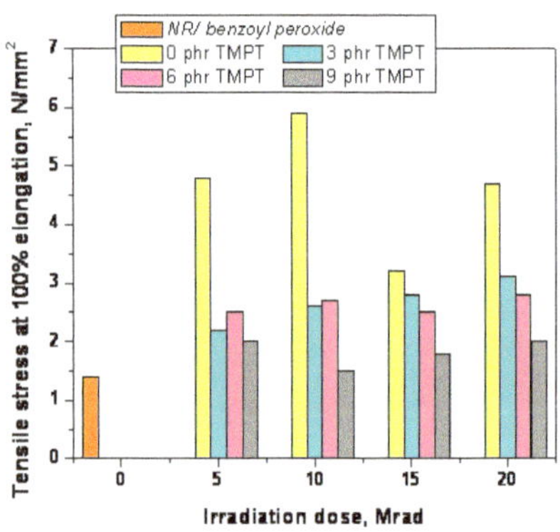

Figure 14. Changes in tensile stress at 100% elongation versus the irradiation dose and TMPT percentage

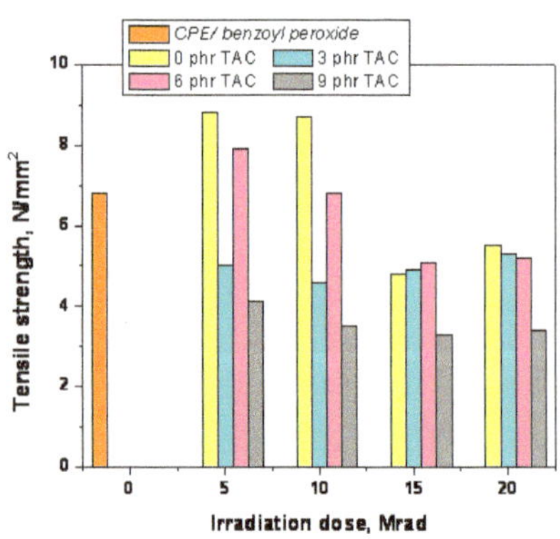

Figure 15. Changes in tensile strength versus the irradiation dose and TMPT percentage

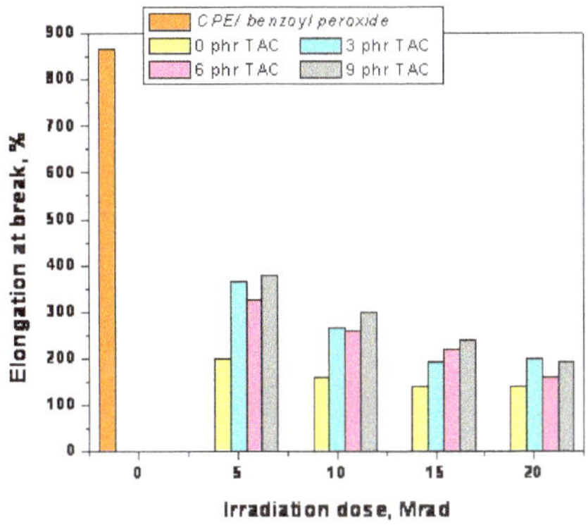

Figure 16. Changes in elongation at break versus the irradiation dose and TMPT percentage

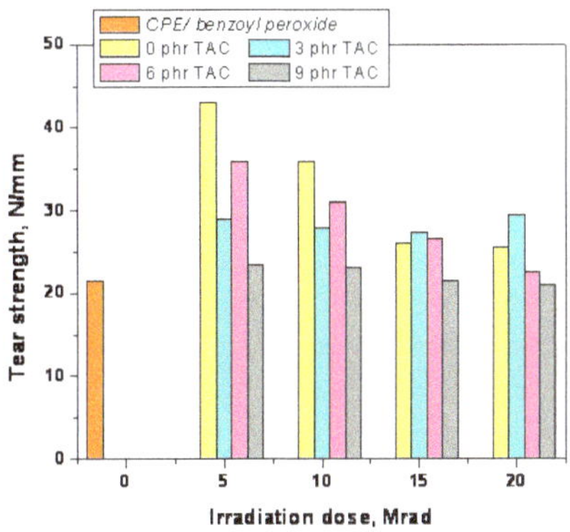

Figure 17. Changes in tear strength versus the irradiation dose and TMPT percentage

Elasticity (Figure 13) has shown a decrease when irradiating the sample with 5 Mrad and thereafter has increased with the in increase in the AE dose. Elasticity has increased as the TAC amount was increased. The elasticity in controls (crosslinked with Percadox) has decreased (from 18 % down to 15%) as the TAC amount was increased. Tensile stress at 100% elongation has shown a decrease as the TAC amount was increased and has shown higher values than those for the non-irradiated blend for all irradiation doses (Figure 14); in controls the tensile stress at 100% elongation has shown increased values (from 1.6 N/mm^2 up to 3 N/mm^2) as the TAC amount was increased. Tensile strength (figure 15) has shown a decrease as the as the irradiation dose was increased from 5 Mrad up to 15 Mrad, and thereafter a slight increase revealed for all the TAC amounts; it has shown values comparable with those for the samples crosslinked with peroxides; Elongation at break have decreased as the irradiation dose and the TAC amount were increased, the resulting values being lower than those for the non-irradiated blend (Figures 16); the variation of this characteristic has revealed the CPE crosslinking with an irradiation dose as low as 5 Mrad.

The control blends have shown a decrease in the elongation at break (from 50 % down to 260 %) as the TAC amount was increased. Tear strength (Figure 17) has increased at irradiation doses of 5 Mrad, and thereafter has decreased as the EB irradiation dose was increased. Tear strength has shown a non-uniform variation as the TAC amount was increased. In controls the tear strength has increased as the TAC amount was increased from 29.5 N/mm up to 38 N/mm.

The rheological characteristics of the controls obtained measured by the Monsanto rheometer are shown in table 12.

Characteristic	*0 phr TAC*	*3 phr TAC*	*6 phr TAC*	*9 phr TAC*
T_{min}	1′30	1′30	1′15	1′15
T_2	2′30	3′	2′15	2′15
M_{min} (dN/m)	10	7.5	6.4	7.5
M_{max} (dN/m)	33.2	36.2	47	45
M_{90}	30.88	33.33	42.94	41.25
T_{90}, min	9′15"	7′15″	7′	10′
I_{vv}	14.81	23.53	21.05	12.9
V_{2c}	0.814	0.830	0.851	0.847

Table 12. Rheological characteristics of CPE blends cross-linked with peroxides

The above data have revealed the following features with the increase in the amount of polyfunctional TAC monomer: the minimum moment of shearing has decreased, and the maximum moment of shearing has increased up to a TAC amount of 6 phr; the best vulcanization time has shown a minimum at 6 phr TAC and thereafter has increased. The index of vulcanization rate has increased by 59 % at 3 phr TAC, then has decreased slightly (by 10 %) when the amount of TAC was increased up to 6 phr, and has decreased significantly when the amount of TAC was increased further up to 9 phr. The highest value (by 4.5 % higher than that for a blend free TAC) of the vulcanization factor was found for 6 phr TAC, and it has decreased (by 0.5 %) as the TAC amount was increased up to 9 phr.

The best physical-mechanical characteristics were obtained by adding 0 – 6 phr TAC to the blends and applying an irradiation dose of 5 Mrad. When comparing the best vulcanization time (T_{90}) with the time required with an irradiation of 5 Mrad, it has shown a decrease of 2.8 times in crosslinking by irradiation that has resulted in a shorter time required to obtain CPE-based finished products. In addition, the irradiation process is in continuous flow.

5.5. Characteristics of ethylene-propylene rubber (EPDM) obtained by cross-linked with electron beam and microwave in the presence of trimethylopropane trimethacrylate (TMPT)

Ethylene-propylene rubbers use the same chemical building blocks or monomers as polyethylene (PE) and polypropylene (PP) thermoplastic polymers. These ethylene (C2) and propylene (C3) monomers are combined in a random manner to produce rubbery and stable polymers. A wide family of ethylene-propylene elastomers can be produced ranging from amorphous, non-crystalline to semi-crystalline structures depending on polymer composition and how the monomers are combined (Karpeles & Grossi, 2001). These polymers are also produced in an exceptionally wide range of Mooney viscosities (or molecular weights). The ethylene and propylene monomers combine to form a chemically saturated, stable polymer backbone providing excellent heat, oxidation, ozone and weather aging. A third, non-conjugated diene monomer can be terpolymerized in a controlled manner to maintain a saturated backbone and place the reactive unsaturation in a side chain available for vulcanization or polymer modification chemistry. The terpolymers are referred to as EPDM (or ethylene-propylene-diene with "M" referring to the saturated backbone structure). An EPDM polymer structure is illustrated in Figure 18. The ethylene-propylene copolymers are called EPM.

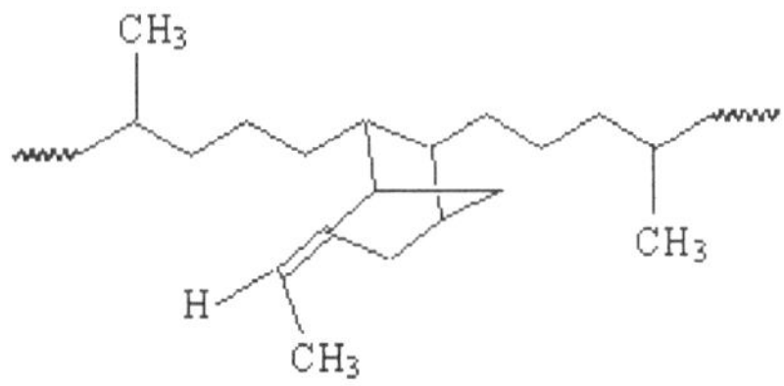

ethylene-propilene elastomers

Figure 18. Chemical structure of ethylene-propylene terpolymer (EPDM)

The most important stage in the EPDM rubber processing technology is vulcanization / crosslinking by sulphur and accelerator or by peroxides. Radiation vulcanization is applicable, but the dose required for EPDM vulcanization is very high (Odian, 1964). To reduce the dose for vulcanization of EPDM, some researches have worked on the introduction of coagents/polyfunctional monomers such as ethylene glycol dimethacrylate, triallyl cyanurate, triallyl isocyanurate, trimethylolpropane trimethacrylate etc (Manaila, Stelescu, Ighigeanu et al., 2011)

The following materials were used in the study: EPDM rubber (Nordel 4760); trimethylopropane-trimethacrylate Luvomaxx TMPT DL 75; dibenzoyl peroxide Perkadox 14-40B for he control blends (Manaila, Stelescu, Ighigeanu et al., 2011).

Blends were prepared on an electrically heated laboratory roller mill. For preparation of EPDM with polyfunctional monomer TMPT, the blend constituents were added in the following sequence and amounts: 100 phr EPDM and 3 phr TMPT. Process variables: temperature 75±5°C, friction 1:1.1, and total blending time 5 min. Plates required for physical-mechanical tests were obtained by pressing in a hydraulic press at 120 ±5°C and 150 MPa. Dibenzoyl peroxide vulcanized samples were prepared similarly to the experimental ones with the following specifications: 8 phr of dibenzoyl peroxide as vulcanizing agent was added and the blend vulcanization was achieved in a hydraulic press at 160°C; the optimum vulcanization time was measured by means of Monsanto Rheometer. The resulted plates were treated by irradiation, using the accelerator ALIN-10.

In Table 13 are presented characteristics of the mixtures cross-linked with dibenzoyl peroxide. These, along with non-irradiated and non-crosslinked samples are considered control samples for mixtures cross-linked by irradiation with EA or EA and microwave.

Rheologic characteristics	*EPDM+TMPT+ dibenzoyl peroxide*
T90 (minutes)	23'30"
M_{min} (dNm)	1
M_{max} (dNm)	46.8
Δ M (dNm)	31.6
M_{90} (dNm)	42.64
T_{90} (minutes)	6'45"
Physical-mechanical characteristics	
Hardness, °ShA	64
Elasticity, %	60
Tensile strength, N/mm^2	2.1
Elongation at break, %	127
Permanent set, %	5
Tear strength, N/mm	13.5

Table 13. Rheological characteristics of EPDM/TMPT blends cross-linked with peroxides

The mechanical properties of samples are summarized in Figures 19-26. The hardness (Figure 19) increases with 7 ºShA by EB irradiation at a dose of 120 kGy and with 5 ºShA by EB + MW irradiation at 30 kGy+83 s and 60 kGy+176 s. For the other treatments, hardness change insignificantly - with 1-2 ºShA; also, there is no big difference between the EB and EB + MW crosslinking. Analyzing the irradiated samples (with EB or EB+MW) and the control sample crosslinked with peroxide (see Table 13), can be noticed that there are no significant differences. The hardness increasing of samples crosslinked by irradiation or by curing with peroxide comparing with non-crosslinked sample is due to the reinforcing effect that occurs as a result of the formation of crosslinking bridges. The elasticity (Figure 20) does not change significantly by irradiation with EB or EB + MW (an increase of only 2%). There is a growing of elasticity with maximum 1% for EB + MW compared with EB only, for the treatment of 30 kGy + 88 s. The same increase in elasticity was observed in the case of cross-linked with peroxide. The tensile stress at 100% elongation (Figure 21) and tensile stress at 300% elongation (Figure 22) increase by irradiation, but the obtained values are small because the material is not with filler.

There are little differences between values obtained by peroxide, EB and EB + MW crosslinking. Considering the fact that the tensile stress at 100% elongation is a measure of the crosslinking degree, the values obtained show that by samples irradiation takes place the increasing of the crosslinking degree (Yasin et al., 2005).

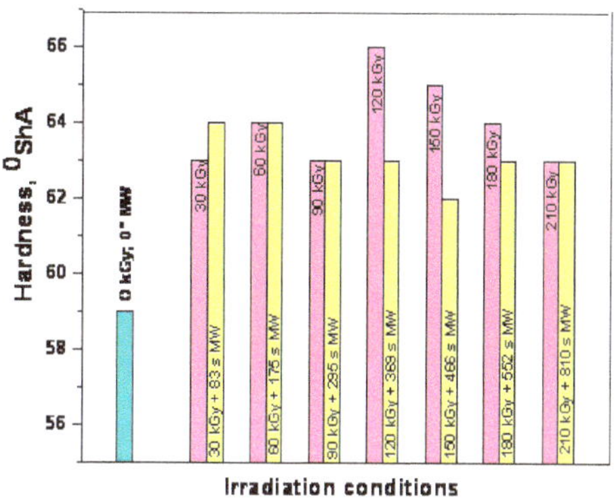

Figure 19. The EB and EB+MW effects on the hardness of the EPDM samples

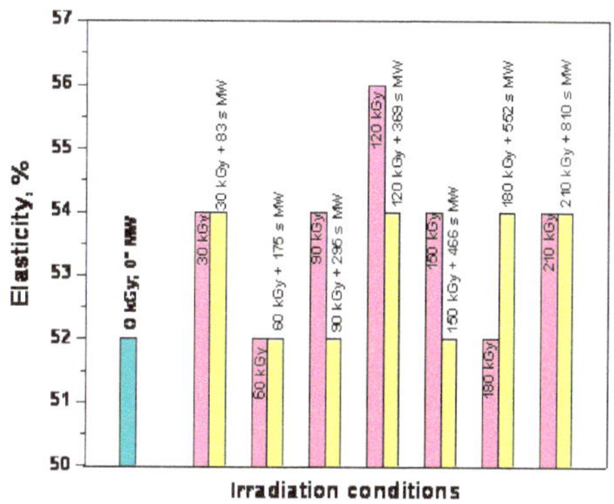

Figure 20. The EB and EB+MW effects on the elasticity of the EPDM samples

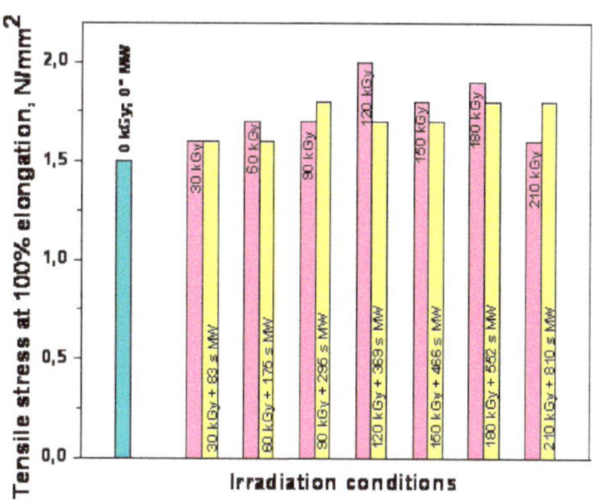

Figure 21. The EB and EB+MW effects on the tensile stress at 100% elongation of the EPDM samples

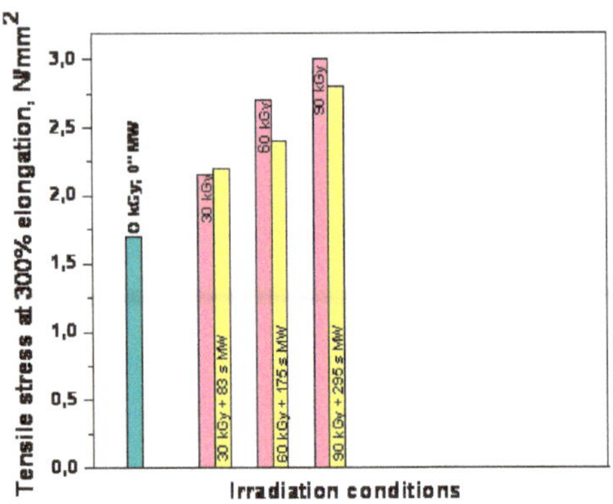

Figure 22. The EB and EB+MW effects on the tensile stress at 300% elongation of the EPDM samples

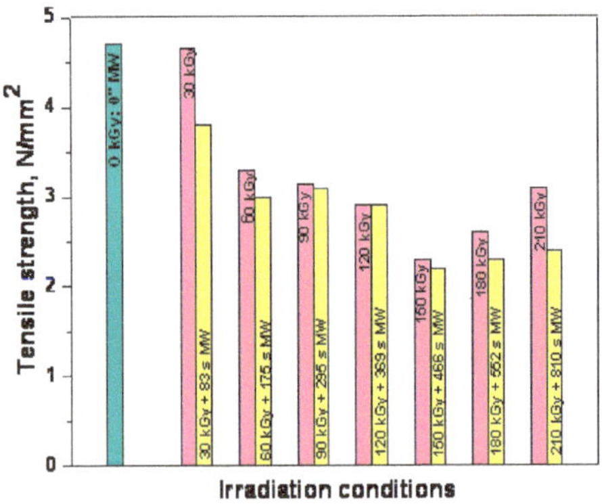

Figure 23. The EB and EB+MW effects on the tensile strength of the EPDM samples

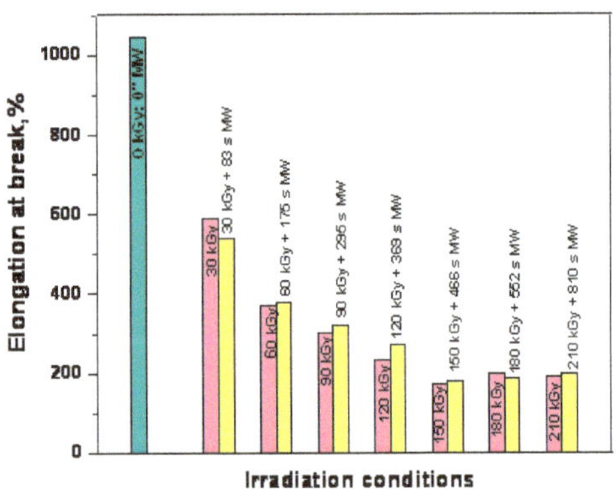

Figure 24. The EB and EB+MW effects on the elongation at break of the EPDM samples

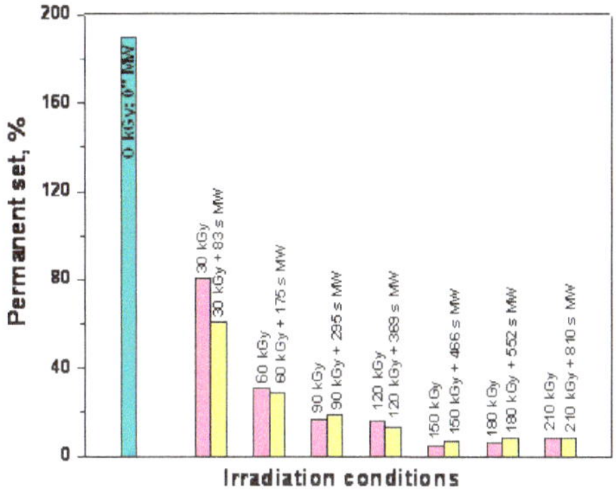

Figure 25. The EB and EB+MW effects on the permanent set of the EPDM samples

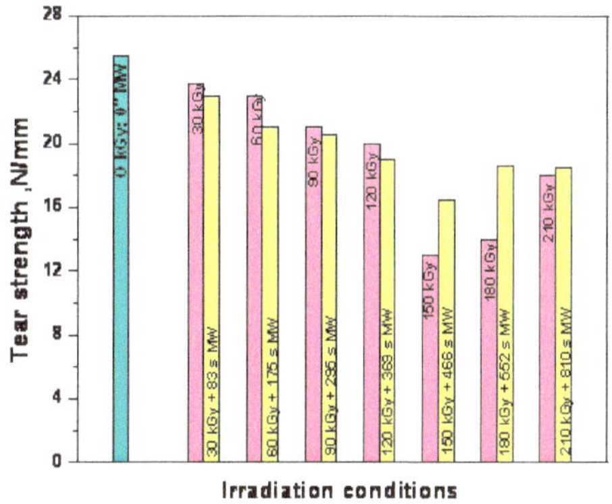

Figure 26. The EB and EB+MW effects on the tear strength of the EPDM samples

Tensile strength (Figure 23), elongation at break (Figure 24), permanent set (Figure 25) and tear strength (Figure 26) decrease with the irradiation dose increasing for both EB and EB + MW irradiation, (comparing with non-irradiated control sample), but the characteristics from crosslinking by irradiation are better compared with crosslinking with peroxide. For all these mechanical features the best results are obtained for a dose of 150 kGy (crosslinking with EB) and 150 kGy + 466 s (crosslinking wit EB + MW). Following the obtained results it can be stated that a dose of maximum 150 kGy + 466s. leads to a good degree of crosslinking, for which there is a return to its original shape after a very good stretching (29%).

6. Conclusion

This review gives an overview about our research on elastomer crosslinking by irradiation with accelerated electrons, a much more ecologic method that does not need to add crosslinking agents into the blend. Crosslinking by EB and EB+MW also shows a series of advantages, like as: reduced crosslinking time, no polymer degradation due to high temperature (as in the classic peroxide curing) because EB and EB+MW crosslinking is performed at room temperature, the process is very fast and can be controlled precisely. At the interaction of ionizing radiation with (co)polymers, breaking of covalent bonds occurs, as well as the emergence of free radicals (transitional chemical species) on the main chain (if the lateral groups break) or in the main chain (if it breaks itself). The final effect is either crosslinking of macromolecular assembly or cutting the main chain of macromolecules and decreasing average molecular weight. In fact, the two effects, crosslinking and degradation, coexist and we need to point out the predominance of one of them. In addition, through appropriate surface treatments, some features of the feet such as appearance, adhesion to different materials or slip resistance can be significantly improved (Zaharescu, 2000).

Author details

Elena Manaila, Maria Daniela Stelescu* and Gabriela Craciun
National Research and Development Institute for Laser, Plasma and Radiation Physics, Magurele, Bucharest, Romania

Maria Daniela Stelescu
National Research and Development Institute for Textiles and Leather – Leather and Footwear Research Institute, Bucharest, Romania

7. References

Alvarez Grima, M.M. (2007). *Novel Co-agents for Improved Properties in Peroxide Cure of Saturated Elastomers*, PhD Thesis, Printed by Print Partners Ipskamp, ISBN: 90-365-2456-3, Enschede, Netherlands

Andreo, P., Cunningham, J.F., Hohlfeld, K., &. Svensson, H. (1997). Absorbed dose determination in photon and electron beams : An International code of practice, *Technical Reports Series No.277*, 2nd edition, International Atomic Energy Agency, Vienna

Bhowmick, A. K. & Vijayabaskar, V. (2006). Electron beam curing of elastomer, *Rubber Chemistry and Technology*, Vol.79, No.3, pp. 402-428

Boye, W.M. (2008). Utilizing Coagents in the Electron Beam Cure of Elastomers, Proceedings of the 57th International Wire & Cable Symposium IWCS, pp.335- 341

Chowdhury, R. & Banerji, M.S. (2005). Electron Beam Irradiation of Ethylene-propylene Terpolymer: Evaluation of Trimethylol Propane Trimethacrylate as a Crosslink Promoter, *Journal of Applied Polymer Science, Vol.97, No.3, pp.* 968–975

Chlorinated polyethylene; CM (CPE). (2009). pp. 23-25, www.hallstar.com;

Dikland, H. G., Ruardy, T., Van der Does, L. & Bantjes, A. (1993). New coagents in peroxide vulcanization of EPM, *Rubber Chemistry and Technology*, Vol.66, No.5, pp.693-711

Dluzneski, P.R. (2001). Peroxide vulcanization of elastomers, *Rubber Chemistry and Technology*, Vol.74, No.3, pp. 451-492

* Corresponding Author

Endstra, W.C. (1990). Application of coagents for peroxide crosslinking, *Kautschuk und Gummi Kunststoffe,* Vol.43, No.9, pp.790-793

Flynn, J. H. & Davis, W. H. (1985). Tyriny Brand CPE Thiadiazole Cure System Studies-Chemistry and Dispersion, presented at a meeting of the *Rubber Division, American Chemical Society*, Los Angeles, California, 23–26 April 1985.

Hafezi, M., Khorasani, S.N. & Ziaei, F. (2006). Application of Taguchi method in determining optimum level of curing system of NBR/PVC blend, *Journal of Applied Polymer Science*, Vol.102, No.6, pp.5358-5362

Henning, S.K. (2008). The Use of Coagents in the Radical Cure of Elastomers, *Proceedings of the 56 th International Wire & Cable Symposium*, pp. 587-593

Henning, S.K. (2008). Use of coagents in the radical cure of elastomers, , *Wire & Cable Technology International*, Vol. XXXVI, No. 3, pp.52-59

Hofmann, H. (1967). *Vulcanization and vulcanizing agents*, MacLaren and Sons Ltd, London, Palmerton Publishing Co. Inc., New York.

Karpeles, R. & Grossi, A.G. (2001). *EPDM Rubber Technology*, Handbook of Elastomers, 2nd Ed., Anil K. Bhowmick and Howard L. Stephens [Editors], Marcel Decker, Inc., New York, pp. 845-876.

Manaila, E., Martin, D., Zuga, D., Craciun, G., Ighigeanu, D. & Matei, C. (2008). Radiation processing of rubber mixtures with polyfunctional monomers, *Proceedings of the 11th International Conference on Optimization of Electrical and Electronic Equipment*, OPTIM, Vol.1, pp. 125-130, Brasov, Romania, 22-23 May 2008

Manaila, E., Martin, D., Stelescu, M.D, Craciun, g., Ighigeanu, D. & Matei, C. (2009). Combined effects of microwaves, electron beams and polyfunctional monomers on rubber vulcanization, *Journal of Microwave Power and Electromagnetic Energy (JMPEE)*, Vol.43, No.3, pp.26-34

Manaila, E., Martin, D., Craciun, G., Zuga, D. & Ighigeanu, D. (2009). Electron beam processing of rubber mixtures with polyfunctional monomers, *Proceedings of the 7th Conference on Nuclear and Particle Physics (NUPPAC'09)*, pp.645-650, Sharm El-Sheikh, Egypt, 11-15 Nov. 2009

Manaila, E., Stelescu, M.D., Ighigeanu. D, Craciun, D. & Berechet, D. (2011), The influence of TMPT coagent on the mechanical properties of the EPDM rubber cross-linked by irradiation *Leather and Footwear Journal,* Vol.11, No.3, pp.201-210

Manaila, E., Stelescu, M.D. & Craciun, G. (2011). Characteristics of natural rubber blends vulcanized with electron beam and microwave, *Leather and Footwear Journal*, Vol.11, No.1, pp. 43-52.

Martin, D., Jianu, A. & Ighigeanu, D. (2001). A method for the 2.45-GHz magnetron output power control, *Transactions on Microwave Theory and Techniques IEEE*, Vol.49, No.3 (2001) 542.

Martin, D., Ighigeanu, D., Mateescu, E., Craciun, G. & Ighigeanu, A. (2002). Vulcanization of rubber mixtures by simultaneous electron beam and microwave irradiation, *Radiation Physics and Chemistry*, Vol.65, No.1, pp. 63-65

MGM Rubber Company-Research and Development, (2007). Electron Beam Radiation Technology for Curing, http://www.mgmrc.com/ebeam-trech.htm

Odian, G. (1964). Radiation crosslinking of polyethylene–polyfunctional monomer mixtures, *Journal of Polymer Science*, Vol.2, No.6, pp.2835-2848

Stelescu, M.D, Manaila, E. & Zuga, N. (2011). The use of polyfunctional monomers in the radical cure of chlorinated polyethylene, *Polymer Journal*, Vol.43, No.9, pp.792–800

Stelescu, M.D., Manaila, E., Craciun, G. & Zuga, N. (2012). Crosslinking and grafting ethylene vinyl acetate copolymer with accelerated electrons in the presence of polyfunctional monomers, *Polymer Bulletin*, Vol. 68, No.1, pp.263-285.

Stelescu, M.D. & Manaila, E. (2007), Crosslinking and grafting the natural rubber by means of accelerated electrons in the presence of trimethylol-propane trimethacrylate (TMPT), *Scientific Bulletin, B Series: Chemistry and Materials Science*, Vol.69, No.4, pp.33-38

Stelescu, M.D., Georgescu, M. & Manaila, E. (2010). Aspects regarding crosslinking of a natural rubber blend, *Proceedings of the 3rd International Conference Advanced Materials and Systems, ICAMS 2010*, pp. 313-318, Bucharest, Romania, 16-18 September 2010

Stelescu, M.D., Niculescu-Aron, I.G. & Manaila, E. (2009). Processing and statistical analysis of the experimental data resulted from EPDM rubber grafting and crosslinking with accelerated electrons in the presence of TMPT, *Materiale plastice*, Vol.46, No.1, pp.48-52

Stelescu, M.D., Vaslan, M. & Manaila, E. (2008). Materials based on chlorinates polyethylene obtained by crosslinking and grafting by accelerated electrons in the presence of trialylcyanurated, *Proceedings of the 2rd International Conference Advanced Materials and Systems, ICAMS 2008*, pp. 96-101, Bucharest, Romania, 23-24 October 2008

U.S. Patent 3 454 544, Process for the Chlorination of Polyolefins, Issued 8 July 1969 to Dow Chemical USA.

Van Duin, M. (2002). Chemistry of EPDM Crosslinking, *Kautschuk und Gummi, Kunststoffe*,Vol.55, No.4, pp.150-156

Vijayabaskar, V. & Bhowmick, A. K. (2005). Electron beam modification of nitrile rubber in the presence of polyfunctional monomer, *Journal of Applied Polymer Science*, Vol. 95, No.2, pp. 435-447

Yasin, T., Ahmed, S., Ahmed, M. & Yoshii, F. (2005). Effect of concentration of polyfunctional monomers on physical properties of acrylonitrile-butadiene rubber under-electron beam irradiation, *Radiation Physics and Chemistry*, Vol.73, No.3, pp. 155-158

Zaharescu, T., Setnescu, R., Jipa, S. & Setnescu, T. (2000). Radiation processing of polyolefin blends. I. Crosslinking of EPDM–PP blends, *Journal of Applied Polymer Science*, Vol. 77, No.5, pp. 982-987

Zuga, M.D., Iovu, H., Trandafir, V., Manaila, E., Martin., D. & Stelescu, M.M. (2007). Study on the preparation of some biocomposites based on silicone elastomers and collagen, *Jurnal of Optoelectronics and Advenced Materials (JOAM)*, Vol.9, No.11, pp.3325-3329

Zuga, M.D., Miu., L., Crudu, M., Bratulescu, V., Iovu, H. & Manaila, E. (2007). Products of ethylene-propylene-terpolymer rubber (EPDM) obtained by an environmentally friendly process, *Advanced Materials Research – Materials and Technologies*, Vol.23, pp. 333-336

Section 2

Liquid Crystal Elastomers

Chapter 2

Polysiloxane Side-Chain Azobenzene-Containing Liquid Single Crystal Elastomers for Photo-Active Artificial Muscle-Like Actuators

Jaume Garcia-Amorós and Dolores Velasco

Additional information is available at the end of the chapter

1. Introduction

Liquid crystals (LCs) are unique materials with amazing properties and uses. Since their discovery in 1889 by F. Reinitzer[1], they have experienced an explosive growth because of their successful application in a wide variety of areas such as information displays[2], cosmetics and health care[3], thermography[4-5], artificial muscle-like actuators[6-7] and enantioselective synthesis[8], among others. Thus, liquid crystals play an important role in modern technology and they are present in the most common devices used in our daily live. Research into this field is growing day by day and new promising applications for such materials are discovered and developed continuously.

The liquid-crystalline state or mesophase is a different state of condensed matter, which is intermediate between solids and liquids (**Figure 1**).[9] Liquid-crystalline materials combine both the molecular order typical of the solid phases and the fluidity characteristic of the isotropic liquid state. Compared with conventional liquids, liquid crystals exhibit long-range orientational order and, in some cases, some partial positional order. Although the degree of order present in a mesophase is lower in comparison to that in conventional solids, it is high enough to induce a great anisotropy in the properties of the system. Hence, liquid crystals are fluids with anisotropic properties because of the mesogens tendency to point to a common direction called director, ***n***.

Liquid-crystalline elastomers (LCEs) consist on high-molecular mass liquid crystals where the mesogenic moieties can be connected either head-to-tail, forming the polymer main chain (main-chain LCEs), or attached as side-chain groups to the main polymer backbone (side-chain LCEs), generally *via* flexible spacers, which allow the polymer main chain to accommodate the anisotropic arrangement of the mesogenic side groups (**Figure 2**).[10-11]

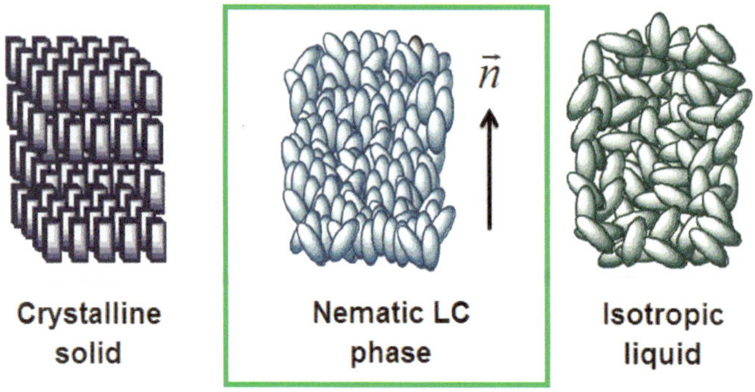

Figure 1. Schematic representation of the molecular distribution in the solid, nematic liquid-crystalline and liquid state, respectively.

Among all liquid-crystalline phases, the nematic one, where the mesogens are approximately oriented parallel to their longest axis, is the less ordered and, therefore, the less viscous. As a consequence, the mesogens alignment can be more easily manipulated than in the more ordered ones. Hence, nematic liquid-crystalline materials are the most commonly used for technical applications. Specifically, this chapter focuses on the photo-actuating properties of nematic liquid-crystalline elastomeric materials.

The properties of liquid-crystalline materials can be easily modulated by applying external perturbations such as light, temperature, electro-magnetic fields, changes of solvent or pH, and so on, which induce changes in the molecular and supermolecular organization of such systems. This feature is the basis of all their further applications.[12-13]

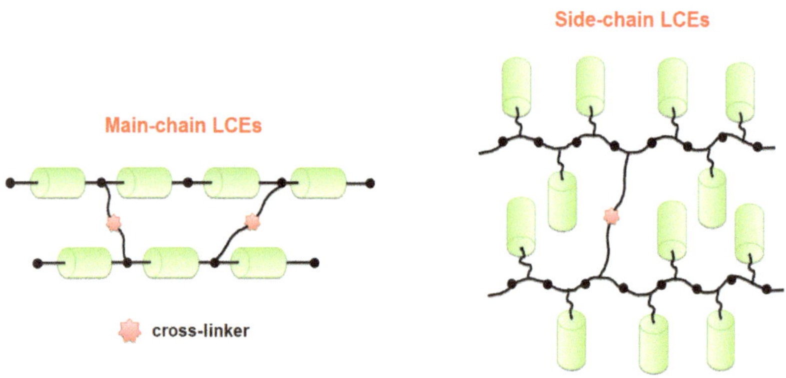

Figure 2. Main-chain (left) and side-chain (right) liquid-crystalline elastomeric materials.

Among all the possible external inputs, light enables a rapid and punctual wireless control of the properties of the material and, moreover, it is a clean, cheap and environmentally-friendly energy source. Several chromophores are known in photochemistry: spiropyranes, diarylethenes, fulgides, stilbenes, viologens, *etc*. However, azobenzenes are doubtlessly the most used ones for designing optically-controlled materials because of their totally clean and reversible isomerisation process between their two isomers of different stability: *trans* and *cis*. The most interesting feature of azobenzenes is that both isomers can be switched back and forward with light of particular wavelengths: UV light ($h\nu_1$), for the *trans*-to-*cis* conversion, and visible light ($h\nu_2$), for the *cis*-to-*trans* isomerisation. Moreover, *cis* isomer is less stable than the *trans* one because it has a bent shape which increases the steric hindrance and somehow breaks the conjugation of the *trans* linear form. Thus, the metastable *cis* isomer will also relax back spontaneously to the thermodynamically stable *trans* form in the dark isothermally (Δ, **Figure 3**).[14]

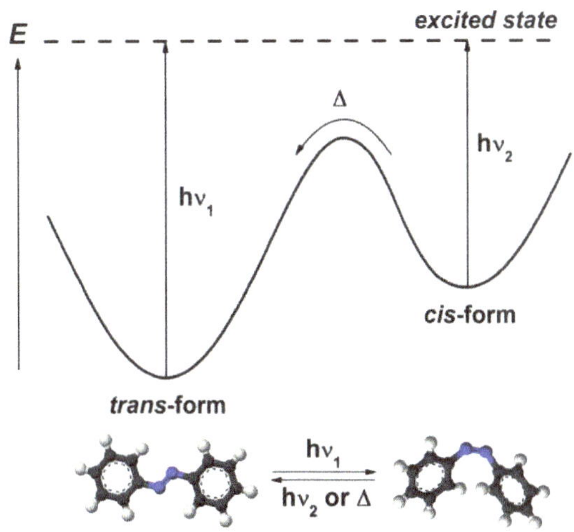

Figure 3. Photochromism of azobenzene and energetic profile for its *trans*-to-*cis* and *cis*-to-*trans* isomerisation processes.

The rod-like structure of *trans*-azobenzene allows its easy introduction in both low- and high-molecular mass nematic liquid crystals without causing the destruction of the host mesophase. When the system is irradiated with light of the appropriate wavelength ($h\nu_1$), the bent *cis* form is generated. As a consequence, the orientation of all the mesogenic molecules of the sample will also change (domino effect). In this way, the *cis* azo-moiety acts as an impurity lowering the nematic order. This effect can be nicely observed experimentally when azobenzene-doped liquid-crystalline mixtures are irradiated at a constant temperature, *T*, within $T_{N\text{-}I\ (trans)}$ and $T_{N\text{-}I\ (cis)}$. In this case, the sample changes from the ordered nematic phase to the disordered isotropic one isothermally, that is, a photo-induced nematic-to-isotropic phase transition occurs (**Figure 4 and 5**).[15-17] The initial state is restored on turning off the irradiation due to the thermal back *cis*-to-*trans* isomerisation of the azo-dye in the dark (Δ).[18] This effect can be exploited for obtaining promising light-controlled liquid-crystalline materials such as photo-active artificial muscle-like actuators and light-driven optical switches.

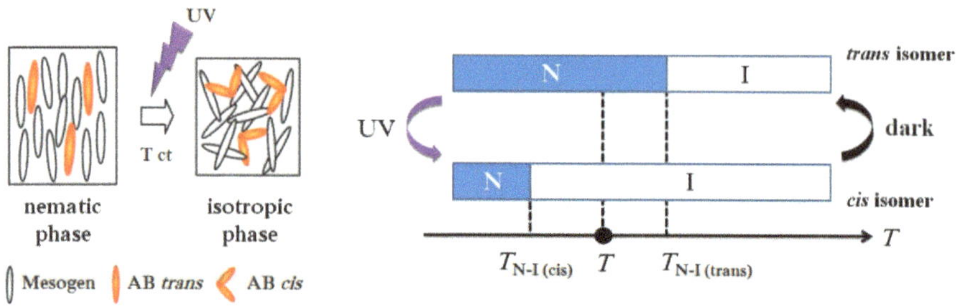

Figure 4. Photo-induced nematic-to-isotropic phase transition in azobenzene-doped nematic liquid crystals.

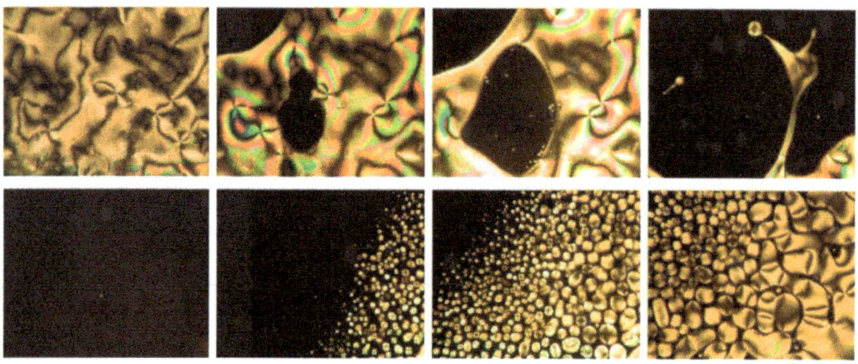

Figure 5. Photo-induced nematic-to-isotropic phase transition in an azobenzene-containing nematic liquid-crystalline mixture. Destruction of the host nematic mesophase by irradiation with UV light at a constant temperature *T* between $T_{N-I\ (cis)}$ and $T_{N-I\ (trans)}$ (up). Regeneration of the nematic mesophase when the thermal *cis*-to-*trans* relaxation of the azobenzene occurs in the dark (down).

2. Liquid-crystalline elastomers for light-induced artificial muscle-like actuation

Photo-active artificial muscle-like actuators convert light into mechanical quantities such as displacement, strain, velocity and stress. Moreover, materials used to produce muscle-like movements should be soft and deform easily upon irradiation. Specifically, polymers are very interesting materials for this purpose due to their many attractive properties and characteristics; they are lightweight, inexpensive, easily manufacturable and implementable, fracture tolerant, pliable and biocompatible.[19] So, light-driven polymer-based actuators are valuable materials to be implemented in a wide range of micro- and macro-scale devices.

Generally, photo-active liquid-crystalline polymers and elastomers are multidomain systems and, therefore, the director changes abruptly from one domain to another. Hence, both LCPs and LCEs deform themselves when non-polarised light falls on them in an isotropic way, that is, there is no preferential direction for the deformation and, as a consequence, the degree of deformation of the material is generally small.[20]

Highly noteworthy photo-actuating properties have been successfully achieved in polydomain azobenzene-based liquid-crystalline elastomeric materials by Ikeda.[21-26] These materials show a great variety of three-dimensional contraction and expansion movements when they are exposed to polarised light of the appropriate wavelength. Within the last few years, it has been reported that it is possible to control the bending direction of the LCE film depending on the role of the azoderivative within the elastomeric network, that is, if the azo-chromophore acts as a cross-linker or as a simply pendant group.[27] Moreover, the first prototypes of real light-controlled plastic rotating motors have been fabricated with laminated films of photo-sensitive LCEs.[28]

However, if non-polarized light is used instead, it is strictly necessary that the director points to a unique direction in the whole sample, that is, a monodomain sample is required.

Liquid single crystal elastomers (LSCEs) are a subclass of liquid-crystalline polymers, which were synthesized for the first time by Küpfer and Finkelmann in the early nineties[29], although their possible use for both thermally- and photo-controlled artificial muscles was predicted earlier theoretically by P. G. de Gennes.[30] LSCEs consist in weakly cross-linked polymer networks with a macroscopic orientation of the director, $\boldsymbol{n}$. These materials combine the elasticity typical of conventional rubbers with the anisotropic properties characteristic of liquid-crystalline systems. As a result of this coupling, the macroscopic dimensions of the elastomeric sample can be easily modified by the appropriate modification of the LC order through the application of different external stimuli such as temperature variations or, when an azo-dye is introduced in the system, by irradiating the probe with light of the appropriate wavelength.[31]

Indeed, LSCEs experiment a spontaneous contraction along the director direction when the system is driven from the ordered nematic phase to the disordered isotropic one by heating the sample over the nematic-to-isotropic phase transition temperature, $T_{N\text{-}I}$, due to the mesogens misalignment. If the temperature is lowered below $T_{N\text{-}I}$, the LSCE expands back thereby recovering its original shape. This effect is the so-called thermo-mechanical effect (**Figure 6a and 7**).[32-36]

The introduction of photo-sensitive azoderivatives in LSCEs, not only as side-chain groups but as cross-linkers as well, affords a new way to control the macroscopic dimensions of the sample isothermally just by applying light. So, azobenzene-containing LSCEs are very attractive materials for the fabrication of light-controlled artificial muscle-like actuators.[37-38] During the last decades, both the preparation and properties modulation of light-driven artificial muscle-like actuators based on azobenzene-containing LSCEs are topics of intensive research. Thus, in this chapter we will describe the last investigations that have been performed in our group towards not only to enhance the mechanical efficiency of such materials but also to optimise the response time of the final artificial muscle.

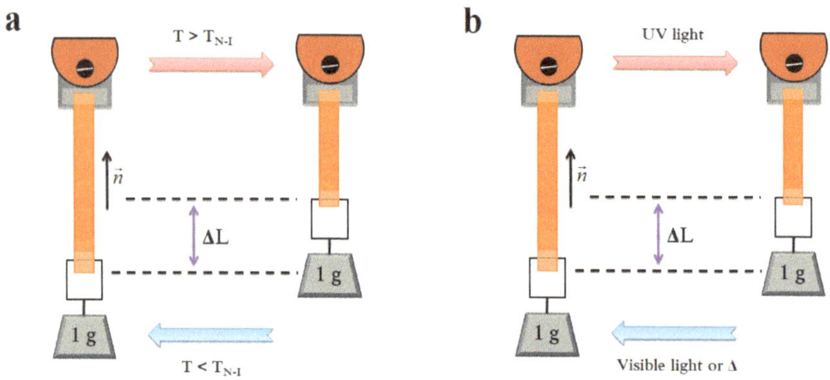

Figure 6. Thermo-mechanical (a) and opto-mechanical (b) effect in nematic LSCEs.

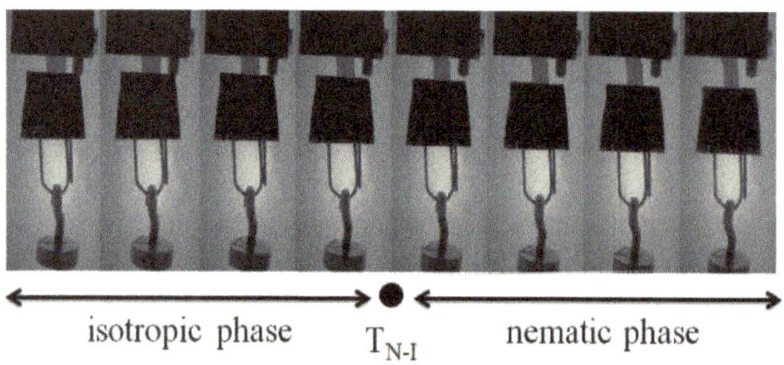

Figure 7. Uniaxial thermal expansion of an LSCE sample on lowering the temperature as a consequence of the isotropic-to-nematic phase transition.

Besides the thermo-mechanical effect commented above, LSCEs that contain isomerisable azobenzenes as light-sensitive molecules also undergo macroscopic contractions in a preferential direction when they are exposed to non-polarised light of the appropriate wavelength (opto-mechanical effect, **Figure 6b**). This happens due to the *trans*-to-*cis* isomerisation of the azo-dye, which drops the nematic order parameter of the elastomeric sample.[38] On turning off the light, the system recovers its initial dimensions due to the thermal back isomerisation of the azo-chromophore. Photo-mechanical effect in nematic LSCEs was first observed by Finkelmann *et al.*[39] in 2001 and it has been deeply investigated during the last decade from both the theoretical and the experimental point of view. A mathematical description of the opto-mechanical effect will be given in Section 4 of this chapter.

3. Preparation and characterization of polysiloxane-based liquid single crystal elastomers

As it has been commented above, both low- and high-molecular mass liquid crystals present different domains and, therefore, there are some places in the sample where the director changes abruptly. However, for many applications like artificial muscle-like actuation, it is strictly necessary that the system should be macroscopically oriented, *i.e.* a monodomain sample used.

Several techniques are used for this purpose with low-molecular mass LCs like rubbing, surface treatments with polymers or the application of external fields in a determined direction. High-molecular mass LCs, like linear LCPs, can be macroscopically oriented by extruding the polymeric mass in order to create very thin oriented fibres. However, this technique is not useful for LCEs. In such systems, the macroscopic orientation of the sample should be carried out during its preparation.

Many smart and functional liquid-crystalline elastomeric materials have been prepared using different polymer backbones. However, silicone is doubtlessly one of the most commonly used polymer backbone for preparing LSCEs due to its great advantages like good thermal stability, constancy of properties over a wide range of temperature which leads to a large operating temperature range – from −100 to 250 ºC –, hydrophobicity, excellent resistance to oxygen, ozone and sunlight, good flexibility, anti-adhesive properties and low toxicity. An additional noteworthy point is that silicone polymers, of which polyhydrogenomethylsiloxane is an example, have very low glass transition temperatures (*ca.* $T_g = -120$ °C). This feature allows polysiloxane-based LSCEs to present photo-actuation even at room temperature.

Generally, polysiloxane-based side-chain liquid single crystal elastomers are prepared following the synthetic methodology in three steps, which was first developed by Küpfer and Finkelmann (**Figure 8**).[29]

In the first step, the different monomers, those are, mesogen/s, cross-linker/s and comonomer/s, react through their terminal olefin with the polysiloxane backbone through a Pt-catalysed hydrosilylation reaction which takes place at 70 °C (**Figure 9**). This reaction is carried out at 5000 rpm by means of the spin-casting technique. The aim of this step is to obtain a stable elastomeric system, which can be handled properly, but partially cross-linked for its further macroscopic orientation. For this purpose, the hydrosilylation reaction is stopped before it is completely finished.

In the second step, the orientation of the different directors of the entire sample is performed. This process is induced by the application of a uniaxial force to the sample along its longest axis. After the suitable time, a monodomain LCE is obtained, that is, a liquid single crystal elastomer (LSCE). The progress of the orientation process can be followed qualitatively *in situ* by checking the optical transparency of the probe. When the orientation of the director of the different domains is successful, one gets an LSCE, which is perfectly transparent (**Figure 10a**). Contrarily, if a macroscopically disordered sample is obtained, it will show opacity due to the scattering of the incoming light through the sample (**Figure 10c**).

As it has been aforementioned, macroscopic changes in shape are coupled to changes at the molecular scale. Hence, the deformation of an elastomer can lead to reorientation of the

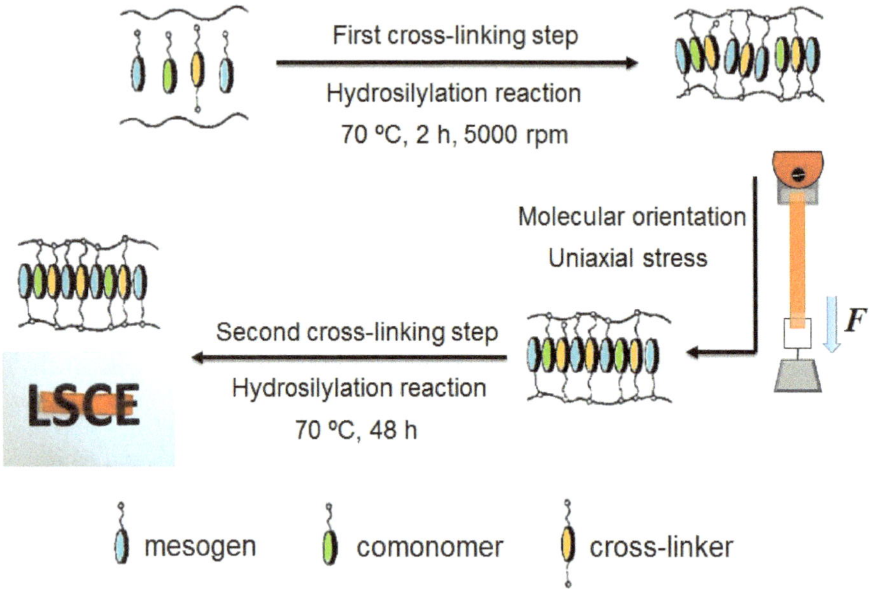

Figure 8. Synthetic methodology in three steps used for preparing photo-active side-chain polysiloxane-based nematic LSCEs.

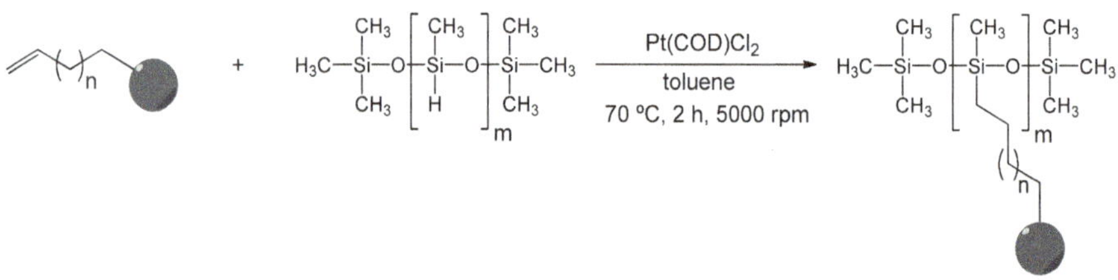

Figure 9. Attachment of an alkene-terminated monomer to the polysiloxane backbone through the Pt-catalysed hydrosilylation reaction. ($Pt(COD)Cl_2$ = dichloro(1,5-cyclooctadiene)platinum(II)).

Figure 10. Qualitative evaluation of the macroscopic orientation of the different directors of the elastomeric sample by means of their optical transparency: monodomain (a), partial monodomain (b) and polydomain (c).

director or vice versa, reorientation of the mesogens can change the shape of the elastomer (see thermo- and opto-mechanical effects in Section 2 of this chapter). In this way, since an LSCE is prepared, its macroscopic dimensions are directly determined by the degree of order created during the second stage of its fabrication. So, the macroscopic orientation of the different directors of the system becomes the key step of the whole synthetic procedure since it will determine the proper actuation of the final material.

In the third stage of the synthesis, the created anisotropy should be fixed by a second cross-linking reaction without removing the applied force. This second hydrosilylation reaction is carried out completely in an oven at 70 °C for 48 h. Afterwards, the LSCE is purified in order to remove both the catalyst and the non-reacted monomers (*ca.* less than 1%) from the network. This cleaning procedure is carried out by a swelling-deswelling process using acetone and hexanes, respectively. Finally, the complete characterisation of the prepared LSCE should be done. The conventional techniques used for characterising nematic LSCEs are briefly described below.

Not only the proper macroscopic orientation but also both the thermal and the mechanical properties of liquid-crystalline elastomers can be studied by means of a wide variety of techniques. Generally, polarised optical microscopy (POM) and X-Ray diffraction (XRD) are used together to get insight into the structure of the mesophase unambiguously but also to evaluate its macroscopic ordering. Moreover, differential scanning calorimetry (DSC) is needed to determine the temperature range of stability of the LC phase. Besides these techniques, additional experiments, such as swelling experiments, are performed on LSCEs to be used further for muscle-like actuation purposes.

The density of cross-linking units present in an LSCE, which is directly related with the mechanical properties of the elastomeric material, can be easily evaluated by means of swelling experiments. Liquid-crystalline elastomers, as a main difference with linear liquid-crystalline polymers and other solids, cannot be dissolved due to the presence of cross-linking points. When an LSCE is immersed in a suitable solvent, it absorbs a large amount of it without dissolving and, as a consequence, it experiments a large deformation producing a small internal stress in the network. Hence, the free energy change that takes place in the elastomer during its swelling process can be separated in two additive contributions, those are, the free energy of mixing, ΔG_{mix}, and the free energy related with the elastic deformation of the network, ΔG_{el}.[40-42]

Traditional swelling experiments are carried out by placing the sample in a vessel containing a suitable solvent (*e.g.* toluene).[43] Then, the system is allowed to reach the thermodynamically equilibrium conditions ($\Delta G_{swelling} = 0$; $\Delta G_{mix} = -\Delta G_{el}$). The swelling parameter, q, is defined as the ratio of volumes of the swollen and original elastomer both measured under thermodynamically equilibrium conditions. Because of the anisotropy of the liquid-crystalline elastomer in its z- direction (**Figure 11**), both x- and y-isotropic dimensions change in the same way but differently from the z- one. Thus, the swelling parameter can be expressed as it is indicated in eq. 1.

$$q = \frac{V}{V_0} = \frac{x \cdot y \cdot z}{x_0 \cdot y_0 \cdot z_0} = \frac{z}{z_0} \cdot \left(\frac{x}{x_0}\right)^2 \tag{1}$$

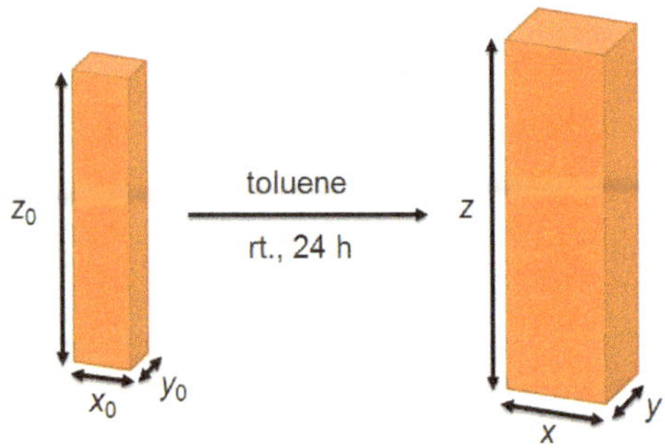

Figure 11. Change in the elastomer volume on reaching thermodynamically equilibrium after swelling.

According to that mentioned above, those samples that contain a high density of cross-linking units will yield a low value of the swelling parameter. However, it is highly remarkable that the magnitude of the swelling parameter depends also greatly on the chemical nature of the cross-linker. In this way, elastomers with more rigid cross-linkers afford lower values of the swelling parameter than those with more flexible ones, although they contain the same density of cross-linking points.[44]

4. Opto-mechanical effect in liquid single crystal elastomers

The macroscopic dimensions of an LSCE are directly related with the degree of order induced during its fabrication. So, any reorientation of the mesogens will lead to changes in the shape of the elastomer. When a photo-active LSCE is illuminated with light of the appropriate wavelength, the azo-chromophore changes its geometry from linear to bent due to its *trans*-to-*cis* photo-isomerisation. This fact produces a microscopic disorganisation of those mesogen molecules that are close to the azo-dye ones producing a decrease in the local order parameter. As a consequence of this molecular disorganisation, a shortening of the LSCE in the director direction is observed (see **Figure 6b**). If the network is fixed by both ends, the system cannot shrink and, as a consequence, the appearance of a retractive force in the elastomer is observed.[38-39]

Opto-mechanical experiments consist in the measurement of the evolution of the internal stress generated inside the LSCE with the time. A typical opto-mechanical experiment for a photo-active liquid single crystal elastomer is shown in **Figure 12**. On turning on the light, the internal stress created in the elastomer, σ, grows until the photo-stationary state is reached. The curve describes a *plateau* which corresponds to the maximum opto-mechanical response produced by the artificial muscle-like actuator, $\Delta\sigma_{max}$. When the irradiation is ceased, the thermal back *cis*-to-*trans* isomerisation of the azo-dye occurs and the stress starts to diminish with the time until the initial stress value is recovered.

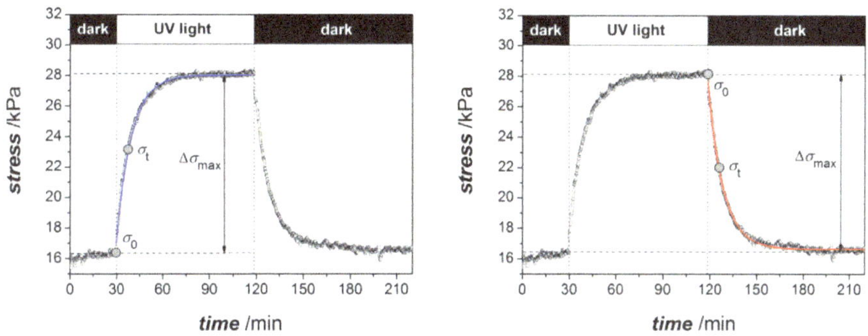

Figure 12. Opto-mechanical experiments: increase of the internal stress generated in the network upon irradiation with UV light (left) and decrease of the stress with the time in the dark at a constant temperature, *T* (right).

Besides the maximum stress that the system is able to generate when UV light falls on it, $\Delta\sigma_{max}$, opto-mechanical experiments allow also determining the time required by the network to produce its maximum mechanical response, τ^{irrad}, and that necessary to recover the initial state thermally, τ^{th}. These three parameters are crucial in the overall performance of LSCEs for artificial muscle-like photo-actuation.

The UV-irradiation process of an LSCE is a bidirectional step, since the *trans*-to-*cis* photo-isomerisation reaction competes with the thermal *cis*-to-*trans* isomerisation simultaneously.

$$\textbf{\textit{trans}} \underset{k^{th}}{\overset{k^{ph}}{\rightleftarrows}} \textbf{\textit{cis}} \tag{2}$$

Both individual isomerisation processes are well known to follow a first order profile for low-molecular weight azoderivatives in isotropic and nematic solution as well as in dense polymer matrixes.[45] Assuming a similar kinetic behaviour for LSCEs, the rate of this process is given by both the disappearance of the *trans* isomer with the time by the photo-induced reaction and its formation through the thermally-activated one (eq. 3).

$$v = -\frac{d[T]}{dt} = k^{ph} \cdot [T] - k^{th} \cdot [C] = k^{ph} \cdot [T] - k^{th} \cdot \left([T]_0 - [T] \right) \tag{3}$$

k^{ph} and k^{th} correspond to the first-order rate constants for the photo-induced *trans*-to-*cis* and the thermal *cis*-to-*trans* isomerisation processes of the elastomeric network, respectively. $[T]$ and $[C]$ stand for the concentration of *trans* and *cis* isomer at each moment of the reaction, respectively, and $[T]_0$ is the total concentration of azo-dye present in the elastomeric system, that is, the total number of *trans* plus *cis* isomers. $[T]_0$ is a constant value and it is given by the initial composition of the material.

Once integrated the differential equation 3, one gets the time-dependence of the concentration of both isomers *trans* (eq. 4) and *cis* (eq. 5). The curve describing this process is an exponential growth with an apparent rate constant, k^{irrad}, which is related with the first-order kinetic constants of both processes ($k^{irrad} = k^{ph} + k^{th}$).

$$[T]=[T]_0 \frac{k^{th} + k^{ph} \cdot \exp\left[-(k^{ph} + k^{th}) \cdot t \right]}{k^{th} + k^{ph}} = [T]_0 \frac{k^{th} + k^{ph} \cdot \exp\left[-k^{irrad} \cdot t \right]}{k^{irrad}} \tag{4}$$

$$[C]=[T]_0 \frac{k^{ph}}{k^{th} + k^{ph}} \left(1 - \exp\left[-(k^{ph} + k^{th}) \cdot t \right] \right) = [T]_0 \frac{k^{ph}}{k^{irrad}} \left(1 - \exp\left[-k^{irrad} \cdot t \right] \right) \tag{5}$$

As we mentioned above, the generation of the *cis*-isomer in the LSCE is the responsible of the mechanical stress observed when the sample is UV-irradiated. In fact, the bent *cis* isomer should be considered as an impurity which shifts the critical temperature of the nematic-to-isotropic phase transition of the LSCE (see Section 1 of this Chapter).[38] Hence, equation 5 can be transformed into equation 6 which stands for the variation of the measured internal stress produced by the elastomer shrinking with the time when it is illuminated with UV light.

$$\sigma_t - \sigma_0 = \Delta\sigma_{max} \cdot \left(1 - \exp\left[-k^{irrad} \cdot t \right] \right) \tag{6}$$

On the other hand, the thermal relaxation process of the LSCE is a unidirectional step.

$$\textbf{\textit{cis}} \xrightarrow{k^{th}} \textbf{\textit{trans}} \tag{7}$$

In this case, the rate of the process is given uniquely by the disappearance of the *cis* isomer with the time. The rate equation for this unimolecular first-order reaction is described by equation 8.

$$v = -\frac{d[C]}{dt} = k^{th} \cdot [C] \tag{8}$$

The integration of the differential equation 8 yields the integrated rate equation 9 which describes the evolution of the *cis* isomer concentration with the time during the thermal relaxation of the network. The curve describing this process is an exponential decrease with a first-order rate constant, k^{th}.

$$[C] = [C]_0 \cdot \exp\left(-k^{th} \cdot t \right) \tag{9}$$

In the same manner than for eq. 6, equation 9 can be transformed into eq. 10 which fits for the variation of the measured internal stress in the elastomer with the time when it is kept in the dark at a constant temperature, *T*:

$$\sigma_t - \sigma_0 = \Delta\sigma_{max} \cdot \exp\left(-k^{th} \cdot t \right) \tag{10}$$

The maximum stress that the system is able to generate upon irradiation, $\Delta\sigma_{max}$, is directly obtained by subtracting the initial stress, σ_0, from the steady-state saturation value reached under irradiation (see **Figure 12**). Both k^{irrad} and k^{th} are determined by fitting equation 6 and 10 to the experimental data, respectively. The characteristic times for both processes, τ^{irrad} and τ^{th}, are calculated from k^{irrad} and k^{th}, respectively, as $\tau = 1/k$.

The thermal activation parameters, the enthalpy ($\Delta H^{\neq}$) and entropy ($\Delta S^{\neq}$) of activation, can be easily determined by studying the evolution of both isomerisation rate constants as a function of the temperature. This behaviour is well described by means of Eyring's equation (eq. 11).[46-47]

$$\ln\frac{k}{T} = \frac{-\Delta H^{\neq}}{R}\cdot\frac{1}{T} + \ln\frac{k_B}{h} + \frac{\Delta S^{\neq}}{R} \qquad (11)$$

R is the universal gas constant (R = 8.314 J·K^{-1}·mol^{-1}), k_B is the Boltzmann's constant (k_B = 1.381 × 10^{-23} J·K^{-1}) and h is the Planck's constant (h = 6.626 × 10^{-34} J·s), respectively.

The characteristic parameters ($\Delta\sigma_{max}$, τ^{irrad} and τ^{th}) for all the artificial muscle-like actuators presented herein have been determined according to the kinetic model described above. The relaxation time registered for the different photo-active LSCEs reported in this chapter ranges from several hours to a few seconds. This fact demonstrates that not only fast-responding artificial muscles but also bistable systems can be obtained easily from polysiloxane-based LSCEs just through the proper substitution of the light-sensitive azo-chromophore.

5. Mechanical efficiency of photo-active liquid single crystal elastomer-based actuators through the variation of the azo-cross-linker flexible spacer

Light-controlled artificial muscle-like actuators based on photo-active polysiloxane LSCEs have been growing interest during the last decade and deeply investigated from both the theoretical and the experimental point of view. As it has been aforementioned, two key parameters are needed to characterise properly the actuation ability of such materials, those are, the maximum mechanical response that they are able to generate by irradiation with light of the appropriate wavelength as well as the time required to produce it and that to recover the initial state further. At the present moment, many efforts are being put forward to improve and optimise both of them.

At this point, we will turn our attention towards the enhancement of the mechanical response produced by those photo-actuators based on polysiloxane azobenzene-containing LSCEs. One of the factors that has been probed to play a clear role on the mechanical response produced by such light-driven actuators is the variation of the photo-active azo-cross-linker spacer length, in other words, the flexible alkyl chain that links the azobenzene core with the main polysiloxane backbone (**Figure 13**).[48]

In order to describe this effect, five different photo-active elastomers containing the nematic mesogen 4-methoxyphenyl-4-(3-butenyloxy)benzoate (**M4OMe**, 90 % mol), the isotropic cross-linker 1,4-di-(10-undecenyloxy)benzene (**V1**, 5 % mol) and one of the light-sensitive cross-linkers shown in **Figure 13** (**AZOX**, 5 % mol) will be considered (**Figure 14**). All elastomers were prepared and characterised according to that described previously in Section 3 and 4 of this chapter. The photo-active azo-cross-linkers present lateral alkoxyl chains of different lengths, bearing 3, 4, 6, 8 and 11 carbon atoms, n, and thereby producing a total number of methylene units in the flexible spacer, n_{total}, of 6, 8, 12, 16 and 22, respectively. All the systems exhibited a broad enantiotropic nematic phase between their glass transition temperature at T_g = 261-276 K and their nematic-to-isotropic phase transition temperature at $T_{N\text{-}I}$ = 336-342 K ($\Delta H_{N\text{-}I}$ = 1.6-2.1 J·g^{-1}). Table 1 displays the temperature range of stability of the nematic liquid-crystalline phase for each LSCE. Moreover, all of them showed a clear macroscopic orientation of the director as it reveals their orientational order parameter which falls between 0.71 and 0.73. The swelling ratio, q, of the different elastomers ranges from 2.5 to 4.3. According to that discussed in Section 3 of this chapter, the slight differences observed in the swelling parameter for the different systems can be related with the different length of their cross-linker spacer. In this way, elastomer **EAZO3**, which has the most rigid cross-linker, shows the lowest q value (q = 2.5) whereas **EAZO11**, with the most flexible one, exhibits the highest swelling parameter (q = 4.3, Table 1) since it can create bigger empty spaces inside the elastomeric system.

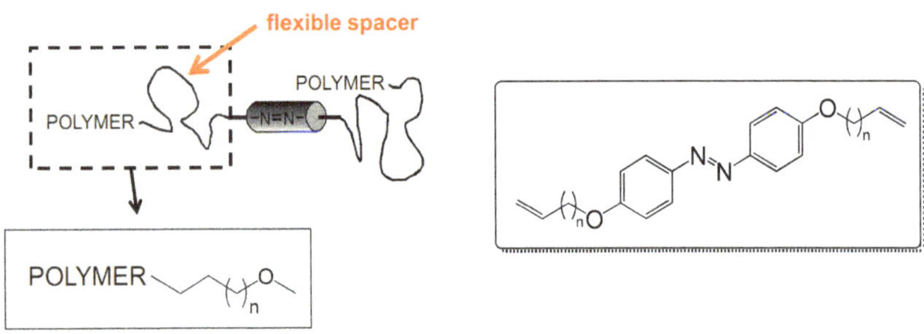

Figure 13. Concept of flexible spacer and general chemical structure of the different photo-active azo cross-linkers used in this study.

Elastomer	S	T_g (K)	$T_{N\text{-}I}$ (K)	$\Delta H_{N\text{-}I}$ (J·g^{-1})	q
EAZO3	0.71	265	336	1.9	2.5
EAZO4	0.73	261	340	2.1	3.0
EAZO6	0.73	264	342	1.6	3.3
EAZO8	0.73	284	348	1.6	3.0
EAZO11	0.72	276	342	1.8	4.3

Table 1. Nematic order parameter, S; glass transition and nematic-to-isotropic phase transition temperatures, T_g and $T_{N\text{-}I}$; nematic-to-isotropic phase transition enthalpy, $\Delta H_{N\text{-}I}$; and swelling parameter, q.

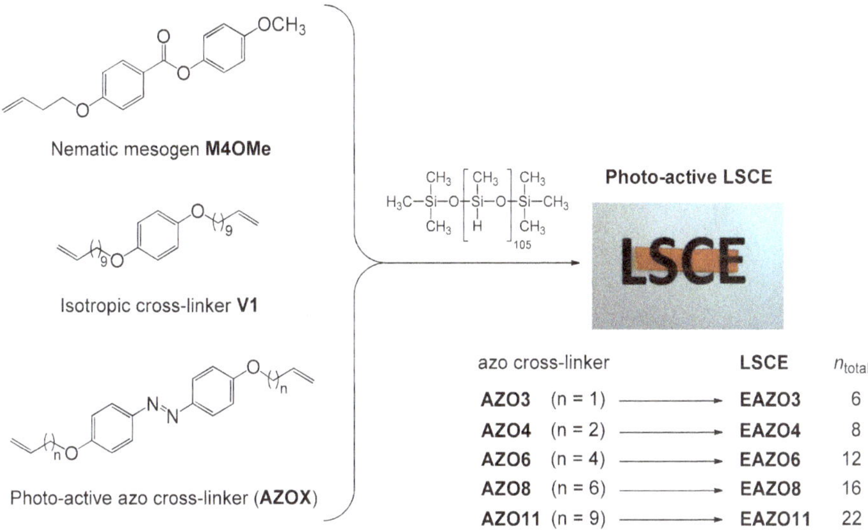

Figure 14. Chemical composition of the different photo-active nematic LSCEs **EAZOX**.

The maximum opto-mechanical response produced by the different LSCEs **EAZOX** under UV-irradiation, $\Delta\sigma_{max}$, depends clearly not only on the number of methylene units of the flexible spacer (**Figure 15**, left) but also on the temperature (**Figure 15**, right). It is highly remarkable that very much efficient light-controlled artificial muscle-like actuators are obtained on using photo-active cross-linkers with long alkoxyl chains in their flexible spacers. Hence, a maximum opto-mechanical response of $\Delta\sigma_{max}$= 19.0 kPa is registered for the LSCE **EAZO11** at 333 K, which bears the longest spacer. Otherwise, a three-fold lower value of $\Delta\sigma_{max}$= 6.4 kPa is obtained for the nematic elastomer **EAZO3** at the same temperature, which contains the shortest one.

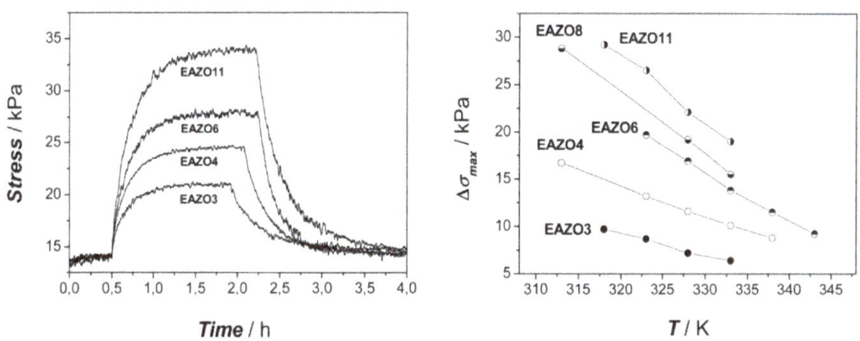

Figure 15. Opto-mechanical experiments for the different nematic liquid single crystal elastomers under irradiation with UV light (λ_{irrad} = 380 nm) at 333 K (left). Evolution of the maximum opto-mechanical response, $\Delta\sigma_{max}$, generated by the different elastomers with the temperature (right).

According to the molecular model presented in **Figure 16**, azo-cross-linkers with long spacers can connect two quite separated points of the main polysiloxane backbone in comparison with their shorter counterparts. As the azo-cross-linker is clamped by both ends, the constraint of the elastomer after the photo-isomerisation to the *cis*-isomer should be greater for those systems which connect two points of the polymeric backbone which are far away separated. Indeed, the cross-linker with the longest spacer experiments the highest opto-mechanical response when the azo moiety adopts its bent geometry.

As a representative example, **Figure 16** shows the calculated distance for the all *trans* conformation of both **AZO3** and **AZO11**, which bear a total number of methylene units in their flexible spacer, n_{total}, of 6 and 22, respectively. For *trans*-**AZO3** and *trans*-**AZO11**, the calculated distance between both terminal carbon atoms are 17.6 and 35.5 Å, while for their corresponding *cis* form, which has a bent shape, it drops until 9.8 and 18.8 Å, respectively. This fact is reflected in a distance change between both terminal carbon atoms that are attached to the main polymeric chain of the network of 7.8 and 16.7 Å, respectively, when the azoderivative is isomerised upon UV-irradiation. Hence, from the calculated distances, it should be expected a greater effect on the macroscopic dimensions of the elastomer when azo-cross-linkers with larger flexible spacers are used. Although the presented model is a qualitative estimation, since the flexible spacer of the azo-cross-linker undergoes a conformational equilibrium, it stands for the different experimental observations.

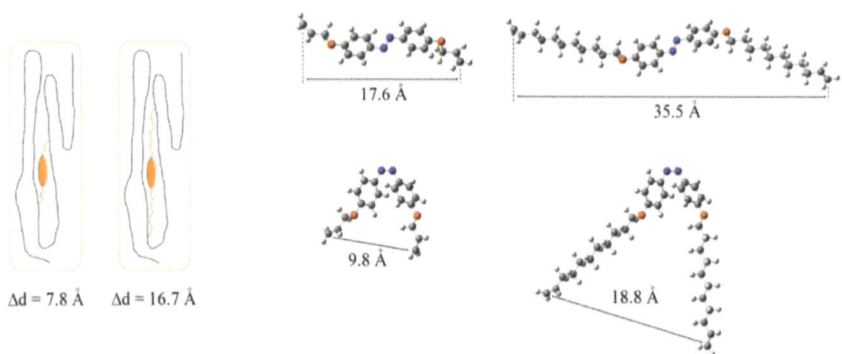

Figure 16. Schematic model of the bonding points of the photo-active azo-cross-linker with the elastomeric network (left). Molecular model for the all *trans* conformation of azo-dyes **AZO3** and **AZO11** and calculated distances between both terminal carbon atoms (right) ($\Delta d = d_{trans} - d_{cis}$).

On the other hand, a marked decrease of the opto-mechanical response of the artificial muscle-like actuator on rising the temperature is detected. This phenomenon is well understood since the thermal back isomerisation process, which competes with the *trans*-to-*cis* photo-isomerisation reaction, takes place faster on increasing the temperature and, consequently, the *cis* isomer concentration at the photo-stationary state, which is the responsible of the observed mechanical response, is lower at higher temperatures.

Moreover, the maximum opto-mechanical response at a determined temperature, T, shows a clear dependence on the total methylene units of the cross-linker spacer, n_{total} ($n_{total} = 2n + 2$, where n is the number of methylene units of the alkoxyl chain). A growing exponential behaviour tending to a *plateau* for a long spacer length is observed at each temperature. It can be nicely seen from **Figure 17**, that there is a threshold length value for the flexible spacer from which the mechanical response of the system is not enhanced anymore. The threshold spacer length value for the selected architecture of the photo-active azoderivative corresponds to *ca.* $n_{total} \approx 30$ methylene units, that is, 15 carbon atoms in each alkoxyl chain of the flexible spacer.

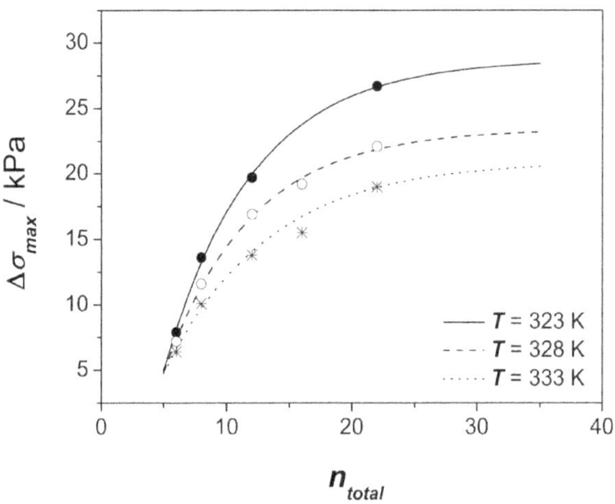

Figure 17. Evolution of the maximum opto-mechanical response, $\Delta\sigma_{max}$, with the total number of methylene units in the spacer, n_{total}, at different temperatures.

The result displayed in **Figure 17** evidences that there is a maximum opto-mechanical response for each type of azo-cross-linker used. In this case, responses up to 30 kPa can be obtained using 4,4′-dialkoxysubstituted azo-dyes as photo-active cross-linkers at 323 K. Considering a linear relationship between $\Delta\sigma_{max}$ and T (see **Figure 15**, right), it can be concluded that this type of azo-cross-linkers can afford responses up to 60 kPa at room temperature.

The analysis of the characteristic times found for both the whole UV-irradiation (τ^{irrad}) and the thermal relaxation (τ^{th}) process when the azo-dye is incorporated as a cross-linker in the nematic LSCE yields that the rate of both processes is independent of the length of the azo-cross-linker spacer. The relaxation time for the irradiation process ranges from 15 to 21 minutes at 323 K. For the reverse process, the thermal isomerisation in the dark, relaxation times ranging from 29 to 35 minutes were registered for all the LSCEs at 323 K, except for **EAZO3** which showed a slightly higher value of 45 minutes, probably related with the lower flexibility of the azo-cross-linker (**Table 2**). The thermal *cis*-to-*trans* isomerisation of the azo-dye takes place faster inside the nematic LSCE than in isotropic media, but it is similar to that obtained in low molar mass liquid crystals. This fact evidences that the kinetic acceleration of the thermal *cis*-to-*trans* isomerisation of the azo-dye in the LSCE is mainly due to the presence of the nematic mean field in the network.[49-51]

Elastomer	323 K		328 K		333 K	
	τ^{irrad}	τ^{th}	τ^{irrad}	τ^{th}	τ^{irrad}	τ^{th}
EAZO3	19.8	44.1	15.7	30.0	12.4	20.0
EAZO4	18.3	34.5	14.1	22.5	10.8	15.9
EAZO6	15.3	29.0	12.8	20.0	10.7	14.7
EAZO8	18.5	30.3	15.1	20.2	12.8	14.4
EAZO11	20.5	33.5	17.3	23.8	14.4	16.4

Table 2. Relaxation time for the *trans* isomer under irradiation with UV light, τ^{irrad}, and for the *cis* isomer on heating in the dark, τ^{th} at different temperatures. All values are given in minutes.

Although all the photo-active LSCEs presented till this moment are mechanically efficient, they need several hours to reach their maximum mechanical response and also to relax back to the initial state.[37,38,48] Hence, the time required by the network to recover its initial dimensions is also another crucial parameter to consider for obtaining functional artificial muscle-like actuators. Likewise, our discussion will be focused in the next section on presenting different strategies to get photo-active artificial muscle-like actuators with a low thermal relaxation time.

6. Response time of artificial muscle-like actuators by using fast thermally-isomerising azoderivatives

While the *trans*-to-*cis* photo-isomerisation can be easily accelerated either by using more powerful light sources or the appropriate modification of the azobenzene core, the thermal *cis*-to-*trans* back reaction depends mainly not only on the chemical functionalization of the azo-dye but also on the environment where the azo-chromophore is located. Hence, it is essential for getting fast azobenzene-based artificial muscle-like actuators that the return to the thermodynamically stable *trans* form of the azo-chromophore in the dark elapses as fast as possible.

Only two examples of dye-doped polysiloxane-based photo-active nematic LSCEs with a fast isomerisation rate have been published so far. These systems use the well-known push-pull azo-dyes 4-amino- and 4-*N,N*-dimethylamino-4′-nitroazobenzene as photo-active molecules, which are doped into host elastomeric networks but not covalently bonded to the polymeric structure[52,53]; this fact decreases the stability of the final photo-actuator. Besides those azoderivatives which bear a push-pull configuration, azophenols are also endowed with a very rapid thermal *cis*-to-*trans* isomerisation process since they are capable to establish azo-hydrazone tautomeric equilibria which allows their thermal back reaction to proceed through the rotational isomerisation mechanism. Hence, these two type of azoderivatives are valuable photo-active molecules to be introduced as covalently bonded comonomers in LSCE to obtain stable and fast responding photo-sensitive elastomeric materials.

6.1. Artificial muscle-like actuators using push-pull azoderivatives as chromophores

Since an azo-dye is introduced into a polymer, the polymer backbone undergoes different motions when the thermal relaxation of the azoderivative occurs. These motions may modify the kinetic parameters of the *cis*-to-*trans* isomerisation of the azo chromophore. In this way, structural factors such as chain flexibility and chain conformation play an important role on the rate of the thermal isomerisation process of the chromophore.[45,54,55] Moreover, very different kinetic behaviour should be expected when the azo-dye is bonded to the polymer backbone than when it is just dissolved in the host polymer matrix as a doped guest. This effect has been already detected in LSCEs that contain the push-pull azo-dyes 4-(5-hexenyloxy)-4'-nitroazobenzene and 4-(5-hexenyloxy)-4′-methoxy-2′-nitro-azobenzene introduced as side-chain covalently-bonded photo-active moieties (**Figure 18**).[56]

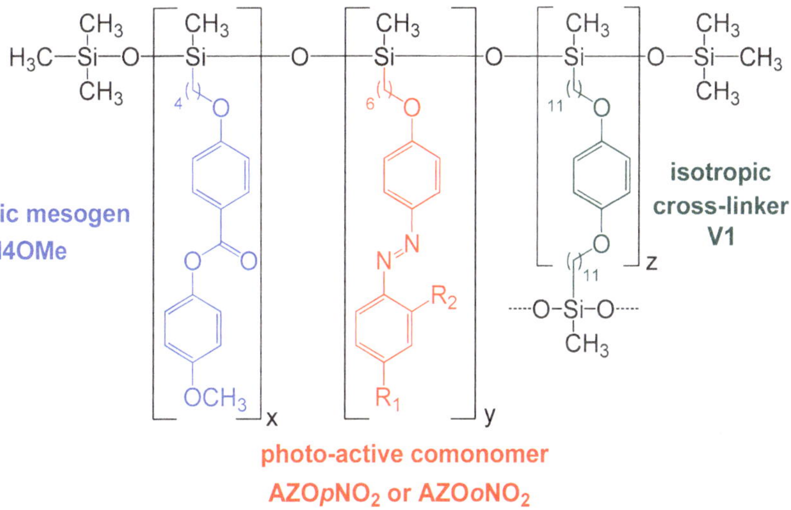

E*p*NO2 R1 = -NO2; R2 = -H (x : y : z 0.85 : 0.05 : 0.10)

E*o*NO2 R1 = -OCH3; R2 = -NO2 (x : y : z 0.85 : 0.05 : 0.10)

Figure 18. Chemical composition of the photo-active nematic LSCEs **E*p*NO2** and **E*o*NO2**.

For describing this effect, two nematic LSCEs **E*p*NO2** and **E*o*NO2** will be considered being composed by the nematogen **M4OMe** (85 % mol), the photoactive co-monomer (**AZO*p*NO2** or

AZO*o*NO$_2$, respectively, 5 % mol) and the isotropic cross-linking agent **V1** (10 % mol) as it is shown in **Figure 18**. Both elastomers, **E*p*NO$_2$** and **E*o*NO$_2$**, which differ only in the placement of the nitro- electron-withdrawing group, showed a broad enantiotropic nematic phase between their glass transition temperature at T_g = 276-277 K and their nematic-to-isotropic phase transition temperature at $T_{N\text{-}I}$ = 331-332 K ($\Delta H_{N\text{-}I}$ = 1.3-1.7 J·g^{-1}), respectively. Completely transparent monodomain nematic samples were obtained for both elastomers exhibiting orientational order parameters of 0.75 and 0.76 for **E*o*NO$_2$** and **E*p*NO$_2$**, respectively.

The determination of the thermal *cis*-to-*trans* isomerisation rate of both azo-dyes acting as covalently-bonded co-monomers in the nematic LSCE by means of opto-mechanical experiments yielded relaxation times for *cis*-**AZO*p*NO$_2$** and *cis*-**AZOoNO$_2$** of 4 and 3 seconds, respectively, at 298 K. The impressive acceleration of the kinetics of the thermal *cis*-to-*trans* isomerisation of both nitro-substituted azoderivatives in the LSCE is highly remarkable. The process was found to be more than 10^3 times faster in the LSCE, in which the azo-dye is chemically bonded to the polymer backbone, than that observed in isotropic solvents (between 16 and 92 min.), and about 10^2 times quicker than the one shown as a guest chromophore doped in a low molar mass nematic liquid crystal (6-7 min.). This acceleration is associated to the covalent attachment of the push-pull azoderivative to a nematic polymer backbone which possesses an anisotropic chain conformation.[56]

The great acceleration exhibited by these nitro-substituted azo-dyes when they are chemically bonded into nematic LSCEs has been already exploited for the preparation of fast photo-active artificial muscle-like actuators. The maximum opto-mechanical response generated by the elastomer **E*p*NO$_2$** was of $\Delta\sigma_{max}$ = 1.4 kPa, while a lower value of 0.7 kPa was obtained for the LSCE **E*o*NO$_2$** (**Figure 19**). The fast thermal relaxation of these systems is the responsible of the low internal stress generated in both elastomers due to the low presence of *cis* isomer in the photo-equilibrium state. The slightly faster thermal isomerisation for **E*o*NO$_2$** accounts for the lower mechanical response produced by this network. Both LSCEs exhibit a reversible behaviour in the time scale of seconds at room temperature, that is, relaxation times of 2-4 s for the whole irradiation process with blue light and of 3-4 s for the *cis*-to-*trans* thermal relaxation in the dark for **E*o*NO$_2$** and **E*p*NO$_2$**, respectively.

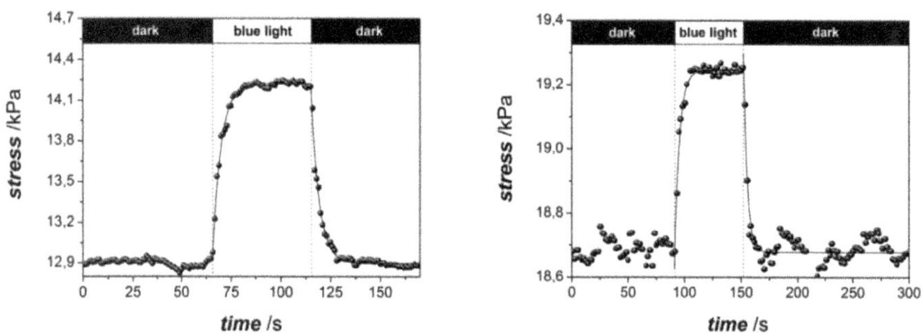

Figure 19. Opto-mechanical experiments for both nematic elastomers **E*p*NO$_2$** (left) and **E*o*NO$_2$** (right) at 298 K (irradiation of the networks was performed with blue light, λ_{irrad} = 450 nm).

6.2. Artificial muscle-like actuators based on azophenol derivatives exhibiting azo-hydrazone taumeric equilibrium

A second strategy to decrease the response time of the artificial muscle-like actuators implies the use of azophenols as photo-active moieties. Azophenols are promising chromophores for designing fast-responding artificial muscle-like actuators since they are endowed with a rapid thermal isomerisation process at room temperature, with relaxation times ranging from 6 ms to 300 ms in polar protic solvents depending on the position of the phenol groups.[57] But, the main drawback is that hydroxyazobenzenes show a fast thermal isomerisation rate only when they are dissolved in polar protic solvents.

A successful strategy to transfer this fast thermal isomerisation to solid elastomeric materials consists in the preparation of a co-elastomer which contains a nematic mesogenic monomer (**M4OMe**) and the photo-active hydroxyazoderivative (**AZO-OH**). The liquid-crystalline co-elastomer obtained contains only a small proportion of the azo moiety (5% mol) in order to not overly disrupt the nematic order of the elastomer. However, in that system, the azo-dye concentration is high enough for hydrogen bonding being established between the hydroxyazobenzene monomers without niether losing the liquid-crystalline properties of the nematic system nor diminishing the temperature range where the nematic phase exists. As a result of this interaction, the resulting LSCE exhibits an isomerisation rate as fast as that of the azo-monomer dissolved in a protic isotropic solvent.[58]

Two different liquid single crystal elastomers will be presented, **EAZO-Ac** and **EAZO-OH**, in order to describe properly the influence of the phenolic group in the opto-mechanical properties of the final photo-actuator (**Figure 20**). The uniaxially oriented acetylated elastomer **EAZO-Ac** was prepared following the well-known spin-casting technique commented above. Elastomer **EAZO-OH** was prepared from **EAZO-Ac** in order to assure an identical composition in both systems. For preparing elastomer **EAZO-OH**, the acetylated elastomer **EAZO-Ac** is swollen in a mixture of CH_2Cl_2 and MeOH (2:1 v/v) with one drop of acetyl chloride at room temperature for 2 days. Acetyl chloride enters the network by diffusion producing a mild and chemoselective cleavage of the ester group of the acetylated side-chain group, **AZO-Ac**, without hydrolysing the other ester group placed in the mesogen.

The characterization of both LSCEs was carried out by using the standard techniques (see Section 3 of this chapter). DSC experiments showed that both elastomers, **EAZO-Ac** and **EAZO-OH**, exhibited a broad enantiotropic nematic phase between their glass transition temperature at T_g = 277-278 K (ΔC_p = 0.3 J·g^{-1}·K^{-1}) and their nematic-to-isotropic phase transition temperature at $T_{N\text{-}I}$ = 335-336 K ($\Delta H_{N\text{-}I}$ = 1.6-1.8 J·g^{-1}), respectively. X-Ray scattering experiments clearly indicate that no change in the degree of order of the LSCE occurred during the cleavage reaction since very close orientational order parameters were obtained for both **EAZO-Ac** and **EAZO-OH** (0.71 and 0.69, respectively). Interestingly, the swelling ratio, q, was determined to be 5.0 and 3.3 for **EAZO-Ac** and **EAZO-OH**, respectively. The lower q value found for **EAZO-OH** evidenced that hydrogen bonding is established between the phenol groups of the azo-dyes in the LSCE acting as additional cross-linking units.

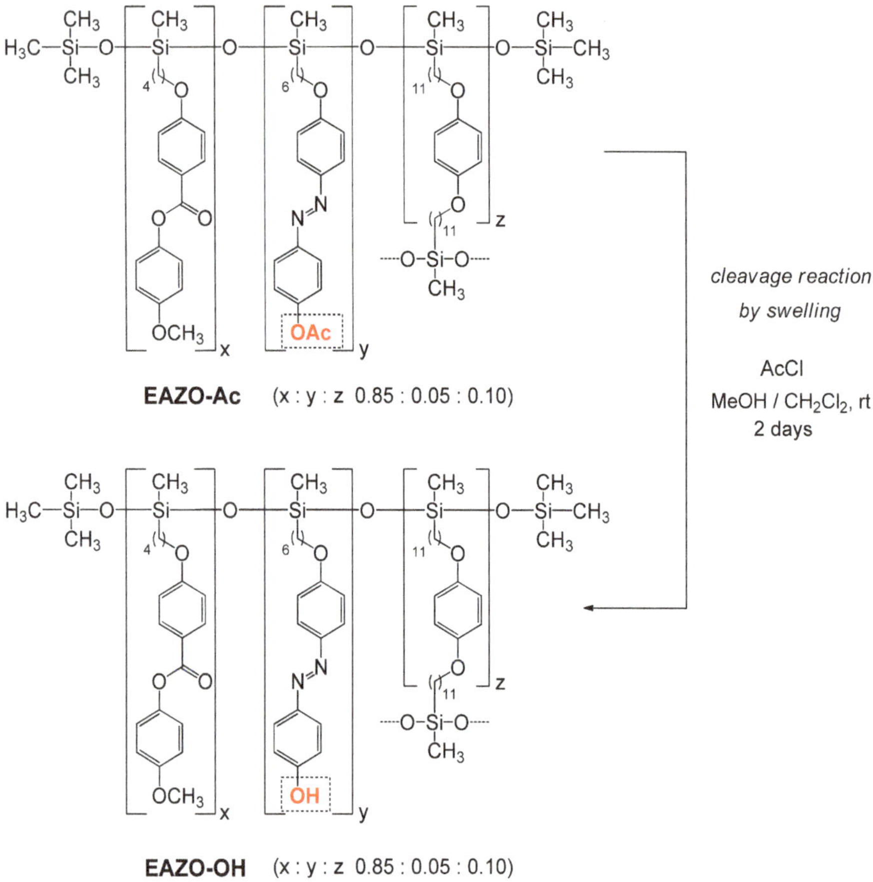

Figure 20. Chemical composition of **EAZO-Ac** and **EAZO-OH** and synthetic concept for the photo-active nematic LSCE **EAZO-OH**.

The analysis of the mechanical response generated by both elastomers as well as their relaxation time by means of opto-mechanical experiments evidenced that both LSCEs, **EAZO-Ac** and **EAZO-OH**, can act as light-controlled actuators when they are irradiated with UV light (λ_{irrad} = 380 nm). **Figure 21** shows the evolution of the internal stress generated inside both LSCEs, **EAZO-Ac** and **EAZO-OH**, with the time at 323 and 298 K, respectively. **EAZO-Ac** produced a maximum opto-mechanical response of $\Delta\sigma_{max}$ = 15 kPa at 323 K (**Figure 21a**). The opto-mechanical response of **EAZO-Ac** was enhanced to *ca.* $\Delta\sigma_{max}$ = 30 kPa on dropping the temperature down to at 313 K (**Figure 22**). When the same experiment is carried out at 298 K for the elastomer **EAZO-OH**, a lower maximum opto-mechanical response of $\Delta\sigma_{max}$ = 0.6 kPa is obtained (**Figure 21b**). Once more, similar results than those aforementioned are obtained. On comparing the non push-pull systems shown in Section 5 with those containing push-pull azoderivatives, which have been presented in Section 6.1, it is clearly seen that the mechanical efficiency, expressed in terms of $\Delta\sigma_{max}$, is greater in the former (up to 60 kPa at 298 K), that is, for those azo-dyes that undergo their thermal *cis*-to-*trans* isomerisation through the inversional mechanism. Otherwise, those azocompounds that isomerises by means of the rotational pathway, those are, push-pull azo-dyes and azophenols, show lower mechanical efficiencies around 0.6 - 1.4 kPa at room temperature.

Besides, the nematic liquid-crystalline elastomer **EAZO-Ac**, which contains the azo-chromophore as a side-chain monomer, is less effective ($\Delta\sigma_{max}$ = 15 kPa at 323 K) than those elastomers **EAZOX**, where the azo-dye is introduced as a photo-active cross-linker in the elastomeric network ($\Delta\sigma_{max}$ = 30 kPa at 323 K).

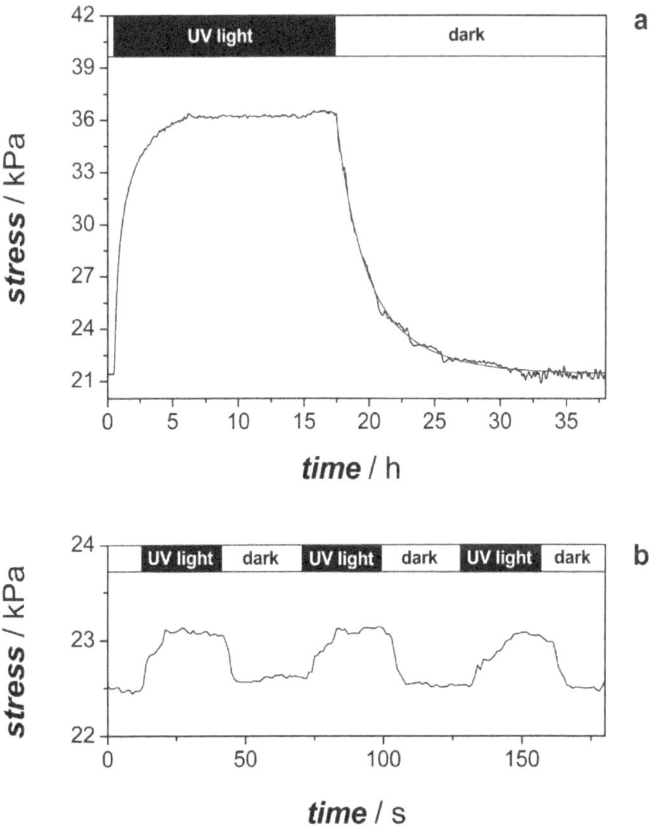

Figure 21. (a) Opto-mechanical experiment for the nematic LSCE **EAZO-Ac** at 323 K.(b) Opto-mechanical experiment for the nematic LSCE **EAZO-OH** at 298 K (λ_{irrad} = 380 nm).

EAZO-Ac exhibits a relaxation time for its thermal back reaction of τth = 21.3 h at room temperature, which was determined by extrapollation of the corresponding Eyring's plot

(**Figure 23**). Opto-mechanical experiments also reveal a completely different kinetic behaviour between both photo-actuators. While **EAZO-Ac** presents a relaxation time of almost 1 day, **EAZO-OH** gives a value of only 1 s at room temperature. The fast thermal relaxation exhibited by **EAZO-OH** evidences clearly the establishment of hydrogen bonding between the different phenol groups due to their spatial proximity within the elastomeric network. This interaction indicates that the azo-dye isomerisation takes place through the rotational mechanism as it has been described for other azophenol derivatives.[57]

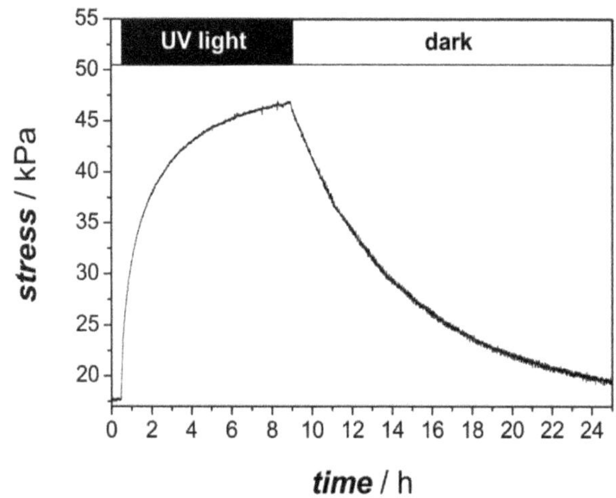

Figure 22. Opto-mechanical experiment for the nematic LSCE **EAZO-Ac** at 313 K (λ_{irrad} = 380 nm).

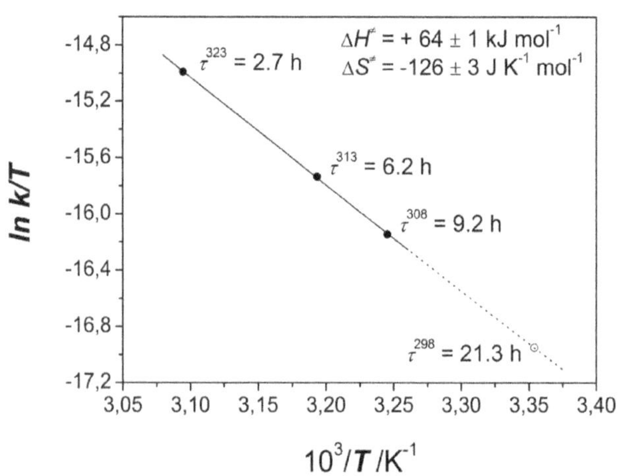

Figure 23. Eyring's plot for the thermal *cis*-to-*trans* isomerisation of **EAZO-Ac**.

The low mechanical response registered for the free-phenol-containing elastomer **EAZO-OH** in comparison with that of **EAZO-Ac** is associated to the low proportion of *cis* isomer present when **EAZO-OH** reaches its photo-equilibrium state at 298 K, as it has been commented above for the nitro-substituted azobenzene-containing elastomers **E*p*NO$_2$** and **E*o*NO$_2$** (see above). Similarly, the slower thermal relaxation of **EAZO-Ac** at 313 K than at 323 K makes the photo-equilibrium state richer in *cis* isomer in the former. This fact accounts for the higher opto-mechanical response exhibited by **EAZO-Ac** at 313 K.

7. Conclusion

Photo-active liquid single crystal elastomers (LSCEs) are valuable materials for artificial muscle-like applications since their macroscopic dimensions can be easily changed by applying light, an environmentally-friendy energy which can be wireless applied to the material. Those actuators based on the azobenzene chromophore are the most commonly used since it presents a clean and totally reversible isomerisation process.

Two different key parameters are crucial for characterising properly the artificial muscle-like actuation of such materials, those are, the maximum mechanical response that they are able to generate by irradiation with light of the appropriate wavelength as well as the time required to produce it and that to recover the initial state further.

The use of non-push-pull azoderivatives, which thermally-isomerise through the inversional mechanism, both as light-sensitive cross-linkers and as side-chain pendant groups in polysiloxane nematic LSCEs produces high mechanically-efficient artificial muscle-like actuators. However, those actuators that bear the photo-active azo-dye as a cross-linker exhibit the highest mechanical efficiency. In such materials, the opto-mechanical response produced by the artificial muscle increases greatly by using larger flexible spacers on the photo-active cross-linker. For 4,4′-dialkoxysubstituted azobenzenes, there is a maximum spacer length, n_{th} ~30, above which the opto-mechanical response of the network reaches a constant value and it is not enhanced anymore. Nevertheless, the relaxation time of the photo-actuator remains unaltered on varying the cross-linker spacer length. Hence, such materials can be used as bistable artificial muscles.

On the other hand, the introduction of both alkoxynitro-substituted azobenzenes and azophenols, which isomerise by means of the rotational pathway, yield fast-responding artificial muscle-like actuators. Indeed these two different types of fast-isomerising azoderivatives have been successfully used as covalently-bonded side-chain chromophores in nematic LSCEs. Alkoxynitro-substituted azobenzenes accelerate up to 10^3 times their thermal *cis*-to-*trans* isomerisation kinetics in isotropic media when they are attached in LSCEs. This rapid thermal isomerisation allows their use as chromophores to get stable and fast artificial muscle-like actuators working in time scale of 1-3 seconds.

Besides, the thermal *cis*-to-*trans* isomerisation process of azophenols depends strongly on the proticity of the environment where they are placed. In this way, their thermal isomerisation rate is accelerated up to 10^4 times if ethanol is used instead of toluene. The fast isomerisation rate exhibited by these azo-dyes in ethanol has been successfully transferred to polymeric and elastomeric materials, where no solvent is present, due to the hydrogen bonding established between the side-chain azophenol monomers. Similar fast and stable artificial muscle-like actuators than those based on alkoxy-nitro-substituted azobenzenes have been described.

Author details

Jaume Garcia-Amorós and Dolores Velasco
Grup de Materials Orgànics, Institut de Nanociència i Nanotecnologia (IN²UB),
Departament de Química Orgànica, Universitat de Barcelona, Barcelona, Spain

Acknowledgement

Financial support from the European project: "*Functional Liquid-Crystalline Elastomers*" (FULCE-HPRN-CT-2002-00169) and from the *Ministerio de Ciencia e Innovación* (CTQ-2009-13797) is gratefully acknowledged. The authors thank Prof. Dr. Heino Finkelmann for its continuous support and helpful discussions.

8. References

[1] F. Reinitzer, *Monatsh. Chem.*, 1998, 9, 421 (Translation in English: *Liq. Cryst.* 1989, 5, 7).
[2] M. Schadt, *Appl. Phys. Lett.*, 1971, 18, 127.
[3] H. Iwai, J. Fukasawa and T. Suzuki, *International Journal of Cosmetic Science*, 1998, 20, 87
[4] M. Shlens, M. R. Stoltz and A. Benjamin, *West J. Med.*, 1975, 122, 367.
[5] J. Stasiek, A. Stasiek, M. Jewartowski and M. W. Collins, *Optics & Laser Technology*, 2006, 38, 243.
[6] Y. Yu and T. Ikeda, *Macromol. Chem. Phys.*, 2005, 206, 1705.
[7] T. Mirfakhrai, J. D. W. Madden and R. H. Baughman, *Materials Today*, 2007, 10, 30.
[8] Y. Ishida, Y. Kai, S. Kato, A. Misawa, S. Amano, Y. Matsuoka and K. Saigo, *Angew. Chem. Int. Ed.*, 2008, 47, 8241.
[9] J. W. Goodby and G. Gray, *Handbook of Liquid Crystals*; Wiley-VCH: New York, Weinheim, 1998, 1, 17.
[10] H. Finkelmann in *Thermotropic Liquid Crystals*; John Wiley & Sons, UK, 1987.
[11] H. Finkelmann, H. Ringsdorf and J. H. Wendorff, *Makromol. Chem.*, 1978, 179, 273.
[12] P. J. Collings and M. Hird, *Introduction to liquid crystals: chemistry and physics*; Taylor and Francis Ltd, Bristol, 1997.
[13] P. J. Collings, *Liquid crystals. Nature's delicate phase of matter*; Princeton University Press, New Jersey, 2002.
[14] H. Rau in *Photochemistry and photophysics*; CRC Press Boca Raton FL, 1990.
[15] C. H. Legge and G. R. Mitchell, *J. Phys. D: Appl. Phys.*, 1992, 25, 492.
[16] S. K. Prasad, G. G. Nair, K. L. Sandhya and D. S. S. Rao, *Current Science*, 2004, 86, 815.
[17] Y. Yu and T. Ikeda, *J. Photochem. Photobiol. C: Photochem. Rev.*, 2004, 5, 247.

[18] J. H. Sung, S. Hirano, O. Tsutsumi, A. Kanazawa, T. Shiono and T. Ikeda, *Chem. Mater.*, 2002, 14, 385.
[19] T. Ikeda, J.-I. Mamiya and Y. Yu, *Angew. Chem. Int. Ed.*, 2007, 46, 506.
[20] C. J. Barrett, J.-I. Mamiya, K. G. Yager and T. Ikeda, *Soft Matter*, 2007, 3, 1249.
[21] H. Yu and T. Ikeda, *Adv. Mater.*, 2011, 23, 2149.
[22] T. Ikeda and T. Ube, *Mater. Today*, 2011, 14, 480.
[23] Y. Yu, M. Nakano and T. Ikeda, *Nature*, 2003, 425, 145.
[24] H. Jiang, S. Kelch and A. Lendlein, *Adv. Mater.*, 2006, 18, 1471.
[25] R. Yin, W. Xu, M. Kondo, C.-C. Yen, J.-I. Mamiya, T. Ikeda and Y. Yu, *J. Mater. Chem.*, 2009, 19, 3141
[26] M. Yamada, M. Kondo, R. Miyasato, Y. Naka, J.-I. Mamiya, M. Kinoshita, A. Shishido, Y. Yu, C. J. Barrett and T. Ikeda, *J. Mater. Chem.*, 2009, 19, 60.
[27] A. Priimagi, A. Shimamura, M. Kondo, T. Hiraoka, S. Kubo, J.-I. Mamiya, M. Kinoshita, T. Ikeda and A. Shishido, *ACS Macro Lett.*, 2012, 1, 96.
[28] M. Yamada, M. Kondo, J.-I. Mamiya, Y. Yu, M. Kinoshita, C. J. Barrett and T. Ikeda, *Angew. Chem. Int. Ed.*, 2008, 47, 4986.
[29] J. Küpfer and H. Finkelmann, *Makromol. Chem. Rapid Commun.*, 1991, 12, 717.
[30] P. G. de Gennes, *C. R. Acad. Sci.*, Paris, 1975, 281, 101.
[31] M. Warner and E. M. Terentjev, *Liquid Crystal Elastomers*, Claredon, 2003.
[32] H. Wermter and H. Finkelmann, *e*-polymers, 2001, 13, 1.
[33] D. L. Thomsen, P. Keller, J. Naciri, R. Pink, H. Jeon, D. K. Shenoy and B. R. Ratna, *Macromolecules*, 2001, 34, 5868.
[34] D. K. Shenoy, D. L. Thomsen, A. Srinivasan, P. Keller and B. R. Ratna, *Sensors and Actuators A: Phys.*, 2002, 96, 184.
[35] C. M. Spillmann, J. Naciri, M.-S. Chen, A. Srinivasan and B. R. Ratna, *Liq. Cryst.*, 2006, 33, 373.
[36] C. M. Spillmann, J. Naciri, B. D. Martin, W. Farahat, H. Herr and B. R. Ratna, *Sensors and Actuators A: Phys.*, 2007, 133, 500.
[37] P. M. Hogan, A. R. Tajbakhsh and E. M. Terentjev, *Phys. Rev. Lett. E*, 2002, 65, 041720.
[38] J. Cviklinski, A. R. Tajbakhsh and E. M. Terentjev, *Eur. Phys. J. E*, 2002, 9, 427.
[39] H. Finkelmann, E. Nishikawa, G. G. Pereira and M. Warner, *Phys. Rev. Lett.*, 2001, 87, 015501-1.
[40] P. J. Flory and J. Rehner, *J. Chem. Phys.*, 1943, 11, 521.
[41] P. J. Flory, *J. Chem. Phys.*, 1950, 18, 108.
[42] N. A. Neuburger and B. E. Eichinger, *Macromolecules*, 1988, 21, 3060.
[43] B. Erman and M. Baysal, *Macromolecules*, 1985, 18, 1696.
[44] A. Greve and H. Finkelmann, *Macromol. Chem. Phys.*, 2001, 202, 2926.
[45] C. D. Eisenbach, *Makromol. Chem.*, 1978, 179, 2489.
[46] M. G. Evans and M. Polanyi, *Trans. Faraday Soc.*, 1935, 31, 875.
[47] H. Eyring, *J. Chem. Phys.*, 1935, 3, 107.
[48] J. Garcia-Amorós, H. Finkelmann and D. Velasco, *J. Mater. Chem.*, 2011, 21, 1094.
[49] J. Garcia-Amorós, M. Martínez, H. Finkelmann and D. Velasco, *J. Phys. Chem. B*, 2010, 114, 1287.

[50] J. Garcia-Amorós, H. Finkelmann and D. Velasco, *Phys. Chem. Chem. Phys.*, 2011, 13, 11233.
[51] P. Beyer and R. Zentel, *Macromol. Rapid Commun.*, 2005, 26, 874.
[52] M. Camacho-López, H. Finkelmann, P. Palffy-Muhoray and M. Shelley, *Nat. Mater.*, 2004, 3, 307.
[53] C. L. M. Harvey and E. M. Terentjev, *Eur. Phys. J. E*, 2007, 23, 185.
[54] C. D. Eisenbach, *Polymer*, 1980, 21, 1175.
[55] C. J. Barrett, A. Natansohn and P. Rochon, *Chem. Mater.*, 1995, 7, 899.
[56] J. Garcia-Amorós, H. Finkelmann and D. Velasco, *Chem. Eur. J.*, 2011, 17, 6518.
[57] J. Garcia-Amorós, A. Sánchez-Ferrer, W. A. Massad, S. Nonell and D. Velasco, *Phys. Chem. Chem. Phys.*, 2010, 12, 13238.
[58] J. Garcia-Amorós, A. Piñol, H. Finkelmann and D. Velasco, *Org. Lett.*, 2011, 13, 2282.

Chapter 3

Monitoring the Contractile Properties of Optically Patterned Liquid Crystal Based Elastomers

Kokou D. (Honorat) Dorkenoo, Emel Sungur,
Hervé Bulou, Grégory Taupier and Alex Boeglin

Additional information is available at the end of the chapter

1. Introduction

Liquid crystal materials offer many opportunities for applications such as data storage [1], image processing [2], or optical modulators [3] calling for media that are sensitive to external stimuli . Today, they are considered to have great potential for the realization of so called "smart materials" through micro- to nano-scale patterning techniques. One interesting material in the flourishing group of smart materials is the Liquid Crystal Elastomer (LCE) which combines the properties of liquid crystals and polymers. An elastomer is formed by a weakly crosslinked network of polymers, which confers it a high elasticity. The polymer network has maximum entropy in his undistorted state, and as a result, it resists deformation [4]. As for liquid crystals, these are materials which present phases intermediate between crystalline and isotropic called mesophases, the molecules responsible for this property are named mesogens.

In this chapter, we will present our results on the optical microstructuring of a LCE and the monitoring of its elastic properties. The liquid crystal based elastomer can be considered as an artificial muscle material. The designation of artificial muscle is attributed to soft actuators which can have muscle like behavior. A muscle is in reality an energy converter which converts chemical energy to mechanical motion. The actuators are converters which accept different types of energy and deliver a mechanical quantity like a displacement, a tension, *etc.*. The LCEs can be considered as artificial muscles in the sense that they may present a distortion, for example a contraction/extension generated by the crystalline-isotropic phase transition. The possibility of using liquid crystal elastomers (LCE) as external stimuli driven size changing materials was predicted by de Gennes in 1997 [5]. Since then, different LCE materials showing reversible macroscopic shape changes under thermal [6][7] or optical [8] stimulations have been produced. Here, we explore the use of this contraction/extension property to create stimulable microsystems.

The LCE used in this study is a nematic thermotropic (sensitive to thermal stimuli) elastomer. A nematic elastomer can be described as crosslinked polymer chains with incorporated mesogens corresponding to rigid rod-like units which can order nematically. The average shape of the material is coupled to the molecular orientation order [4][9]: the nematic-isotropic phase transition results in the change of shape of the macroscopic sample. When the elastomer is judiciously prepared to obtain a monodomain sample [10], meaning that all the mesogens are oriented uniformly, the sample is elongated in the nematic phase where all the mesogens are parallel to each other, and has a contracted shape in the isotropic phase where the nematic order is lost.

First, we will show that advantage can be taken of the crosslinking properties of the material to inscribe refractive index modulations by optical means as in a common photoresist [11]. We present two approaches based respectively on one- and two-photon photopolymerization, that can be implemented using an optical microscopy setup. Such a method provides an accurate way to locally modify the properties of the material. Compared to its one-photon relative, the two-photon excitation mode will yield finer patterns because of its higher intrinsic spatial resolution. This technique relies on the fact that the simultaneous absorption of two low energy photons is equivalent to the absorption of a high energy photon with an energy equal to the sum of these two photons energy. The two-photon absorption (TPA) process is a quadratic phenomenon: the probability of the simultaneous absorption of two photon is not proportional to the photon flux as in the case for the linear absorption, but to the square of the flux. TPA is then significant only for high incident fluences. Thus, the use of ultra-short pulse lasers combined with microscopy techniques allows light-matter interaction to be confined within a micrometric volume around the focal point.
After the presentation of the microstructuring process, we will describe how the contractile properties of the material can be measured by detecting the thermally induced step size changes of an inscribed grating. Finally, these results will be interpreted with the help of Molecular Dynamics (MD) simulations. MD is a powerful tool to elucidate the structure and the behavior of the molecules. It has been used to simulate the behavior of polymers [12] and has proved its efficiency to describe the liquid crystalline phases [13], thus it also allows the description of elastomer behavior [14]. MD consists of a computational method which calculates time dependent atomic motions by applying the laws of classical mechanics. We have used a combination of Lennard-Jones and Gay-Berne potentials to represent the anisotropic mesogens and the calculations have been carried out in the Parrinello-Rahman-Nosé-Hoover ensemble.

2. Materials

2.1. Photopolymerizable liquid crystals

LCE are materials combining the elastic properties of the elastomers and the anisotropic properties of liquid crystals. They are formed by reticulated liquid crystalline polymers, which, in general, at low reticulation degrees, conserve the mesomorphic properties of the polymer before reticulation as well as the nature of the phases and the transition temperatures. These elastomers, when subjected to external stimuli like temperature or electromagnetic fields, can mimic the actions of muscles (contraction/extension). They are, thereby, considered as artificial muscle materials. Different types of actuators have been developed using polymers as parent material, like for example polymer gels [15] [16] [17], conductive polymers [18] [19] [20], carbon nanotubes [21] [22] [23] [24] or dielectric elastomers [25]. Soft and highly flexible, the polymers are well adapted to be used as artificial muscle materials due to their high strength, deformation capabilities, and their capacity to keep intact their properties after several operating cycles. Recently, LCE took over the field of artificial muscles. They are robust and don't need any solvent to operate (unlike the gels or conductive polymers). The concept of LCE was proposed by P. G. de Gennes in 1975 [9]. In his studies on reticulated liquid crystal polymers, he mentioned the possibility of the material deformation without constraint. In 1981 Frinkelmann *et al.* have realized the first mesomorphic oriented side-chain LCE [26] and afterwards the first main-chain LCE was synthesized by Bergmann *et al.* [27]. Indeed, according to the insertion topology of the mesogen (liquid crystal phase-forming unit)

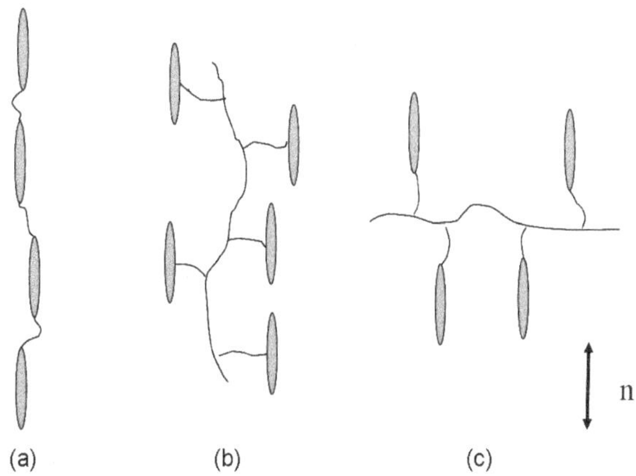

Figure 1. Schematic representation of different types of liquid crystal polymers: (a) main-chain, (b) side-chain side-on, (c) side-chain end-on.

in the elastomer, there exist two main families of LCEs: side-chain and main-chain elastomers. In the case of main-chain elastomers, the mesogens are directly integrated in the polymer chain, while in the case of side-chain elastomers, they are laterally attached to the polymer chain, either orthogonal to the chain, attached by one end (end on), or parallel to the chain, attached by the side (side on), as shown in figure 1.

Since the monomer used in this study is thermotropic, we will study the contraction properties of the resulting elastomer as a function of temperature. The studied elastomer is a nematic elastomer. The nematic phase is favored by the side-chain conformation. In nematic crystals, the mesogens have a random position but an orientational order: they are, on average, parallel to each other. This behavior is encouraged by the elongated form of these units. For a nematic crystal, the amount of order can be characterized by the order parameter P_2 defined as the average value of a second-order Legendre polynomial [28]:

$$P_2 = \frac{1}{2} < 3cos^2(\theta) - 1 > \quad (1)$$

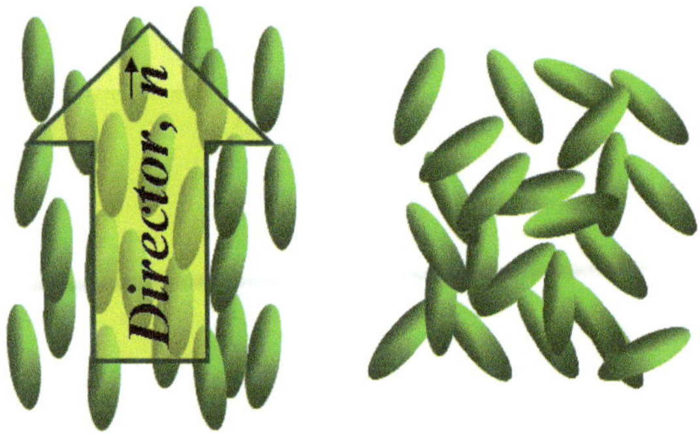

Figure 2. Schematic representation of a liquid crystal in the nematic phase (left) and in the isotropic phase (right). The average orientation is represented by the nematic director **n**.

where θ corresponds to the angle between the long axis of the molecule and the nematic director. For a completely ordered phase P_2=1, while P_2=0 corresponds to a completely disordered phase (isotropic). The mesogens interact *via* long range dipolar or Van der Walls interactions. Below a given temperature, the interactions are sufficiently strong to orient the director axis and organize the system. Beyond this temperature, the preferential orientation of the mesogens is destroyed by the effect of entropy and an isotropic phase appears.

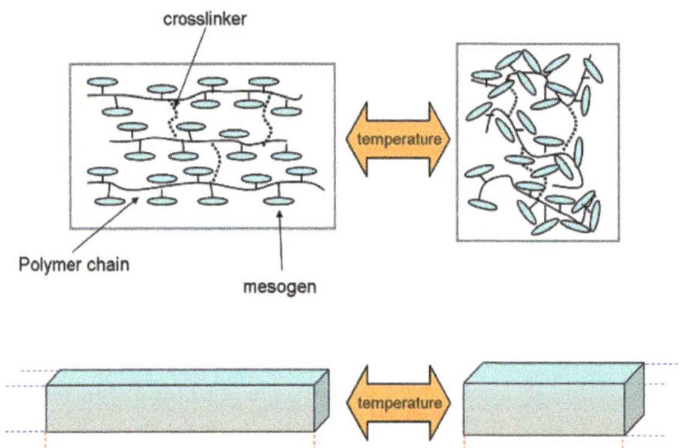

Figure 3. Contraction of a nematic thermotropic elastomer. Microscopic point of view (top), macroscopic point of view (bottom).

In the case of the elastomer, the mesomorphic properties (as the nature of phases of the initial monomers) are conserved. The mesogens attached to the sides of the main chain are free to move. During the nematic-isotropic phase transition the mesogens are disoriented and the main chain drag behind them. These microscopic form changes are transferred to the macroscopic sample. The destruction of the nematic order thus generates a contraction of the system along the mesogens alignment direction and an extension in the orthogonal directions as schematized in figure 3.

2.2. Elastomer thin film preparation

The precursor of the LCE used in this study is a mixture of three compounds: a liquid crystalline acrylate monomer, a crosslinker and an ultraviolet (UV) photoinitiator. The monomer whose structure is represented in figure 4 is the 4′-acryloyloxybutyl 2,5-(4′-butyloxybenzoyloxy)benzoate [29]. The part containing the aromatic chains is the rigid part of the monomer and the attached carbon chains correspond to soft parts. The photoinitiator is the 2-benzyl-2-(dimethylamine)-4′ morpholinobutyrophenone (Irgacure 369) added at a concentration of 1 mol %, the crosslinker is the 1,6-hexanediol diacrylate at a concentration of 10 mol %. The elastomer is obtained by the photopolymerization of the monomers and the crosslinking of the polymeric chains. Let us recall that in the case

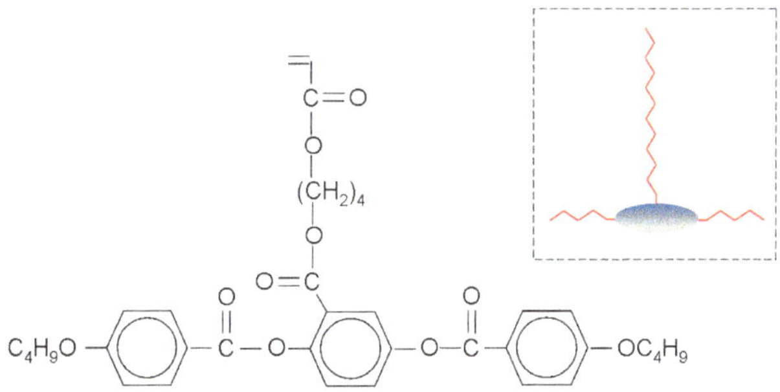

Figure 4. Chemical structure of the acrylate monomer. Inbox: schematics representation.

of a thermotropic nematic elastomer, the destruction of the nematic order by an increase in temperature will cause the contraction of the material along the nematic director. For the contraction to be maximum, the initial order parameter of the sample should be high. The samples were prepared in several micrometer thick glass cells filled by capillarity with the material in its isotropic phase (above 81.5 °C, the nematic-isotropic phase transition temperature). The cell surfaces were coated with rubbed poly-vinyl alcohol to insure the alignment of the liquid crystal moieties by guiding them during the filling process and allowing the production of a fully monodomain aligned sample. Thus the contraction can be unidirectional and along the original mesogens′ alignment. To produce the elastomer the filled cell was slowly cooled down (1 °C/min) to the nematic phase (63 °C) conserving the alignment of the liquid crystal moieties. The elastomer film was then obtained by radical photopolymerization and photocrosslinking under UV light, fixing in this way the nematic alignment of the mesogens. The resulting monodomain elastomer film is named "Liquid Single Crystal Elastomer" (LSCE). The radical photopolymerization starts with the excitation of the photoinitiator Irgacure 369 which is a benzoyl type chromophore with a large absorbance and a high initiation efficiency in the range of 300-400 nm. The initiator generates a radical (a chemical species possessing one or several unpaired electrons in the outer shell) which can react with the monomer and form the macroradical which is also able to react with other monomers. The propagation of the polymerization proceeds by successive addition of monomers to this macro-radical. The monomers are bound to each other by the $CH{=}CH_2$ groups located at the extremities of the molecules. It is the breaking of the double bond that allows the bonding between monomers. The polymerization takes place in such a way as to produce a form of "sheathing" of mesogen around the principal chain, thus avoiding steric hindrance (see figure 5). The crosslinker present in the mixture helps to create a crosslinked network and the elastomer so obtained can be extracted from the cell.

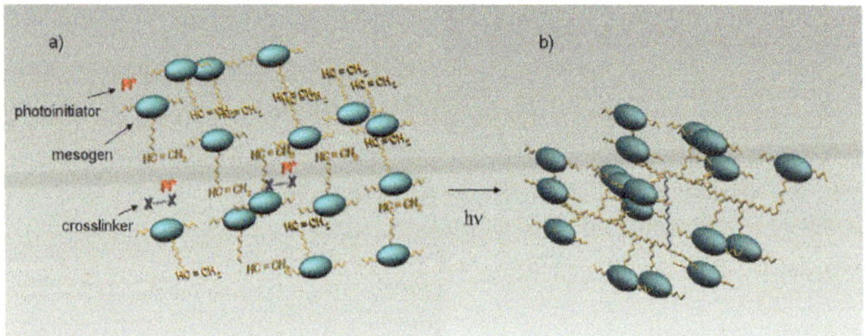

Figure 5. Elastomer formation by photopolymerization and photocrosslinking a) the initial mixture, b) the elastomer: the radical generated by the photoinitiator will activate the monomer which can then make a C-C bond with the following monomer; hence, the pôlymerization is accompanied by the formation of a sheathing around a main chain.

3. Experimental setup

Two different light sources were used to photo-structure the elastomers. We have used 365 nm UV light from an Argon Ion (Ar^+) laser to polymerize the mixture by one-photon photopolymerization or 780 nm infrared (IR) femtosecond laser pulse for two-photon photopolymerization. The advantage of two-photon absorption is the possibility to spatially confine the process by limiting the light-matter interaction to volumes which can be smaller than 1 μm^3. The probability of TPA is very low, its cross section is about 31 orders of magnitude lower than a one-photon absorption cross section. To be efficient TPA therefore necessitates a high density of photons. Thus, we have to spatially and temporally confine the photons. This can be achieved by combining the use of high numerical aperture objectives and ultra-short pulsed lasers such as titanium-sapphire lasers emitting in the near-IR region. When approaching the focal point of the objective, the photon density will sharply increase. The low probability TPA process will thus be automatically confined around this focal point in a volume called "voxel". If the sample is transparent to the IR radiation of the laser, but absorptive for the UV light, only the confined volume around the focal point will be excited, while the rest of the sample will remain transparent to the incident radiation. This microscopy technique is named "two-photon microscopy". Figure 6 presents a comparison between the excitation volumes in the case of conventional microscopy and two-photon microscopy. From the Rayleigh criterion, in optimal conditions, the radial resolution of a conventional microscope is given by the radius of the Airy disc:

$$r_{x,y} = \frac{0.61\lambda}{NA} \tag{2}$$

In the case of two-photon microscopy, Webb *et al.* have proposed an estimation of the ω_{xy} and axial ω_z dimensions of the voxel [30]:

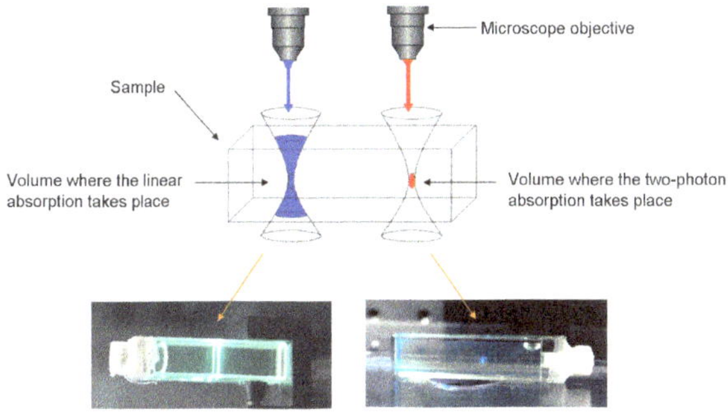

Figure 6. Comparison of the excitation volumes for the linear absorption (left) and two-photon absorption (right): in the linear case all the volume of the sample crossed by the beam is excited, in the case of two-photon absorption the excitation is confined to the voxel around the focal point. The photographies show the fluorescence signal emitted by the excited part of the sample.

$$\omega_{xy} = \begin{cases} \dfrac{0.32\lambda}{\sqrt{2}\mathrm{ON}} & \text{if ON} \leq 0.7 \\ \\ \dfrac{0.325\lambda}{\sqrt{2}\mathrm{ON}.^{0.91}} & \text{if ON} > 0.7 \end{cases} \quad \text{and} \quad \omega_z = \frac{0.532\lambda}{\sqrt{2}} \left(\frac{1}{n - \sqrt{n^2 - \mathrm{ON}^2}} \right) \tag{3}$$

The excitation is given by:

$$V_{2photon} = \pi^{3/2} \omega_{xy}^2 \omega_z \tag{4}$$

The experimental setup used for the microstructurating and the realization of the elastomers is based on a confocal microscope. The cell containing the elastomer precursor mixture is placed on a heating plate mounted on a motorized stage that can execute computer-controlled 3D translations along the X, Y, and Z axes. The observation is done by reflection with a CCD camera. The sample is illuminated in transmission with a white light lamp; a filter is used to stop any UV photons. The sample is also placed between polarizers allowing the monitoring of isotropic and nematic zones through the polarization of transmitted light induced by the mesogen alignment. The photo-patterning of the desired structures is realized in the nematic phase by moving the sample under the focus of the objective with the translation stage.

4. Experimental results

4.1. Photostructuration of the material

We have first studied the creation of a patterned elastomer by photopolymerization initiated by linear or two-photon absorption. Different sets of experiments have been performed depending on the sample alignment state (using treated or untreated glass cells) and the excitation source (UV or IR light). The first set of experiments was made using UV excitation (Argon laser, λ= 365 nm, power = 8 μW, objective N.A = 0.45) and a non-aligned nematic monomer sample. In a first step the photopolymerizable mixture was introduced in an untreated cell. Then the cell was heated beyond the isotropic-nematic

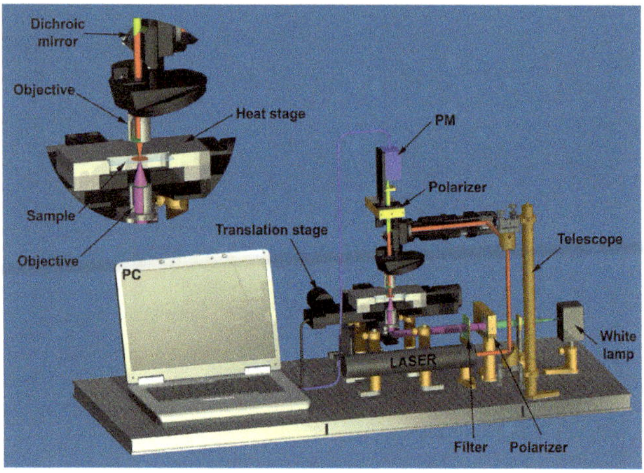

Figure 7. Experimental setup.

phase transition temperature (> 81.5 °C) for the mixture initially in the form of a powder to liquefy and penetrate by capillarity. The mixture was then cooled down until the nematic phase could form (63 °C). A shape representing the capital letter E has been patterned by photopolymerization with the focused UV beam using the translation stage with a displacement velocity of 200 $\mu m/s$. The resulting polymerized and crosslinked part draws the letter "E" as can be seen in figure 8(a). The second set of experiments was performed under the same conditions as described above, except that an aligned nematic monomer sample was used figure 3(b). This polymerization was also successful. The polymerized letter "E" presents thinner (30 μm) and smoother lines than those obtained for the non-aligned sample (50-60 μm). The patterned lines are undistorted in the aligned sample. Therefore, using aligned nematic samples results in considerably improved micro-pattern formation. In the third set of experiments performed again on aligned samples, the IR source (Ti:Sa laser, pulses duration = 100 fs, repetition rate = 80 MHz , λ= 780 nm, power = 500 mW, objective N.A = 0.45) was used instead of the Argon laser. The results presented in figure 3(c) show that such two-photon excitation processes produce a smooth pattern with even thinner lines measuring 5 to 10 μm across. The high resolution achieved in this case is indeed better than UV illumination.

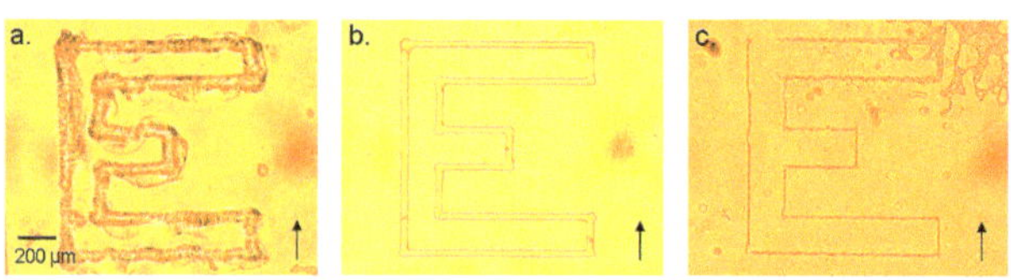

Figure 8. Letter "E" patterns (a) by one-photon photopolymerization (UV irradiation) in a nonaligned sample, (b) by one-photon photopolymerization (UV irradiation) in an aligned sample, (c) by two-photon photopolymerization (IR irradiation) in an aligned sample. The nematic director orientation is indicated by the arrow.

4.2. Obtaining the thermo-active elastomer

In order to stabilize the patterns, a post-photopolymerization of the whole sample can be performed using UV light of weak intensity. Under these conditions, the pattern written by photopolymerization using a high intensity beam is preserved as it can be seen on figure 9. The resulting rubber-like film is then removed from the cell and placed on a heating plate to observe the shape changes as a function of temperature. Figure 10 represents the heating of an elastomer with an inscribed grating of 170 μm step size. The elastomer lies on a film of glycerol oil to minimize the constraints during the contraction. By heating this sample up from room temperature to 120 °C, a clear deformation of the film is observed in the direction of the mesogen alignment. An elastomer with a circular pattern was also produced to highlight the

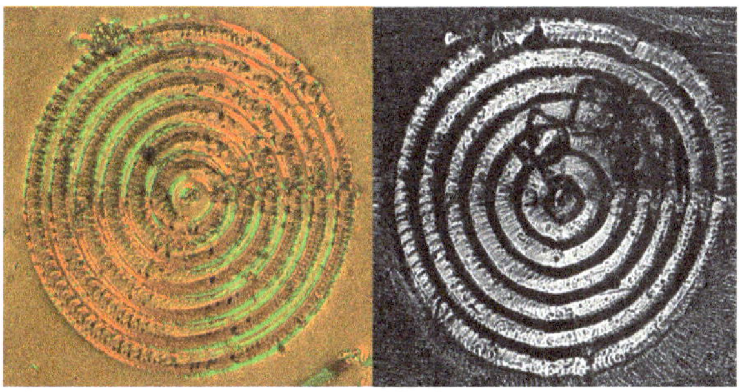

Figure 9. Concentric figure inscribed by photopolymerization before postpolymerization of the whole sample (left), after postpolymerization (right), the zone surrounding the pattern has a weaker polymerization degree.

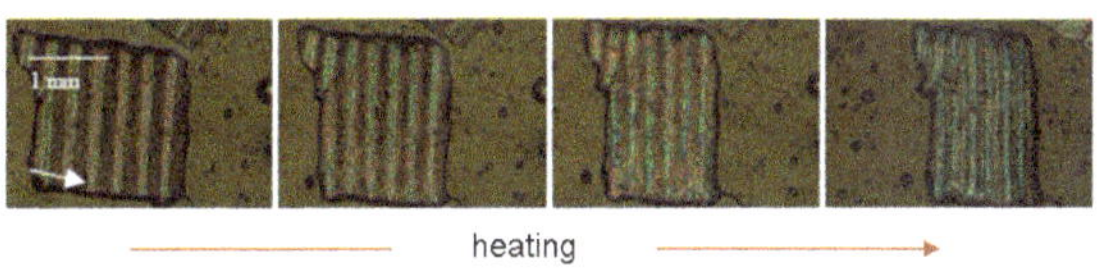

Figure 10. Contraction of a 20 μm thick elastomer as a function of temperature (from 60 °C to 120 °C).

uniaxial contraction and the reversible aspect of the phenomenon as can be seen in figure 11.

5. Discussion

5.1. Contraction properties

Direct observation of the micropatterned elastomer's contraction can be readily achieved by optical microscopy. Figure 12 shows the temperature dependent deformation of the concentric pattern inscribed in a 13 μm thick film of elastomer. The value of the width perpendicular to the nematic director (denoted a) shows a small variation (less than 2% of extension) whereas the perpendicular direction (denoted b) experiences about 17% of contraction. This demonstrates that the shape change consists mostly in a contraction along the nematic director and proves that a well aligned monodomain sample was achieved. This contraction factor varies with the sample thickness. About 50% contraction is observed in 30 μm thick samples

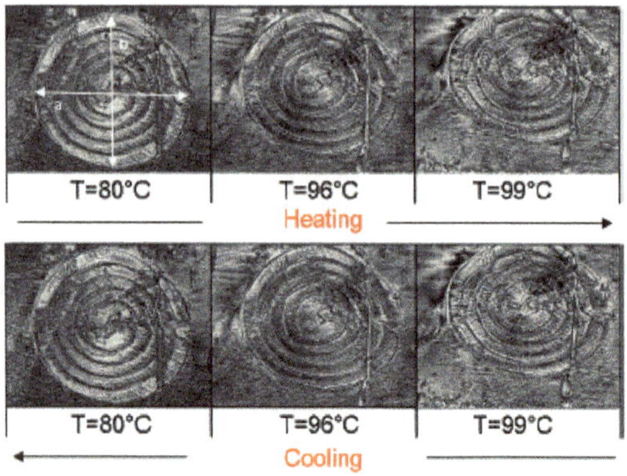

Figure 11. Deformation of concentric cercle patterned in the elastomer as a function of temperature. "a" and "b" correspond to the resulting ellipse axis, "b" is in the mesogen alignment direction.

but as the thickness decreases, the contraction drops to about 30% for 20 μm thick samples and to about 17 % for 13 μm thick samples. This behavior can be attributed to mechanical stress. Indeed, the size of the artificial muscle with respect to the surface of adhesion is larger for a thick film than for a thin film. The contraction forces thus become prevalent (as the samples become thicker) over the constrains resulting from the friction of the film on its substrate. It is noteworthy that no dependency of the rate of contraction on the pattern structure has been observed, suggesting that any diffraction element can be inscribed without altering the physical properties of the material. The contraction of the elastomer starts around 90 °C. It stops when the phase becomes isotropic, towards 100 °C. Both of these temperatures are higher than the phase transition temperatures of the initial liquid crystal which is 81.5 °C. The mesogens being free to move within the elastomer, the thermal characteristics of the mesophases should be preserved. This is, however, only partially true since the mesogens are not completely decoupled from the main chain. However it is reasonable to assume that the longer the spacer, the higher the degree of this decoupling. This aspect may thus bear on the phase transition temperature but there is no doubt that the primary cause behind the observed contraction temperatures lies with the fact that the response of the material to the drop in nematic order is not immediate. Contraction will only occur once the degree of disorientation of the mesogens reaches levels that are enough to overcome the frictional forces between the elastomer and its substrate. In addition, with the monitoring technique that is being put to use, the contraction becomes only apparent beyond a certain level. All this may be the origin of the gap between the phase transition temperature and the temperature at which contraction is being observed.

5.2. Diffraction properties

The contraction of the material can be monitored indirectly by the observation of the diffraction pattern proceeding from a grating inscribed in the elastomer. Temperature induced contraction of the material will result in a change in the grating period which will modify the diffraction pattern. We present the study of the diffraction pattern obtained for a 13

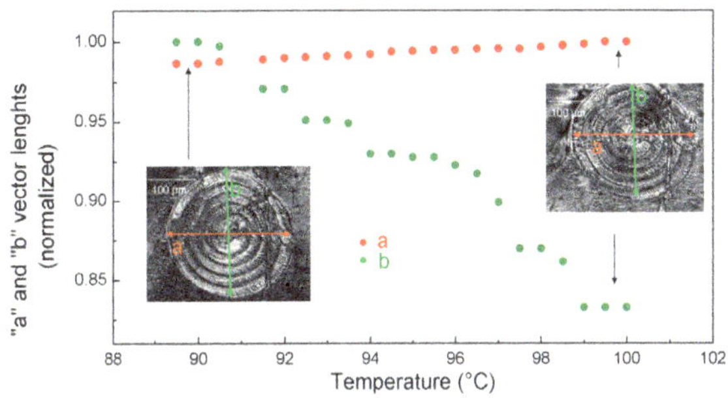

Figure 12. Deformation of concentric circles separated by 25 μm as a function of temperature; "a" and "b" correspond to the axes of the resulting ellipse. The insets show the samples at the temperatures indicated by the arrows.

μm thick sample with a 30 μm period linear grating using a linearly polarized He-Ne laser beam at 633nm. Before testing the properties of the obtained gratings, we have examined

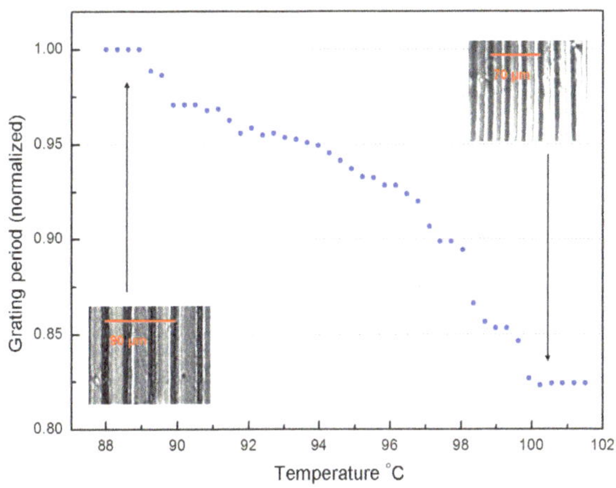

Figure 13. Grating step as a function of temperature, in the case of a 13 m thick sample. The internal figures represent the grating at the corresponding temperatures.

the surface of the elastomer with a profilometer. In some cases when the sample is removed from the cell, its surface exhibits ridges corresponding to the grating. Such observations are commonly made in photopolymerization experiments [31]. The surface corrugation comes from the relaxation of the mechanical stresses introduced by the cross-linking process and which have been maintained by the glass plates. These ondulations are not observed in two-photon photopolymerized samples where the patterns are much thinner and inscribed inside the volume of the elastomer. In the cases where inscribed gratings led to surface ridges, we have checked that the addition of a compensating liquid did not modify the diffraction properties. Thus we may pretend that we are dealing with refractive index gratings. Index inhomogeneity stems from spatial variations in material density between parts that have been strongly polimerized and the remainder of the sample which has only been subjected to a light post-polymerization. The diffraction regime is indicated by the Klein-Cook parameter $Q = 2\pi\lambda_0 L/\Lambda^2 n_0$, where L is the grating thickness, λ_0 is the vacuum wavelength of light, Λ is the grating step size, and n_0 is the mean refractive index. High values of Q (Q>10) correspond to the Brag regime while low values (Q<1) correspond to the Raman-Nath regime. One usually ascribes the Bragg regime, with essentialy only one diffracted wave being produced, to gratings that are thick with respect to step size and the Raman-Nath regime, with several diffracted waves being produced, to thin gratings. In the case of a grating with a step size of 30 μm and a thickness of 13 μm yielding the diffraction pattern shown in figure 14, the Q parameter is about 0.4, placing the diffraction in the Raman-Nath regime as demonstrated by the multiple order diffraction pattern. Figure 15 shows the variation of the angle of diffraction for the +1 order as a function of temperature. It clearly demonstrates the decrease

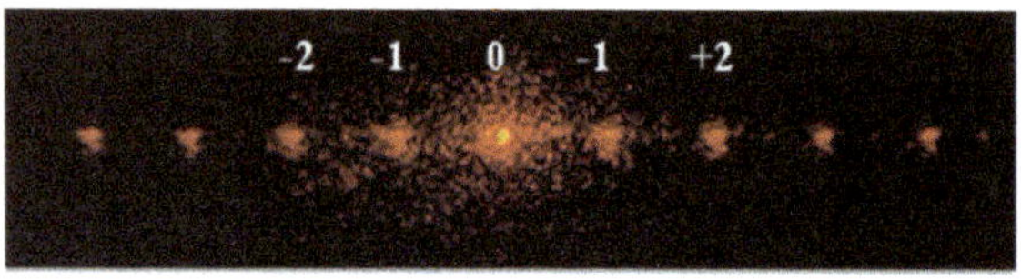

Figure 14. Diffraction figure of 30m step gratings inscribed into the elastomer.

of the grating step with the temperature rise stemming from the unidirectional contraction accompanying the nematic/isotropic transition. In figure 15, the black dots correspond to measurements of the diffraction angle, while the red dots represent values calculated following the diffraction grating formula from the data shown in figure 13 for the 13 μm thick sample. The agreement between both sets of data shows the consistency between the grating period and the diffraction angle measurements (discrepancies between data points may come from the fact that they have been gathered over several heating cycles). In conclusion, the contraction induced by the temperature is easily monitored by the widening of the diffraction figure. This allows a feedback on the deformation of the material through direct monitoring of the resulting diffraction.

Since we work with a cross-linked polymer, the contraction/extension process will affect all three directions in space. The initial homogeneous nematic order makes the film behave as a biaxial material with an optical axis oriented along the nematic director. Thus, the material may change the polarization state of the light upon diffraction. To illustrate the birefringence properties of the gratings, we have measured the diffraction efficiency dependence on the orientation of the incident linear polarization. The resulting angular distribution for the polarization of the diffracted light is presented in figure 16b. Here, the simplest configuration is adopted: the incident beam is normal to the surface and its polarization is linear. As one can see in figure 16a, the diffraction efficiency is higher when the direction of the polarization of the incident beam and the grating vector are parallel. The diffracted light intensity has been fitted by :

$$I = \eta_{\parallel} \cos^2 \phi + \eta_{\perp} \sin^2 \phi \tag{5}$$

where the angle ϕ corresponds to the rotation of the polarisation and I is the light intensity of the first diffracted order. Thus we can determine the parallel and the perpendicular diffraction

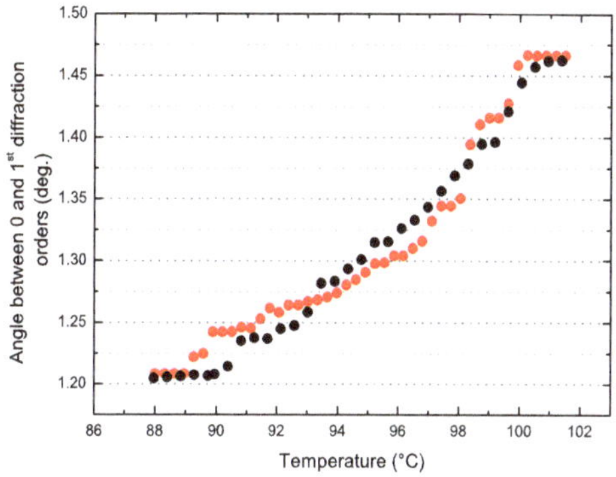

Figure 15. Variation of the first order diffraction angle of a 30 μm period grating as a function of temperature: black dots are the measured diffraction angles, red dots are calculated from the data on figure 13.

efficiencies which amount respectively to $\eta_{\parallel} = 1$ and $\eta_{\perp} = 0.53$.
In the case of figure 16b the incident beam polarization is at 45^0 of the grating vector. This time a light intensity dependence on the rotation of an analyzer placed after the sample is observed. Equation (6) represents the theoretical function which will fit the intensity of light measured after the analyser.

$$I = \left| \sqrt{\eta_{\parallel}} \cos\phi \, \frac{\sqrt{2}}{2} \, e^{(i\psi/2)} + \sqrt{\eta_{\perp}} \sin\phi \, \frac{\sqrt{2}}{2} \, e^{(-i\psi/2)} \right|^2 \tag{6}$$

For $\eta_{\parallel}$ and $\eta_{\perp}$ we take the diffraction efficiencies determined in figure 16a. The phase difference ψ is introduced by the birefringence and ϕ is the analyser angle. Figure 16b represents experimental data and the fits for the zero and the first diffraction orders. We observe the slight depolarization of the zero order due to the intrinsic material birefringence when comparing it to the polarization of the incident beam added as a reference. Diffraction order +1 shows a more elliptic polarization and a rotation of the major axis. Calculations give a rotation of 29.7^0. We have been able to make these observations on various samples and are led to consider that this behavior finds its origin in two simultaneous effects. The first is obviously the birefringence of the material which introduces a phase difference between the parallel and the perpendicular components of the field and causes the depolarization of the diffracted beam. The other effect originates in the diffraction grating: the refractive index modulation seen by the beam is different for the direction along the grating vector and for the perpendicular to it. Therefore, the diffraction efficiencies for polarizations along these two directions are not identical. Hence, the decomposition of the polarization of the diffracted beam along these directions will be different with respect to that of the incoming light. As temperature increases, the material tends to the isotropic configuration and the observed birefringence tends to get attenuated.

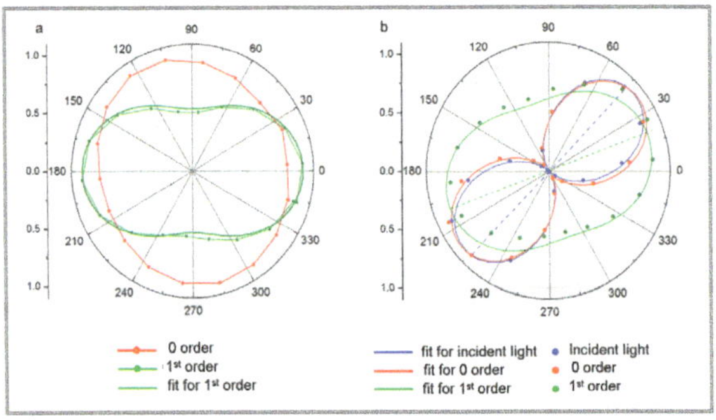

Figure 16. Polarization of diffracted light. (a) zero- and first-order diffraction intensities versus incident beam linear polarization orientation. (b) Incident beam, zero-, and first-order diffraction intensities versus analyzer angle. The direction of the grating vector corresponds to the angle zero.

5.3. Molecular dynamics simulations

Molecular dynamics (MD) simulations are a powerful tool for understanding the properties of a sample in terms of molecular collective phenomena. We used classical MD method to simulate the contraction of the elastomer. First, the monomer molecules composed of 95 atoms was fully optimized by the density functional theory (DFT). The chosen functional was B3LYP, a hybrid functional obtained by linear combination of exchange-correlation energy functionals (Local-density approximations LDA, Generalized gradient approximations GGA) and the Hatree-Fock exchange [32]. The orbital basis set was the contracted Gaussian 6-311** set. All calculations were carried out with NWChem [33], executed on the IBM Power 6 at the "Institut du Développement et des Ressources en Informatique Scientifique" of Orsay. The obtained minimum energy configuration is represented in figure 17. This model presents a dipole localized on the mesogen corresponding to the aromatic core. This will allow

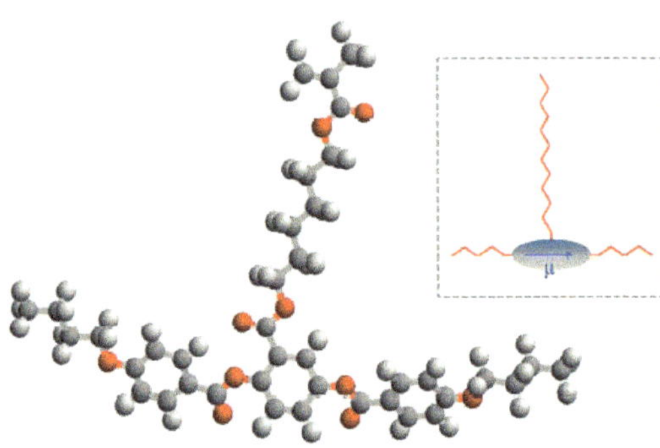

Figure 17. Model of the monomer: minimum energy configuration. Inbox: schematical representation; a dipole is localized on the rigid mesogenic site corresponding to the aromatic core.

dipole-dipole interactions between the monomers. Then to simulate the contraction of the elastomer resulting from the collective behavior of the monomers the interaction between pairs of anisotropic rigid mesogenic sites was modeled using Gay-Berne (GB) potential. The Gay-Berne potential corresponds to a modified form of Lennard-Jones potential which can be easily adjusted to modify the shape of the studied system [34]. In that sense it has proven its efficiency to model mesogenic systems [35][36][37].

The force field modeling intra and inter-molecular contribution to the interaction energy is given by the following expression:

$$\begin{aligned} U = & \sum_{i=1}^{N_{angles}} \frac{k_i^{ang}}{2} \left(\theta_i - \theta_i^0\right)^2 \\ & + \sum_{i=1}^{N_{GB}} \frac{k_i^{GB}}{2} \left(\nu_i - \nu_i^0\right)^2 + \\ & + \sum_{i=1}^{N_{dihedrals}} \left(a_{i,1}\left(1+\cos\phi_i\right) + a_{i,2}\left(1-\cos 2\phi_i\right) + a_{i,3}\left(1+\cos 3\phi_i\right)\right) \\ & + \sum_{i=1}^{N_{LJ}} \sum_{j>i}^{N_{LJ}} U_{LJ} \\ & + \sum_{i=1}^{N_{LJ}} \sum_{j=1}^{N_{GB}} U_{LJ/GB} \\ & + \sum_{i=1}^{N_{GB}} \sum_{j>i}^{N_{GB}} U_{GB} \end{aligned} \quad (7)$$

where N_{angles}, $N_{dihedrals}$, N_{GB}, N_{LJ}, are respectively the angles number of the dihedral angles of the numbers, and the number of GB and LJ sites numbers. The variations in dihedral angles ϕ, which represent torsions, are modeled by a truncated Fourier series ($a_{i,1,2,3}$ coefficients). The alkyne groups of the chain are condensed into single atom sites and represented as spheres. All the chain segment parameters are modeled by resorting to universal force field [38]. U_{LJ}, U_{GB} et $U_{LJ/GB}$ are respectively Lennard-Jones, Gay-Berne and the mixte Lennard-Jones/Gay-Berne potentials. Bond lengths between the mesogen and the flexible chains were set from *ab initio* results. They are 1.495 Å for the C-mesogen bond and 1.353 Å for the O-mesogen bond. The bond lengths between chain segments and GB unit were constrained by using the SHAKE procedure [39] during molecular dynamics calculations. ν_i corresponds to the angle between the long axis of the GB site and the bond between this site and the adjacent one. ν_i^0 is the equilibrium value and k_i^{GB} corresponds to the force constant given by DFT calculations (ν_{GBC}^0 =112.62 °C and k_{GBC}^0=6.521 eV/rad^2).

For a separation distance of r_{ij} between two particles, the standard LJ potential is written as:

$$U_{LJ} = \sum_i \sum_{j>i} 4\epsilon_{ij} \left(\left(\frac{\sigma_{ij}^{(0)}}{r_{ij} - \sigma_{ij} + \sigma_{ij}^{(0)}} \right)^{12} - \left(\frac{\sigma_{ij}^{(0)}}{r_{ij} - \sigma_{ij} + \sigma_{ij}^{(0)}} \right)^{6} \right) \quad (8)$$

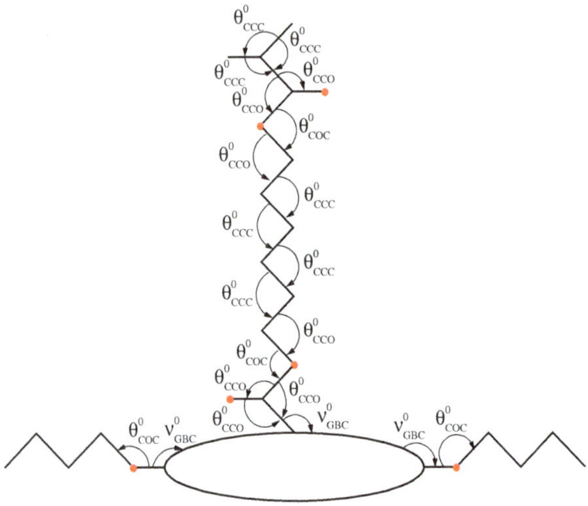

Figure 18. Representation of different bonding angles.

where ϵ_{ij} corresponds to the energy well depth, σ_{ij} is the contact parameter, it corresponds to the separation distance at which the inter-particle potential is zero.

U_{GB} represents the orientation dependent interaction energy for two GB particles:

$$U_{GB} = \sum_{i}\sum_{j>i} 4\epsilon_{ij}\left(\hat{\mu}_i, \hat{\mu}_j, \hat{r_{ij}}\right)\left(\left(\frac{\sigma_{ij}^{(0)}}{r_{ij} - \sigma_{ij}\left(\hat{\mu}_i, \hat{\mu}_j, \hat{r_{ij}}\right) + \sigma_{ij}^{(0)}}\right)^{12} - \left(\frac{\sigma_{ij}^{(0)}}{r_{ij} - \sigma_{ij}\left(\hat{\mu}_i, \hat{\mu}_j, \hat{r_{ij}}\right) + \sigma_{ij}^{(0)}}\right)^{6}\right) \tag{9}$$

where $\hat{\mu}_i$ and $\hat{\mu}_j$ are unitary vectors along the molecular axis of GB particles, ϵ_{ij} and σ_{ij} correspond respectively to the energy well depth and the contact parameter depending on the relative orientation of the particles. These terms are defined as [36] [40]:

$$\sigma_{ij}\left(\hat{\mu}_i, \hat{\mu}_j, \hat{r_{ij}}\right) = \frac{\sigma_{ij}^{(0)}}{\sqrt{1 - \frac{\chi}{2}\left[\frac{\left(\hat{\mu}_i \cdot \hat{r}_{ij} + \hat{\mu}_j \cdot \hat{r}_{ij}\right)^2}{1 + \chi\left(\hat{\mu}_i \cdot \hat{\mu}_j\right)} + \frac{\left(\hat{\mu}_i \cdot \hat{r}_{ij} - \hat{\mu}_j \cdot \hat{r}_{ij}\right)^2}{1 - \chi\left(\hat{\mu}_i \cdot \hat{\mu}_j\right)}\right]}} \tag{10}$$

$$\epsilon_{ij}\left(\hat{\mu}_i, \hat{\mu}_j, \hat{r_{ij}}\right) = \epsilon_0 \epsilon_1^{\gamma}\left(\hat{\mu}_i, \hat{\mu}_j, \hat{r_{ij}}\right) \epsilon_2^{\nu}\left(\hat{\mu}_i, \hat{\mu}_j\right) \tag{11}$$

$$\epsilon_1\left(\hat{\mu}_i, \hat{\mu}_j, \hat{r_{ij}}\right) = 1 - \frac{\chi\prime}{2}\left[\frac{\left(\hat{\mu}_i \cdot \hat{r}_{ij} + \hat{\mu}_j \cdot \hat{r}_{ij}\right)^2}{1 + \chi\prime\left(\hat{\mu}_i \cdot \hat{\mu}_j\right)} + \frac{\left(\hat{\mu}_i \cdot \hat{r}_{ij} - \hat{\mu}_j \cdot \hat{r}_{ij}\right)^2}{1 - \chi\prime\left(\hat{\mu}_i \cdot \hat{\mu}_j\right)}\right] \tag{12}$$

$$\epsilon_2\left(\hat{\mu}_i,\hat{\mu}_j\right) = \frac{1}{\sqrt{1-\chi^2\left(\hat{\mu}_i\cdot\hat{\mu}_j\right)^2}} \tag{13}$$

$$\chi = \frac{(\sigma_E/\sigma_s)^2-1}{(\sigma_E/\sigma_s)^2+1}$$

$$\chi' = \frac{1-(\epsilon_E/\epsilon_s)^{1/\gamma}}{1+(\epsilon_E/\epsilon_s)^{1/\gamma}} \tag{14}$$

with σ_E and σ_S corresponding to the contact parameter along the GB particle length and width and ϵ_E and ϵ_S corresponding to the energy well depth for two GB particles with an "end to end" and a "side by side". r_{ij} is the distance between the two particles i and j, $\hat{r}_{ij}$ corresponds to the unitary vector along $\vec{r}_{ij}$, $\hat{\mu}_i$ and $\hat{\mu}_j$ are unitary vectors along the molecular axis of the particles i and j. γ and ν are adjustable parameters which were set to their canonical values of 1.0 and 2.0, respectively. The ratio was σ_E/σ_s=3.46 and ϵ_0=21.2 meV.
$U_{LJ/GB}$ corresponds to the potential representing the mixte interaction between LJ and GB sites [40]:

$$U_{GB/LJ} = \sum_i\sum_j 4\epsilon_{ij}\left(\hat{\mu}_i,\hat{r_{ij}}\right)\left(\left(\frac{\sigma_{ij}^{(0)}}{r_{ij}-\sigma_{ij}\left(\hat{\mu}_i,\hat{r_{ij}}\right)+\sigma_{ij}^{(0)}}\right)^{12} - \left(\frac{\sigma_{ij}^{(0)}}{r_{ij}-\sigma_{ij}\left(\hat{\mu}_i,\hat{r_{ij}}\right)+\sigma_{ij}^{(0)}}\right)^{6}\right) \tag{15}$$

We have considered the molecular dynamics in a modified isothermal-isobaric (NPT) ensemble to allow an anisotropic deformation of the simulation cell. The ensemble used is the Parrinello-Rahman-Nosé-Hoover ensemble based on the H_{PRNH} Hamiltonian [41] [42] [43]. Thanks to the introduction of a thermostat and a barostat, this ensemble allows to control the temperature and the constraints acting on the simulation cell. It allows the study of anisotropic deformation of a system as a function of temperature. It is in that sense well adapted to the study of systems containing liquid crystal molecules. The H_{PRNH} Hamiltonian is given by:

$$H_{PRNH} = \frac{1}{2}\sum_{i=1}^{N}\frac{1}{m_i}\vec{P}_i^T\bar{\bar{G}}\vec{P}_i + U \tag{16}$$

$$+\frac{1}{2Q}P_\eta^2 + gk_BT\eta$$

$$+\frac{1}{2W}Tr\left(\bar{\bar{P}}_h^T\bar{\bar{P}}_h\right) + P_{ext}\Delta\bar{\bar{h}}$$

$$+\frac{1}{2}\sum_{i=1}^{N_{GB}}\frac{1}{m_{\vec{\rho}_i}}\vec{P}_{\vec{\rho}_i}^T\vec{P}_{\vec{\rho}_i} \tag{17}$$

with $\bar{\bar{G}} = \bar{\bar{h}}^T\bar{\bar{h}}$.
$\bar{\bar{h}}$ is the matrix containing the dimensions of the unit cell L_x, L_y and L_z, $\bar{\bar{P}}_h$ is the conjugated moment and together they form the dynamical variables of the system relating to the control of

the constraints acting on the simulation cell. W is a fictive mass which allows the adjustments of the barostat's reactions to the constraints' changes. m_i is the i particle mass. $\vec{P}_i$ is its momentum:

$$\vec{P}_i = m_i \frac{d\vec{S}_i}{dt} \tag{18}$$

where $\vec{S}_i$ is the reduced position of the particle in the simulation cell

$$\vec{S}_i = \bar{\bar{h}}^{-1} \vec{q}_i \tag{19}$$

and $\vec{q}_i$ corresponds to the absolute position. η and P_η are dynamical variables relating to the control of the temperature. Q is a fictive mass which allows to adjust the thermostat reactions to temperature changes in the system. $\vec{\rho}$ and $\vec{P}_{\vec{\rho}}$ are dynamical variables relating to the long range interactions between GB sites. $m_{\vec{\rho}}$ is the mass relating to this type of interactions and controlling its dynamics.
The forces acting on the system are obtained by deriving the field U. The system's deformations are obtained by integrating the equations of motion using Beeman's algorithme [44]:

$$\vec{S}_i(t+\delta t) = \vec{S}_i(t) + \frac{\delta t}{m_i}\vec{P}_i(t) + \frac{\delta^2}{6m_i}\left(4\frac{d\vec{P}_i(t)}{dt} - \frac{d\vec{P}_i(t-\delta t)}{dt}\right) \tag{20}$$

$$\vec{P}_i(t+\delta t) = \vec{P}_i(t) + \frac{\delta t}{6}\left(5\frac{d\vec{P}_i(t)}{dt} + 2\frac{d\vec{P}_i(t+\delta t)}{dt} - \frac{d\vec{P}_i(t-\delta t)}{dt}\right) \tag{21}$$

$$\bar{\bar{h}}(t+\delta t) = \bar{\bar{h}}(t) + \frac{\delta t}{W}\bar{\bar{P}}_h(t) + \frac{\delta^2}{6W}\left(4\frac{d\bar{\bar{P}}_h(t)}{dt} - \frac{d\bar{\bar{P}}_h(t-\delta t)}{dt}\right) \tag{22}$$

$$\bar{\bar{P}}_h(t+\delta t) = \bar{\bar{P}}_h(t) + \frac{\delta t}{6}\left(5\frac{d\bar{\bar{P}}_h(t)}{dt} + 2\frac{d\bar{\bar{P}}_h(t+\delta t)}{dt} - \frac{d\bar{\bar{P}}_h(t-\delta t)}{dt}\right) \tag{23}$$

$$\eta(t+\delta t) = \eta(t) + \frac{\delta t}{Q}P_\eta(t) + \frac{\delta^2}{6Q}\left(4\frac{dP_\eta(t)}{dt} - \frac{dP_\eta(t-\delta t)}{dt}\right) \tag{24}$$

$$P_\eta(t+\delta t) = P_\eta(t) + \frac{\delta t}{6}\left(5\frac{dP_\eta(t)}{dt} + 2\frac{dP_\eta(t+\delta t)}{dt} - \frac{dP_\eta(t-\delta t)}{dt}\right) \tag{25}$$

$$\vec{\rho}_i(t+\delta t) = \vec{\rho}_i(t) + \frac{\delta t}{m_{\vec{\rho}_i}}\vec{P}_{\vec{\rho}_i}(t) + \frac{\delta t^2}{6m_{\vec{\rho}_i}}\left(4\frac{d\vec{P}_{\vec{\rho}_i}(t)}{dt} - \frac{d\vec{P}_{\vec{\rho}_i}(t-\delta t)}{dt}\right) \tag{26}$$

$$\vec{P}_{\vec{\rho}_i}(t+\delta t) = \vec{P}_{\vec{\rho}_i}(t) + \frac{\delta t}{6}\left(5\frac{d\vec{P}_{\vec{\rho}_i}(t)}{dt} + 2\frac{d\vec{P}_{\vec{\rho}_i}(t+\delta t)}{dt} - \frac{d\vec{P}_{\vec{\rho}_i}(t-\delta t)}{dt}\right) \tag{27}$$

An autocoherent process is necessary to calculate the conjugated moments. The time derivatives of the moments are calculated with the Parrinello-Rahman-Nosé-Hoover

Hamiltonian.

$$
\begin{aligned}
\frac{d\vec{P}_i}{dt} &= -\frac{1}{Q}P_\eta \vec{P}_i - \bar{\bar{G}}^{-1}\frac{d\bar{\bar{G}}}{dt}\vec{P}_i - \bar{\bar{h}}^{-1}.\vec{\nabla}_{\vec{q}_i}U && (28)\\
\frac{d\vec{P}_{\vec{\rho}_i}}{dt} &= -\vec{\nabla}_{\vec{\rho}_i}U && (29)\\
\frac{dP_\eta}{dt} &= \sum_{i=1}^{N}\frac{1}{m_i}\vec{P}_i^T\bar{\bar{G}}\vec{P}_i + \frac{1}{W}Tr\left(\bar{\bar{P}}_h^T\bar{\bar{P}}_h\right) - gk_BT && (30)\\
\frac{dP_{h,\alpha\beta}}{dt} &= -\frac{1}{Q}P_\eta P_{h,\alpha\beta} + \frac{1}{2}\sum_{i=1}^{N}\frac{1}{m_i}\vec{P}_i^T\frac{d\bar{\bar{G}}}{dh_{\alpha\beta}}\vec{P}_i - \frac{\partial U}{\partial h_{\alpha\beta}} - P_{ext}\frac{\partial\Delta\bar{\bar{h}}}{\partial h_{\alpha\beta}} && (31)\\
&\alpha,\beta = x,y,z
\end{aligned}
$$

$$
\begin{aligned}
\frac{\partial U}{\partial \mu_{k,\alpha}} = \frac{1}{\mu_k}\sum_{i\neq k}^{N_{GB}}\Bigg\{&(\hat{\mu}_{i,\alpha} - \hat{\mu}_{k,\alpha}\hat{\mu}_k\cdot\hat{\mu}_i)\cdot\Bigg[\nu\chi^2\varphi_{ki}\Omega_{ki}^2\hat{\mu}_i\cdot\hat{\mu}_k\\
&+\mu\chi'^2\frac{\varphi_{ki}}{\Theta_{ki}}\left[\frac{(\hat{\mu}_k\cdot\hat{r}_{ki}+\hat{\mu}_i\cdot\hat{r}_{ki})^2}{(1+\chi'\hat{\mu}_k\cdot\hat{\mu}_i)^2} - \frac{(\hat{\mu}_k\cdot\hat{r}_{ki}-\hat{\mu}_i\cdot\hat{r}_{ki})^2}{(1-\chi'\hat{\mu}_k\cdot\hat{\mu}_i)^2}\right]\\
&-\frac{\chi^2}{\sigma_{ki}^{(0)^3}}\epsilon_{ki}\sigma_{ki}^3\left(12\rho_{ki}^{13}-6\rho_{ki}^7\right)\left[\frac{(\hat{\mu}_k\cdot\hat{r}_{ki}+\hat{\mu}_i\cdot\hat{r}_{ki})^2}{(1+\chi\hat{\mu}_k\cdot\hat{\mu}_i)^2} - \frac{(\hat{\mu}_k\cdot\hat{r}_{ki}-\hat{\mu}_i\cdot\hat{r}_{ki})^2}{(1-\chi\hat{\mu}_k\cdot\hat{\mu}_i)^2}\right]\Bigg]\\
&+2\left(\hat{r}_{ki,\alpha}-\hat{\mu}_{k,\alpha}\hat{\mu}_k\cdot\hat{r}_{ki}\right)\Bigg[\frac{\chi}{\sigma_{ki}^{(0)^3}}\epsilon_{ki}\sigma_{ki}^3\left(12\rho_{ki}^{13}-6\rho_{ki}^7\right)\\
&\cdot\left[\frac{\hat{\mu}_k\cdot\hat{r}_{ki}+\hat{\mu}_i\cdot\hat{r}_{ki}}{1+\chi\hat{\mu}_k\cdot\hat{\mu}_i}+\frac{\hat{\mu}_k\cdot\hat{r}_{ki}-\hat{\mu}_i\cdot\hat{r}_{ki}}{1-\chi\hat{\mu}_k\cdot\hat{\mu}_i}\right]\\
&-\mu\chi'\frac{\varphi_{ki}}{\Theta_{ki}}\left[\frac{\hat{\mu}_k\cdot\hat{r}_{ki}+\hat{\mu}_i\cdot\hat{r}_{ki}}{1+\chi'\hat{\mu}_k\cdot\hat{\mu}_i}+\frac{\hat{\mu}_k\cdot\hat{r}_{ki}-\hat{\mu}_i\cdot\hat{r}_{ki}}{1-\chi'\hat{\mu}_k\cdot\hat{\mu}_i}\right]\Bigg]\Bigg\} \qquad (32)
\end{aligned}
$$

Calculations have been carried out for 100 molecules and for simulation times up to $\delta t = 1$ fs. We considered that the molecules are bonded by the $CH{=}CH_2$ groups located at their extremity as represented in figure 19. The bonding of the molecules (the polymerization) takes

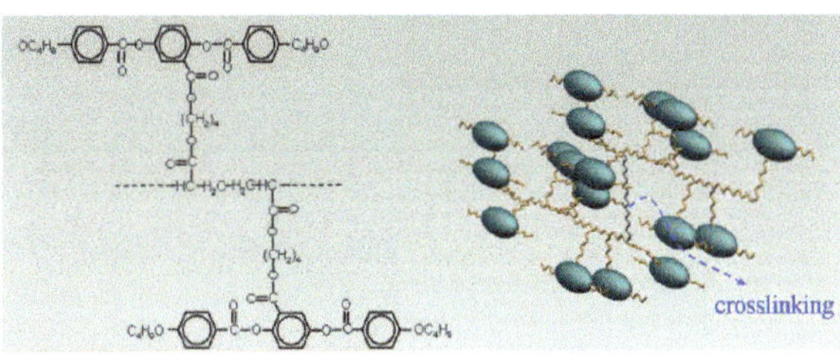

Figure 19. Schematical representation of the modeled elastomer.

place in such a way to produce a form of sheathing around the principal chain. Crosslinking between principal chains was modeled by resorting to alkyne chains.

To measure the contraction of the simulated system we have monitored the obtained dimensions of the simulation cell vectors L_x and L_y. In figure 20 and 21 we have represented the results of the simulation and the measured contraction. We can notice a difference between the curvature of the two curves. Figure 22 represents the behavior of the order parameter

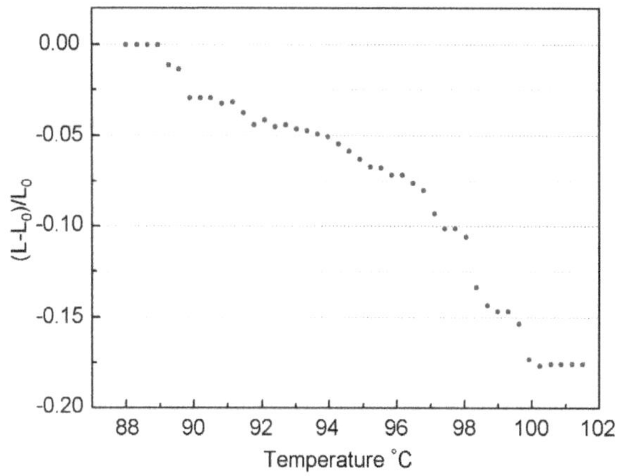

Figure 20. Contraction of the elastomer: fractional change in length $(L - L_0)/L_0$ as a function of temperature.

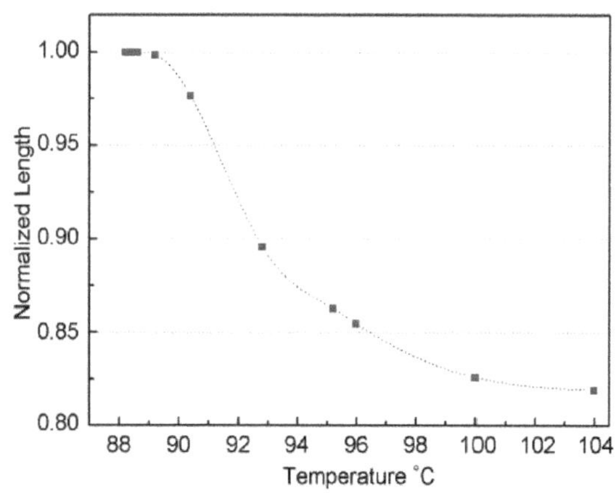

Figure 21. Simulation result. The curve is a guide to the eye.

obtained by polarized Fourier transform spectroscopy (FTIR) in the case of the same elastomer [45]. If we compare the simulation curve with figure 22 we can see that they have the same apparence. This similarity was to be expected, in fact it is the decrease of the order parameter which generates the contraction of the material. But differences between the behavior of the order parameter and the contraction can be identified: the contraction is not immediate, it

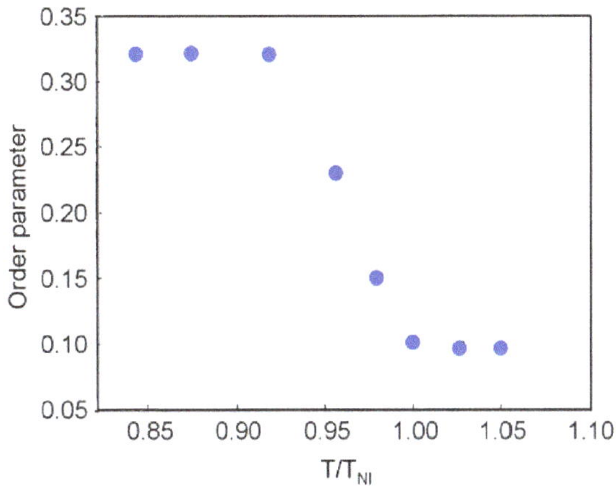

Figure 22. Order parameter as a function of reduced temperature T/T_{NI}, T_{NI} is the phase transition temperature [45].

starts above a given temperature which is higher than the phase transition temperature. In reality a sufficient contraction force should be achieved before observing the phenomena and these contraction forces should also exceed the friction forces with the glass plate. The fact that the simulation curve looks like the order parameter curve more than the contraction curve encourages us to consider external mechanical forces acting on the elastomer in the future.

6. Conclusion

We have demonstrated that one photon "UV" photopolymerization as well as two-photon "IR" photopolymerization can be used to microstructure artificial muscle materials made of nematic liquid crystalline elastomers without losing the contraction/extension properties. We have shown that the use of two-photon absorption allows to achieve 65 % greater spatial resolutions. A major advantage of the TPA consists in creating shape-changing volume objects, a property particularly interesting for the domain of microfluidics. We have established the possibility of generating a grating design in the sample which can be used as a step changing grating when subject to a temperature increase. We have shown that the contraction induced by the temperature is easily monitored by the widening of the diffraction figure. We can then consider the use of this kind of grating for temperature adjusted feedback devices. In addition, the birefringence properties of the gratings can open the path for polarization dependent diffractive elements.

Molecular dynamics simulations have been used to describe the contraction of the elastomer as a function of temperature. The proposed model allows the simulation of anisotropic molecules made from a combination of Gay-Berne potentials, representing the rigid mesogenic parts, and Lennard-Jones sites representing the alkyne groups of the flexible chains. Even if experimental and numerical contractions behave somewhat differently as functions of the temperature, a fact that may be attributed in part to the neglect of external mechanical forces, a rather good agreement between the behavior of the simulation curve and the order parameter has been achieved and a reasonable agreement with the contraction curve has been obtained.

Author details

Kokou D. (Honorat) Dorkenoo, Hervé Bulou, Grégory Taupier, and Alex Boeglin
IPCMS, UMR 7504, 23 rue du Loess, BP 43, F-67034 Strasbourg Cedex 2, France

Emel Sungur
Bilkent University, Department of Physics, Advanced Research Laboratories, 06800, Ankara, Turkey

7. References

[1] Ikeda, T. and Tsutsumi, O. (1995) Optical switching and image storage by means of azobenzene liquid-crystal films. Science 268: 1873–1875.
[2] Lagerwall, S. T. (1999) Ferroelectric and Antiferroelectric Liquid Crystals. Wiley-VCH.
[3] Drzaic, P. S. (1999) Liquid Crystal Dispersions. World Scientific.
[4] Warner, M. and Terentjev, E.M. (2003) Liquid crystal elastomers. Oxford University Press, USA.
[5] de Gennes, P.G. (1997) A semi-fast artificial muscle. CR Acad. Sci. Ser. II B 324: 343-348.
[6] Yu, Y. and Ikeda, T. (2006) Two-Photon Excitation in CaF2Eu2+. Phys. Rev. Lett. 45: 5416-5418.
[7] Naciri, J. and Srinivasan, A. and Jeon, H. and Nikolov, N. and Keller, P. and Ratna, B.R. (2003) Nematic elastomer fiber actuator. Macromolecules 36: 8499-8505.
[8] Ikeda Y. Yu T., and Mamiya J.(2007) Soft actuators based on liquid-crystalline elastomers. Angaw. Chem. Int. Ed, 46: 506.
[9] De Gennes, P.G(1975) Réflexions sur un type de polymères nématiques. ACR Acad. Sci. Paris, Ser. B, 46: 101-103.
[10] Kupfer, J. and Finkelmann, H. (1991) Nematic liquid single crystal elastomers. Makromol Chem., Rapid Commun. 12: 717-726.
[11] Lessard, R. A. Gurusamy, M. (1995) Photoreactive Polymers in Advance Applications. Chapman and Hall, New York.
[12] McCammon, J.A. and Gelin, B.R. and Karplus, M. (1977) Dynamics of folded proteins. Nature. 267: 585–590.
[13] Allen, M.P. and Warren, M.A. and Wilson, M.R. and Sauron, A. and Smith, W. (1996) Molecular dynamics calculation of elastic constants in Gay-Berne nematic liquid crystals. J. Chem. Phys. 105: 2850-2858.
[14] Darinskii, A.A. and Zarembo, A. and Balabaev, N.K. (2007) Molecular Dynamic Simulation of Side-Chain Liquid Crystalline Elastomer Under Load. Macromolecular Symposia. 252: 101-109.
[15] Osada, Y. and Okuzaki, H. and Hori, H.(1992) A polymer gel with electrically driven motility. Nature 355: 242-244.
[16] Osada, Y. and De Rossi, D.E.(2000)Polymer sensors and actuators. Springer Berlin.
[17] Osada, Y. and Khokhlov, A.R. (2002)Polymer Gels and Networks. Marcel Dekker, Inc, 270 Madison Avenue, New York, NY 10016, USA, 1-381.

[18] Osada, Y. and Khokhlov, A.R. (1996)Conducting polymer artificial muscles. Synth. Met. 78: 339-354.
[19] Lu, W. and Fadeev, A.G. and Qi, B. and Smela, E. and Mattes, B.R. and Ding, J. and Spinks, G.M. and Mazurkiewicz, J. and Zhou, D. and Wallace, G.G. and others (2002)Use of ionic liquids for pi-conjugated polymer electrochemical devices. Science 297: 983.
[20] Pyo, M. and Bohn, C.C. and Smela, E. and Reynolds, J.R. and Brennan, A.B.(2003)Direct strain measurement of polypyrrole actuators controlled by the polymer/gold interface. Chem. Mater. 15: 916-922.
[21] Baughman, R.H. and Cui, C. and Zakhidov, A.A. and Iqbal, Z. and Barisci, J.N. and Spinks, G.M. and Wallace, G.G. and Mazzoldi, A. and De Rossi, D. and Rinzler, A.G. and others(1999)Carbon nanotube actuators. Science 284: 1340.
[22] Kim, P. and Lieber, C.M.(1999)Nanotube nanotweezers. Science 286: 2148.
[23] Zhang, Y. and Iijima, S.(1999) Elastic response of carbon nanotube bundles to visible light. Phys. Rev. Lett. 82: 3472-3475.
[24] Spinks, G.M. and Wallace, G.G. and Fifield, L.S. and Dalton, L.R. and Mazzoldi, A. and De Rossi, D. and Khayrullin, I.I. and Baughman, R.H.(2002) Pneumatic carbon nanotube actuators. Adv. Mater. 14: 1728-1732.
[25] KPelrine, R. and Kornbluh, R. and Pei, Q. and Joseph, J.(2000)High-speed electrically actuated elastomers with strain greater than 100%. Science 287: 836.
[26] Finkelmann, H. Kock, H.J. and Rehage, G.(1981)Investigations on liquid crystalline polysiloxanes 3. Liquid crystalline elastomers-a new type of liquid crystalline material. Makromol. Chem., Rapid Commun. 2: 317-322.
[27] Bergmann, G.H.F. and Finkelmann, H. and Percec, V. and Zhao, M.(1997) Liquid-crystalline main-chain elastomers. Macromol. Rapid Commun. 18: 353-360.
[28] C. Zannoni (1979) The molecular physics of liquid crystals. Rijeka: G. R. Luckhurst and G. W. Gray.
[29] Thomsen, D.L. and Keller, P. and Naciri, J. and Pink, R. and Jeon, H. and Shenoy, D. and Ratna, B.R.(2001)Liquid crystal elastomers with mechanical properties of a muscle. Macromolecules 34: 5868-5875.
[30] Zipfel, W.R. and Williams, R.M. and Webb, W.W.(2003)Nonlinear magic: multiphoton microscopy in the biosciences. Nature Biotechnology 21: 1369-1377.
[31] Zipfel, Naydenova, I. and Mihaylova, E. and Martin, S. and Toal, V.(2005) Holographic patterning of acrylamide–based photopolymer surface. Opt. Express 13: 4878-4889.
[32] Becke, A.D.(1993)Density-functional thermochemistry. III. The role of exact exchange. Chem. Phys. 98: 5648–5652.
[33] Valiev, M. and Bylaska,E.J. and Govind, N. and Kowalski, K. and Straatsma, T.P. and van Dam, H.J.J. and Wang, D. and Nieplocha, J. and Apra, E. and Windus, T.L. and de Jong, W.A.(2010) NWChem: a comprehensive and scalable open-source solution for large scale molecular simulations. Comput. Phys. Commun. 181: 1477.
[34] Gay, J.G. and Berne, B.J.(1981) Modification of the overlap potential to mimic a linear site–site potential. J. Chem. Phys. 74: 3316.
[35] Miguel, E. and Rull, L.F. and Chalam, M.K. and Gubbins, K.E. and Van Swol, F.(1991) Location of the isotropic-nematic transition in the Gay-Berne model. Mol. Phys. 72: 593-605.

[36] Luckhurst, G.R. and Stephens, R.A. and Phippen, R.W.(1990) Computer simulation studies of anisotropic systems. XIX. Mesophases formed by the Gay-Berne model mesogen. Liq. Cryst. 8: 451.
[37] Allen M. P., M. A. Warren, M. R. Wilson, A. Sauraon, and w. Smith(1996) molecular dynamics calculation of elastic constants in Gay-Berne nematic crystals. J. Chem. Phys. 105: 2850-2858.
[38] Rappe, AK and Casewit, CJ and Colwell, KS and Goddard Iii, WA and Skiff, WM (1999) UFF, UFF, a full periodic table force field for molecular mechanics and molecular dynamics simulations. JOSA 114: 10024-10035.
[39] Allen M. P., M. A. Warren, M. R. Wilson, A. Sauraon, and w. Smith(1985) molecular dynamics calculation of elastic constants in Gay-Berne nematic crystals. Mol. Phys. 55: 549.
[40] Cleaver, D.J. and Care, C.M. and Allen, M.P. and Neal, M.P.(1996) Extension and generalization of the Gay-Berne potential. Phys. Rev. E 54: 559-567.
[41] Parrinello, M. and Rahman, A.(1981) Polymorphic transitions in single crystals: A new molecular dynamics method. J. Appl. Phys. 52: 7182.
[42] Nose, S.(1984) A unified formulation of the constant temperature molecular dynamics methods. J. Chem. Phys. 81: 511.
[43] Hoover, W.G.(1985) Canonical dynamics: Equilibrium phase-space distributions. Phys. Rev. A 31: 1695-1697.
[44] Beeman, D.(1976) Some multistep methods for use in molecular dynamics calculations. J. Comput. Phys. 20: 130.
[45] Li, M.H. and Keller, P. and Li, B. and Wang, X. and Brunet, M.(2003) Light-driven side-on nematic elastomer actuators. Adv. Mater. E 15: 5569-572.

Section 3

Elastomer Nanocomposites

Chapter 4

Pigment and Dye Modified Fillers as Elastomeric Additives

Anna Marzec and Marian Zaborski

Additional information is available at the end of the chapter

1. Introduction

Pigments are a dynamically developinged area of chemical technology. This interesting scientific concept involves preparation of hybrid pigments by the formation of a stable dye sphere surrounding an inorganic substrate grain, such as silica, titanium dioxide, or aluminosilicates. The concept of combining properties of organic and inorganic pigments has been employed for a long time. At the end of the 19th century, this concept was applied to the production of a pigment lake where freshly precipitated aluminium hydroxide was employed as the support. These lakes were used for the production of graphic dyes due to their averaged properties. Silica, zinc oxide, carbon black, and titanium dioxide may be employed in rubber compounds as pigments and fillers. Fillers boast a wide range of industrial applications, such as in the pharmaceutical and textile industries, and are important as additives for paints and varnishes. Fillers are relatively inexpensive, solid substances that are added in fairly large volumes to polymers to adjust the volume, weight, cost, surface, colour, expansion coefficient, conductivity, permeability, and mechanical properties. They can be roughly divided into inactive or extender fillers and active or functional or reinforcing fillers. One of the most popular fillers is synthetic silica, which is used in the polymer industry as an active filler (Liu et al., 2008, Wang et al., 2009). The rubber industry utilises fine silicas, which have a large specific surface leading to favourable effects on the strength of the composites. One important processing challenge is that silica surfaces are hydrophilic in nature, which impedes their dispergation in polymer media. Nearly all of the silica forms tend to agglomerate, which is an undesirable behaviour from the point of view of utilising their functional properties in industrial applications. On-going research on this topic focuses on altering the characteristics of the filler surfaces through some type of modification (Ciesielczyk et al., 2007; Ciesielczyk et al., 2011; Klapiszewska et al., 2003; Kohno et al., 2011; Ladewig et al., 2012; Kishore et al., 2012; Natinee et al., 2011). For example, industry has developed hydrophobic silicas that easily dispergate in polymers. The production of these silicas involves a silica silanisation reaction with alkoxysilane containing functional groups (Jesionowski et al., 2001, Krysztawkiewicz et al., 2000; Elham et al., 2012; Kister et al., 2012; Jianzhong et al., 2011; Wunpen et al., 2011; Jung-Woo et al., 2011). Over the past decade, numerous studies have employed a dye to modify the surface of silica to obtain hydrophobic silica surfaces and visually appealing end product colours (Ansarifar et al., 2004; Al Dwayya et al., 2012). The presence of amino groups in these supports enables the formation of ionic bonds with acid and direct dyes as well as the formation of covalent bonds with reactive dyes (Donia et al., 2009). Dyes, which were applied for this purpose, belonged to the textile class. Although they were combined with silica, they could not fulfill the requirements of modern pigments, such as light-resistance or resistance to solvents and to high temperature. High temperature during polymer processing can damage or destroy the dye causing changes in the shade or loss of colour. The present study was directed at obtaining composite pigments of indigothiazine and silicas with different surface areas. Indigothiazine consists of two thiazine residues conjugated to each other by a double bond. Because their cyclic carbonamide groups , thiazine can form hydrogen bond to silanol groups present on the silica surface. The thiazine pigments exhibited good heat stability, insolubility and exposure durability. An alternative method of producing coloured fillers is via a mechanical approach in which a coloured organic system is mechanically applied to the surface of a non-organic support. This method is a simple and cheap method of producing an organic-inorganic pigment encasing silica in the coloured chromophore with the additional benefit of eliminating the undesirable solvents from the process. The anti-ageing properties of indigothiazine were confirmed by oxidation-reduction potential measurements. The composites were produced by mechanically applying a coloured organic system (indigothiazine pigment) to a non-organic support (silica). The prepared pigment composites were characterised by the following properties: agglomerate size distribution, zeta potential, and specific surface area of the silica fillers before and after application of indigothiazine. The composite pigments were evaluated for their functional properties and subsequently employed as fillers in an acrylonitrile butadiene rubber (NBR). The mechanical properties as well as the spectrophotometric and DSC data of the vulcanisates produced with the composite pigment fillers were also studied before and after being aged under UV radiation.

2. Colourants in polymer composites

2.1. Fillers modification

Compared to inorganic pigments, organic pigments exhibit greater vividness, higher colour intensity and higher staining potential. Their drawbacks include greater sensitivity to the action of chemical compounds, temperature and solvents (Binkowski et at., 2000). Several studies were conducted to obtain organic pigments that were permanently attached to a silica core to improve the resistance of the pigments to chemical and thermal treatments. Numerous studies have shown that the reaction mechanism that occurs on the surface of inorganic substances is extremely difficult to identify. For example, the reaction that occurs on the surface of silica with fluorescein results in the formation of covalent bonds between the reagents. The interactions may involve the formation of hydrogen bonds as well as van der Waals or electrostatic interactions (Jesionowski 2003; Wu et al., 1997). Silica is one of the most popular fillers employed as a pigment carrier. Silicas are extensively modifiable materials. Their modification results in products that have new functional groups on their surface and are capable of interacting with various organic compounds. For example, modification with aminosilane provided functional amino groups that can react with the carbonyl groups of aldehydes, ketones or esters. The aminosilane-modified silicas find multiple applications as polymer fillers in industry. Recently, they have been used with increasing frequency as coupling agents in pigment or organic systems. Until recently, studies of organic pigments involved modification of their basic properties, such as stability or intensity of the colours. However, in many cases, the pigments must satisfy additional requirements (e.g., well defined particle diameters). Pigments with silica cores can satisfy these requirements (Krysztafkiewicz et al., 2003). Bińkowski (Bińkowski et al., 2000) proposed a simple procedure for obtaining coloured silicas, which involves attaching the dye to a modified silica surface. The procedure is possible due to the availability of a wide range of commercial dyes. The resulting pigment exhibits a structure in which the silanol groups on the surface of the silica are chemically coupled to an organic dye (e.g., azo dye) through an aminosilane coupling agent. They represented the pigment structure using the following formula: $S\text{-}(O\text{-}Si\text{-}R^1\text{-}Z\text{-}D)_n$ where R^1 is the alkyl group, Z is the amine bridge fragment , D is the dye, n is a number not less than 1 and $\text{-}Si\text{-}R^1\text{-}Z\text{-}$ represents a group originating from the silane coupling agent. In this work, they employed precipitated silica, 3-aminopropylotriethoxysilane (U-13), 2-(aminoethyl)-3-aminopropylotriethoxysilane (U-15) and a dye obtained from Boruta-Kolor, C.I. Direct Red 81. The modifications resulted in uniform pigments. A subsequent report by the authors (Jesionowski et al., 2011) showed the adsorption of C.I. Mordant Red 3 on the surface of silica that was both unmodified and modified with *N*-2-(aminoethyl)-3-aminopropyltrimethoxysilane. They proposed two mechanisms for the dye addition to the silane modified silica surface including via a hydrogen bond between the hydroxyl group of C.I. Mordant Red 3 and the amine groups of the silane coupling agents and/or via electrostatic interaction between the dissociated anion of the dye and the cation appearing on the silica surface modified with silane. Sudam (Sudam et al., 2003) described the adsorption behaviour of selected styryl pyridinium dyes on silica gel. The adsorbents were styryl pyridinium dyes with a monochromatic group containing an alkyl chain at the pyridinium nitrogen and a bischromatic group containing methylene groups bridging at the pyridinium nitrogen atoms. They observed that the bischromophoric dyes anchor on the silica surface in a flat-on position in which the methylene units also come in contact with the silica surface. Andrzejewska (Andrzejewska et al., 2004) performed studies to obtain pigments by adsorption of organic dyes on the modified surface of a titanium dioxide pigment system. The titanium oxide surface was modified with silane coupling agents, such as 3-aminopropyltriethoxysilane (U-13), *N*-2-(aminoethyl)-3-aminopropyltriethoxysilane (U-15D), in various solvents (methanol, toluene, acetone and methanol-water mixture). C.I. Acid Orange 7 and C.I Reactive Blue 19 were employed to prepare the pigments. The organic dyes were deposited on the surface of the modified titanium white. The modification of the titanium white with both silanes (U-13 and U-15D) boosted the efficiency of adsorption of C.I. Acid Orange 7 dye on the surface. The improved adsorption efficiency of C.I Reactive Blue 19 on the titanium was obtained following modification with silane U-13. For the titanium modified with U-15D, the efficiency of C.I. Reactive Blue 19 dye adsorption decreased slightly. A subsequent report (Raha et al., 2012) proposed a new type of inorganic carrier, Na-montmorilllonite (MMT). If the carrier is a smectite clay or layered silicate, the most effective pathway found involved using the ion exchange process. For that reaction, the dye must exist in a cationic form, which is the case for basic or cationic dyes. Example for that intercalated products of Rhodamine dyes with hectorite, montmorilllonite and other smectite silicates. The synthesis of the water-soluble cationic dyes was performed using two azo dyes (i.e., Solvent Red 24 and Solvent Yellow 14) and one disperse dye (i.e., Red 60) intercalated with Na^+-MMT to create nano-structured pigments. The dye-intercalated montmorillonite was successfully obtained. The modification of the dyes into their respective cationic species was confirmed by NMR study, and XRD indicated that these nanopigments had a nanostructured morphology. The authors also observed that the use of the nanopigments in PP reduced the migration of the dye, which results from the intercalation /adsorption of the dye within the clay-based pigments. The most common methods of modifications are presented in Table 1.

Type of supporter	The type of coupling agent	Type of dye	Reference
Silica	-	D&C Red 6, Acid Yellow 1, Acid Blue 25, Guinea Green B	Wu et al., 1997
Silica	-	Monochromatic and bischromatic styryl piridinium dyes	Parida et al., 1997
Silica	3-aminopropyltriethoxysilane, *N*-2-(aminoethyl)-3-aminopropyltriethoxysilane	C.I. Direct Red 81	Binkowski et al., 2000
Sodium-aluminium silicate	Vinyltris(β-methoxyethoxy)silane; γ-methacryloxypropyltrimethoxysilane	-	Werner et al., 2001
Silica	*N*-2-(aminoethyl)-3-aminopropyltrimethoxysilane	C.I. Reactive Blue 19	Jesionowski et al., 2002
Silica	*N*-2-(aminoethyl)-3-aminopropyltrimethoxysilane	C.I. Reactive Blue 19, C.I. Acid Red 18	Jesionowski 2003
Silica	*N*-2-(aminoethyl)-3-aminopropyltrimethoxysilane	C.I. Acid Red 18	Jesionowski et al., 2003
Silica	3-aminopropyltriethoxysilane, *N*-2-(aminoethyl)-3-aminopropyltriethoxysilane, 3-ureidopropyltriethoxysilane	C.I. Acid Red 18, C.I. Acid Violet 1	Krysztafkiewicz et al., 2003
Silica	*N*-2-(aminoethyl)-3-aminopropyltrimethoxysilane	C.I. Reactive Blue 19, C.I. Acid Green 16, C.I. Acid red 18, C.I. Acid Violet 1, C.I. Direct Red 81	Jesionowski et al., 2004
Titanium dioxide	3-aminopropylotriethoxysilane, *N*-2-(aminoethyl)-3-aminopropyltrimethoxysilane	C.I. Acid Orange 7, C.I. Reactive Blue 19	Andrzejewska et al., 2004

Type of supporter	The type of coupling agent	Type of dye	Reference
Silica	*N*-2-(aminoethyl)-3-aminopropyltrimethoxysilane	C.I. Basic Blue 9, C.I. Basic Blue 9, C.I. Acid Orange 52	Jesionowski et al., 2005
Magnesium silicates	*N*-2-(aminoethyl)-3-aminopropyltrimethoxysilane; octyltrimethoxysilane; 3-chloropropyltrimethoxysilane; vinyltrimethoxysilane	-	Ciesielczyk et al., 2007
Montmorillonite	-	Thioindigo	Ramirez et al., 2011
Magnesium silicate	*N*-2-(aminoethyl)-3-aminopropyltrimethoxysilane; 3-glicydoxypropyltrimethoxysilane	C.I. Basic Blue 9, C.I. Acid Yellow 17	Ciesielczyk et al., 2011
Silica	Imidazolium ionic liquids	Indigothiazine	Marzec et al., 2010
Silica	Imidazolium ionic liquids	Berberine chloride; 2-hydroxy naphthol quinone	Marzec et al., 2011
Silica	*N*-2-(aminoethyl)-3-aminopropyltrimethoxysilane	C.I. Mordant Red 11	Jesionowski et al., 2011
Silica	*N*-2-(aminoethyl)-3-aminopropyltrimethoxysilane	C.I. Mordant Red 3	Jesionowski et al., 2011
Na-Montmorillonite	-	Solvent Yellow 14, Solvent Red 24, Disperse Red 60	Raha et al.,2012
Silica	3-aminopropyl trimethoxysilane	C.I. Acid Blue 113, C.I. Acid Red 114, C.I. Acid Green 28, C.I. Acid Yellow 127, C.I. Acid Orange 67	Aniba et al., 2012

Table 1. Common methods for the preparation of pigments.

2.2. Organic and inorganic pigments

Colourants used in polymers are pigments or dyes (Drobny, 2007). Dyes are organic compounds that are soluble in the polymers forming a molecular solution. They produce bright, intense colours and are transparent and easy to disperse and process. In general, pigments are insoluble in polymers. They produce opacity or translucence in the final product. Pigments can be inorganic or organic compounds and are available in a variety of forms, such as dry powders, colour concentrates and liquids. Pigments and dyes produce colour in a resin due to the selective absorption of visible light with wavelengths ranging from ~380 (violet) to 760 nm (red). Because the dyes are in solution, the colour is produced only from light absorption, and the material is transparent. The colour shade is dependent on the particle size of the pigment. Ultramarine blue pigments are nonreflective due to their refractive index, which is similar to that of the polymer. Pigments must be adequately dispersed in the polymer for optimum scattering. Specks and uneven colouration can result from incomplete dispersion. If agglomerates are present, they may adversely affect the mechanical properties (i.e., tensile strength, impact strength, and flex fatigue) of the resulting product. The pigments should be compatible with the polymer because poor compatibility may result in part failure. High temperature during processing can damage or destroy the pigment resulting in shade variation or colour loss. The thermal sensitivity is related to both temperature and duration of exposure. Long cycles during injection moulding and rotational moulding can have a more adverse effect than high-speed extrusion. Some pigments can act as nucleating agents altering the mechanical properties and improving the clarity of the polymer. The most common inorganic pigments include oxides, sulphides, hydroxides, chromates, and other complexes based on metals, such as cadmium, zinc, titanium, lead, and molybdenum. In general, they are more thermally stable than organic pigments and are more opaque and resistant to migration, chemicals, and fading. They can cause wear on the processing equipment (e.g., extrusion machine screws and barrels). The use of heavy metal compounds (e.g., cadmium) has been restricted due to toxicity issues. The most widely used white pigment is titanium oxide (rutile),which is used either alone or in combination with other colourants to control the opacity and produce pastel shades. Other white pigments include zinc oxide, zinc sulphide, and lead carbonate (lead white).The most widely used black pigment is carbon black, which is essentially composed of pure carbon. When combined with white pigments, it produces various shades of grey depending on the particle size and tinting strength of the carbon black grade. Iron oxide, Fe_3O_4, is another black pigment that has a lower thermal stability and tinting strength. A variety of inorganic compounds are used as pigments to colour polymers. Drobny (Drobny, 2007) classified the most common types of pigments. The yellow pigments include chrome yellow, chrome-titanium yellow, iron oxides, and lead chromates. The orange pigments include molybdate orange and cadmium orange. The brown pigments include iron oxide or a combination of chrome/iron oxides. The red pigments include iron oxide and cadmium sulphide/selenide. The blue pigments include ultramarine (aluminosilicate with sodium ion and ionic sulfur groups) and mixed metal oxides that are primarily based on cobalt aluminate. The green pigments include chrome oxide and cobalt-based mixed oxides. In general, the organic pigments are brighter, stronger, and more transparent than inorganic pigments but are not as light resistant. They may be partially soluble in many polymers but exhibit a much greater tendency to migrate. The largest group of organic pigments are the azo pigments, which contain one or more azo chromophoric groups and form yellow, orange, and red pigments. Monoazo pigments, which have only one chromophore, exhibit low thermal and light stability and have a tendency to bleed. In addition, the monoazo pigments are not typically employed in plastics. Polyazo pigments, which have more than one chromophore, do not tend to bleed and have better thermal stability and excellent chemical stability. Nonazo pigments have a variety of structures including polycyclic and metal complexes. Phtalocyanine blues and greens, most of which are complexed with copper, are highly stable to light, heat, and chemicals and form highly transparent, intense colours with a high tinting strength. Other organic pigments include quinacridones (red, violet, orange), dioxazines (violet), isoindolines (yellow, orange, red), perylenes, flavanthrones, and anthraquinones. In addition, thiazine pigments are suitable for almost all organic pigment applications (Smith, 2002). In low concentrations, thiazine pigments provide high value-in-use due to their heat stability (up to 290 -300 ^{0} C) in polyolefins enabling warpage-/deformation-free HDPE mouldings, high opacity coupled with very high colour saturation, ease of incorporation (dispersion, flow), chemical inertness, high insolubility in nearly all solvents, high performance in the intended applications (processability and durability), ability to form supramolecules with synergistic values, potential environmental compatibility, and cheap cost.

3. Experimental

3.1. Characterization and preparation pigment and pigment composites

The silicas used as the support medium of composite pigments in these studies are as follows: Aerosil 380 (Degussa S.A.) – a hydrophilic fumed silica with a specific surface area of 340 - 380 m^2/g; Zeosil 175 (Rhodia) – a precipitated silica with a specific surface area of 175 m^2/g; Silica Gel 60 (Merck) – a silica with a specific surface area of 470 – 540 m^2/g. The indigothiazine pigment was used as the coloured chromophore. Below 180 ºC, the *cis* form of this pigment is yellow in colour and transforms into the highly durable *trans* form characterised by the red colour upon heating (Figure 1). The latter form is highly resistant to organic solvents, high temperatures, and light. The *cis* form of indigothiazine was obtained by according to the description in the patent Bansi Lal (Bansi Lal et al., 2002).

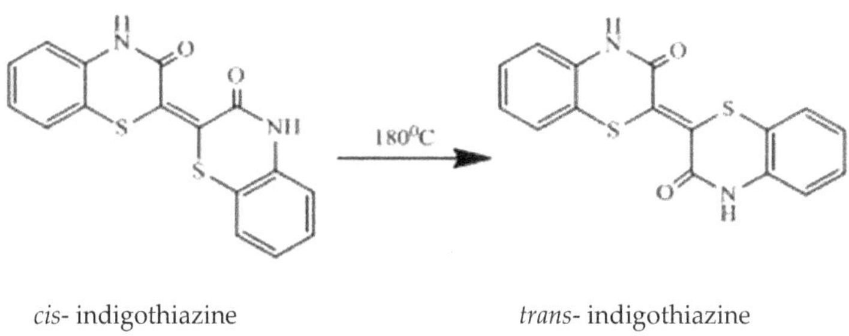

Figure 1. Isomeric conversion of *cis*- to *trans*–indigothiazine – ([2,2']-bi(1,4-benzothiazynylidene)-3,3'-(4H, 4'H)-dione.

3.2. Characterisation and preparation pigment and pigment composites

The composite pigments were prepared by mechanically applying *cis*-indigothiazine on silica surfaces and subjecting them 180 ºC for 2 h to induce their transformation into the highly durable *trans* form. The mechanical application of the *cis* form on silica was accomplished by means of a high-speed blender (Warning Commercial) at ~14,000 rpm. The quantity of the pigment was measured relative to the weight of silica, and the modification required 15 min (Fig. 2).

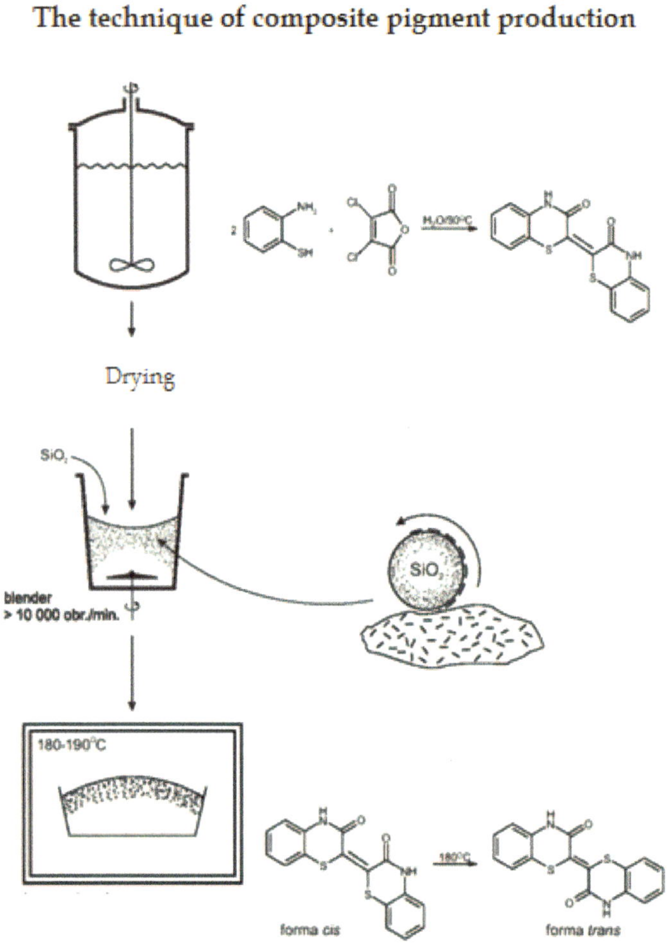

Figure 2. Production process scheme of composite pigments (Copyright Lipińska at al.2011).

To assess the mechanism and kinetics of the electrochemical oxidation of the compounds under investigation, cyclic voltammetry (CV) and differential pulse (DPV) methods were employed on an Autolab analytical unit (EcoChemie, Holland). A three-electrode system was employed for the measurements. Platinum was used as the anode and the auxiliary electrode. The potential of the tested electrode was measured versus a ferricinium/ferrocene reference electrode (Fc+/Fc) where the standard potential is defined as zero and independent of the solvent used. Prior to the measurements, all of the solutions were deoxygenated with argon. During the measurements, an argon atmosphere was maintained over the solution. The effect of the scan rate on the electrooxidation of indigothiazine in an anhydrous medium was investigated. The differential scanning calorimeter (DSC) measurements of *cis*– and *trans*–indigothiazine and vulcanisates were performed on a DSC1 calorimeter (Mettler Toledo) with a heating rate of 10 ^{0}C/min. The morphology of the indigothiazine particles and the dispersion in the elastomer matrix containing either unmodified silica or the composite pigments was estimated using scanning electron microscopy with a LEO 1530 SEM microscope. The NBR vulcanisates were broken down in liquid nitrogen, and the fracture surfaces of the vulcanisate were examined. Prior to the measurements, the samples were coated with carbon. The prepared pigment composites were characterised to evaluate the agglomerate size distribution in the water medium using dynamic light scattering (Zetasizer Nano S90, Malvern) for a dispersion concentration of 0.2 g/l (the powders were pre-treated with ultrasound for 0.5 h). The specific surface area of the silica fillers before and after application of indigothiazine was determined. The measurements were performed using the Brunauer-Emmet-Teller (BET) nitrogen adsorption method (GEMINI 2360 V2.01). The zeta potential of the silicas and of pigment composite were evaluated (ZETASIZER 2000, the water dispersion with concentration 2 g/l)). The powders were pre-treated with ultrasound for 1 h.

3.3. Preparation and vulcanization of rubber compounds

The composite pigments were used as the rubber mixture fillers. The rubber mixture composition was made up of the following: NBR (Perbunan 28-45F containing 28 wt % acrylonitryle groups from Lanxess, 100 phr); sulphur crosslinking system (See Table 2) and optional: composite pigments or silica (30 phr).

The mixture composition	[a]phr
NBR rubber	100
Zinc oxide	5
Sulphur	2
Mercaptobenzothiazole	2
Stearic acid	1
Filler	30

[a]phr = parts per hundred of rubber by weight

Table 2. Rubber formulations.

The mixture was prepared in a laboratory rolling mill (Roll dimensions: D = 200 mm, L = 450), and the vulcanisation process was carried out in a hydraulic press (160 ^{0}C) in line with the curing time specifications (See Table 2). The vulcanisation kinetics was measured with evaluated using a Monsanto vulcameter with an oscillating rotor (ZACH Metalchem) in compliance with the ISO-4317standard. The tests performed on the vulcanizates are summarised as follows: the mechanical properties were characterised using a Zwick 1435 strength testing machine and according to ISO-37. The crosslink density of vulcanizates was determined by equilibrium swelling in toluene, based on the Flory-Rehner (Flory et al., 1943) equation using the Huggins parameter of elastomer–solvent interaction $\mu = 0.381 + 0.671Vr$ [eq. (1)].

$$v_e = -\frac{\ln(1-V_r)V_r+\mu V_r^2}{V_0(V_r^{\frac{1}{3}}-\frac{V_r}{2})} \tag{1}$$

where: v_e - crosslink density, Vr - volume fraction of elastomer in swollen gel, Vo - molar volume of solvent [mol/cm^3].

spectrophotometric data were taken using a KONICA–MINOLTA CM–3600 spectrophotometer under computer control with spectra analysis software (Colour Data Software CM – S 100 in Spectra Magic) according to the following standards: CIE No. 15; PN-EN ISO 105- J01; ASTM E1 164; and DIN 5033 Tell.7, resistance to ageing under UV radiation was tested by exposing the vulcanizates to a UV source (Atlas UV200) for a total of 5 days, equivalent to the ageing by approximately 3 months under normal conditions. The test timeline was as follows:day segment: UV power 0.7 W/m^2; temperature 60 ºC; duration 8 h, night segment: UV power 0.0 W/m^2; temperature 50 ºC; duration 4 h. The vulcanizates subjected to UV ageing were tested for tensile strength. Spectrophotometric measurements were performed before and after UV irradiation. For each vulcanizate being tested, the ageing coefficient, K, was computed using Equation 2:

$$K = \frac{(TS \cdot Eb)_{after\ ageing}}{(TS \cdot Eb)_{before\ ageing}} \tag{2}$$

where: K – ageing coefficients; TS – tensile strength; Eb – relative elongation.

4. Results

The electrode reactions characterising the electrochemical oxidation of indigothiazine at the platinum electrode were studied by cyclic pulse voltammetry. The half-wave potential of the peak in the cyclic voltammogram is characteristic of each subsequent step in the investigated electrode reaction. Selected cyclic voltammograms recorded in a solution consisting of the indigothiazine and the supporting electrolyte are presented in Fig. 3. The cyclic voltammograms recorded for the indigothiazine solution exhibit two peaks, which involve at least three electrode steps of the pigment electrooxidation in the potential range prior to the potential at which electrolyte decomposition is initiated. The supporting electrolyte (in *N*-methyl-2-pyrrolidone) has no characteristic peaks except for charging the electrical double layer. The electrooxidation potentials were determined fallowing: the first peak - 0.71 V, second peak - 0.82 V, third peak - 1.01 V. In addition, the electrode reactions of the examined connection are becoming more and more irreparable. Polarisation towards the negative potentials of the reduction reaction was not observed (Fig. 4). The recorded voltammograms, under the linear diffusion of the first electrooxidation step, were employed to determine the half-wave potential (E1/2). Based on the results provided in Table 3, indigothiazine, E1/2 = 0.36 V, was easily oxidised. The energy of the highest filled molecular orbital (EHOMO) was estimated to determine the ease of electron back-donation (potential of the ionisation) and the enthalpy (ΔH). The EHOMO energy (-7.65 eV) and enthalpy (22.83) confirmed the good antioxidant properties and stability.

Compound	$E_{1/2}$,V	E_{HOMO}, eV	ΔH, kcal/mol
Indigothiazine	0.85	- 7.65	22.83

Table 3. Cyclic voltammetry and molecular orbital parameters for the studied compound

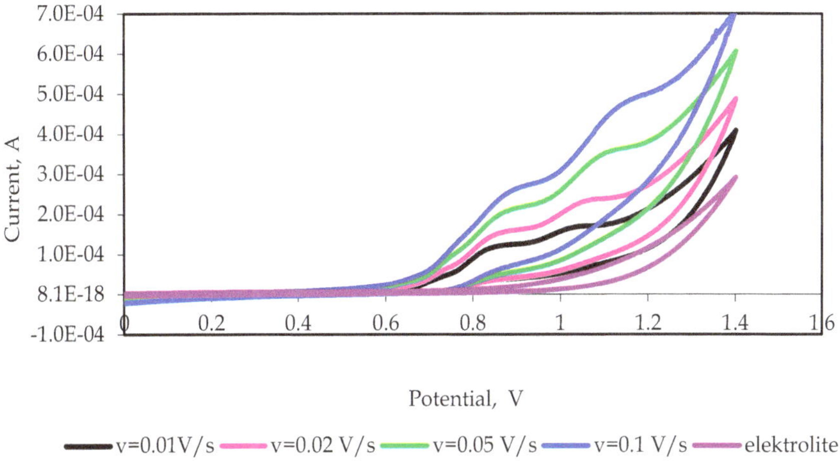

Figure 3. Voltammogram of indigothiazine oxidation on a Pt electrode; c = 1 mmol/dm³ in 0,1 mol/dm³ in *N*-methyl-2-pyrrolidone at various polarisation speeds.

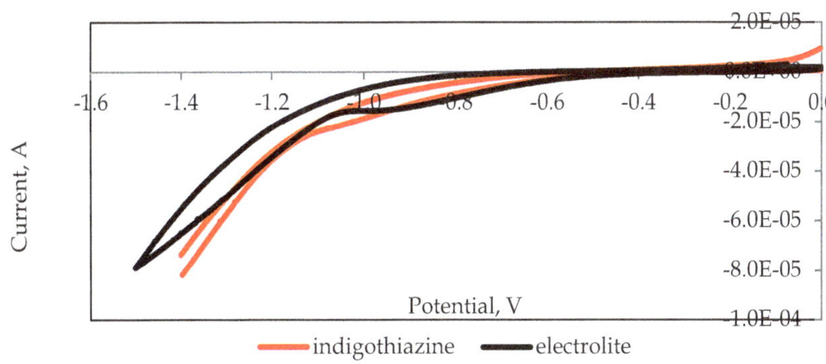

Figure 4. Voltammogram of indigothiazine for negative potential on the Pt electrode; c = 1 mmol/dm³ in 0,1 mol/dm³ in *N*-methyl-2-pyrrolidone.

The thermal strength of *cis* – and *trans*– indigothiazine (Fig.6) was measured applying the DSC method (DSC 1 Mettler Toledo). The resulting diagrams show the melting bands of the pigment's crystalline phase at around 440° C (Fig. 5). The exothermic peak on the *cis*–indigothiazine curve is in the regime where the *cis*- to *trans*- conversion took place most intensively.

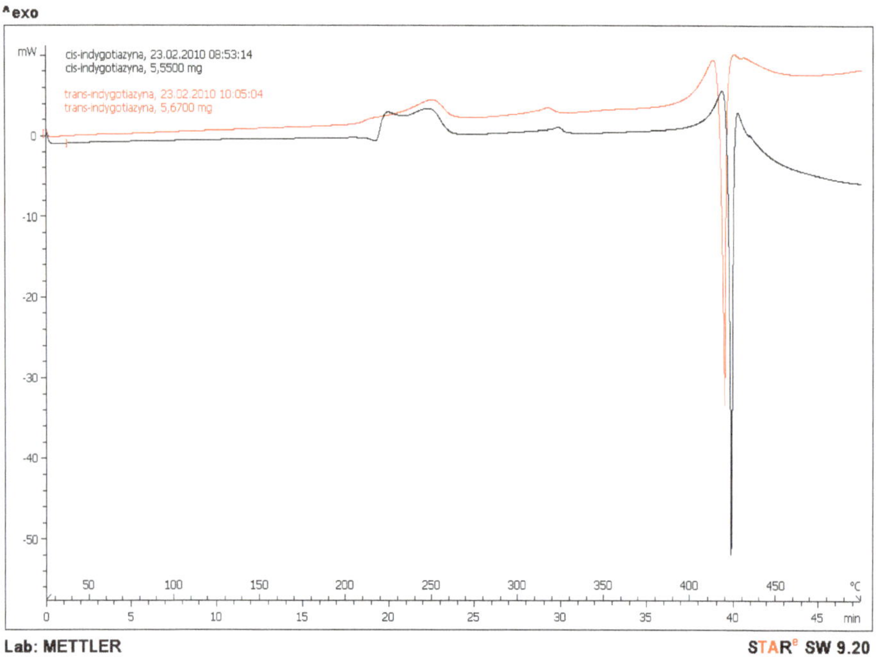

Figure 5. DSC curves of indigothiazine pigment – *cis* and *trans* conformation.

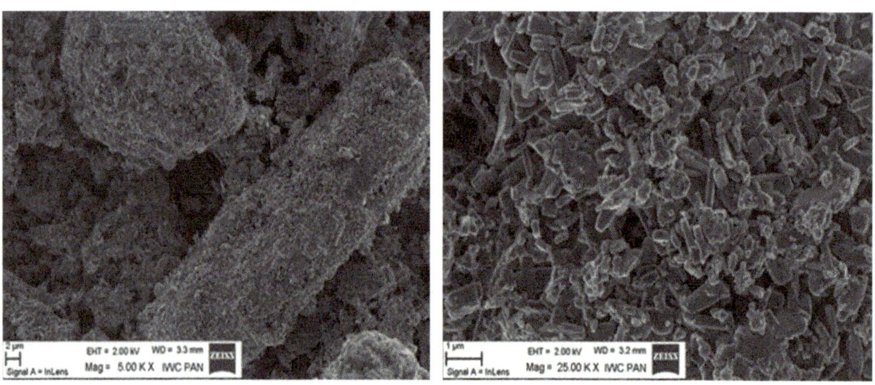

Figure 6. SEM image of trans-indigothiazine pigment .

The fumed silica, Aerosil 380, used in the tests is characterised by the high numerical percentage (99.67 %) of aggregates in the range of 116-330 nm in size; however, these aggregates account for only 28.83 % of the filler's volume in the water medium. The outstanding fraction (71.17 %) consists of agglomerates whose sizes are in the range of 1056-3000 nm. The application of indigothiazine pigment on the surface of Aerosil 380 silica reduced the average particle size and fragmented agglomerates, effectively producing the composite pigments with sizes ranging from 82-131 nm (100 %). For the precipitated silica, Zeosil 175, the application of pigment helped produce the composite pigments with 99.92 % of the aggregates in the size range from 116-262 nm and the rest the size range of 2379-3000 nm. Although the modification of Zeosil 175 reduced the average particle size and increased the particles uniformity, the particles still formed agglomerates in the order of a few μm in size. For Silica Gel 60, applying indigothiazine only slightly increased the particles tendency to agglomerate; that is, the aggregates size increased from 104-147 nm (unmodified) to 116-208 nm (modified). The results are summarized in Table 4.

Aggregates name	Aggregates size [nm]	Numerical percentage [%]	Volumetric percentage [%]	BET [m²/g]
A 380 (Aerosil 380)	116 - 330	99.67	28.83	246
	1056 - 3000	0.33	71.17	
A 380/ indigothiazine	82 - 131	100	100	185
Z 175 (Zeosil 175)	147 - 234	89.24	61.68	187
	330 - 527	10.75	33.13	
	2379 -3000	0.01	5.19	
Z175/ indigothiazine	116 - 262	99.92	34.85	131
	2379 - 3000	0.08	65.15	
S 60 (Silica Gel 60)	104 - 147	100	100	526
S60/ indigothiazine	116 - 208	100	100	396

Table 4. Aggregates size and specific surface area of the silicas and composite pigments.

The specific surface area was measured using the BET nitrogen adsorption method before and after applying the indigothiazine pigment. Each composite pigment was characterised by reduced specific surface area as compared with the unmodified silicas. The decrease in the specific surface proceeded to a larger extent in the case of Aerosil 380 and Silica Gel 60 with a large specific surface.

Figures 7-9 shows the change in zeta potential as a function of pH for both unmodified silicas and composite pigments. The largest change in the surface charge was caused by the application of indigothiazine to Zeosil 175 silica. The difference in zeta potential between the composite and unmodified silica Zeosil 175 was approximately 30 mV (pH 4), indicating that the modification reduced the acidic characteristics of the Zeosil 175 surface, which is a favourable result for its use as a polymer composite filler. For the A 380/ indigothiazine pigment, the zeta potential was slightly higher than for the unmodified Aerosil 380. For Silica Gel 60 and the Silica Gel 60/ indigothiazine composite, the two zeta potential curves overlapped one another, indicating that the application of the pigment induced practically no change to the surface charge in this filler.

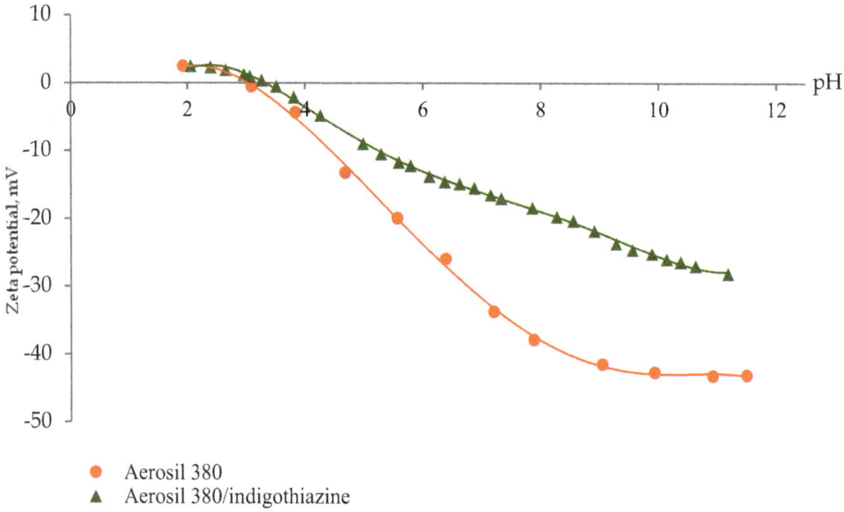

Figure 7. Zeta potential diagram for the composite pigments in which the supports was Aerosil 380.

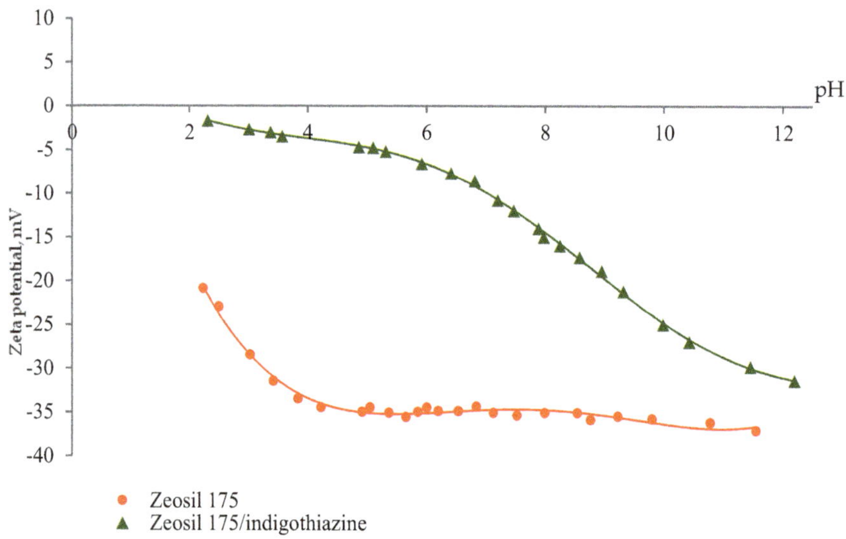

Figure 8. Zeta potential diagram for the composite pigments in which the support was Zeosil 175.

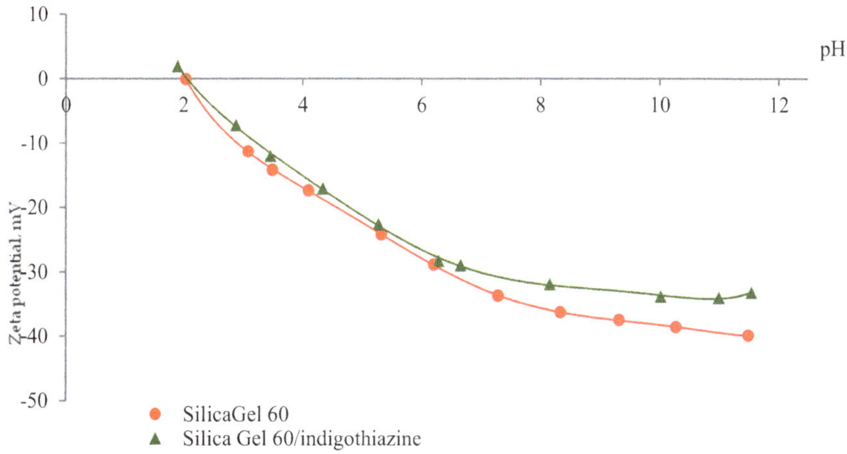

Figure 9. Zeta potential diagram for the composite pigments in which the support was Silica Gel 60.

The effects of silicas and the composite pigments on the properties of rubber mixtures were evaluated with respect to the curing kinetics and tensile strength of the vulcanizates (Table 5). Rheometric tests showed that adding composite pigments to the rubber mixture did not significantly affect the vulcanisation time, t_{09}, torque increase, ΔM, and the initial viscosity values (M_{min}) in comparison to the mixtures containing unmodified silicas. The only exception was the Zeosil 175/ indigothiazine composite, which showed a lower torque increase than the original Zeosil 175. Tg values from DSC curves for vulcanizates filled modified and unmodified silicas have similar glass transitios temperature in range -28 ^{0}C – 30 ^{0}C (Table 5) The vulcanisation process produced red-coloured vulcanizates, which were then subjected to tensile strength tests. The results showed that the application of the composite pigments had a similar effect on the mechanical properties of the vulcanizates as did the application of the unmodified fillers, as indicated by the tensile strength TS (Table 5) and the Scanning Electron Microscope SEM observations. The SEM images revealed agglomerates in the vulcanizates filled with unmodified silicas (Figs. 11, 13, 15) and composite pigments (Figs. 12, 14, 16). Even though the modification of the test fillers reduced the average size of agglomerates in the water medium, it did not significantly affect the dispersion level of silicas in the elastomer. The vulcanizates containing modified silicas had also similar values of curing density, module values at 100% stress and elongations at break EB in comparison to composites filled unmodified silicas.

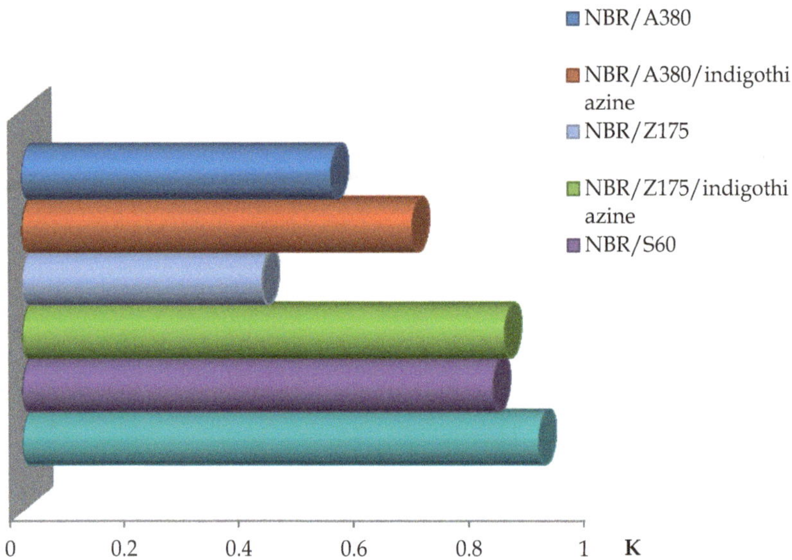

Figure 10. Ageing coefficient value (K) of vulcanizates filled with composte pigments and unmodified silicas.

Next, the functional properties of the vulcanizates after undergoing the 120 h of UV ageing were re-examined. After ageing, all samples exhibited lower TS and EB values and a higher curing density; however, the vulcanizates containing either the A 380/ indigothiazine, Z 175/ indigothiazine and S60/ indigothiazine composites exhibited less UV degradation. Based on the K values (Fig.10), it is possible to conclude that the pigment protects against ageing most effectively when applied to the A380 and Z175 silica supports. The closer the K value is to unity, the less ageing is considered to have taken place. Colorimetric tests of the aged vulcanizates were also performed using the non-aged samples as benchmarks (Table 6). The colorimetric measurements were taken in the CIE-Lab space that expressed the colours in the Cartesian L-a-b coordinate system of the brightness L, the red-greenness a, and the yellow-blueness b. The L value ranges from 0 (black) to 100 (white), and all shades of grey are expressible by the values in between. The colorimetric data reveal that the UV radiation not only degrades the functional properties but also changes the colour of the vulcanizates (Table 6). The parameter dE*ab depicts the degree of change in colour as a result of ageing. All the vulcanizates containing composite pigments exhibited a lesser degree of change in colour than those containing unmodified silica. The vulcanizates with either the A 380/ indigothiazine composite or the Silica Gel 60/ indigothiazine composite exhibited three times less the degree of change in colour than those with ordinary silicas.

Mixture symbol	M_{min} [dNm]	ΔM [dNm]	t_{09} [min]	Sample condition	SE_{100} [MPa]	TS [MPa]	Eb [%]	v_e*10^{-5} [mol/cm^3]	DSC [^{0}C]
NBR/ A 380	31	112	27.0	non-aged	1,74	17,6	614	5,41	-29
				UV aged	2,87	10,5	564	6,48	
NBR/ A 380/ indigothiazine	31	109	30.0	non-aged	1,87	17,1	677	5,44	-28
				UV aged	2,52	12,12	652	6,34	
NBR/ Z 175	32	134	25.0	non-aged	1,69	15,3	619	5,75	-30
				UV aged	2,38	7,85	510	6,54	
NBR/ Z 175/ indigothiazine	25	132	22.8	non-aged	1,53	16,7	628	5,69	-28
				UV aged	2,22	14,2	617	6,41	
NBR/ S 60	29	89	11.9	non-aged	1,73	4,39	466	6,12	-28
				UV aged	1,94	4,06	413	7,05	
NBR/ S 60/ indigothiazine	28	98	14.6	non-aged	2,42	4,52	248	7,89	-29
				UV aged	2,92	4,46	226	9,05	

$M_{min.}$ – minimum torque [dNm], ΔM – torque increase [dNm], t_{09} – vulcanisation time [min], SE_{100} – stress at 100 % elongation [MPa], TS – tensile strength [MPa], EB – relative elongation at break [%], v_T – curing density of vulcanizates calculated using the Flory Rehner equation [mol/cm^3]

Table 5. Effects of composite pigments and unmodified silicas on selected properties of NBR.

Sample name Rubber/silica/pigment	dE*ab	L*	a*	b*	dl*	da*	db*
NBR/ A380 (benchmark)	-	71.77	2.31	16.80	-	-	-
A 380 UV	30.70	53.03	12.82	38.73	-18.74	10.51	21.93
NBR/A 380/ indigothiazine (benchmark)	-	46.09	46.97	35.58	-	-	-
NBR /A 380/ indigothiazine UV	9.48	38.05	43.03	38.68	-8.05	-3.93	3.10
NBR/Z 175 (benchmark)	-	73.09	1.65	21.84	-	-	-
NBR/Z 175 UV	29.52	52.77	13.48	39.71	-20.31	11.83	17.86
NBR/Z175/ indigothiazine (benchmark)	-	46.66	52.13	42.95	-	-	-
NBR/Z175/ indigothiazine UV	17.40	40.50	40.32	31.76	-6.16	-11.81	-11.19
NBR/Silica Gel 60 (benchmark)	-	66.86	3.08	15.76	-	-	-
NBR/Silica Gel 60 UV	30.44	52.99	14.07	40.52	-13.88	10.99	24.76
NBR/Silica Gel60/ indigothiazine (benchmark)	-	48.28	43.64	29.66	-	-	-
NBR/Silica Gel 60/ indigothiazine UV	9.19	40.75	40.78	34.07	-7.54	-2.86	4.41

L* – brightness parameter, a* – redness – greenness reading, b* – yellowness – blueness reading, dl*, da*, db* – the difference in colour between trial and standard readings; dE*ab – vector of overall errors in colour and colour darkness;

Table 6. Colorimetric data for the vulcanizates containing unmodified silicas or composite pigment.

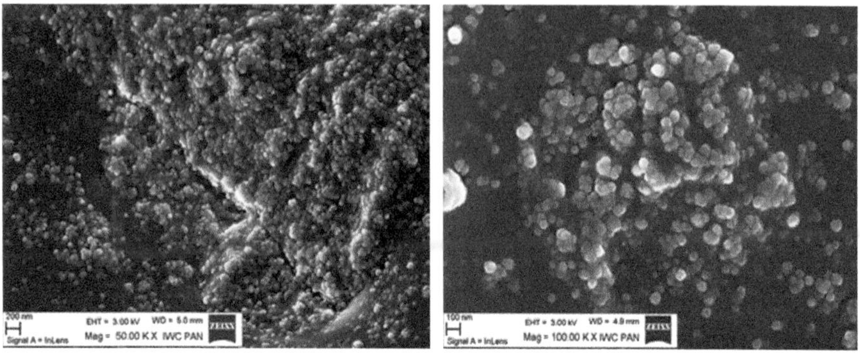

Figure 11. SEM image of NBR vulcanizate filled with Aerosil 380.

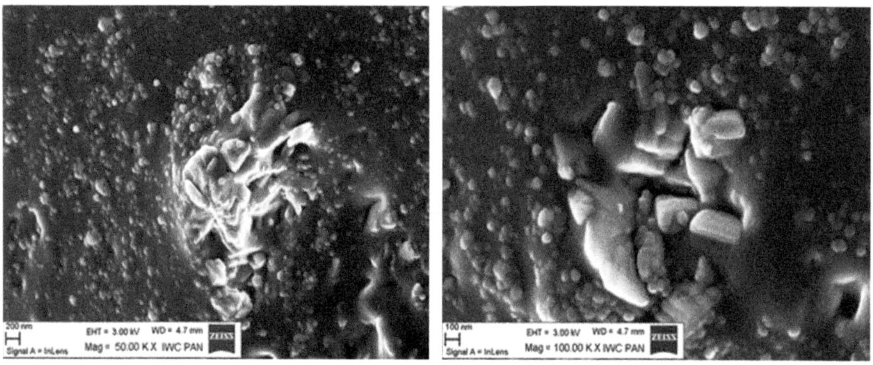

Figure 12. SEM image of NBR vulcanizate filled with A 380/indigothiazine.

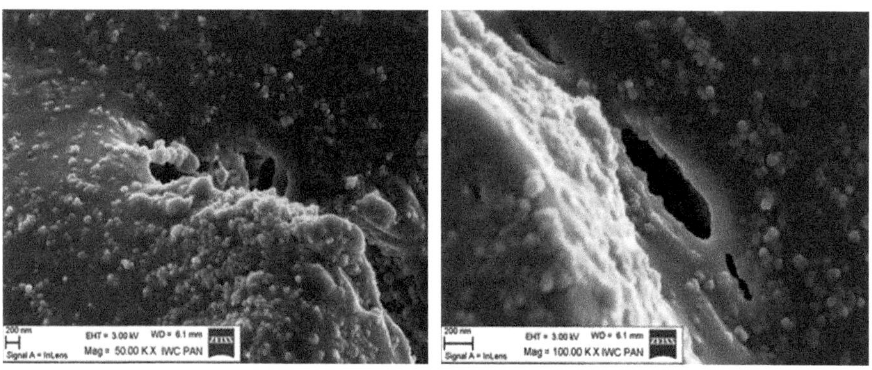

Figure 13. SEM image of NBR vulcanizate filled with Zeosil 175.

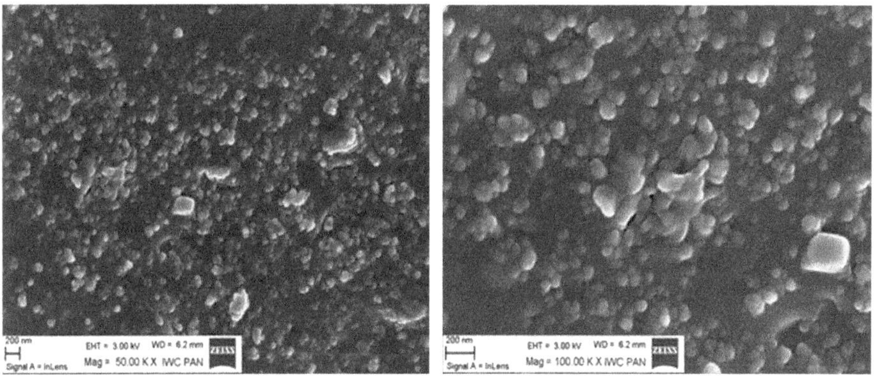

Figure 14. SEM image of NBR vulcanizate filled with Z 175/indigothiazine.

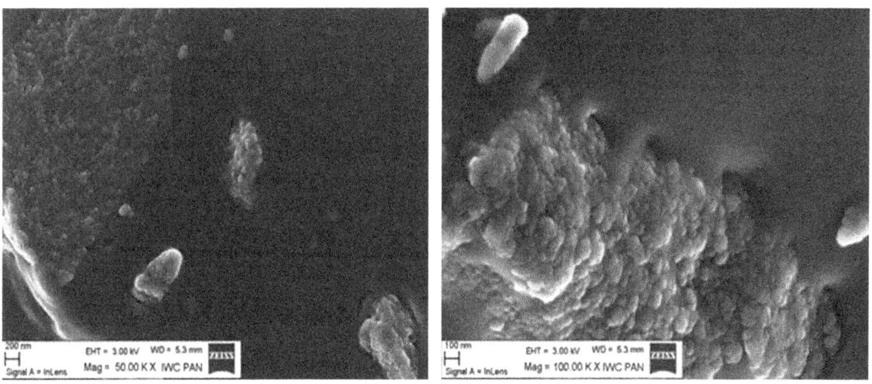

Figure 15. SEM image of NBR vulcanizate filled with Silica Gel 60.

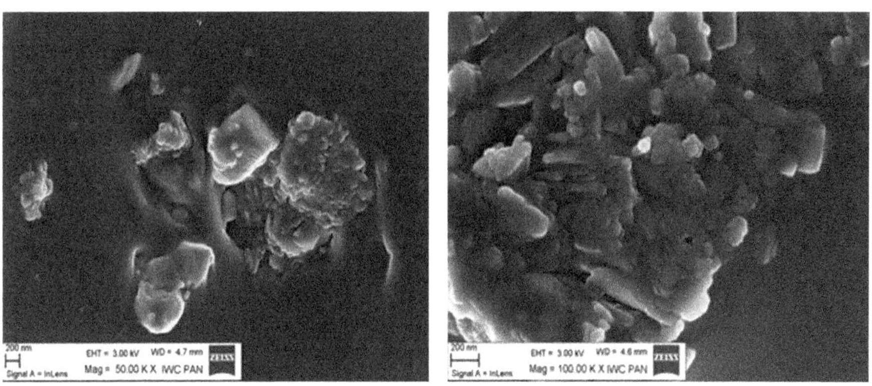

Figure 16. SEM image of NBR vulcanizate filled with S60/indigothiazine.

5. Conclusions

Indigothiazine exhibited antioxidant properties, which allows them to be successfully employed as anti-ageing substances. Red-colour composite pigments were produced by modifying silica Aerosil 380, Zeosil 175, and Gel 60. The modification method employed in the present work eliminated harmful solvents employed in the conventional process. The newly developed method is much simpler, cheaper, and easier to use on an industrial scale compared to the conventional process, which requires a solvent. The silica modification process using the blender reduced the specific surface area of the tested fillers. For the composites with Aerosil 380, Zeosil 175 and Silica Gel 60, this process reduced the agglomerate sizes. For the samples with Aerosil 380 and Zeosil 175, the zeta potential also increased, which indicated that their surface became less polarised and their acceptor characteristic was partially lost. The addition of the composite pigments to the polymer mixtures did not significantly change their curing kinetics or processing properties. The vulcanisation of NBR in the presence of the modified silicas resulted in vulcanisates with a vivid red colour were produced. Vulcanisates filled with the composite pigments and those filled with unmodified silicas exhibited similar mechanical strengths. The strength tests after UV ageing of the vulcanisates showed that the A 380/ indigothiazine composite and the Zeosil 175/ indigothiazine composite possess anti-ageing properties and have a favourable impact on the plastic's mechanical properties and colour stability. The present work led to the successful production of vulcanisates with enhanced ageing resistance.

Author details

Anna Marzec and Marian Zaborski
Institute of Polymer and Dye Technology,
Technical University of Lodz,
Poland

6. References

Al Dwayyan, A.S., Qaid, M.H., Khan, M.A. & Al Salhi M.S. (2012). Structural and spectral investigations of Rhodamine (Rh6G) dye-silica core–shell nanoparticles. *Opt. Mater.*, 34, 761–768.

Andrzejewska, A., Krysztafkiewicz, A. & Jesionowski, T. (2004). Adsorption of organic dyes on the aminosilane modified TiO_2 surface. *Dyes Pigm.*, 62, 121-130.

Anbia, M. & Salehi, S. (2012). Removal of acid dyes from aqueous media by adsorption onto amino-functionalized nanoporous silica SBA-3 *Dyes Pigm.*, 94, 1-9.

Ansarifar, A., Azharb, A., Ibrahima, N., Shiaha, S.F. & Lawtonc, J.M.D. (2005). The use of a silanised silica filler to reinforce and cross link natural rubber. *Int. J. Adhes. Adhes.*, 25, 77–86.

Bansi Lal, K. & Bruno, P. (2002). Production of thiazine-indigo pigments. 6339084, United States.

Binkowski, S., Jesionowski, T. & Krysztafkiewicz A. (2000). Preparation of pigments on modified precipitated silicas. *Dyes Pigm.*, 47, 247-257.

Ciesielczyk, F., Krysztafkiewicz, A. & Jesionowski, T. (2007). Physicochemical studies on precipitated magnesium silicates. *J. Mater. Sci.*, 42, 3831-3840.

Ciesielczyk, F., Nowacka, M., Przybylska, A. & Jesionowski, T. (2011). Dispersive and electrokinetic evaluations of alkoxysilane-modified $MgO{\cdot}SiO_2$ oxide composite and pigment hybrids supported on it. *Colloids Surf.*, 376, 21-30.

Donia, A.M., Atia, A.A., Al-amrani, W.A. & El-Nahas, A.M. (2009). Effect of structural properties of acid dyes on their adsorption behaviour from aqueous solutions by amine modified silica. *J. Hazard. Mater.*, 161, 1544–1550.

Drobny, J. G. (2007). *Handbook of Thermoplastic Elastomers,* Plastics Design Library.

Elham, E. & Behzad, P. (2012). One-Pot Synthesis Of Sub-50 Nm Vinyl- And Acrylate-Modified Silica Nanoparticles., *Powder Technol.*, 219, 276-283.

Flory, P.J.& Rehner, J. (1943). Statistical mechanics of cross-linked polymer networks. I. Rubberlike elasticity. *J. Chem. Phys.*, 1-52.

Jianzhong, S., Yan, Y. & Rui, Z. (2011). Preparation and characterization of silica/PMMA composites with highly dispersed silica nanoparticles dispersion of nanosilica in MMA. *Acta Polym Sci.*,7, 745-751.

Jesionowski, T., Przybylska, A., Kurc, B., Ciesielczyk, F. (2011). Hybrid pigments preparation via adsorption of C.I. Mordant Red 3 on both unmodified and aminosilane – functionalised silica supports. *Dyes Pigm.*, 89, 127-136.

Jesionowski, T. & Krysztafkiewicz, A. (2001). Influence of silane coupling agents on surface properties of precipitated silicas. *Appl. Surf. Sci.*, 172,18-32.

Jesionowski, T. (2002). Synthesis of organic-inorganic hybrids via adsorption of dye on an aminosilane-functionalised silica surface. *Dyes Pigm.*, 55, 133-141.

Jesinowski, T. (2003). Influence of aminosilane surface modification and dyes adsorption on zeta potential of spherical silica particles formed in emulsion system. *Colloids Surf.*, 222, 87-94.

Jesionowski, T., Pokora, M., Tylus, W., Dec, A. & Krysztafkiewicz, A. (2003). Effect of *N*-2-(aminoethyl)-3-aminopropyltrimethoxysilane surface modification and C.I. Acid Red 18 dye adsorption on the physicochemical properties of silica precipitated in an emulsion route, used as a pigment and a filler in acrylic paints. *Dyes Pigm.*, 57, 29–41.

Jesionowski, T., Binkowski, S. & Krysztafkiewicz, A. (2005). Adsorption of the selected organic dyes on the functionalized surface of precipitated silica via emulsion route. *Dyes Pigm.*, 65, 267-279.

Jesionowski, T. (2005). Characterisation of pigments obtained by adsorption of C.I. Basic Blue 9 and C.I. Acid Orange 52 dyes onto silica particles precipitated via the emulsion route. *Dyes Pigm.*, 67, 81-92.

Jesionowski, T., Przybylska, A., Kurc, B. & Ciesielczyk, F. (2011). The preparation of pigment composites by adsorption of C.I. Mordant Red 11 and 9-aminoacridine on both unmodified and aminosilane-grafted silica supports. *Dyes Pigm.*, 88, 116-124.

Kishore, P.N.R. & Jeevanandam, P. (2012). A novel thermal decomposition approach for the synthesis of silica-iron oxide core–shell nanoparticles. *J. Alloy.Comp.*, 52, 51–62.

Kister, O. & Roessner F. (2012). Synthesis and characterization of mesoporous and amorphous silica modified with silica-organo-sulfogroups. *J. Porous Mat.*,19, 119-131.

Klapiszewska, B., Krysztafkiewicz, A. & Jesionowski, T. (2003). Highly dispersed green silicate and oxide pigments precipitated from model systems of postgalvanic waste. *Environ. Sci. Technol.*, 37, 4811-4818.

Kohno, Y., Senga, M., Shibata, M., Yoda, K., Matsushima, R., Tomita, Y., Maeda, Y. & Kobayashi, K. (2011). Stabilization of flavylium dye by incorporation into Fe-containing mesoporous silicate. *Rev. Mineral. Geochem.*, 141, 77–80.

Krysztafkiewicz, A., Jesionowski, T. & Binkowski, S. (2000). Precipitated silicas modified with 3-aminopropyltriethoxysilane. *Colloids Surf.*, 173, 73–84.

Krysztafkiewicz, A., Binkowski, S. & Jesionowski, T. (2002). Adsorption of dyes on a silica surface. *Appl. Surf. Sci.*, 199, 31–39.

Krysztafkiewicz, A., Binkowski, S. & Wysocka, I. (2003). Pigments on amorphous silica carriers. *Powder Technol.*, 132, 190– 195.

Ladewig, K., Seifert, A., Hahn, H., Hietschold, M., Moszner, N., Burtscherd P. & Spange, S. (2012). apara-Nitroaniline- functionalized chromophoric organic–inorganic hybrid. *J. Mater. Chem.*, 22, 3839-3844.

Liu, X. & Zhao, S. (2008). Study on structure and properties of SSBR/SiO2 co-coagulated rubber and SSBR filled with nanosilica composites. *J. Appl. Polym. Sci.*,109, 3900-3907.

Lipińska, M., Sokołowska, J., Boruszczak, Z. & Zaborski, M. (2011).Synthesis and use of composite silica/trans-benzothiazine indigo pigment in elastomeric systems. *Przem. Chem.*, 90, 1225-1231.

Marzec, A., Lipińska, M., Sokołowska, J. & Zaborski M. (2010). Pigment-modified silicas as fillers of elastomers. *Przem. Chem.*, 89, 1475-1478.

Natinee, L., Dolmalik, J. & Manus, S. (2011). Hybridized reinforcement of natural rubber with silica-modified short cellulose fibers and silica. *J. Appl. Polym. Sci.*, 120, 3242-3254.

Park, S-J. & Cho K-S. (2003). Filler–elastomer interactions: influence of silane coupling agent on crosslink density and thermal stability of silica/rubber composites. *J. Colloid Interface Sci.*, 267, 86–91.

Parida, S.K. & Mishra, B. K. (1998). Adsorption of styryl pyridinium dyes on polyethylene-glycol-treated silica. *Colloids Surf.*, 134, 249-255.

Raha, S., Quazi, N., Ivanov, I. & Bhattacharya, S. (2012). Dye/clay intercalated nanopigments using commercially available non-ionic dye. *Dyes Pigm.* 93, 1512-1518.

Rampazzo, E., Bonacchi, S., Montalti, S., Prodi, L. & Zaccheroni, L. (2007). Self - Organizing Core - Shell Nanostructures: Spontaneous Accumulation of Dye in the Core of Doped Silica Nanoparticles. *J. Am. Chem. Soc.*,129, 14251-14256.

Ramirez, A., Sifuentes, C., Manciu, F.S., Komarneni, S., Pannell, K.H. & Chianelli, R.R. (2011). The effect of Si/Al ratio and moisture on an organic/inorganic hybrid material: Thioindigo/montmorillonite. *Appl. Clay Sci.*, 51, 61-67.

Siwińska, D., Kołodziejczak-Radzimska A., Krysztafkiewicz, A. & Jesionowski, T. (2009). Adsorption of octylamine on titanium dioxide. *Appl. Surf. Sci.*, 255, 7337-7342.

Smith, H. M.,(2002). *High performance pigments*, WILEY-VCH.

Sójka-Ledakowicz, J., Lewartowska, J., Kudzin, M., Leonowicz, M., Jesionowski, T., Siwińska-Stefańska, K. & Krysztafkiewicz, A. (2009). Functionalization of textile materials by alkoxysilane-grafted titanium oxide. *J. Mater. Sci.*, 44, 3852-3860.

Wieczorek, M., Krysztafkiewicz, A. & Jesionowski, T. (2004). Influence of modification by N-2-(aminoethyl)-3-aminopropyltrimethoxysilane on physicochemical properties of bentonite. *J. Phys. Chem. Solids*, 65, 447-452.

Wang J-X, Kang, L. & Chen., J-F. (2009). Fabrication of antibacterial monodispersed Ag–SiO2 core–shell nanoparticles with high concentration. *Mater.Lett.* ,63, 31–33.

Werner, R., Krysztafkiewicz, A., Dec, A. & Jesionowski, T. (2001). Effect of surface modification on physicochemical properties of precipitated sodium-aluminium silicate, used as a pigment in acrylic dispersion paints. *Dyes Pigm.*, 50, 41-54.

Wu, G., Koliadima, A., Her Y-S. & Matijevic, E. (1997). Adsorption of dyes on nanosize modified silica particles. *J. Colloid Interface Sci.*, 195, 222-228.

Wunpen, Ch., Withawat, M. & Ulchulee, M. (2011). Vulcanization characteristics and dynamic mechanical behavior of natural rubber reinforced with silane modified silica. *J. Nanosci Nanotechno.*, 11, 2018-2024.

Zaborski, M. & Marzec, A. (2011). Effect of some dyes on antiageing properties of the acrylonitrile-butadiene rubber. *Przem Chem.*, 90, 1083-1087.

Jung-Woo, P., Jun, P. Y. & Chul-Ho, J. (2011). Post-grafting of silica surfaces with pre-functionalized organosilanes: new synthetic equivalents of conventional trialkoxysilanes. *Chem Commun.*, 47, 4860-4871.

Chapter 5

Rubber Clay Nanocomposites

Maurizio Galimberti

Additional information is available at the end of the chapter

1. Introduction

RCN Nomenclature and classification of clays are first summarized, highlighting the most important clay features that affect their behaviour as fillers for rubbers. The modification of clays with organophilic compensating cations, to promote their compatibilization with the polymer matrix, is then presented. The processing methods for the preparation of RCN is then discussed and the main aspects of RCN are reviewed, namely: rheology, vulcanization, barrier and mechanical properties. Finally, commercial applications of RCN are presented.

A large scientific and patent literature is available on RCN and it was taken into consideration for the preparation of this Chapter. As the RCN applications are treated as well, press releases and news available on web sites were considered and commercial products were examined. Patents, when cited, are regarded just as publications, without considering if they were already granted or if, viceversa, they are still patent applications. Literature reporting on polymer clay nanocomposites (PCN) is available [1-13] and it has to be considered, to fully understand structure and properties of RCN. Reviews are also available, dealing specifically with rubbers [14-21].

2. Clays

2.1. Nomenclature and classification

The Joint Nomenclature Committee of the AIPEA (Association Internationale pour l'Etude des Argiles) and the CMS (Clay Minerals Society) say that a clay is a naturally occurring material composed primarily of fine-grained minerals, which is generally plastic at appropriate water contents and hardens when dried or fired. Hence, the definition of clay refers essentially to the macroscopic clay properties. Precisely determined crystallographic structures should be named as clay minerals [22].

Clays are layered: this is their most important feature. They are inorganic compounds made by stacked layers, whose atoms are joined together by iono-covalent bonds, that are bound to each other in the perpendicular direction through weaker forces. This implies that layers can be separated from each other by applying a minor amount of energy, whereas remarkable energy is required to break the layers.

Clays are a sub-family belonging to the larger family of layered oxides (or oxyhydroxides). They are not only silicates, as some of them do not contain any silicon atom, though those applied for the preparation of polymer nanocomposites are indeed essentially silicates.

Clays can be classified according to the electrical charge of the layer, as summarized in Table 1. Clays have either (i) neutral layers or (ii) negatively charged layers or (iii) positively charged layers. In the last two cases, the layer charge is exactly compensated by an equal amount of opposite charges located in the interlayer space. Clays with negatively charged layers or positively charged layers are respectively called cationic clays and anionic clays.

Type of layers	Type of clay	Main features
neutral layers	pyrophyllite, talc, kaolinite	*neutral clays* layers joined together by van der Waals interactions and/or hydrogen bonds
negatively charged layers	phyllosilicates: e.g. bentonites (main component: montmorillonite)	*cationic clays* the negative layer charge is exactly compensated by compensating cations are located in the interlayer space.
positively charged layers	hydrotalcite (HT). layered double hydroxides (HT-like family)	*anionic clays* the positive layer charge is exactly compensated by compensating anions located in the interlayer space

Table 1. Classification of clays as a function of the electrical charge of the layer

2.2. Organization and structure of clays

Clays give rise to a multiscale organization. Examining the upper level of said organization, clays particles of micrometric size form millimetric-size agglomerates. In a polymer matrix, the dispersion of these agglomerates is one of most important feature of the polymer nanocomposite. Figure 1 reports the structure of a montmorillonite (Mt) (Figure 1a), the most applied clay for the preparation of polymer nanocomposites, thanks to its large availability, low cost and high surface area. Moreover, purified Mt, with less than 1% of crystalline silica, is considered safe and is handled as a standard powder, as its platy nanoparticles, with only one dimension at the nanoscale, appear to have little chance to cross biological barriers [23].

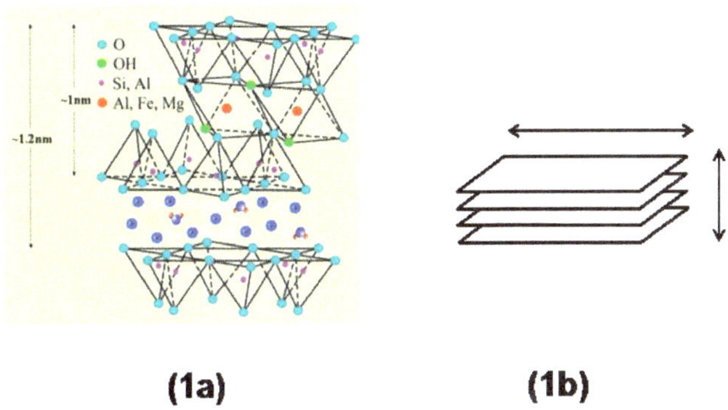

Figure 1. Structure of montmorillonite (Figure 1a) and a scheme (Figure 1b) to highlight the significant difference in length with respect to thickness of the layers

By examining a clay such as Mt at the lower level of its organization, it can be seen that a single clay layer is characterized by lateral dimensions from 100 to 1000 nm and by a thickness of about 1 nanometer. Mt is a TOT type clay mineral, with two tetrahedral (T) sheets linked to both sides to a central octahedral (O) sheet. These sandwiches are held together by weak interforces and alkaline and alkaline-earth cations are located in the interlayer space. Each silicate layer is terminated on its faces by oxygen atoms and on its periphery by oxygen atoms and hydroxide groups. The hydrophilic nature of Mt implies that a compatibilizer has to be used in order to disperse Mt in a lipophilic matrix, such as the one of most diffused rubbers. To modify a pristine Mt with a compatibilizer, the cations located in the interlayer space are exchanged with organophilic cations, preferentially with long chain quaternary alkyl ammonium cations, an organoclays (OC) are formed [24]. OC are discussed in the next paragraph. The high cationic exchange capacity (CEC) of Mt further pushes its use in polymer nanocomposites.

The term nanocomposites was many times used in previous lines. In fact, nanocomposites are defined as composite materials characterized by the presence of dispersed particles whose size is in the "nanoscale", defined as "having one or more dimensions of the order of 100 nm or less" [25]. Mt can be considered as a nanomaterial and, as it will be discussed later on, as a nanofiller and its polymer composites are thus nanocomposites.

Figure 1 shows as well a scheme (Figure 1 b) that highlights the peculiar shape of the layers, that have a significant difference in length with respect to thickness. Considering Mt as a reinforcing filler, this means that Mt has a high aspect ratio. As it will be discussed in Paragraph 10, the mechanical reinforcement of a polymer matrix strongly depends on the aspect ratio of the filler, i.e. on the ratio between its longest and shortest dimensions. This ratio is often indicated as *shape factor f*. The higher is the shape factor f and the higher are the values obtained for the dynamic-mechanical moduli. It is worth commenting here that to have a high shape factor, the Mt layers have to be brought apart, that means Mt has to be exfoliated. The highest aspect ratio for a clay would be given by the ratio between the longest lateral dimension and the thickness of a single layer.

2.3. X-ray diffraction of clays

Clays are crystalline materials and X-Ray Diffraction (XRD) analysis is a technique largely used to assess clay features. In the Bragg model of X-rays diffraction, that indicates the plane spacing with d index and the plane orientation with three Miller indexes (h, k, ℓ), a constructive interference is obtained for X-rays scattered from adjacent planes when the angle θ, between the plane and both the incident and reflected beams, and the plane spacing is related with the X-ray wavelength λ. through the equation known as the Bragg law [26]:

$$2\,d\sin\theta = n\,\lambda \tag{1}$$

From the values of θ angles detected in the XRD pattern, it is possible to determine clay features such as: $d_{hk\ell}$ interlayer distance, $D_{hk\ell}$ correlation length in a crystallographic direction, for example in the plane of the layers and in the direction orthogonal to the layers, and hence the number of stacked layers giving rise to a crystalline unit.

Figure 2 shows the XRD patterns of two pristine Mt available in the market: Cloisite® Na from Southern Clay and Dellite® HPS from Laviosa Chimica Mineraria.

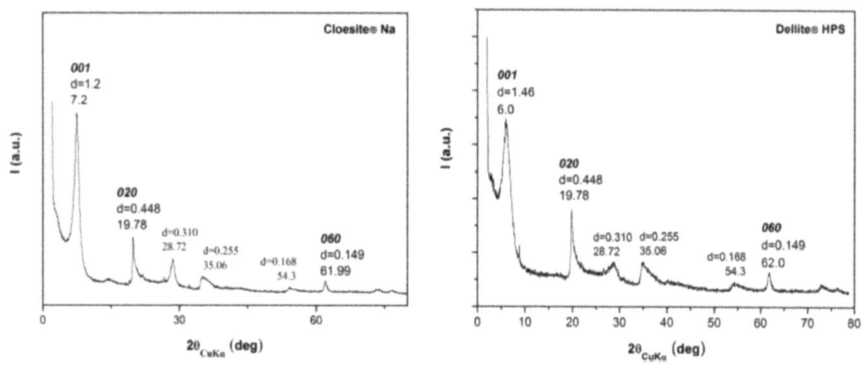

Figure 2. X-Ray diffraction patterns in the 2theta range 0-80° for pristine Mt samples: Cloisite® Na and Dellite® HPS

The d_{001} interlayer distance was determined to be about 1.2_0 and 1.4_6 nm for Cloisite® Na and Dellite® HPS, respectively. As mentioned above, from XRD data it is also possible to calculate the $D_{hk\ell}$ correlation length of the crystalline domain in the direction orthogonal to the structural layer, through the Scherrer equation:

$$D_{hk\ell} = 0.9\,\lambda \;/\; \beta_{hk\ell}\cos\theta_{hk\ell} \tag{2}$$

where: λ is the wavelength of the irradiating beam (1.5419 Å, CuKα), $\beta_{hk\ell}$ is the width at half height, and $\theta_{hk\ell}$ is the diffraction angle [27]. With reference to the (001) reflection and by introducing the correction factor, to be used in case $\beta_{hk\ell}$ is lower than 1°, a D_{001} value of about 9 nm and about 4 nm was calculated for Cloisite® Na and Dellite® HPS, respectively. Taking into account that the d_{001} interlayer distances reported above, a number of about 8 and about 4 stacked layers was calculated for Cloisite® Na and Dellite® HPS, respectively. This elaboration demonstrates that the available pristine Mt are characterized by a pretty typical interlayer distance and by a relatively low number of stacked layers.

3. Organoclays

As mentioned in the previous paragraph, OC are prepared through the exchange reaction of a pristine clay with an organophilic ion. In the case of cationic clays, such as Mt, organophilic ammonium cations are mostly used for preparing OC for polymer nanocomposites. The intercalation of ammonium cations leads to an expansion of the interlayer distance. In the XRD spectrum, (00ℓ) indexes refer to a crystalline order in the direction perpendicular to the structural layers: the (001) reflection is due to regularly stacked layers and higher order reflections, e.g.. (002) and (003), originate from the regular arrangement of intercalants. An expansion of d interlayer spacing corresponds to a shift of (001) reflection towards lower 2 θ angle values.

Figure 3 shows two types of ammonium ions largely used as Mt compensating cations and the expansion of interlayer distance that occurs as a consequence of their intercalation. The ammonium cations are: dimethyl-talloyl-dihydroxyethyl and dimethylditalloyl and the talloyl (T) group can be optionally hydrogenated (HT).

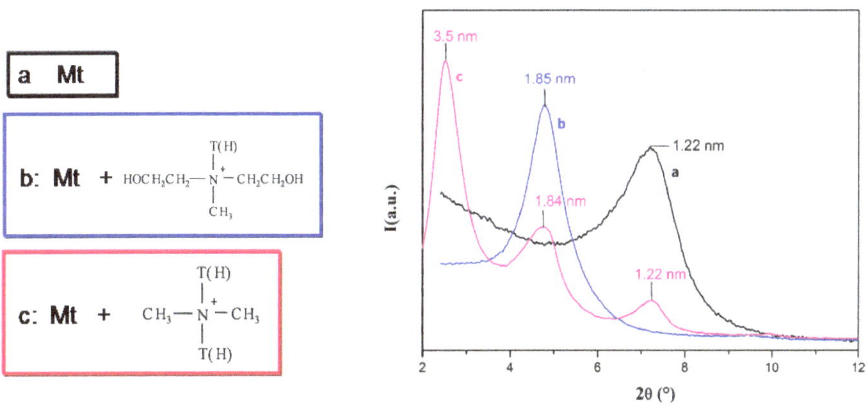

Figure 3. Expansion of Mt interlayer distance as a consequence of ammonium cations intercalation. 2(H)T and (H)Tdihydroxyethyl as the ammonium cations

It is worth underlying that the same ammonium cation can give rise to different interlayer distances, as a function of its weight percent and of its bond with the inorganic layer. In fact the ammonium can be ionically bonded or can be only absorbed on the layer surface. Table 2 reports the values of interlayer distances detected for commercially available OC with 2HT as the compensating ammonium cation.

Organoclay	% wt organic	2θ (deg)	Hkl	d (nm)	d_{001} (nm)
Dellite® 72T	36 – 38	3.42	001	2.59	~ 2.5
		7.12	002	1.24	
Closite® 20A	38	3.57	001	2.48	~ 2.5
		7.1	002	1.25	
Closite® 15A	43	2.98	001	2.97	~ 3.0
		7.15	00?	1.24	
Dellite® 67G	45	2.50	001	3.53	~ 3.6
		4.66	002	1.89	
		7.23	003	1.22	
		9.8	004	0.90	

Table 2. Organoclays with 2HT as the compensating ammonium cation

4. Clays and organoclays used for RCN

Literature available on RCN essentially refers to cationic clays. Those used for RCN preparation were: Na-Mt, Na-bentonite, Na-fluorohectorite, rectorite, vermiculite and a fibrillar silicate such as attapulgite. Clays of the bentonite family and, among them, Mt in particular, were the most applied ones. This chapter refers thus to data based on cationic clays. As discussed in paragraphs 2 and 3, a lipophilic compensating cation promotes the compatibilization of the cationic clay with the polymer matrix. Those used for the preparation of RCN are listed as follows, indicating for the ammonium cation the substituents of the nitrogen atom: (i) cations from primary alkenylamines, (ii) ammonium cations with three methyls and one long chain alkenyl, (iii) ammonium cations with two methyls, an hydrogenated tallow and a benzyl group, (iv) ammonium cations with two methyls and two ethylhexyl or short chain alkenyl groups, (iv) ammonium cations with a methyl, a tallow and polar groups such as 2-hydroxyethyl, (v) ammonium cations with two methyls and two (hydrogenated) tallow groups.

5. Processing methods for the preparation of RCN

In this paragraph, distribution and dispersion of clays in a rubber matrix are discussed. They will be shown as depending on the chemical nature of the clay (pristine or organically modified) and on the selection of the appropriate processing technology.

Different methods have been developed for the preparation of RCN. A classification can be attempted, based on the type of clay used for the preparation of the nanocomposite: pristine or organically modified, as it is summarized in Table 3.

A pristine clay can be blended with rubber latexes (NR and E-SBR as the rubbers) through what is known as the emulsion blending. It can also be blended with the rubber in the melt state, performing the exchange reaction with an ammonium salt and using the rubber as the reaction medium.

For an organically modified clay, melt and solution blending can be applied. In the melt blending, the OC is mixed directly with the rubber in the melt state. In the solution blending, the rubber is first dissolved in a good solvent, adding the OC swollen in the same solvent.

Methods that use pristine clay and the melt blending of an OC are suitable for an industrial development, whereas the solution blending can be adopted in a laboratory. In this paragraph, the impact of mixing methods on clay distribution and dispersion is examined. An even clay dispersion is a crucial property for a PCN, as it allows to best exploit the properties of the layered filler.

Type of Clay	Ammonium cation	Type of blending	State of the rubber
pristine clay	=	emulsion blending	in a latex
pristine clay	added during blending	melt blending	in the melt state
organoclay	in the organoclay	melt blending	in the melt state
organoclay	in the organoclay	solution blending	in solution

Table 3. Processing methods of clays and organoclays with rubber

5.1. Emulsion blending from a pristine clay

Zhang first introduced the emulsion compounding: a mixture obtained from an aqueous clay suspension and a rubber latex was coagulated in an electrolite solution. This approach was applied for blending Mt in rubbers such as NR [28, 29], SBR [28-32], NBR [28, 29, 33], XNBR [29]. Moreover, bentonite was blended with SBR [34-37] and SVBR (V = vinylpiridine) [34, 35] and rectorite was blended in SBR [38, 39]. Different electrolytes were used: triethylenetetrammonium chloride (2% wt solution) [29, 38, 39], diluted sulphuric acid solution [28, 30-32, 36], 1% calcium chloride aqueous solution [32], dilute hydrochloric acid solution [34, 35, 37], dilute dichloroacetic acid solution [33]. A scheme for blending a pristine clay with a rubber latex, as reported by Zhang, is shown in Figure 4.

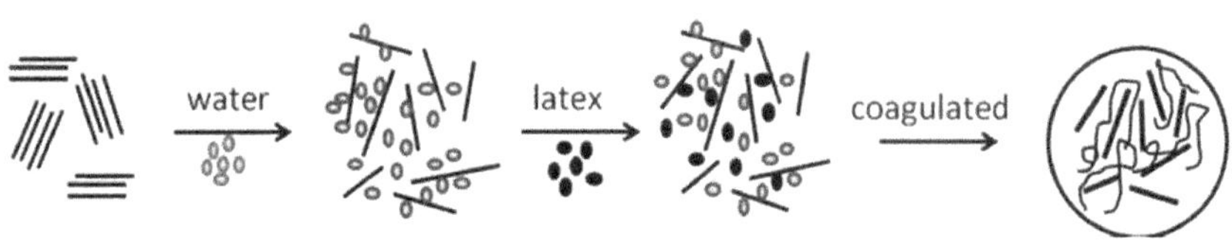

Figure 4. Emulsion blending of a pristine clay with a rubber latex

It was reported that the emulsion compounding led to nanocomposites, whereas conventional microcomposites were obtained when a pristine clay was melt blended with a rubber [28]. The Mt dispersion was commented to be excellent, up to 20 phr as concentration, and best results were obtained with diluted sulphuric acid.

Modifications of this procedure, leading to satisfactory silicate dispersion, were adopted by:

i. adding a slurry containing further ingredients prior to coagulation, with NBR as the rubber and a sodium salt of methylene-bis-naphthalene sulphonic acid as the electrolyte.

ii. adding further ingredients by melt mixing and avoiding the use of coagulating agents, with NR as the rubber [40]
iii. avoiding both coagulating agents and the addition of further ingredients, with NR as the rubber [41].

Bentonite and fluorohectorite (10 phr) dispersions in NR were observed to be to some extent worse when further ingredients were added prior to coagulation and coagulating agents were not used [42, 43].

When a prevulcanized NR latex was used, a minor amount of fluorohectorite layers and stacks with an increased distance between opposite layers were observed [44]. Fluorohectorite revealed a better swelling ability than bentonite and thus gave a more exfoliated structure [45].

5.2. Melt blending from clays and organoclays

Melt blending was performed with the help of the typical processing technologies of the rubber industry, from internal mixers such as brabender® and banbury® to open mills to twin screw extruders.

The melt blending of a pristine Mt with a rubber leads to composites with undispersed agglomerates, as demonstrated for NR [46-49], IR [50], BR [51-53], E-SBR [48, 54], E-SBR/NBR blend [55], EPDM [56] as the rubbers. Analogous results were presented with fluorhectorite and attapulgite as the clays.

A satisfactory dispersion was achieved when the inorganic clays contained a lipophilic modifier [57]. With primary alkenylamines, exfoliated platelets and aggregates composed of different number of platelets were observed in the final composites, commenting that the C_{18} alkenyl group promoted a better dispersion with respect to the shorter ones.

An even dispersion of the organoclay in the rubber matrix was achieved in any the composites, though a fully exfoliated morphology was not achieved, when an ammonium cation with either three methyls and one long chain alkenyl substituent or two methyls and two long chain substituents was used. The polar hydroxyl group as ammonium cation substituent led to some layers agglomeration. These results were independent of the type of the rubber matrix.

5.3. Solution blending from clays and organoclays

Microcomposites were prepared when a bentonite or a Mt were mixed with the help of an organic solvent with many different types of rubbers: NR [58], BR [55], SBR [55, 59-61], (H)NBR [55, 62], BIMS [63]. Big lumps of clay aggregates and agglomerates were formed. Viceversa, an even dispersion of organoclays was achieved, when a clay containing a lipophilic ammonium as compensating cation was used. As reported in paragraph 4, different types of ammonium cations were used, in many types of rubber. A rationalization is proposed in ref. [57]. To summarize, the mentioned even dispersion was achieved in NR,

BR, SBR, NBR, BIMS and EPDM with a primary alkenyl amine as the clay modifier. In NR, IR, BR, SBR, (H)NBR when ammonium cations having either two or three methyl groups and longer chain substituents were adopted as clay compensating cations. Results appeared very similar to those obtained with melt blending.

5.4. Formation of organoclays in situ in the rubber matrix

The preparation of the organoclay *in situ* in the rubber matrix was reported by Galimberti [57, 64-67], by adopting the melt blending approach, with all the processing technologies mentioned in paragraph 5.2. An even clay dispersion was achieved in the rubber matrix and the organoclays revealed a lower number of stacked layers. This approach was adopted for many types of rubbers: IR, NR, SBR, NBR, XIIR, EPDM.

6. Structure of RCN

In this paragraph, the overall picture of RCN structure is presented, discussing in particular the interlayer distance of clays and the type of intercalants present in the interlayer space, of low molecular mass or polymeric.

This aspect is of the outmost importance for polymer clay nanocomposites. In fact, it is largely reported [1-22] that the intercalation of polymer chains in the interlayer space is at the origin of the remarkable properties of PCN. The intercalation of polymer chains is considered not only to favour the improvement of the polymer matrix but also to promote the clay exfoliation, leading to the ultimate clay dispersion. The overall picture of RCN, that is proposed for PCN is shown in Figure 5.

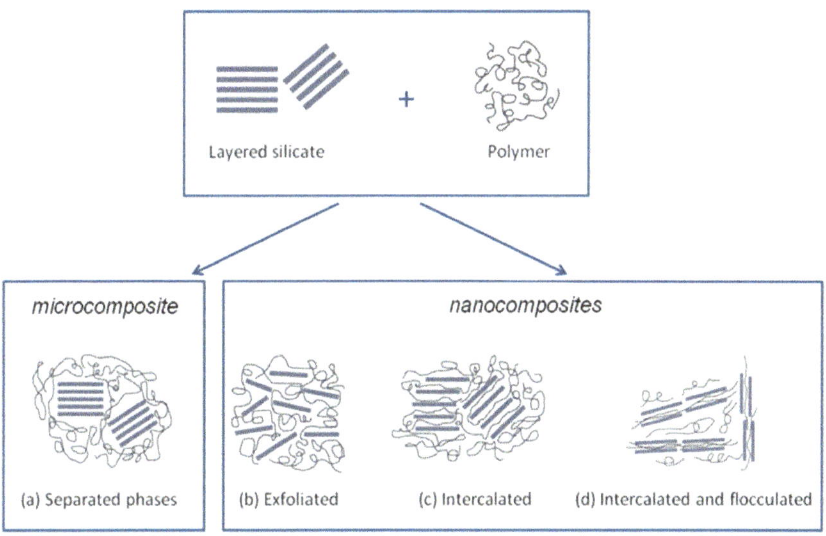

Figure 5. Overall picture of the structure of polymer (rubber) clay nanocomposites

The study of the interlayer distance is performed through XRD and TEM analysis [57]. In particular, as discussed in paragraph 3, the modification of the interlayer distance, determined from XRD patterns, is taken as the evidence of the occurring of modifications in the interlayer space. It is worth summarizing here the two mechanisms proposed in the literature to explain the occurring of clay exfoliation and the modification of the interlayer distance.

6.1. Two mechanisms for clays intercalation and exfoliation

The two mechanisms can be visualized as reported in Figure 5 and are summarized as follows.

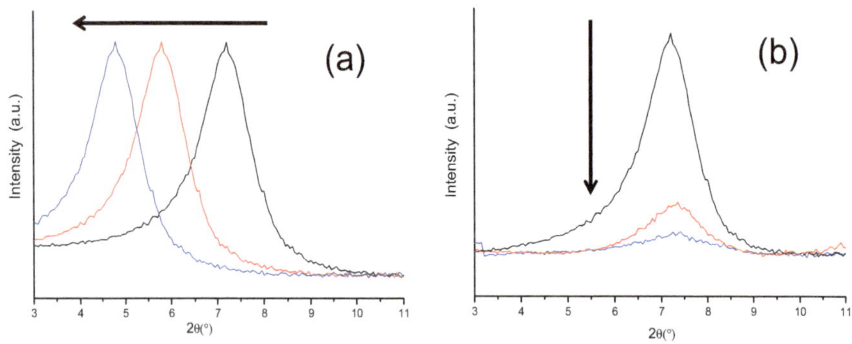

Figure 6. Schemes to visualize the two mechanisms proposed for clay intercalation and exfoliation: first mechanism (a), second mechanism (b) (see text for the explanation)

First mechanism. Polymer chains are intercalated in the interlayer space, with a consequent expansion of the interlayer distance. In Figure 6a, the position of the (00ℓ) reflection progressively shift towards lower values of the 2θ angle, from black to red to blue curve, as a consequence of the progressive intercalation of the polymer chains. The intercalated polymer chains are not only responsible for the expansion of the interlayer distance but also cause the separation of clay layers, promoting the ultimate exfoliation. This first mechanism, originally proposed for polystyrene as the polymer, is widely accepted in the prior art.

Second mechanism. The intercalation of polymer chains is regarded as unlikely and only low molecular mass substances are considered to be present in the interlayer space, when crystalline OC are formed. In particular, it is commented that it is hard to suppose that highly crystalline structures (as shown by the presence of several (00ℓ) reflections in the XRD pattern), can be generated in the presence of a polymer chain. The variation of the interlayer distance is seen as a consequence of: (i) the different arrangements of the substituents of the compensating cations, that indeed can give rise to different d_{001} values (as shown in Table 2) and/or (ii) of intercalation of guests in the interlayer space and/or (iii) of molecules absorption on clay surfaces. The clay exfoliation occurs through a progressive peeling off of the clay stacks thanks to the shear mixing. In Figure 6b, the (00ℓ) reflection remains at the same 2θ value and its intensity decreases, passing from the black to the red to the black curve, as a consequence of the progressive exfoliation of the clay stacks. Authors presenting this mechanism propose the scheme reported in Figure 7, where the XRD pattern of nanocomposites in the low 2θ region (peaks are due to (00ℓ) reflections are shown as well the mechanism hypothesized for their formation.

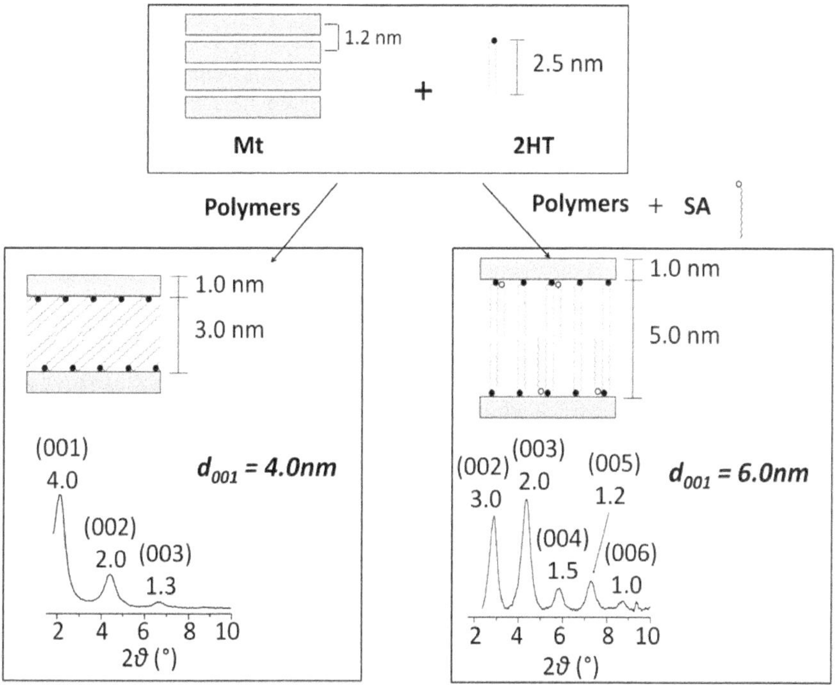

Figure 7. Schematic presentation for the formation of an organoclay by blending Na-Mt with di(hydrogenated tallow)-dimethylammonium chloride (2HT), in the absence or in the presence of stearic acid (SA) (Scheme reproduced from ref. 66)

7. Rheology of RCN

Clays compatibilized and evenly dispersed in a polymer matrix tend to build networks at low concentration. Rheological measurements, performed on RCN based on various types of rubbers, revealed the pronounced rubber-clay interaction, when measurements were taken at zero shear. The storage modulus in the low frequency region was investigated as a function of clay content and the clay percolation threshold (as wt%) was found to be about 4 (OC was Mt/methyl tallow bis-2-hydroxyethyl ammonium cation), above 5 (OC was Mt/dimethyl-dialkylammonium halide (70% C18, 26% C16 and 4% C14)) and above 7.5 (OC was Mt/ dimethyl dehydrogenated tallow quaternary ammonium chloride) for IR, EPR and EVA, respectively [68-71]. The filler networking phenomenon was observed as well in matrices based on IR, ENR [72], SBR [73] and EPR [69]. At zero shear, the viscosity of RCN is thus higher than the one of the neat elastomer.

However, organophilic clays were shown to reduce the steady shear viscosity of RCN. Most evident results were a pronounced shear-thinning behaviour, increasing with the clay content, a higher extent of extrudate, a lower swelling and a better surface smoothness, by increasing the shear rate. These findings are of great importance, as they indicate that OC have a positive effect on the processability of rubber compounds. Data were reported for RCN based on various rubbers, such as BR [60], SBR [60], NBR [60], BIMS [74] and fluoroelastomer [75]. A compensating cation with longer alkyl chains led to a reduction of Mooney viscosity, that was not observed with a short chain substituent [76].

This behaviour of OC, that is opposite to that of traditional fillers, is attributed to the orientation, occurring at high shear rates, of clay platelets along flow direction, and to the slippage of platelets on the chains, thanks to the organophilic clay substituents. Data to demonstrate that were provided with an OC (Mt/octadecyltrimethylammonium) in BR [77] and with another OC (Mt/dimethyl hydrogenated-tallow (2-ethylhexyl) quaternary ammonium methylsulfate) in poly(epichlorohydrine) [78].

The improvement of processability brought about by OC was found to be higher when the matrix was apolar rubber such as NBR [79].

8. Vulcanization of RCN

OC promote a fast sulphur based crosslinking of unsaturated polymer chains. Moreover, they increase the delta torque value indicating a higher value of crosslinking density [80].

Many data were reported in the literature, supporting these conclusions, in rubbers such as NR (hexadecyltrimethyl ammonium, octadecyltrimethyl ammonium, tetraoctyl phosphonium, triphenyl vinylbenzyl phosphonium, octadecylamine chloridrate were used as intercalants) [81-86], SBR (Mt/octadecyltrimethylamonium) [87], (NBR (octylamine; dodecylamine and octadecylamine chloridrates were the intercalants) [76]. A reduction of the activation energy for sulphur based NR crosslinking was found by using a bentonite modified with octadecylamine chloridrate [83, 88]. OC could thus present a warning, as they could lead to premature scorching but, at the same time, they could present the chance of performing vulcanization reactions at lower temperatures.

To explain the behaviour of OC in sulphur based crosslinking, the formation of tertiary amines from the thermal degradation of ammonium cations and the enhanced mobility of sulphur accelerating anionic species were proposed [89].

9. Barrier properties of RCN

The reduction of air permeability thanks to clays dispersed in a poly(isobutene) matrix led to the first commercial application of RCN, that will be discussed in Paragraph 11.2.

9.1. The tortuous path model

Platelets with a high aspect ratio substantially reduce the diffusion of penetrating molecules in polymer matrices. The tortuous path model, represented in Figure 8, is generally accepted to explain this result.

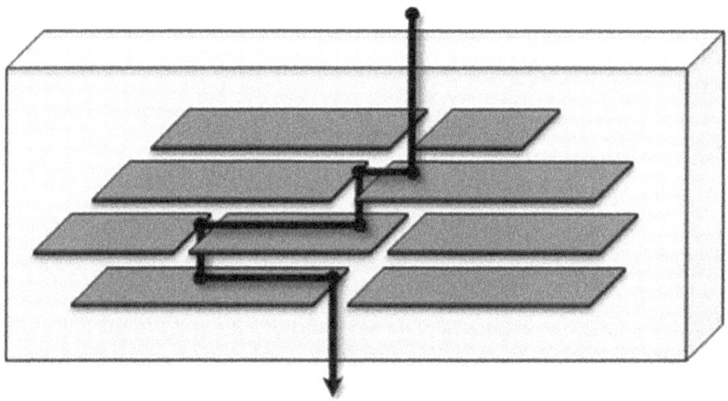

Figure 8. Tortuous path for molecules in a matrix containing platelets

Various continuum models were developed and applied to polymer clay nanocomposites, assuming in most cases a random dispersion of plates, parallel to each other and perpendicular to the direction of molecules diffusion, as summarized in refs. [90] and [91]. According to these models, reduction of permeability is enhanced by the increase of the platelet aspect ratio. These models are able to interpret the reduction of permeability in a polymer matrix, without necessarily having the presence of nano-platelets. This means that a nano-effect should not be invoked. It was commented [91] that permeating molecules have a sub-nano level and their permeation is slowed down by platelets, whatever is their size.

9.2. Improvement of barrier properties of rubber matrices thanks to the use of clays

Better barrier properties of a rubber matrix, thanks to clay addition, were demonstrated for many different rubbers. In NR as the matrix, the barrier properties were enhanced at low clay (Mt) content, preparing the nanocomposite through the emulsion blending: 1, 2 and 3 phr of clay led respectively to more than 35% and to about 45% and 50% reduction of oxygen permeability [92]. 3 phr of OC (Mt/ didodecyl methyl amine) gave a 50% reduction of the oxygen permeability and a 40% reduction of toluene absorption at 20 °C [93]. 5 and 10 phr of OC led, respectively, to about 10% and 15% reduction of the oxygen permeability, and, at 15% OC, to a 30% reduction of toluene absorption at 30°C [94]. In BR as the matrix, 5 phr OC (Mt/dimethyl ditallow-ammonium) dispersed from solution blending, led to a reduction of about 80% of water vapour permeability, whereas 10 phr OC gave about 20% reduction of toluene uptake at equilibrium [95]. In NBR as the matrix, 1 phr of OC (Mt/dimethyl ditallow-ammonium) dispersed from solution blending brought to about 80% decrease of water vapour permeability [96]. Better barrier properties were observed by increasing the AN content of NBR: the relative nitrogen permeability was reduced by 11.5, 10.4, and 9.0% for 42NBR, 35NBR, and 26NBR (the figures indicate the wt% of AN), respectively, with 10 phr OC (Mt/dimethyl dialkyl (C14–C18) ammonium) [97].

A clay with a high aspect ratio such as rectorite was found to give better barrier than carbon black N330 [38].

10. Mechanical properties of RCN

10.1. The origin of reinforcement: nano-structured and nano-fillers

Particulate fillers are used to reinforce traditional rubber compounds: they are carbon black and silica and are made by spherical individual particles of few tents of nanometers as the diameter, fused together to form aggregates extending up to few hundreds of nanometers. These fillers are called nano-structured as the aggregates can not be separated into individual particles by thermomechanical mixing.

Theories on the origin of reinforcement [98], developed on carbon black based compounds, led to present the modulus of a filled rubber compound as due to the sum of different contributions, as it shown in Figure 8.

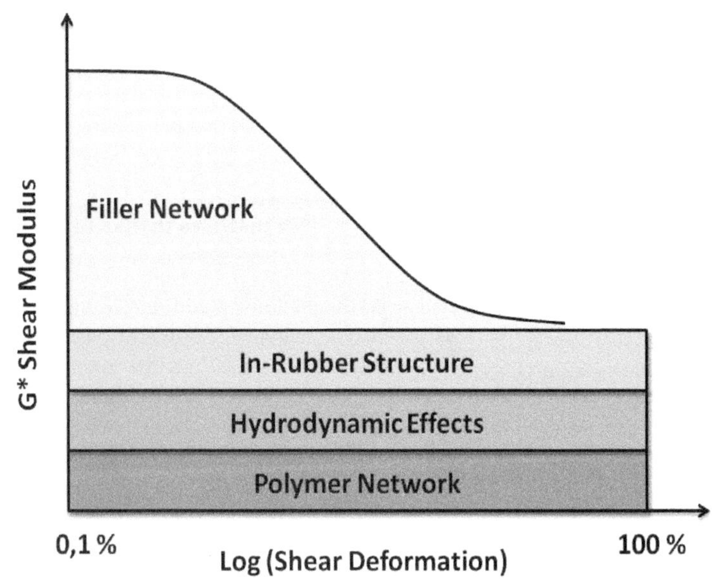

Figure 9. Contributions to the modulus of a rubber compound filled with a particulate filler

Contributions that do not depend on the strain amplitude are due to the polymer network (entanglements, physical and chemical crosslinks), hydrodynamic effects (related to the filler volume fraction and implying the strain amplification mechanism and thus the enhancement of the modulus), immobilization of rubber on filler particles (that transform a highly viscous liquid in a solid). A contribution to the modulus that strongly depends on the strain amplitude is due to the so called filler network: filler particles are joined together either directly or through polymer layers.

The Guth Equation is used to correlate the compound initial modulus with the filler volume fraction (equation 1 as follows).

$$E = Em\,(1 + 0.67\, f\phi + 1.62\, f^2 \phi^2) \tag{3}$$

This equation holds up to a threshold of filler content, known as the percolation threshold, at which a continuous network is established in the rubber matrix and thus accounts for the sum of the three contributions commented above and shown in Figure 8 that do not depend on the strain amplitude.

In equation 1, appears the *shape factor f*, that was already commented in Paragraph 2. The *shape factor f* was introduced in the Guth model in order to account for the fact that particle aggregation (clustering) has a significant impact on stiffness at high volume fractions (higher than 0.15). These clusters are formed by the filler aggregates and the impact of the *f* value on the quadratic term of the Guth equation makes them much important for the total value of the modulus.

In the case of a clay, as it was shown in Figure 1b, the *shape factor f* is given by the ratio between the longest lateral side of the layer and the height of clay stacks, the highest *f* value being obtained in the case of a single layer. Therefore, the exfoliation of a clay not only favours a better clay dispersion but also improves the modulus of the compound. To take into account the lower contribution to the modulus of a platelet like filler, a modulus reduction factor of about 0.7 was determined [99] by fitting experimental data.

An important contribution to the reinforcement is given by the so called "in rubber structure": fillers are able to accomodate polymer chains in the voids of their structure. This mechanism is well known in the case of nanostructured fillers such as carbon black and silica. In the case of clays, the immobilization of polymer chains would require their intercalation in the interlayer space, phenomenon that is however, as explained in paragraph 6.1, still a matter of debate.

OC give an important contribution to the polymer network. In fact, as discussed in paragraph 8, they promote a higher crosslinking density, thanks to the cation-thiolate interaction. Clay layers and aggregates could be represented as framed in a cage formed by short sulphur bonds, that prevents their slip on the polymer chains, phenomenon that occurs in uncrosslinked samples, as commented in paragraph 7.

In the same paragraph, it was commented that a low clay concentration is needed for achieving the percolation threshold in a hydrocarbon rubber matrix. This conclusion was drawn also by determining the dependence of the Young modulus (obtained from stress-strain curves) on the OC concentration, in a SBR matrix (with Nanomer I.42E from Nanocor as OC) [100] or in an isoprene rubber matrix (OC was Mt/2HT), either synthetic [67] or naturally occurring [101]. The excess of the Young modulus was plotted versus the filler fraction [102], determining the percolation threshold, that was: 2.7 vol % in SBR [100], 2.9 vol % in IR [67] and 4 vol % in NR [101]. The excess modulus scales with a power law with an exponent between 1.8 and 2.5 above the percolation threshold, lower with respect to the one typical of carbon black (about 4). The structural difference of the filler is proposed to justify this difference. As mentioned above, the filler network is responsible for the contribution to the modulus that depends on the strain amplitude. Thanks to the easy formation of a filler network, clays promote a remarkable non linearity of the dynamic-mechanical behaviour of rubber nanocomposites, a phenomenon known as Payne effect [103]. Data that the show the remarkable Payne effect of RCN, increasing with the clay content, were reported for NR [104] and IR [67] as the rubber matrices. A strong reduction of the dynamic modulus was observed also by increasing the temperature [50]. The 2-D melting of the paraffinic chains substituents of the compensating ion [105] was commented as a possible explanation [66, 101].

10.2. The improvement of rubber mechanical properties thanks to the use of clays

Many data are available in the literature to show that both pristine clays and OC bring about the improvement of the mechanical properties of a rubber matrix, preparing the nanocomposites from latex, melt and solution blending. Some examples are reported as follows.

Pristine Mt was dispersed in a NR latex, observing an increase of modulus for the obtained nanocomposite up to 30 phr of clay, in the presence of worse ultimate properties [41], as well as higher 300% stress, shore A hardness, tensile strength and tear strength, when Mt was at 20 phr level [29]. In a SBR matrix as well, tensile and tear strengths were improved by adding a pristine Mt via emulsion blending, up to 20 phr [29], without any improvement for a further addition [35].

OC were typically used at lower levels. In NR as the matrix, 5 phr of Mt modified by tetraoctyl phosphonium bromide led to a 3-fold increase of tensile strength [81] and best ultimate properties were obtained with 10 phr of OC (Mt/octadecyltrimethyl ammonium) [106]. In BR as the matrix, best mechanical properties were obtained with a Mt having dimethyl dihydrogenated tallow quaternary ammonium as the compensating cation [107], an ammonium cation largely diffused in RCN field. In a SBR matrix, mechanical properties improved up to 10 phr of OC (Mt/octadecyltrimethylamine, Nanomer I.28E) [87] (Mousa and Karger-Kocsis, 2001), decreasing for a further addition. In a nanocomposite based on maleic anhydride modified EPM (EPM-MA, with 0.42 wt% of maleic anhydride) about 5% of OC (Mt/octadecyl amine) gave a modulus three times higher than that of EPM-MA. Modulus increased and elongation at break decreased with clay content [108].

Reinforcement was obtained from OC also in rubber blends. 2 phr of OC (Mt/octadecyltrimethylamine) in an NR/ENR blend (10 phr ENR50) gave an increase of both strain and stress at break values [109] (Teh et al., 2004).

It was commented in Paragraph 7 the importance of sulphur based vulcanization to prevent the OC slip on the polymer chains. OC (Mt modified with hydrogenated tallow ammonium) was reported to promote a remarkable reinforcement also of a peroxide cured rubber matrix. However, in this case, the rubber was NBR, with acrylonitrile content as high as 50% [110]. It was also shown that higher values of tensile strength and better tear resistance were obtained, with a worse elongation at break, by increasing the AN content, with 10 phr of OC (Mt/dimethyl dialkyl (C14–C18) ammonium) [111].

10.3. Clays in a rubber matrix for lighter weight of the compound

The most important application of rubber compounds is in tires. Lighter rubber compounds are thus pursued, in the light of their remarkable impact on the environment. Direct comparisons were reported between OC and CB in promoting the mechanical reinforcement of rubber compounds.

In NR as the matrix, a comparable mechanical reinforcement was obtained with 10 phr OC (Mt/octadecylamine) and 40 phr CB, with almost twice elongation at break, a lower compression set and similar abrasion loss with OC [112]. With the same OC, a much lower heat build up was obtained with respect to 50 phr of CB [113].

In BR as the matrix, better mechanical performance was obtained with 3 phr OC (Mt/dimethyl dihydrogenated tallow quaternary ammonium) compared with 10 phr of CB [114]. In NBR as the matrix, a comparable tensile strength was obtained with either 10 phr OC or with 40 phr CB [115]. In EPM-MA, 5% of OC (Mt/octadecyl amine) gave the same reinforcement as 30 wt% CB [108].

10.4. Synergistics effects between clays and nanostructured fillers

An increasing number of studies is becoming available in the literature on RCN based on clays and a nanostructured filler such as silica and, in particular, CB. A large scale application of RCN could reasonably imply the use of hybrid filler systems, with a minor amount of OC added to a major part of a traditional filler. In SBR as the matrix, a hybrid sepiolite-silica filler system was adopted [116]. Hybrid OC/CB filler systems were used in the following matrices: SBR [117-119], brominated poly(isobutylene-co-paramethylstyrene) rubber [117], IR [50, 57, 120], NR [67, 121] NR/SBR blend [122], chlorobutyl rubber [123], ENR [124], EPDM [125] (Malas and Das, 2012). In some of these works [50, 57, 120-122], CB was used at a level typical of commercial applications (higher than 50 phr). A remarkable enhancement of material dynamic-mechanical properties was reported, thanks to the use of the hybrid filler system. The most important finding from these study was the synergism developed by the two fillers [118, 120, 121]. In particular, it was shown [126] that initial modulus values obtained with the hybrid CB-OMt filler system were much higher than those calculated through the simple addition of the two initial moduli of composites with only CB or only OC. Figure 9, taken from ref. [127], demonstrates the synergistic effect between OC and CB. The dashed line indicates the initial modulus calculated by simply adding the initial moduli of composites with only CB and only OC, whereas points refer to experimental data taken from composites containing the hybrid filler system.

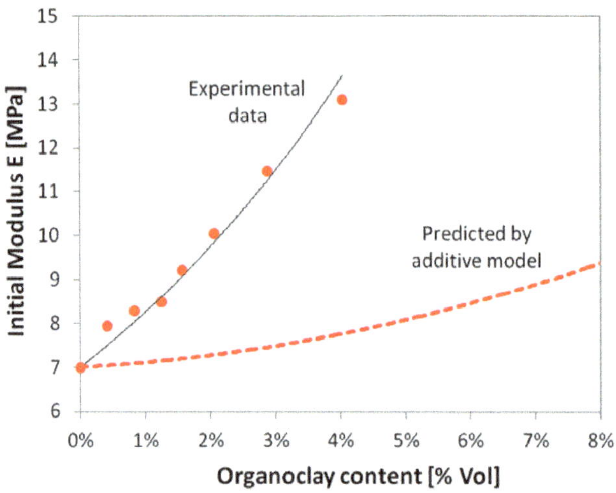

Figure 10. Initial modulus values (•) of IR based composites containing the hybrid OC/CB filler system. The dashed line refers to values predicted through the addition of initial moduli values of composites with only CB and only OC.

11. Applications of RCN

In this paragraph, applications of RCN are examined, moving from products already available on the commercial scale. The most meaningful patents were taken as the reference literature for the considered applications. RCN applications, diffused on the commercial scale, well documented and advertised, are essentially in two fields: tires and sport balls.

11.1. Applications of RCN for tyres

The most abundant application for rubbers is in tire compounds. NR is the most diffused rubber, with almost 11 million ton in 2010 [128] and SBR is the most produced synthetic rubber, with about 6 million ton. The application of both these rubbers is for about 75% in tire compounds. The world demand for tires is rising, at a pace of about 4.7 per year through 2015, achieving a number of produced tires of about 3.3 billion units. In the same period, the tire market is projected to increase, in terms of value, of about 6.5 percent annually, achieving a total value level of $220 billion 129]. It is thus evident that the application of RCN in tire compounds can lead to a real commercial diffusion as well as to an important economic impact.

A tire can be defined as a toroidal high performance composite behaving as a flexible membrane, able to contain gases under pressure and with the following capabilities: load carrying, cushioning, road handling. Figure 9 reports the structure of a tire, with the name of constitutive parts.

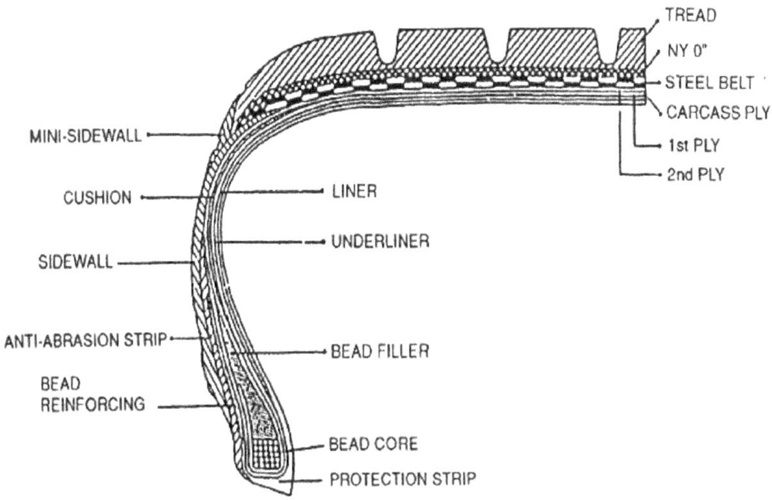

Figure 11. Structure of a tire

RCN have been so far essentially applied in three parts of a tire: tread, base (i.e. a compound immediately before the tread in contact with the road), liner. Table 4 explains the position in a tire of such compounds and clarifies their role.

RCN in tire compounds are used to achieve the following properties and performances: reduced weight, reduced energy dissipation, enhanced air retention and extension of the balance of the so called magic triangle performances for a tire tread: rolling resistance, traction, wear.

RCN in tire treads. By examining the patent literature [130-132], it appears that clays were used in tire tread compounds aiming at longer life of the compound and lower fuel consumption without negatively affecting the tread grip on the road at low temperature. Clays were used with or without an ammonium cation modifier, in the former case in the presence of a large amount of silica. Compound properties claimed in patents are: better ultimate properties, lower abrasion and an hysteresis that remains analogous at low temperature and is lower at medium – high temperature.

RCN in base compounds. OC were used in base compounds of tires commonly referred to as "HP" (High Performance) or "UHP" (Ultra high performance) tires, belonging to classes "V" or "Z", designed to experience extreme driving conditions, as they achieve maximum speeds in a range from 210 to 240 km/h and higher than 240km/h, respectively. In this type of tires, a low thickness base compound is used to favour performances such as tread block stability, road grip, steering stability, cornering stability and ride comfort. This base compound should have high tensile and dynamic-mechanical properties. OC were reported in the patent literature to give a remarkable improvement of the compound dynamic mechanical properties [133-136] and Pirelli Tire launched in 2007 P Zero tires for the HP and UHP segments, with a base compound containing an OC. In particular, it was reported that OC was able to give a much more isotropic behavior (i.e. an equal performances in longitudinal and lateral directions), with respect to traditional reinforcing fibers, such as the aramide ones.

Tyre compound	Position in the tyre	Performances
Tread	In contact with road	Dry, wet, ice and snow traction. Stability at high speed. Abrasion resistance. Protection of compounds below.
Base (tread cushion)	Below the tread. Between tread and nylon 0°	Low hysteresis. Good adhesion. Good fatigue, tear and durability. Compatibility
Liner or innerliner Thin layer of impermeable rubber.	Under the carcass	Impermeability to air and moisture. Prevention of degradation of tire structure (due to air and moisture). Good flex fatigue, crack, long term aging resistance.

Table 4. Tyre compounds explored for RCN application

RCN in innerliner. Most research efforts were dedicated to develop OC based innerliner compounds, with the aim to exploit the barrier property provided by clay platelets, trying in particular to achieve a high clay dispersion. In order to have the best barrier, the following technical solutions were adopted: use of clays with a high aspect ratio, control of the clay organization (stacks of exfoliated layers), use of reactive rubbers to promote the clay exfoliation, use of coating layers with a high impermeability.

A kaolin [137] and a mica [138] with high aspect ratio (at least 50) were used in butyl and in butyl, butadiene and natural rubber, respectively.

Stacks of OC were used in BR [139], in BIMS [140] and in SBR [141]. With an OC content as low as 9 phr, the oxygen transmission was reduced to about one fourth [139].

The clay exfoliation was promoted either by dispersing clays in the emulsion were polymerization of monomers such as isoprene and styrene was performed [142], or by performing the exchange reaction with a cationic polymer latex [143],or by mixing the clay (in a BIMS matrix) in the presence of a tertiary amine. In this latter case, with only 3 phr of clay, a reduction of 20% permeability was detected [144].

An amine terminated oligomer (e.g. a butadiene-acrylonitrile copolymer) was used as the reactive polymer to prepare a 25/75 clay/SBR blend and the oxygen transmission was reduced to less than one half [145]. A clay was mixed with a aqueous dispersion of a diene-based elastomer, with one or more functional groups such as acid or an anhydride [146].

OC was melt blended with poly(isobutylene-co-p-alkylstyrene) and poly(isobutylene-co-isoprene) elastomers and at a concentration of 10 phr the permeability was reduced to one half [147]. OC was melt blended with an halogenated copolymer of isobutylene and p-methyl-styrene and, at an OC level of 5 phr, the air retention was improved up to 30% [148]. Poly(isobutylene-co-N,N-dimethylvinylbenzylamine) copolymer was blended with an OC and an improvement of impermeability up to 70% was reported [149].

Barrier coating mixtures were prepared by mixing an elastomer (butyl rubber) an exfoliated clay with a high aspect ratio (preferably a vermiculite) and a surfactant, adopting a polymer/clay ratio from 1:1 to 20:1 and obtaining a permeability reduction of about 25%. This technology found a large commercial success in tennis ball and it will be discussed in the following paragraph [150].

11.2. Applications of RCN for sport balls

As mentioned above, a clay based barrier coating technology was developed [151-154], and applied to sport balls, with the aim to drastically improve the pressure retention without negatively affecting properties such as the bounce, the feel and the reproducibility of performances. A relatively thin coating layer (about 10-30 micron) was deposed, made by an elastomeric nanocomposite prepared by combining aqueous dispersions of a vermiculite (having a aspect ratio up to 1000) and butyl rubber. In a laboratory study [90] (Takahashi et al., 2006) a coating layer containing vermiculite (20 - 30 wt%) in butyl rubber led to a reduction of the diffusion coefficients by two orders of magnitude and of gas permeability by 20–30 fold. This technology was developed by InMat company and Air D-Fense products were commercialized.

Author details

Maurizio Galimberti
Politecnico di Milano, Dipartimento di Chimica,
Materiali e Ingegneria Chimica G. Natta, Milano, Italia

12. References

[1] Giannelis E.P., Krishnamoorti R., Manias E. (1999) Polymer-Silicate Nanocomposites: Model Systems for Confined Polymers and Polymer Brushes. Adv. Polym. Sci. 138: 107-147.

[2] Alexandre M., Dubois P. (2000) Polymer-layered silicate nanocomposites: preparation, properties and uses of a new class of materials. Materials Science and Engineering. 28: 1-63.

[3] Ray S.S., Okamoto M. (2003) Polymer/layered silicate nanocomposites: a review from preparation to processing. Progr. Polym. Sci. 28: 1539 - 1641.

[4] Gao F. (2004) Clay/polymer composites: the story. Materials Today 7(11): 50-55.

[5] Utracki L.A. (2004) Clay-Containing Polymeric Nanocomposites. Rapra Technology Ltd, Shawbury, Shrewsbury, Shropshire UK

[6] Ray S.S., Bousmina M. (2005). Biodegradable polymers and their layered silicate nanocomposites: In greening the 21st century materials world. Progress in Materials Science. 50: 962–1079

[7] Okada A., Usuki A. (2006). Twenty Years of Polymer-Clay Nanocomposites. Macromol. Mater. Eng. 291: 1449–1476

[8] Kotsilkova R. (2007) Thermoset nanocomposites for engineering applications. Ismithers Rapra Publishing, UK, 344.

[9] Utracki L.A., Sepehr M., Boccaleri E. (2007) Synthetic, layered nanoparticles for polymeric nanocomposites (PNCs), Polym. Adv. Technol., 18: 1-37.

[10] Chen B., Evans J.R.G., Greenwell H.C., Boulet P., Coveney P.V., Bowden A.A., Whiting A. (2008) Chem. Soc. Rev. 37: 568-594.

[11] Garces J.M., Moll D.J., Bicerano J., Fibiger R., McLeod D. G. (2000) Polymeric Nanocomposites for Automotive Applications. Adv. Mater., 12:1835-39

[12] Mittal V. (2010) Advances in Polyolefin Nanocomposites. CRC Press Taylor and Francis Group, London (UK).

[13] Kato M., Usuki A., Hasegawa N., Okamoto H., Kawasumi M. (2011) Development and applications of polyolefin– and rubber–clay nanocomposites. Polymer Journal 43: 583–593.

[14] Karger-Kocsis J., Wu C.M. (2004) Thermoset Rubber/Layered Silicate Nanocomposites. Status and Future Trends. Polym. Eng. Sci. 44: 1083-1093.

[15] Varghese J., Karger-Kocsis, J. (2004) Rubber World 4: 32-38.

[16] Bandyopadhyay A., Maiti M., Bhowmick A. K. (2006) Synthesis, characterisation and properties of clay and silica based rubber nanocomposites, Mater. Sci. Technol., 22(7): 818-828.

[17] Sengupta R., Chakraborty S., Bandyopadhyay S., Dasgupta S., Mukhopadhyay R., Auddy K., Deuri A.S. (2007) A Short Review on Rubber/Clay Nanocomposites With Emphasis on Mechanical Properties, Polym. Eng. Sci., 47(11): 1956-1974.

[18] Maiti M., Bhattacharya M., Bhowmick A.K. (2008) Elastomer Nanocomposites, Rubber Chemistry and Technology 81: 384-469.

[19] Sabu T., Ranimol S., (Eds.) (2010) Rubber Nanocomposites. Preparation, Properties and Applications. Wiley and Sons, New York. 705.

[20] Galimberti M. (2011) Rubber Clay Nanocomposites-Science, Technology and Applications, Wiley and Sons, New York.

[21] Mittal V., Kim J.J., Pal K., (Eds.) (2011) Recent Advances in Elastomeric Nanocomposites, Springer-Heidelberg Dordrecht London New York

[22] Bergaya F., Jaber M., Lambert J.-F. (2011) Clays and Clay Minerals. In: Galimberti M. (Ed.), Rubber Clay Nanocomposites - Science, Technology and Applications, Wiley and Sons, New York, Chapter 1: 3-44.

[23] http://www.bsi-global.com/en/Standards-and-Publications/Industry-Sectors/ Nano technologies/ PAS-71/Download-PAS-712005-Vocabulary-Nanoparticles
[24] Bergaya F., Jaber M., Lambert J.F. (2011) Organophilic Clay Minerals. In: Galimberti M. (Ed.), Rubber Clay Nanocomposites - Science, Technology and Applications, Wiley and Sons, New York, Chapter 2: 45-86.
[25] SCENIHR (Scientific Committee on Emerging and Newly-Identified Health Risks) 21-22 June 2007.
http: //ec.europa.eu/health/ph_risk/committees/04_scenihr/docs/scenihr_o_010.pdf
[26] Klug H.P., Alexander L.E. (1959) in *X-ray diffraction procedure* Wiley NY. Chapter.3 Fundamental Principles of X-ray Diffraction pp.120-174
[27] Klug H.P., Alexander L.E. (1959) in X-ray diffraction procedure Wiley NY, Chapter 9
[28] Wu Y., Huang H., Zhao W., Zhang H., Wang Y., Zhang L. (2008) Flame retardance of MMT/rubber composites. Journal of Applied Polymer Science. 107: 3318-3324
[29] Wu Y., Wang Y., Zhang H., Wang Y., Yu D., Zhang L., Yang J. (2005) Rubber–pristine clay nanocomposites prepared by co-coagulating rubber latex and clay aqueous suspension, Compos. Sci. Technol. 65: 1195-1202
[30] Ray S.; Bhowmick A. K.; Sarma K. S. S.; Majali A. B.; Tikku V. K. (2002) Characterization of electron-beam modified surface coated clay fillers and their influence on physical properties of rubbers. Radiation Physics And Chemistry. 65: 627-640
[31] Tian M.; Cheng L.; Liang W.; Zhang L. (2005) The anisotropy of fibrillar silicate/rubber nanocomposites, Macromolecular Materials and Engineering. 290(7): 681-687
[32] Tian M.; Cheng L.; Liang W.; Zhang L. (2006) Overall properties of fibrillar silicate/styrene-butadiene rubber nanocomposites. Journal of Applied Polymer Science. 101(5): 2725-2731
[33] Kader M. A.; Kim K.; Lee Y. S.; Nah C. (2006) Preparation and properties of nitrile rubber/montmorillonite nanocomposites via latex blending, J. Mater. Sci. 41: 7341-7352
[34] Wang Y.; Zhang L.; Tang C.; Yu D. (2000) Preparation and characterization of rubber-clay nanocomposites, J. Applied Polymer Science. 78: 1879-1883
[35] Zhang L.; Wang Y.; Wang Y.; Sui Y.; Yu D. (2000) Morphology and mechanical properties of clay/SBR nanocomposites, J. Applied Polymer Science. 78: 1873-1878
[36] Chakraborty S.; Sengupta R.; Dasgupta S.; Mukhopadhyay R.; Bandyopadhyay S.; Joshi M.; Ameta S. C. (2009) Synthesis and Characterization of Styrene Butadiene Rubber-Bentonite Clay Nanocomposites, Polymer Engineering and Science, 49(7): 1279-1290
[37] Kim W. S.; Lee D. H.; Kim I. J.; Son M. J.; Kim W.; Cho S. G. (2009) SBR/Organoclay Nanocomposites for the Application on Tire Tread Compounds, Macromolecular Research. 17(10): 776-784
[38] Wang Y.; Zhang H.; Wu Y.; Yang J.; Zhang L. (2005) Preparation, structure and properties of a novel Rectorite/Styrene Butadiene copolymer nanocomposite, J. Applied Polymer Science. 96: 324-328
[39] Wang, Z. F.; Wang, B.; Qi, N.; Zhang, H. F.; Zhang, L. Q. (2005) Influence of fillers on free volume and gas barrier properties in styrene-butadiene rubber studied by positrons, Polymer. 46: 719-724

[40] Rattanasom N.; Prasertsri S.; Ruangritnumchai T.; (2009) Comparison of the mechanical properties at similar hardness level of natural rubber filled with various reinforcing-fillers. Polymer Testing. 28: 8-12
[41] Valadares L. F.; Leite C.A.P.; Galembeck F. (2006) Preparation of natural rubber–montmorillonite nanocomposite in aqueous medium: evidence for polymer–platelet adhesion, Polymer. 47: 672–678
[42] Karger-Kocsis J.; Wu C.M. (2004) Thermoset Rubber/Layered Silicate Nanocomposites. Status and Future Trends, Polym. Eng. Sci., 44(6): 1083-1093
[43] Varghese S.; Karger-Kocsis J. (2003) Natural rubber-based nanocomposites by latex compounding with layered silicates, Polymer, 44: 4921-4927
[44] Varghese S.; Gatos K.G.; Apostolov A.A.; Karger-Kocsis J. (2004) Morphology and Mechanical Properties of Layered Silicate Reinforced Natural and Polyurethane Rubber Blends Produced by Latex Compounding, J. Appl. Polym. Sci., 92: 543-551
[45] Stephen R.; Ranganathaiah C.; Varghese S.; Joseph K.; Thomas S. (2006) Gas transport through nano and micro composites of natural rubber (NR) and their blends with carboxylated styrene butadiene rubber (XSBR) latex membranes, Polymer, 47(3): 858-870
[46] Sharif J.; Zin W.; Wan M.; Mohd D.; Khairul Z.H.; Ahmad M.H. (2005) Preparation and properties of radiation crosslinked natural rubber/clay nanocomposites, Polym. Testing, 24, 211-217
[47] Sun Y.; Wo Y.; Jia D. (2008) Preparation and properties of natural rubber nanocomposites with solid state organomodified montmorillonite, Journal of Applied Polymer Science. 107: 2786-2792
[48] Wu Y.; Huang H.; Zhao W.; Zhang H.; Wang Y.; Zhang L. (2008) Flame retardance of MMT/rubber composites, Journal of Applied Polymer Science, 107: 3318-3324
[49] Hernandez M.; Carretero-Gonzalez J.; Verdejo R.; Ezquerra T. A.; Lopez-Manchado M. A.. (2010) Molecular Dynamics of Natural Rubber/Layered Silicate Nanocomposites As Studied by Dielectric Relaxation Spectroscopy. Macromolecules. 43(2): 643-651
[50] Galimberti M.; Senatore S.; Lostritto A.; Giannini L.; Conzatti L.; Costa G.; Guerra G. (2009) Reinforcement of diene elastomers by organically modified layered silicates. e-Polymers. 57: 1-16
[51] Wang S.; Zhang Y.; Ren W.; Zhang Y.; Lin H. (2005) Morphology, mechanical and optical properties of transparent BR/clay nanocomposites, Polym. Testing. 24: 766-774
[52] Wang S.; Zhang Y.; Peng Z.; Zhang Y. (2005) New Method for Preparing Polybutadiene Rubber/Clay Composites, Journal of Applied Polymer Science. 98: 227-237
[53] Wang S.; Zhang Y.; Peng Z.; Zhang Y. (2006) Morphology and Thermal Stability of BR/Clay Composites prepared by a New Method, J. Appl. Polym. Sci. 99: 905-913
[54] Zhang H.; Wang Y.; Wu Y.; Zhang L.; Yang J. (2005) Study on Flammability of Montmorillonite/Styrene-Butadiene Rubber (SBR) Nanocomposites, J. Appl. Polym. Sci. 97: 844-849
[55] Sadhu S.; Bhowmick A. K. (2004) Preparation and properties of nanocomposites based on acrylonitrile-butadiene rubber, SBR and polybutadiene rubber, J. Polymer Science B. 42: 1573-1585

[56] Morlat-Therias S.; Mailhot B.; Gardette J.; Da Silva C.; Haidar B.; Vidal A. (2005) Photooxidation of ethylene-propylene-diene/montmorillonite nanocomposites, Polym. Degrad. Stab. 90: 78-85
[57] Galimberti M., Cipolletti V., Giudice S. (2011) Morphology of rubber clay nanocomposites. In: Galimberti M. (Ed.), Rubber Clay Nanocomposites - Science, Technology and Applications, Wiley and Sons, New York, Chapter 7: 181-240.
[58] Joly S.; Garnaud G.; Ollitrault, R.; Bokobza L. (2002) Organically modified layered silicates as reinforcing fillers for natural rubber, Chemistry Of Materials. 14: 4202-4208
[59] Sadhu S.; Bhowmick A. K. (2004) Preparation and characterization of SBR based nanocomposites and study of their mechanical properties, Advanced Engineering Materials. 6(9): 738-742
[60] Sadhu S.; Bhowmick A.K. (2005) Unique rheological behaviour of rubber based nanocomposites, J. Polymer Science B. 43: 1854-1864
[61] Sadhu S.; Bhowmick A.K. (2005) Morphology study of rubber based nanocomposites by transmission electron microscopy and atomic force microscopy, J. Mater. Sci. 40: 1633-1642
[62] Sadhu S.; Bhowmick A.K. (2005) Effect of nanoclay on the dynamic mechanical properties of styrene butadiene and acrylonitrile butadiene rubber vulcanizates, Rubber Chemistry & Technology.78: 321-335
[63] Maiti, M.; Sadhu, S.; Bhowmick, A. K. (2004) Effects of Heat and Pressure on Intercalation Structures of Isobutylene-Isoprene Rubber/Clay Nanocomposites. I. Prepared by Melt Blending, J. Polym. Sci. Part B: Polym. Phys. 42: 4489
[64] Galimberti M., Lostritto A., Spatola A., Guerra G., (2007) Clay delamination in hydrocarbon rubbers. Chemistry of Materials.19: 2495–2499
[65] Galimberti M., Senatore S., Conzatti L., Costa G., Giuliano G., Guerra G. (2009) Formation of clay intercalates with organic bilayers in hydrocarbon polymers. Polym. Adv. Technol. 20: 135-142
[66] Galimberti M., Giudice S., Cipolletti V., Guerra G. (2010) Control of Organoclay Structure in Hydrocarbon Polymers. Polymers for Advanced Technologies. 21: 679–684.
[67] Galimberti M., Riccio P., Giudice S., Citterio A., Riccò T., Pandini S., Ramorino G. (2009) Nanofillers for elastomers. Proceedings of the 176th Technical Meeting, Rubber Division of American Chemical Society, Pittsburgh, PA October 13-15.
[68] Austin J.R., Kontopoulou M. (2006) Effect of organoclay content on the rheology, morphology, and physical properties of polyolefin elastomers and their blends with polypropylene. Polym. Eng. Sci. 46: 1491-1501.
[69] Jeon H.S., Rameshwaram J.K., Kim G. (2004) Structure-property relationships in exfoliated polyisoprene/clay nanocomposites. J. Polym. Sci., Part B: Polym. Phys. 42: 1000.
[70] Jeon H.S., Rameshwaram J.K., Kim G., Weinkauf D.H. (2003) Characterization of polyisoprene-clay nanocomposites prepared by solution blending. Polymer 44: 5749-5758.
[71] Prasad R., Pasanovic-Zujo V., Gupta R.K., Cser F., Bhattacharya S.N. (2004) Morphology of EVA based nanocomposites under shear and extensional flow. Polym. Eng. Sci. 44: 1220-1230.

[72] Vu Y.T., Mark J.E., Pham L.H., Engelhardt M. (2001) Clay nanolayer reinforcement of cis-1,4-polyisoprene and epoxidized natural rubber. J. Appl. Polym. Sci., 82: 1391-1403.
[73] Tian M., Qu C., Feng Y., Zhang L. (2003) Structure and properties of fibrillar silicate/SBR composites by direct blend process. J. Mater. Sci. 38: 4917-4924.
[74] Maiti M., Sadhu S., Bhowmick A.K. (2004) Brominated poly(isobutylene-co-para-methylstyrene) (BIMS)-clay nanocomposites: Synthesis and characterization. Journal of Polymer Science Part B: Polymer Physics 42: 4489-4502.
[75] Kader M.A., Lyu M.Y., Nah C. (2006) A study on melt processing and thermal properties of fluoroelastomer nanocomposites. Compos. Sci. Technol. 66: 1431-1443.
[76] Kim J.T., Oh T.S., Lee D.H. (2003) Morphology and rheological properties of nanocomposites based on nitrile rubber and organophilic layered silicates. Polymer International 52, 1203-1208.
[77] Wang X., Gao Y., Mao K., Xue G., Chen T., Zhu J., Li B., Sun P., Jin Q., Ding D., Shi A.C., (2006) Unusual Rheological Behavior of Liquid Polybutadiene Rubber/Clay Nanocomposite Gels: The Role of Polymer-Clay Interaction, Clay Exfoliation, and Clay Orientation and Disorientation. Macromolecules 39, 6653-6660.
[78] Lim S.K., Kim J.W., Chin I.J, Choi H.J. (2002) Rheological properties of a new rubbery nanocomposite: Polyepichlorohydrin/organoclay nanocomposites. J. Appl. Polym. Sci. 86: 3735-3739.
[79] Maiti M., Bhattacharya M., Bhowmick A.K. (2008) Elastomer Nanocomposites, Rubber Chemistry and Technology. 81:, 384-469.
[80] Verdejo R., Hernandez M., Bitinis N., Kenny J.M., Lopez-Manchado M.A. (2011) Vulcanization Characteristics and Curing Kinetic of Rubber/Organoclay Nanocomposites in: Galimberti M. (Ed.), Rubber Clay Nanocomposites - Science, Technology and Applications, Wiley and Sons, New York. Chapter 9: 275-304.
[81] Avalos F., Ortiz J.C., Zitzumbo R., López-Manchado M.A., Verdejo R., Arroyo M. (2008) Effect of montmorillonite intercalant structure on the cure parameters of natural rubber. European Polymer Journal. 44: 3108–3115.
[82] Lopez-Manchado M.A., Arroyo M., Herrero B., Biagiotti J. (2003) Preparation and characterization of organoclay nanocomposites based on natural rubber. Polymer International. 52: 1070-1077.
[83] Lopez-Manchado M.A., Arroyo M., Herrero B., Biagiotti J. (2003) Vulcanization kinetics of natural rubber-organoclay nanocomposites. Journal of Applied Polymer Science. 89: 1–15.
[84] Magaraphan R., Thaijaroen W., Ochakun R.L., (2003) Structure and properties of natural rubber and modified montmorillonite nanocomposites. Rubber. Chem. Technology. 76: 406- 418.
[85] Teh P.L., Mohd Ishak Z.A., Hashim A.S., Karger-Kocsis J., (2006) Physical properties of natural rubber/organoclay nanocomposites compatibilized with epoxidized natural rubber. Journal of Applied Polymer Science. 100: 1083–1092.
[86] Varghese J., Karger-Kocsis J. (2004) Journal of Applied Polymer Science. 91: 813–819.
[87] Mousa A., Karger-Kocsis J. (2001) Rheological and thermodynamical behavior of styrene/butadiene rubber-organoclay nanocomposites. Macromol. Mater. Eng. 286: 260–266.

[88] Lopez-Manchado M.A., Arroyo M., Herrero B. (2004) Organoclay-natural rubber nanocomposites synthesized by mechanical and solution mixing methods. Polymer International. 53: 1766-1772.
[89] Giannini L., Citterio A., Galimberti M. (2011) Chemistry of rubber/organoclay nanocomposites, in: Galimberti M. (Ed.), Rubber Clay Nanocomposites - Science, Technology and Applications, Wiley and Sons, New York, Chapter 5: 127-146.
[90] Takahashi S., Goldberg H.A., Feeney C.A., Karim D.P., Farrell M., O'Leary K., Paul D.R., (2006) Gas Barrier properties of butyl rubber /vermiculite nanocomposite coatings. Polymer. 47: 3083–93.
[91] Paul D.R., Robeson L.M. (2008) Polymer nanotechnology: Nanocomposites. Polymer. 49: 3187–3204.
[92] Jacob A., Kurian P., Aprem A.S. (2008) Transport properties of natural rubber latex layered clay nanocomposites. Journal of Applied Polymer Science. 108: 2623-2629.
[93] Li P., Wang L., Song G., Yin L., Qi F., Sun L. (2008) Characterization of high-performance exfoliated natural rubber/organoclay nanocomposites. Journal of Applied Polymer Science. 109: 3831-3838.
[94] Jacob A., Kurian P., Aprem A. S. (2007) Cure characteristics and mechanical properties of natural rubber-layered silicate nanocomposites. International Journal of Polymeric Materials. 56: 593-604.
[95] Gu Z., Song G., Liu W., Wang B., Li J. (2009) Preparation and properties of organo-montmorillonite/cis-1,4-polybutadiene rubber nanocomposites by solution intercalation. Applied Clay Science. 45: 50-53.
[96] Hwang W. G., Wei K.H., Wu C.M. (2004) Mechanical, thermal and barrier properties of NBR/organosilicate nanocomposites. Polymer Engineering and Science. 44: 2117-2124.
[97] Liang Y.R., Cao W. L., Zhang X. B., Tan Y. J., He S. J., Zhang L. Q., 2009. Preparation and properties of nanocomposites based on different polarities of nitrile-butadiene rubber with clay. Journal of Applied Polymer Science 112: 3087-3094
[98] Donnet J.B., Custodero E., (2005) in The Science and Technology of Rubber Third Ed.; Mark, J.E.; Erman, B.; Eirich, F.R. Eds. Elsevier Academic Press Chapter 8: 367-400
[99] Wu Y., Jia Q., Yu D., Zhang L., (2004) Modeling Young's modulus of rubber–clay nanocomposites using composite theories. Polymer Testing 23 (2004) 903–909
[100] Schön F., Gronski W. (2003) Filler networking of silica and organoclay in rubber composites: reinforcement and dynamic mechanical properties. Kautschuk und Gummi Kunststoffe. 56: 166-171.
[101] Ramorino G., Bignotti F., Pandini S., Riccò T. (2009) Mechanical reinforcement in natural rubber/organoclay nanocomposites. Composites Science and Technology. 69: 1206-1211.
[102] Huber G., Vilgis T.A. (1999) Universal properties of filled rubbers: mechanisms for reinforcement on different length scales. Kautschuk und Gummi Kunststoffe. 52: 102-107.
[103] Payne A.R. (1962) The dynamic properties of carbon black loaded natural rubber vulcanizates. Part II. Journal of Applied Polymer. Scence. 21: 368-372
[104] Ramorino G., Bignotti F., Conzatti L., Riccò T. (2007) Dynamic and viscoelastic behavior of natural rubber/layered silicate nanocomposites obtained by melt blending. Polymer Engineering and Science. 47: 1650-1657.

[105] Osman M.A., Ploetze M., Skrabal P. (2004) Structure and properties of alkylammonium monolayers self-assembled on montmorillonite platelets. Journal of Physics Chemistry B. 108: 2580-2588.

[106] Hrachová J., Komadel P., Chodák I. (2008) Effect of montmorillonite modification on mechanical properties of vulcanized natural rubber composites. Journal of Materials Science. 43: 2012-2017.

[107] Kim M.S., Kim G.H., Chowdhury S.R., (2007) Polybutadiene rubber/organoclay nanocomposites: effect of organoclay with various modifier concentrations on the vulcanization behavior and mechanical properties. Polymer Engineering Science. 47: 308-313.

[108] Hasegawa N., Okamoto H., Usuki A. (2004) Preparation and properties of ethylene propylene rubber (EPR)-clay nanocomposites based on maleic-anhydride modified EPR and organophilic clay. Journal of Applied Polymer Science. 93: 758-764.

[109] Teh P.L., Mohd Ishak Z.A., Hashim A.S., Karger-Kocsis J. (2004) Effects of epoxidized natural rubber as a compatibilizer in melt compounded natural–organoclay nanocomposites. European Polymer Journal. 40: 2513-2521

[110] Bhowmick A.K., Ganguly A., Maiti M. (2006) Factors influencing the morphology and the properties of clay-rubber nanocomposites. Kautschuk und Gummi Kunststoffe. 9: 437-446.

[111] Liang Y.R., Cao W.L., Zhang X.B., Tan Y.J., He S. J., Zhang L.Q. (2009) Preparation and properties of nanocomposites based on different polarities of nitrile-butadiene rubber with clay. Journal of Applied Polymer Science. 112: 3087-3094.

[112] Arroyo M., Lopez-Manchado M.A., Herrero B. (2003) Organo-montmorillonite as substitute of carbon black in natural rubber compounds. Polymer. 44: 2447 –2453.

[113] Jurkowska B., Jurkowski B., Oczkowski M., Pesetskii S.S., Koval V., Olkhov Y.A. (2007) Properties of montmorillonite-containing natural rubber. J. Appl. Polym. Sci. 106: 360-371.

[114] Kim M.S., Kim D.W., Chowdhury S.R., Kim G.H. (2006) Melt-compounded butadiene rubber nanocomposites with improved mechanical properties and abrasion resistance. J. Appl. Polym. Sci. 102: 2062-2066.

[115] Kojima Y., Usuki A., Kawasumi M., Okada A., Fukushima Y., Kurauchi T., Kamigaito O. (1993) Mechanical properties of nylon 6–clay hybrid. Journal of Materials Research 8: 1185–1189.

[116] Bokobza L., Leroy E., Lalanne V. (2009) Effect of filling mixtures of sepiolite and a surface modified fumed silica on the mechanical and swelling behavior of a styrene-butadiene rubber. Eur. Polym. J. 45: 996-1001.

[117] Maiti M., Sadhu S., Bhowmick A.K. (2005) Effect of Carbon Black on Properties of Rubber Nanocomposites. Journal of Applied Polymer Science 96: 443-451.

[118] Praaven S., Chattopadhyay P.K., Albert P., Dalvi V.G., Chakraborty B.C., Chattopadhyay S. (2009) Synergistic effect of carbon black and nanoclay fillers in styrene butadiene rubber matrix: development of dual structure. Composites: Part A. 40: 309-316.

[119] Gopi J.A., Patel S.K., Chandra A.K., Tripathy D.K. (2011) SBR-clay-carbon black hybrid nanocomposites for tire tread application. Journal of Polymer Research. 18: 1625–1634.

[120] Galimberti M., Giannini L., D'Oria F., Lostritto A., Ponta T., Riccio P., Senatore S. (2008) Tailor made nanostructures for improved properties of tyre compounds. Proceedings from 174th Technical Meeting, Rubber Division of American Chemical Society, Louisville, KY (USA), October 14-16,
[121] Galimberti M., Giudice S., Coombs M., Giannini L., Buttafava A., Dondi D., Faucitano A., Pandini S., Riccò T., Bracco S., Cattaneo A., Sozzani, P., Conzatti, L., Cipolletti, V., 2010. Filler matrix interface characteristics and mechanical reinforcement in elastomeric nanocomposites. Proceedings of the Fall 178th Technical Meeting of the Rubber Division of the American Chemical Society, Inc. Milwaukee (WI) October 12-14.
[122] Cataldo F. (2007) Preparation and properties of nanostructured rubber composites with montmorillonite Macromolecular Symposia. 247: 67-77.
[123] Sridhar V., Shanmugharaj A.M., Kim J. K., Tripathy D.K. (2009) Optimization of Carbon Black and Nanoclay Filler Loading in Chlorobutyl Vulcanizates Using Response Surface Methodology. Polymer Composites. 30(6): 691-701
[124] Chattopadhyay P.K., Das N.C., Chattopadhyay S. (2011). Influence of interfacial roughness and the hybrid filler microstructures on the properties of ternary elastomeric composites. Composites: Part A. 1049-1059
[125] Malas A., Das C.K. (2012) Carbon black–clay hybrid nanocomposites based upon EPDM elastomer. Journal of Material Science. 47: 2016–2024.
[126] Galimberti M., Coombs M., Cipolletti V., Giannini L., Conzatti L., Riccò T., Mauro M., Guerra G. (2011) Nano and nanostructured fillers and their synergistic behavior in rubber composites such as tyres. Proceedings of the Fall 180th Technical Meeting of the Rubber Division of the American Chemical Society, Inc. Cleveland (OH) October 11-13.
[127] M. Galimberti, M. Coombs, V. Cipolletti, P. Riccio, T. Riccò, S. Pandini, L. Conzatti (2012) Accepted for publication on Applied Clay Science
[128] source: International Rubber Study Group
[129] source: The Freedonia Group
[130] Heinrich G., Herrmann W., Kendziorra N., Pietag T., Recker C. (2002). Rubber compound for tyre tread rubber. US 6818693 B2 to Continental
[131] Romani F., De Cancellis P. (2006) Studded Tyre WO 2006/002672 to Pirelli Tyre
[132] Zanzig, D.J., Yang X., Cohen M.P. (2010). Tyre with component comprised of a blend of polybutadiene rubber and composite of styrene/butadiene elastomer which contains exfoliated clay platelets. US 7,714,055 B2 to Goodyear.
[133] Galimberti M., Fino L., Verona M. (2003) Tyre for vehicle wheels with tread band of cap and base construction. WO 05/002883 to Pirelli Tyre
[134] Galimberti M., Fino L., Dessanti F. (2002) Tyre for two-wheeled vehicles. EP1597093 to Pirelli Tyre
[135] Giannini L., Fino L., Galimberti M. (2004) Tyre and crosslinkable elastomeric composition. WO 05/042278 to Pirelli Tyre
[136] Giannini L., Fino L., Galimberti M., Bizzi S. (2003) High performance tyre for vehicle wheels. EP 1677996 B1 to Pirelli Tyre
[137] Wada I., Ohara M., Uchino O., Koura T., Ezawa N., Kanoh M., Nakamura S. (2006) Rubber composition for inner liner. US 7.019.063 B2 to Bridgestone Corporation.

[138] Miyazaki T. (2006) Rubber composition for inner liner. EP 1726620 A1 to Sumitomo Rubber Ind.

[139] Elspass W.C., Peiffer G.D., Kresge E.N., Wright P.J., Wang H.C. (1998) Tactoidal elastomer nanocomposites. US 5,807,629 to Exxon Mobil

[140] Elspass C.W., Peiffer D.G. (2000) Nanocomposite materials formed from inorganic layered materials dispersed in a polymer matrix. US 6,034,164 to Exxon Research and Engineering Company.

[141] Ishida, K., Fujiki, K., 2003. WO 03/087214 A1 to Bridgestone

[142] Elspass C.W., Peiffer D.G., Kresge E.N., Hsieh D.T., Chludzinski J.J., Liang K.S., (1999) Nanocomposite materials. US 5,883,173 to Exxon Research and Engineering Company

[143] Parker D.K., Larson B.K.F., Yang X. (2003) Preparation and use of a nanocomposite of elastomer and exfoliated clay platelets formed in situ within an elastomer host and articles of manufacture, including tires, having at least one component comprised thereof. EP 1321489 A1 to Goodyear Tyre and Rubber Company.

[144] Dias A.J., Tsou A. H., Chung D.Y.L. Weng W. (2005). Low permeability nanocomposites. US 2005/0027058 to Exxonmobil.

[145] Kresge E.N., Lohse D.J. (1997) Tire inner-liners comprising a solid rubber and a complex of a reactive rubber and layered silicate clay. US 5,665,183 to Exxonmobil

[146] Ajbani M., Geiser J.F., Parker D.K. (2003) Nanocomposite of elastomer and dispersion therein of intercalated clay prepared in an aqueous medium with funcional elastomer. US 2003/0144401 A1 to Goodyear Tyre and Rubber Company

[147] Gong C., Dias A.J., Tsou A. H., Poole B.J., Karp K.R. (2004) Functionalized Elastomer nanocomposite. WO 04/005387 to Exxonmobil.

[148] Grah M.D. (2004) Tire with improved inner liner. US 2004/0194863 A1 to Michelin

[149] Wang X., Fudemoto H., Hall J., Araki S., Hogan T., Foltz V., Sadhukhan P., Bohom, G.G.A. (2004) Method for clay exfoliation, compositions therefore, and modified rubber containing same. WO 058874 A1 to Bridgestone Corporation.

[150] Feeney C.A., Balzer, R.J. (1998) Barrier coating of an elastomer and a dispersed layered filler in a liquid carrier and coated compositions, particularly tires. WO 98/56598 to Michelin and Herberts Gmbh.

[151] Feeney C. A., Goldberg,H. A., Farrell M., Karim D. P., Oree K. R. (2006) US Patent 7,078453 to InMat Inc

[152] Feeney C. A., Goldberg H. A., Farrell M., Karim D. P., Oree K. R. (2006). US Patent 7,119,138 to InMat Inc

[153] Feeney C.A., Farrell M., Tannert K., Goldberg H.A., Lu M., Grah M.D. (2001) US Patent 6,232,389. to InMat and Michelin Recherche et technique

[154] Feeney C.A., Farrell M., Tannert K., Goldberg H.A., Lu M., Grah M.D. (2000) US Patent 6,087,016 to InMat and Michelin Recherche et technique

Section 4

Smart Elastomers

Chapter 6

Microstructure and Properties of Magnetorheological Elastomers

Anna Boczkowska and Stefan Awietjan

Additional information is available at the end of the chapter

1. Introduction

Magnetorheological elastomers (MREs) belong to the new group of the functional materials called "smart". Although smart materials are known since long time, their intensive development started in the end of the XXth century. The term *smart materials, intelligent materials* or less frequently used *adaptive materials* or *multifunctional materials,* was introduced in the eighties of the twentieth century, when some materials, which were included in the group were already known. Till today there is no accepted universal definition of smart material, it is also not included in the encyclopedia devoted to these materials, published in 2002 [1, 2].

The term smart material generally refers to material which changes its properties under the influence of various external stimuli [3]. However, it seems that this approach is too simplistic. Much more precise description of smart material is given by Takagi in his work [4]. According to him, smart material is capable of reacting to external stimuli by changing its material properties for the desired way and effectively responds to these stimuli. Such material should therefore be some kind of sensor, processor, and actuator. These attributes should be looped, and the effect of changing the properties of the material should be done in real time.

Among smart materials, magnetorheological materials (MR) are an important group. They are a class of materials with rheological properties rapidly varied by the application of a magnetic field (Fig. 1). The change in their properties is in the proportion to the magnitude of the magnetic field applied and is immediately reversible. The Bingham plastic model is often used to model the behaviour of MR materials [1].

Research focussing on magnetorheological materials was initiated by Jacob Rabinow at the National Bureau of Standards, now the National Institute for Science and Technology, in the

United States in the 1940s and early 1950s. His early works on magnetorheological fluids led to a host of devices and products based on dry magnetic powders, for example the magnetic, powder brake. In addition to MR fluids, the MR materials also contain magnetic field responsive gels [5], foams, powders, and elastomers. The comparison of the most typical magnetorheological materials is shown in Table 1.

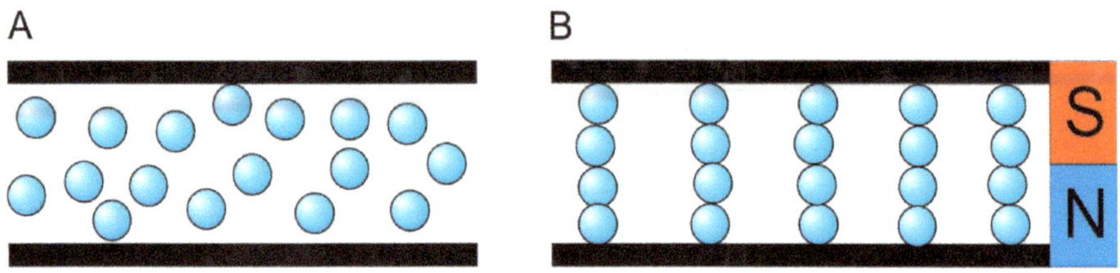

Figure 1. Magnetorheological material A- before, B – and after the application of an external magnetic field.

	Magnetorheological fluid MRF	**Ferrofluid**	**Magnetorheological elastomer MRE**
Particles type	iron	magnetite	iron
Particles size	0,1 – 10 μm	2 –10 nm	10 – 50 μm
Carrier	oils	oils, water	elastomers
Volume fraction	0,1 – 0,5	0,02 – 0,2	0,1 – 0,5
Additives	surfactants, thicsotropic agents	surfactants	none
Viscosity without external magnetic field [mPa·s]	100 - 1000	2 - 200	none
Change of properties in an external magnetic field	Yield stress τ_y up to ~100 kPa	Relative viscosity $\Delta\eta/\eta$ ~1	Storage modulus $\Delta G'$ ~20 kPa

Table 1. Composition and properties of typical magnetorheological materials [6].

The interest in magnetorheological elastomers (MREs) has recently increased because of their prospects for application in various smart systems. They are still much less known than magnetorheological fluids. The MREs are smart materials, analogues of the magnetorheological fluids (MRFs), in which the fluid component is replaced by a cross-linked material like rubber or silicone. They consist of micron sized magnetically permeable particles in non-magnetic matrix. In a similar way, as in the case of MRFs, the particles tend to align themselves in the direction of the magnetic field, but after curing of the matrix such microstructure is fixed. Magnetorheological materials change their rheological properties under the influence of an external magnetic field. Their rheological properties can be changed continuously, rapidly and reversibly by changing an external magnetic field. This behaviour is described by the magnetorheological effect.

In this chapter the overview on magnetorheological elastomers is given, as well as basic knowledge about rheology and magneto-rhelogy. Also the effect of the amount, size and orientation of the particles on the microstructure and properties of MREs is discussed, as well as the influence of magnetic field on the compressive characteristics and rheological properties of developed MREs.

2. Magnetorheological elastomers

The MREs consist of magnetically permeable particles (such as iron or other ferromagnetic particles) added to a viscoelastic polymeric material prior to crosslinking. The MREs contain ferromagnetic particles having sizes from few to few hundreds of μm. Pure iron has the highest saturation magnetization of known elements and it has also high permeability and low remanent magnetization, providing high, short-term inter-particle attraction. It is known from available literature that the amount and the shape of the particles, as well as the type of the polymer matrix influence the MREs properties.

MREs can be classified according to several parameters like: particles type, matrix, structure and distribution of particles [7].

MREs classification:

1. Particles magnetic properties
    a. Soft magnetic particles
    b. Hard magnetic particles
    c. Magnetostrictive particles
    d. Magnetic shape-memory particles
2. Matrix structure
    a. Solid matrix
    b. Porous matrix
3. Matrix electrical properties
    a. Isolating matrix
    b. Conductive matrix
4. Distribution of particles
    a. Isotropic
    b. Anisotropic

Usually magnetic field is applied to the polymer composites during crosslinking of the matrix. Such treatment locks the columnar particle chain structures during the final cure giving special anisotropic properties. The formation of columnar particle structures within the elastomers corresponds to a low dipolar energy state. Shearing of the cured composite under applied magnetic field requires additional energy because of the particles displacement from this low energy state. The amount of work required and the field dependent shear modulus increase steadily with the applied magnetic field [8, 9, 10].

Changes of the properties in MREs under the influence of the magnetic field depend strongly on the microstructure formed during the curing of elastomer matrix, also as a result of magnetic field. Interactions between particles in a magnetic field bring them closer, resulting in increased stiffness of the material. This changes the stress-strain curves [11]. The magnetic field causes a shift towards higher stresses. The same mechanism is responsible for the effect of "magnetostriction", shortening of length of the cylindrical sample in the presence of an external magnetic field [12], as schematically shown in Figure 2A, or its extension, the MRE when the particles are distributed isotropically [13]. Under the influence of the field the particles move in the direction of the field, as shown in Figure 2B. This effect is possible when the adhesion between the particles and the matrix is large enough so that the movement of particles leads to deformation of the elastomeric matrix, which is has such low stiffness that such deformations can occur. A similar phenomenon occurs in the MREs, in which the particles are spherical and have an elongated shape [14]. Then, under the influence of the magnetic field they become dipoles, rotating in the direction of the magnetic field vector, which also leads to deformation of the elastomer and thus increase in the length of the sample (Fig. 2C). In all cases, these changes are rapid and fully reversible.

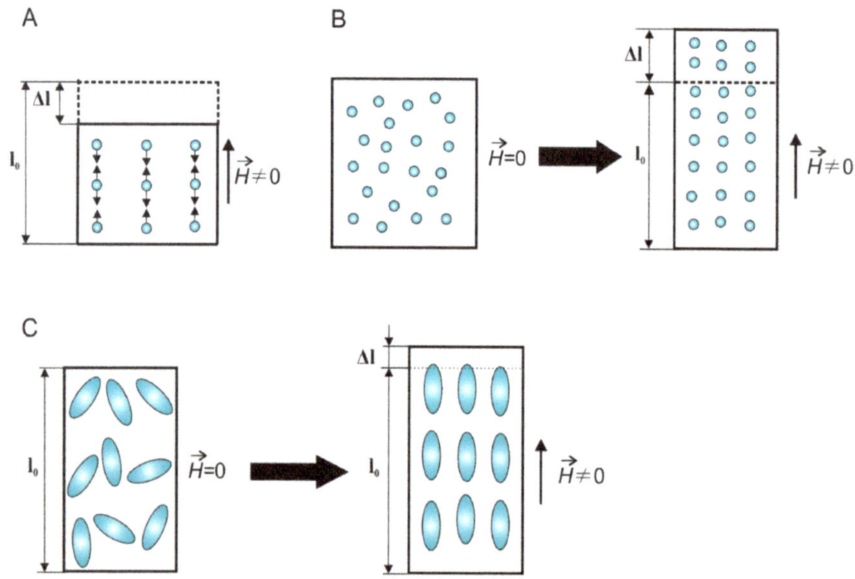

Figure 2. Magnetostriction effect in MREs under external magnetic field.

Under the influence of magnetic field rheological properties of MREs change [15], mainly shear modulus, but also the modulus of elasticity determined in a compression test [11]. These properties are strongly dependent on a magnetic field strength. Shearing of MREs in the presence of a magnetic field displaces the particles from the position of minimum energy state, which requires additional work increasing monotonically with increasing magnetic field strength. Therefore, shear modulus depends on the field and is a characteristic feature of the MREs [10, 16]. Changes of the module in the magnetic field also depend on the content of ferromagnetic particles in the elastomer matrix.

The vast majority of MREs described in the literature contain soft magnetic particles, mainly iron, cobalt and their oxides [17,18,19,20]. There is a group of magnetorheological elastomers with hard magnetic fillers like $BaFe_{12}O_{19}$ or $SrFe_{12}O_{19}$ [21]. These materials, like permanent magnets remain magnetized after turning off the external magnetic field [22,23]. For the fabrication of MREs also magnetostrictive particles are used, usually Terfenol D - a material with giant magnetostriction [24,25] and the particles with magnetic shape memory (MSM), such as Ni-Mn-Ga [26,27,28].

The matrix of magnetorheological elastomers can be solid or porous. MREs with porous matrix, also called magnetorheological foams, have foamed matrix in order to increase the abilities to change their properties. Majority of magnetorheological elastomers have a matrix with electrically insulating properties. In some cases magnetorheological elastomers, which were doped with electrically conductive particles such as graphite or silver and their percolation threshold has been reached, become electric conductors [29]. Conductive polymers such as polypyrole, polyacetylene, polyaniline can also be used for matrix [30,31]. However, using a conductive matrix has no significant effect on changes in the rheological or mechanical properties in a magnetic field, and raises costs.

The spatial distribution of the particles is determined during curing process. In the presence of a magnetic field during curing, the obtained elastomer has anisotropic, oriented structure, consisting of chains of particles [10,6,7,8,32,33,34,35,36] as schematically shown in Figure 3B. Magnetorheological elastomer cured without the magnetic field allows to receive isotropic structure, which is shown on a schematic diagram in Figure 3A [11,20,21,37,38,39,40,41,42].

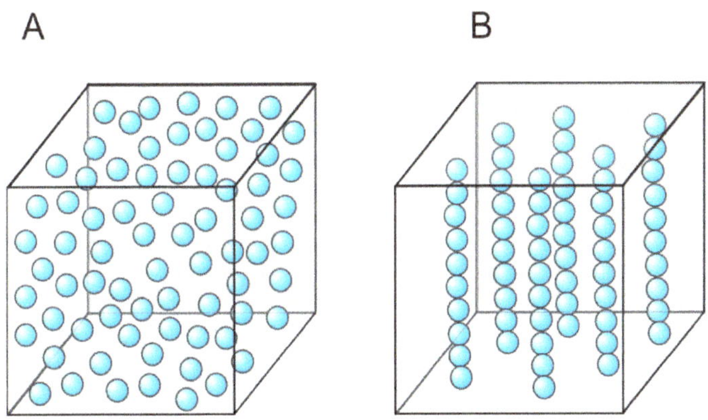

Figure 3. Schematic microstructure of MRE: A - isotropic, B - anisotropic spatial distribution of particles.

The distribution of particles in MREs microstructure is influenced by many factors, such as the magnetic interaction forces between particles, orientation and magnetic field strength, sample size, the volume fraction of particles, temperature. Understanding the influence of all these factors on the MREs microstructure is important, but extremely difficult for both experimental and computational methods.

3. The rheological properties and magnetorheological effect of magnetorheological elastomers

Many materials, especially polymers and their composites are characterized by viscoelastic properties [43]. This means that they combine the features of elastic solids and viscous liquids, as schematically shown in Figure 4. Their behavior is between the ideal solid described by Hooke's law, in which the stress is always directly proportional to the strain and is independent of strain rate, and a viscous liquid, in which according to Newton's law, stress is always directly proportional to the strain rate and does not depend on the strain. Viscoelastic materials under rapid deformation behave more like elastic body, and under very slow - as viscous liquid. Rheology describes the flow and deformation of solids and liquids under the influence of an external forces.

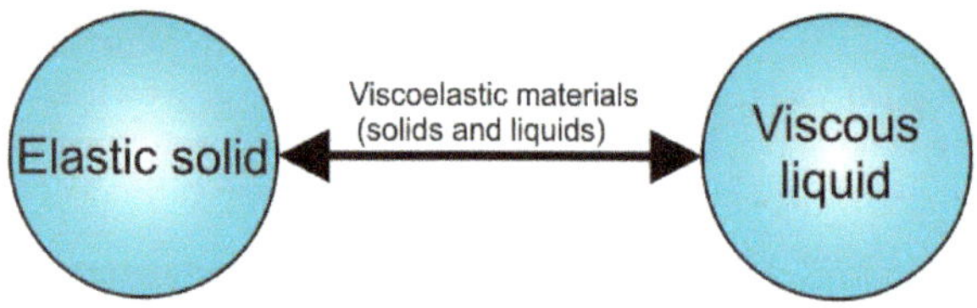

Figure 4. Viscoelastic properties of materials.

For a perfectly elastic solid when the force is applied, strain occurs immediately, and it is linearly proportional to the applied force. The ratio of stress to strain is a measure of elasticity of the material. After unloading deformed body immediately returns to its initial state. The applied force can cause the shear stress (τ) and shear modulus (G). G modulus determines the resistance of the solid to deformation and is expressed by the ratio of shear stress (τ) to the shear strain (γ):

$$G = \frac{\tau}{\gamma}. \quad (1)$$

For an elastic solid, both stress and strain are independent on time. For the viscoelastic solid, rheological parameters are dependent on the time and described by Kelvin-Voight model for linear viscoelasticity:

$$\tau = G \cdot \gamma + \eta \cdot \frac{d\gamma}{dt}, \quad (2)$$

where:

η – dynamic viscosity,

t – time.

Deformed material can undergo relaxation when the applied force is maintained, which results in a decrease in the stress in time, until its complete disappeared, as schematically shown in Figure 5. When the force is removed, the disappearance of deformation is delayed. This delay is given by the relaxation time λ:

$$\lambda = \frac{\eta}{G} \tag{3}$$

In the area of linear viscoelasticity delay times during creep and recovery are the same. Usually solids are more complicated and to describe the viscoelastic behavior during the creep and recovery it is necessary to use the whole spectrum of relaxation times. In most testing methods of viscoelastic materials, instead of constant stress, dynamic strain measurements in the form of an oscillating sinusoidal function of time is used (Fig. 6):

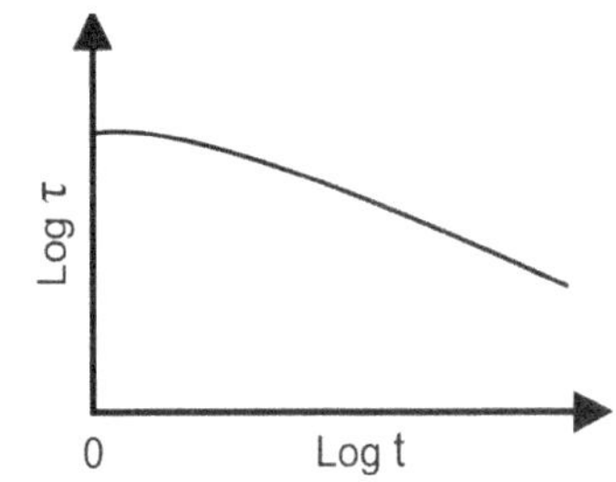

Figure 5. Stress relaxation (τ) at time (t).

$$\tau = \tau_0 \cdot \sin(\omega \cdot t) \tag{4}$$

where: τ_0 - applied stress,
ω - angular velocity [1/s or rad/s],

$$\omega = 2\pi \cdot f, \text{ f} - \text{frequency} \left[\text{Hz}\right]. \tag{5}$$

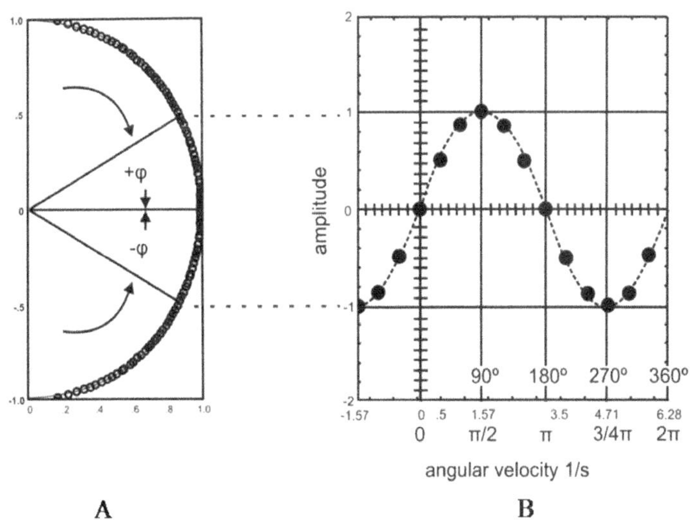

Figure 6. Dynamic measurements: a - deformation of the angle φ, b - sinusoidal strain or stress.

The course of deformation and the induced stress in the perfectly elastic solid is in-phase, as shown schematically in Figure 7. However, in the viscoelastic material phase shift angle δ occurs in the range 0 - 90°, between the stress and strain that induces it (Fig. 8). It is the result of delayed reaction of the material to an applied strain. The response to an applied strain consistent with the phase is called the elastic (material entirely elastic - behaves according to Hooke's law), shifted in phase by 90° - viscous (material entirely viscous - behaves like a Newtonian fluid), and between 0 and 90° - viscoelastic (viscoelastic material).

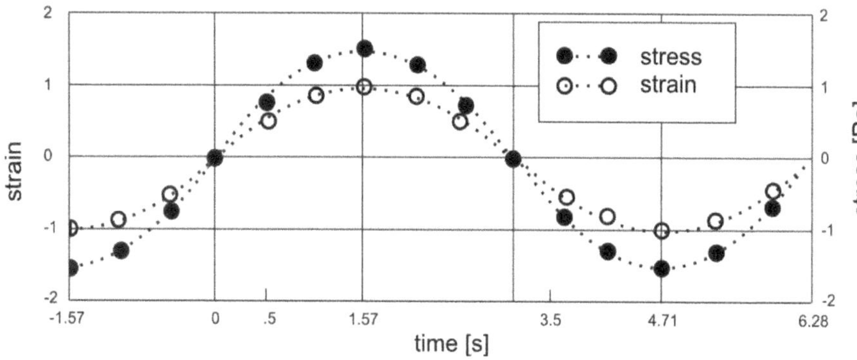

Figure 7. Dynamic measurement: strain and stress in an elastic solid.

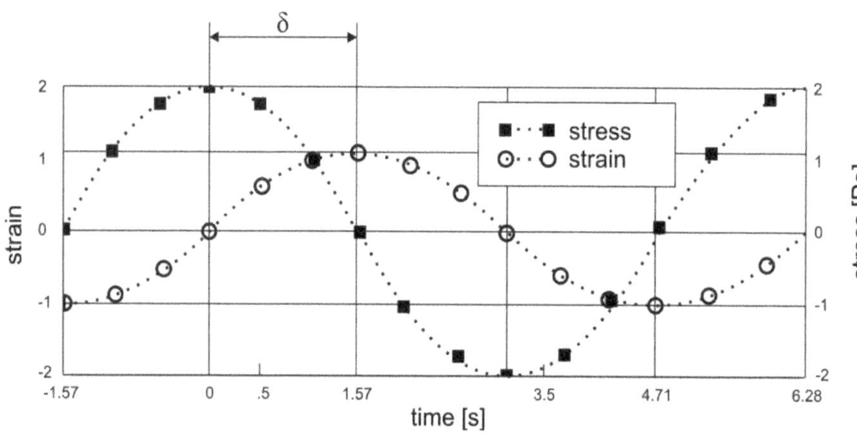

Figure 8. Dynamic measurement: strain and stress in the viscoelastic body, δ - phase shift angle.

To combine the elastic and viscous properties the complex shear modulus G* has been introduced. It is the total resistance of the solid to the applied strain and is defined as:

$$G^* = \frac{\tau_0}{\gamma_0}. \tag{6}$$

For viscoelastic materials, both complex modulus and phase angle is frequency dependent. For easy distinction between elastic and viscous behavior of the material complex numbers were introduced, so that the complex shear modulus can be described as follows:

$$G^{*} = G^{'} + iG^{''}, \tag{7}$$

where:
G'- storage modulus, or elastic (the real part of complex modulus G*),
G''- loss modulus, or viscous (imaginary part of complex modulus G*).

The term storage modulus indicates that the strain energy is temporarily stored during deformation, and can be later recovered, and the loss modulus term means that the energy is irretrievably lost and converted to shear heat. Modules G' and G'' are described by the following relations:

$$G^{'} = G^{*} \cdot \cos\delta = \frac{\tau_0}{\gamma_0} \cdot \cos\delta, \tag{8}$$

$$G^{''} = G^{*} \cdot \sin\delta = \frac{\tau_0}{\gamma_0} \cdot \sin\delta. \tag{9}$$

Storage modulus G' and loss modulus G'' are parameters describing the rheological properties of viscoelastic polymeric materials. Storage modulus G' defined by the ratio of elastic stress to strain, describes the amount of stored (saved) energy during shear, and refers to the elastic properties of the material. Loss modulus G'' expressed by the ratio of viscoelastic stress to strain, shows how much energy was dissipated in a cycle of deformation in the form of heat and defines the viscous properties of the material [44].

If the complex modulus value is equal to the storage modulus G* = G', the loss modulus G'' is equal to zero and the material is completely elastic - the whole cycle of deformation energy is stored and then released, and the deformation is completely reversible. When the complex modulus is equal to the loss modulus G* = G'', storage modulus G' is equal to zero and the material is completely viscous. Modules G' and G'' are dependent on the frequency of oscillation and the temperature. The higher the frequency of oscillation, the higher the value of the modules G' and G''. A rise in temperature causes a decrease in storage modulus G' and increase of loss modulus G'' [44].

The ratio of loss modulus G'' to the storage modulus G' is called the loss angle (damping) and is expressed by the tangent of the phase angle δ [45].

$$\tan(\delta) = \frac{G^{''}}{G^{'}} \tag{10}$$

The loss angle tan(δ) is the ratio between the energy lost and stored during the deformation and determines the viscoelastic material's ability to dissipate (damp) energy [45]. In the case of completely viscous material value of tan (δ) is greater than 1, and the loss modulus values are higher than the storage modulus G" > G'. When the value of tan(δ) is less than 1 it means supremacy of the elastic properties over the viscous, which means higher values of storage modulus in relation to the loss modulus G' > G". The lower the value of the tan(δ), the more elastic the material. A zero value of tan (δ) indicates that the material is perfectly elastic [46].

To describe the damping properties of the material parameters such as the loss angle tan (δ) and the complex shear modulus G* are used [47,48]. Damping of vibration or the damping of cyclic deformation is related to a non-reversible absorption of the deformation energy and its conversion into heat that is dissipated by the material [49].

Magnetorheological elastomers are viscoelastic materials, which because of their potential applications are subjected to dynamic loads. Characteristic MREs work area is in the range of small deformations below the yield stress. Under the influence of the magnetic field MREs change their stiffness over a wide range, allowing customization for effective vibration damping. MREs properties are studied and described by the equations typical for viscoelastic materials, with the parameter describing the influence of the external magnetic field on the rheological properties of MREs.

Change in the properties of MREs under the magnetic field is called the magnetorheological (MR) effect and is the most important and most extensively studied property of magnetorheological composites.

The term "magnetorheological" is used only in relation to a group of magnetorheological materials. It describes the reversible properties changes under the magnetic field in comparison to the material properties without the presence of the field. Changes in the rheological properties are usually described by the shear modules: storage G' and loss G" [10,50,51,52,53,54,55]. The magnetorheological effect is also associated with increased stiffness (modulus of elasticity, measured in compression test for example) and the change in damping of magnetorheological materials [11].

Changes in the properties of MREs under the magnetic field, described as the MR effect are related to the magnetic particles tendency to change their position under the influence of an applied magnetic field. The magnetic field induces dipole moments in the ferromagnetic particles, which tend to obtain the positions of minimum energy state. Particles movement introduce deformations in the elastomer matrix, resulting in the increase of shear modulus and stiffness of the MREs. Secondly, the interactions between particles in a magnetic field causes their attraction to each other, which also introduces a deformation in the elastomeric matrix and, consequently, increases the stiffness of the material and shear modules [56]. We can conclude that due to efforts of magnetic particles to change the position in a magnetic field, the elastic matrix is deformed, which leads to an increase in the shear modules (G*) [20].

The creation of oriented chains of particles in the elastomer leads to a state of minimum magnetic energy [10]. Such MREs under shear deformation in the presence of an external magnetic field losses the low-energy state and requires additional work associated with the movement of entire chains of particles that oppose the deformation and tend to return to equilibrium state. Deformation of the matrix leads to an increase in the value of shear modules G' and G" and increase in the stiffness of the elastomer. MREs properties strongly depend on the magnetic field strength. Studies of the MREs magnetorheological effect are usually carried out at relatively small frequencies (from 1 to 20 Hz) [8,10,57,58,59,60].

Changes of the properties of magnetorheological composites in the magnetic field are usually described by absolute and relative magnetorheological effect [61].

The absolute magnetorheological (MR) effect is the difference between the maximum value of shear modulus G_{max} achieved in a magnetic field, and the value obtained without a magnetic field G_0 (called zero field modulus). The absolute magnetorheological effect is described by the equation:

$$\Delta G = G_{\max} - G_0 \left[\mathrm{MPa}\right]. \tag{11}$$

The relative magnetorheological (MR) effect is the ratio of the absolute magnetorheological effect to the zero field modulus G_0, expressed in percents:

$$\Delta G_r = \frac{\Delta G}{G_0} \cdot 100\%. \tag{12}$$

As it is shown in the literature the absolute and relative MR effects depend on the content of magnetic particles, the frequency of oscillations and the magnetic field strength [20,32,40,42,57,61].

According to reports relative magnetorheological effect can be of 30-40% on average [62], and even 60% for MRE containing 30% vol. iron particles [63]. With the increase in the frequency in the range from 100 to 1400 Hz, shear modulus increases up to four times for MRE containing 27% vol. iron particles. An increase in shear modulus (about 2 MPa) under the influence of the magnetic field of 0.56 T, as well as significant MR effect has been obtained [8,58,60,64].

In most studies, the magnetic field is applied parallel to the chains with iron particles, in order to increase the interaction between particles inside chains [8,57,62].

The MR effect is also influenced by the amplitude of the applied strain, and this is because forces of the magnetic interaction strongly depend on the distance between the dipoles (the iron particles) [20,58]. Magnetorheological elastomers are characterized by a greater strain at which the MR effect is significant than the corresponding MR fluids [10,20,57].

In [65], an increase in storage modulus of 100% in a magnetic field of 670 mT for the MRE based on silicone rubber matrix containing 20 vol.% of iron particles has been reported. The authors of [66] recorded the relative MR effect reaching 878% under 1 T magnetic field (800 kA/m), the MRE was based on silicone rubber (10 wt.%), plasticized with silicone oil (10 wt.%), containing 80 wt.% of particles and cured in 1.5 T magnetic field (1200 kA/m).

Beneficial effects of plasticizers, causing a reduction of material matrix stiffness, on the magnetorheological effect is shown in other papers [20].

Most of the results presented in the literature describes the MREs with isotropic distribution of particles. In [20,61] it was found that larger relative MR effect can be achieved in materials with softer matrix. Therefore, plasticizers are often added in order to increase the relative MR effect. The increase in the relative MR effect is obtained in this case, but the zero field modulus (without magnetic field) is lower.

Also reports can be found about the influence of the anisotropic microstructure on the properties of magnetorheological elastomers. The static and dynamic tensile testing of MREs with particles oriented in the form of chains have shown that they exhibit much higher Young's modulus in comparison to MREs with the isotropically distributed particles [67], which is explained by the presence of column-like structure formed by the particles. Stiffness of the chains depends on the distance between the particles within them, and increases with the applied magnetic field due to the dipole formation and dipole-dipole interactions.

There is also a significant effect of particle size on the MR effect. In order to achieve a significant MR effect, particles should contain at least several magnetic domains [10]. The shape of particles can also have influence on the MR effect, it is better if they have an asymmetric shape with the main axis of anisotropy [63]. Few works have reported positive effect of surface modification of particles on the growth of magnetorheological effect [68], although others show clearly that due to the better adhesion particle-matrix boundary region occurs and the ability of particles to move is reduced, which leads to a reduction of MR effect [69,70].

4. Experimental results

4.1. Materials characterisation

In our study, the selection of composite matrix was carried out to produce MREs with mechanical properties, which can be varied widely under the influence of the magnetic field.

Magnetorheological elastomers were manufactured using:

- soft polyurethane (PU), denoted here as PU 70/30, obtained from polyether polyols VORALUX® HF 505 used in a blend with 14922, with the average molecular weight respectively 3596 and 4350 g/mol, and isocyanate compound HB 6013, supplied by Dow Chemical Company,
- segmented urea-urethane elastomer (EPU), obtained from 4,4′-diphenylmethane diisocyanate (MDI), ethylene oligoadipate (OAE) with the average molecular weight about 2000 g/mol and dicyandiamide (DCDA) as a chain extender.

PU substrates were mixed in the weight ratio, respectively 30:70:23. Mixing and curing process were conducted at room temperature.

EPU with molar ratio of MDI:(OAE+DCDA) equal to 2.5 was synthesized by one-shot method. The curing process was carried out at temperature of 150°C, what makes technology aspects of MRE manufacturing more complicated. The existence in every short hard segment of strong polar urea group and strong polar nitrilimide side-group increases the urea-urethane thermal and mechanical properties, as well as stiffness and hardness in comparison to soft polyether polyurethane (see Table 2).

Material	Viscosity (mPas)	Density (g/cm³)	Glass transition temperature (°C)		Hardness (°ShA)	Young's modulus (MPa)
			Soft segments	Hard segments		
PU 70/30	8000	1.03	-64	41	< 10	0.1
EPU 2.5	1400	1.26	-23	165	87	14

Table 2. Selected physical and mechanical properties of elastomers used for MREs fabrication.

As shown in Table 2 polyurethane PU 70/30 is characterized by lower density and mechanical properties than EPU 2.5. Low hardness and stiffness of the polyurethane matrix can lead to the higher relative property changes of the MRE under an external magnetic field. On the other hand, the EPU 2.5 is distinguished by lower viscosity of reactive mixture of substrates and good mechanical properties. Low viscosity during the processing of the MRE makes the arrangement of the particles into aligned chains very easy.

During the fabrication of MRE, two polyols were mixed first and then carbonyl iron was added. This mixture was subsequently put under vacuum to remove trapped air bubbles. After de-gassing the isocyanate compound was added and the reactive mixture was poured into moulds. The mixing of substrates and curing processes were carried out at room temperature, without or with application of magnetic field.

Three iron powders, which differed in size and shape of the particles, were used as magnetoactive component of MREs:

- carbonyl–iron powder, produced by Fluka, containing spherical particles with average diameters in the range 6-9 μm, Fe > 99,5 wt.%,
- carbonyl-iron powder HQ, produced by BASF, containing spherical particles with average diameter 1.4 μm, Fe > 97,5 wt.%.
- iron powder, produced by PYRON, containing irregular particles of a diameter of 70 μm.

Images of the particles and particle size distribution histograms are presented in Fig. 9, particles characteristics are shown in Table 3.

All the particles used for the study are characterized by their high saturation magnetization, which has a positive effect on the properties of MREs by increasing the strength of interaction between particles in a magnetic field. Moreover, they are commercially available and relatively inexpensive. A variety of MRE samples were produced:

a. pure elastomers (without particles),
b. elastomers with randomly dispersed iron particles,

c. elastomers with aligned iron particles.

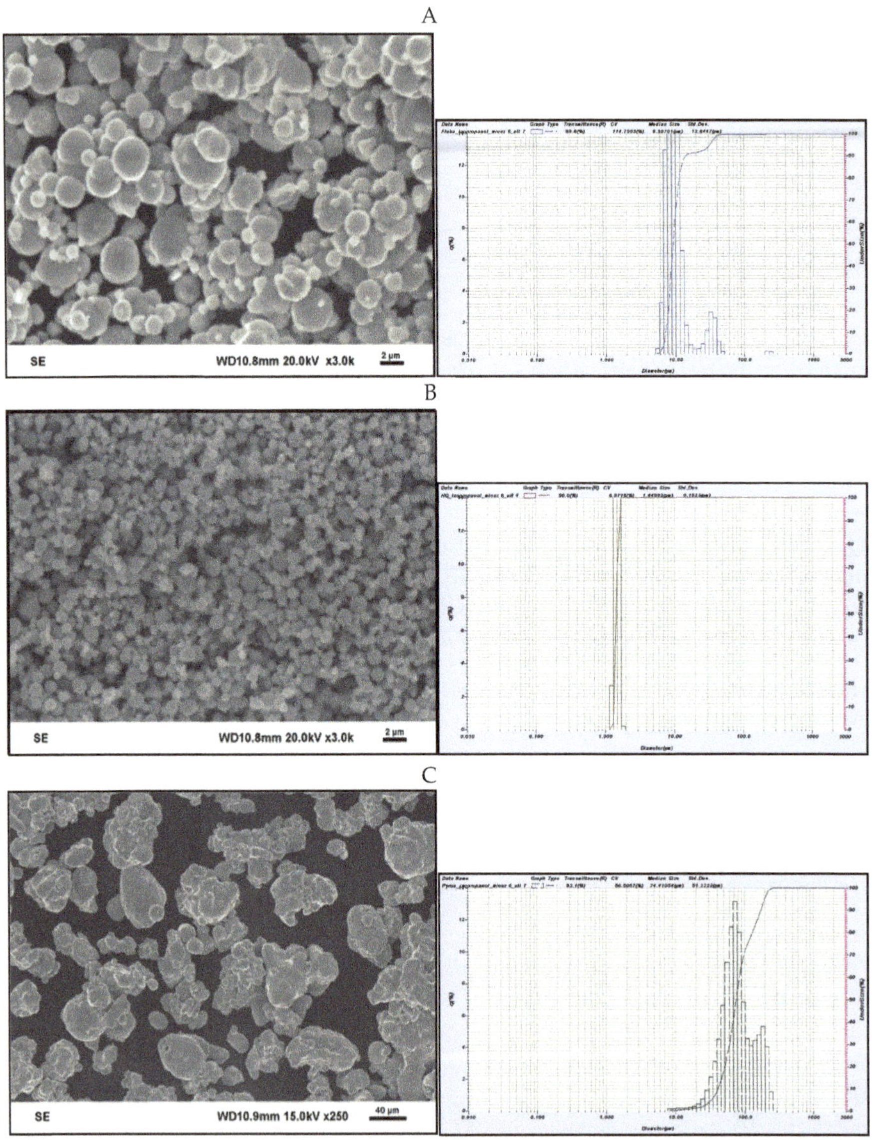

Figure 9. SEM images and particle size distribution histograms A - carbonyl iron Fluka, B - carbonyl iron HQ, C - Iron PYRON.

Particle's type	Manufacturer	Designation	Shape	Particle size (manufacturer's data) [μm]	Average particle size (Horiba LA-950) [μm]	Saturation magnetization [emu/g]
carbonyl iron	Fluka	Fe	spherical	6-9	9	217
carbonyl iron	BASF	Fe HQ	spherical	1-2	1,4	215
iron	PYRON	Fe PYRON	irregular	no data	74	209

Table 3. Characteristics of ferromagnetic particles used for MREs fabrication.

The carbonyl iron particle volume fractions were 1.5, 11.5, 18, 25 and 33.0 vol. %. The samples were subjected to a magnetic field during curing, to produce aligned chains in the elastomer matrices. Two magnetic field strengths were used: 80 and 240 kA/m. Samples with different orientation of the particle chains to the long sample axis (30, 45, 60 and 90 degree) were produced.

4.2. Microstructure-properties relationship in MRE

The spatial distribution of particles in the magnetorheological elastomer matrix depends on the curing conditions. It is known [6,7,8,10,32,33,34,35,36] that curing in the presence of a magnetic field, allows to obtain magneto-anisotropic structure, consisting of chains of particles, as shown in the example of MRE with PU matrix and 11.5 vol.% carbonyl iron particles (Fig. 10A and C). The cross-section parallel to the direction of the magnetic field shows the arranged paths (chains) of Fe particles, clearly indicating the orientation of the particles in a magnetic field (MF). The curing of magnetorheological elastomer without application of magnetic field, leads to the isotropic structure of Fe particles uniformly distributed throughout the volume of the matrix (Fig. 10D).

As it is seen in the cross-section of the composite, perpendicular to the direction of the magnetic field (Fig. 10B), the distribution of chains is relatively uniform, moreover, most of them consist of more than one row of particles.

The viscosity of the reactive mixture is a technological parameter, which turned out to have an significant influence on the ability of particles to create a structure oriented along the magnetic field direction. The viscosity is not constant over time, but increases as the reaction follows, to the principle of doubling the length of the molecules in each act of addition, often starting during mixing of liquid substrates. The lifetime of the mixture is the period in which it is liquid, until the gel formation. The movement of particles in a reactive mixture of substrates is possible until it is liquid.

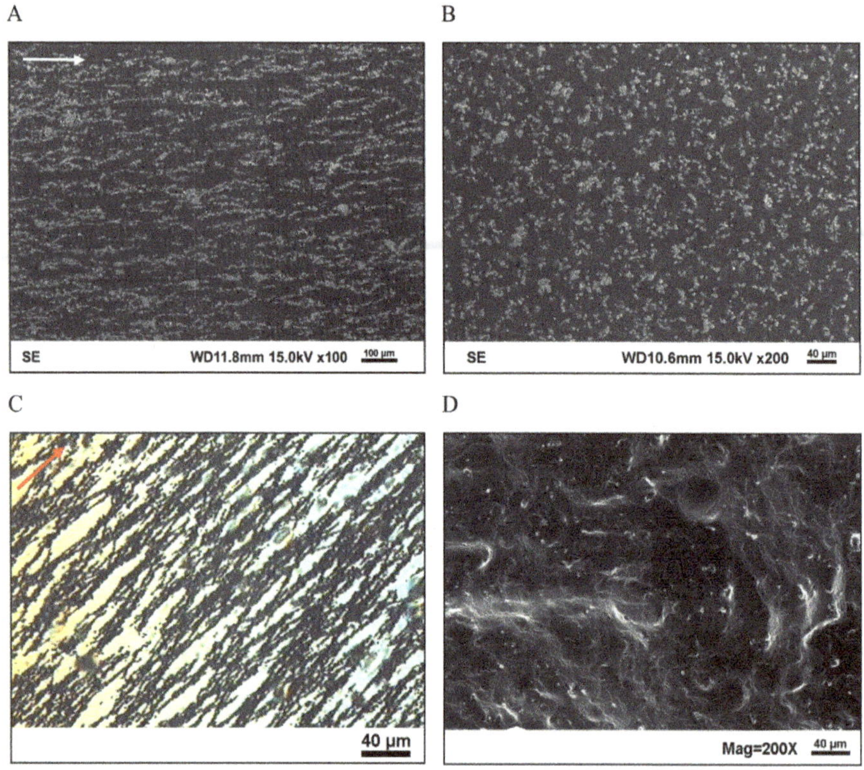

Figure 10. Microstructure of MRE obtained from PU matrix containing 11.5 vol. % Fe particles: A - anisotropic, cross-section parallel to the direction of the MF, SEM image, B - anisotropic cross section perpendicular to the direction of the MF, SEM image, C - anisotropic, cross-section parallel to the field, the image in polarized light, D - isotropic, SEM image. Arrows show the direction of the magnetic field during the curing.

Observations of the brittle fractures of produced MREs, as shown in Figure 11, show that too high viscosity of the reactive mixture inhibits the movement of particles and the applied magnetic field does not allow to obtain the oriented structure of particles chains. This is the case for the silicone rubber matrix (Fig. 11C), which has a viscosity before curing of 30 000 mPa·s and is nearly four times greater than the viscosity of the PU 70/30 mixture and up to 20 times than EPU. Even a relatively short time of gelling (about 10 min for the PU 70/30) is not an obstacle for obtaining oriented structure of particle chains. It is sufficient for particles to shift the position to obtain the lowest Zeeman energy and arrange along the lines of magnetic field.

Another important parameter that affects the microstructure and consequently the properties of MREs is the magnetic field strength during curing. At higher magnetic field strengths the formed chains are thicker, consisting of more than 1-2 rows of particles.

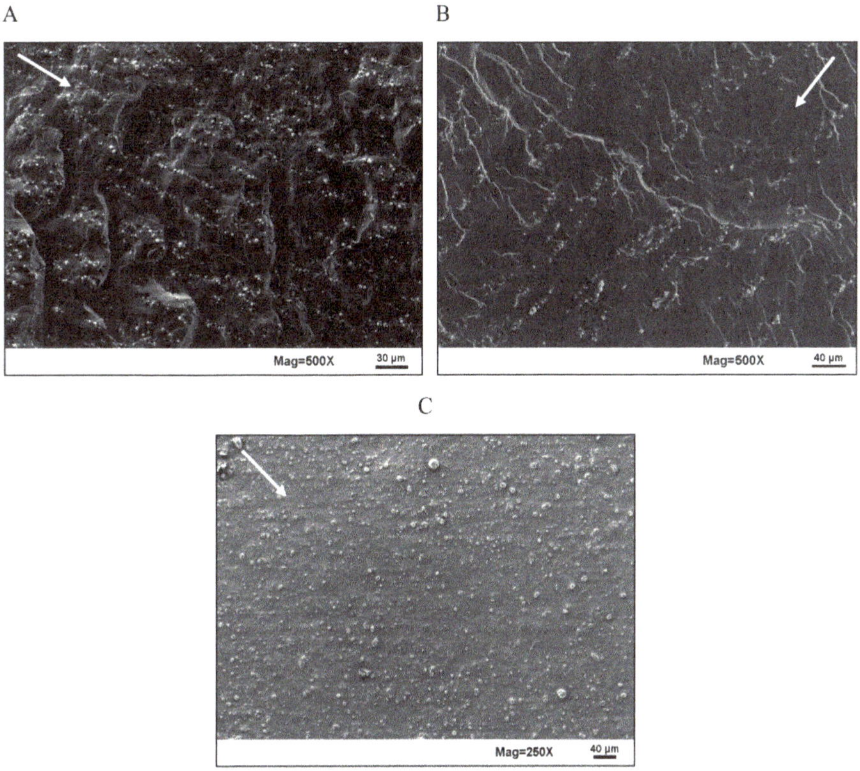

Figure 11. Microstructure of MREs with 11.5 vol.% of Fe particles, cured in the 80 kA/m, matrix: A - EPU 2.5, B - PU 70/30, C - silicone rubber. Arrows show the direction of the magnetic field during curing.

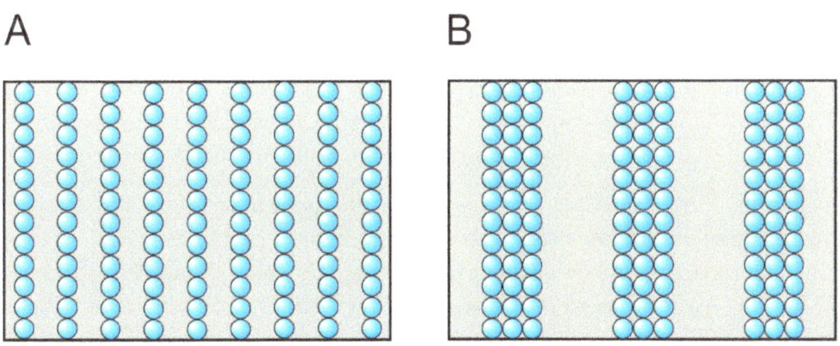

Figure 12. Model of MRE microstructure dependently on the magnetic field strength during the matrix curing: A - 80 kA/m, B - 240 kA/m.

The structural anisotropy was confirmed by the Vibration Sample Magnetometer (VSM) studies. Tests were carried out parallel (II) and perpendicular (L) to the direction of particles chains, corresponding to the magnetic field direction during curing. The example of the hysteresis loops obtained for the MREs is shown in Fig. 13A. From the hysteresis loops the anisotropy coefficient (A_b) was calculated at the selected value of magnetic field strength of 160 kA/m. It was expressed by the ratio of magnetization measured at 160 kA/m, respectively parallel and perpendicular to the particles alignment direction. The A_b values are shown in Fig. 13B.

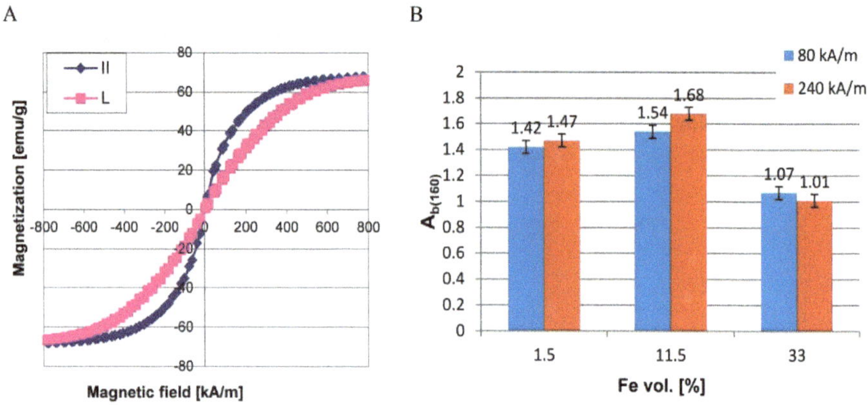

Figure 13. (A) Magnetization curves as a function of a magnetic field in the direction parallel (II) and perpendicular (L) to the chains of particles for the MRE containing 11.5 vol. % Fe particles, cured in the 240 kA/m magnetic field. (B) A change in the magnetic anisotropy $A_{b(160)}$ for MRE with different particle amount and magnetic field during curing (80 and 240 kA/m).

It was found that there is a maximum value of the anisotropy coefficient corresponding to 11,5 vol.% of carbonyl-iron. The anisotropy coefficient decreases with further increase in the volume fraction of particles. Moreover, for the sample with the highest A_b a significant effect of the aligning magnetic field is visible; higher magnetic field leads to a better alignment. Structural and magnetic anisotropy has not been found in the specimens having 33 vol. % of carbonyl-iron, therefore, the anisotropy coefficient is equal to 1 for these specimens. The result can be attributed to the formation of the network of particles, in contrast to the chains of particles that characterize anisotropic composites. This was also observed on SEM images.

The anisotropy of MRE microstructure, obtained due to the magnetic field during curing has a significant influence on the changes of their mechanical properties in the magnetic field. Also the field strength applied during manufacture influences the mechanical properties of the MREs. This was confirmed by the results of mechanical and rheological tests presented in this chapter.

4.2.1. *Compression tests*

Compression tests were performed using a specially designed magnetic coil device for the MTS tensile tester. The study was conducted on samples with different particle volume content (1.5, 11.5 and 33%), produced in a magnetic field of 80 and 240 kA/m. Samples with dimensions of ϕ = 20 mm and h = 25 mm, with particle chains parallel to the axis of the sample, were compressed uniaxially without and with a magnetic field strength of 240 kA/m. The obtained stress-strain curves are shown in Figure 14.

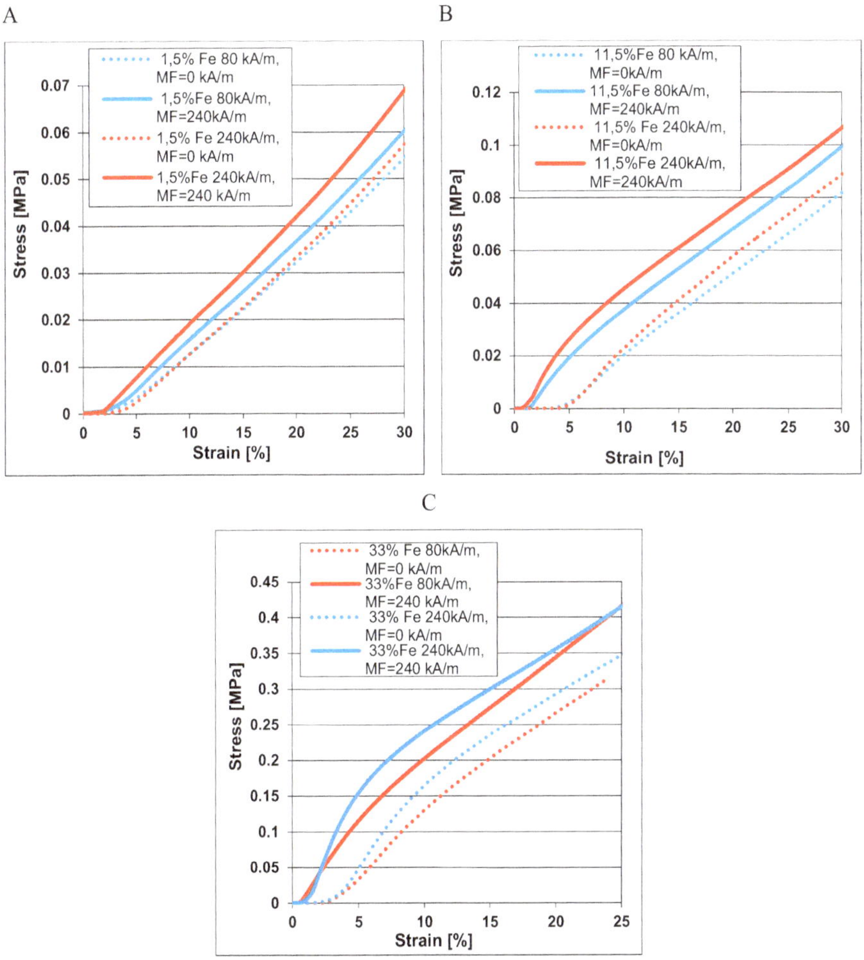

Figure 14. Stress-strain curves during the compression of MREs cured in a field intensity of 80 and 240 kA/m, for various Fe particle volume fractions: A - 1.5%, B - 11.5%, C - 33%.

The compressive curves recorded without magnetic field are marked with a dashed line, while with MF using the solid line. It is clear that in each case MREs in a magnetic field are characterized by higher stiffness and compressive strength than without magnetic field. MREs with a higher content of particles are characterized by higher stiffness and compressive strength both without and with the magnetic field. Moreover, in a magnetic field, obtained stress-strain curves are at a higher level for MREs with higher structural and magnetic anisotropy, those cured in a magnetic field of greater intensity, even when no field curve was at a similar level, as for the MRE containing 1.5% vol. Fe particles (Fig. 14A).

Changes of the properties of magnetorheological elastomers in the magnetic field are described by absolute and relative MR effect, which in this case is expressed by the relative and absolute change in compressive modulus. Based on the compressive curves, the absolute and relative change of the compressive modules in the field of 240 kA/m, at the deformation of 0.1 mm/mm has been calculated and presented in Table 4.

Fe particles fraction [vol. %]	**Magnetic field during curing [kA/m]**	**$\Delta E_{c(240)}$ [MPa]**	**$\frac{\Delta E_{c(240)}}{E_{c(0)}} * 100\%$ [%]**
1,5	80	0,03	25
	240	0,05	51
11,5	80	0,17	86
	240	0,23	100
33	80	0,72	56
	240	0,75	55

Table 4. Comparison of absolute (ΔEc (240)) and relative ($\frac{\Delta E_{c(240)}}{E_{c(0)}}$) changes in the compressive modules in the magnetic field of 240 kA/m, under the deformation of 0.1 mm/mm.

The absolute MR effect is greater with the higher content of iron particles and depends on the applied magnetic field strength during curing. The higher the magnetic field during curing, the larger the absolute MR effect. The relative MR effect is not directly proportional to the particle content and reaches the highest value for contents of 11.5 vol.% Fe. Relative MR effect for the largest particle content, as well as magnetic anisotropy coefficient is practically the same regardless of the magnetic field during manufacture. Moreover, it is worth noting that the relative MR effect reaches its maximum for the MRE with the highest values of the coefficient of magnetic anisotropy.

4.2.2. Rheological properties

The knowledge of dynamic properties is needed especially for materials that are used in equipment subjected to vibration such as in aerospace and automotive industries. To describe the efficiency of damping, parameters such as the loss angle tan (δ) and shear modulus G*, consisting of the storage modulus G' and loss modulus G" are used, determined by rheological tests [47,48].

A

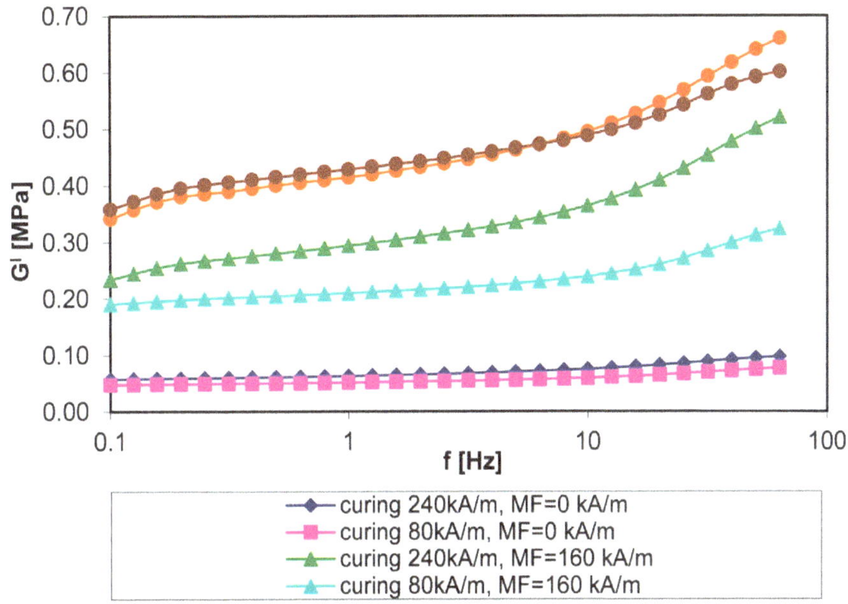
0.70
0.60
0.50
0.40
0.30
0.20
0.10
0.00
G' [MPa]
0.1
1
10
100
f [Hz]
curing 240kA/m, MF=0 kA/m
curing 80kA/m, MF=0 kA/m
curing 240kA/m, MF=160 kA/m
curing 80kA/m, MF=160 kA/m

B

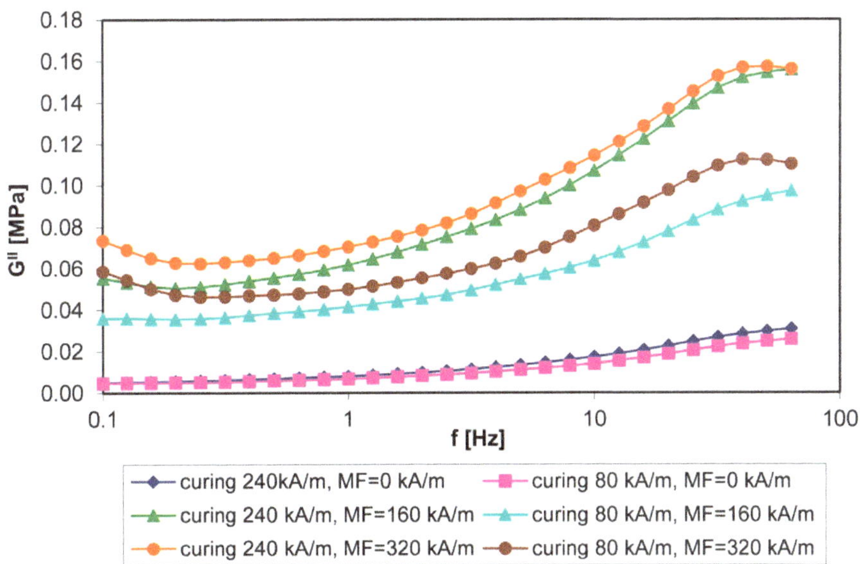
0.18
0.16
0.14
0.12
0.10
0.08
0.06
0.04
0.02
0.00
G'' [MPa]
0.1
1
10
100
f [Hz]
curing 240kA/m, MF=0 kA/m
curing 80 kA/m, MF=0 kA/m
curing 240 kA/m, MF=160 kA/m
curing 80 kA/m, MF=160 kA/m
curing 240 kA/m, MF=320 kA/m
curing 80 kA/m, MF=320 kA/m

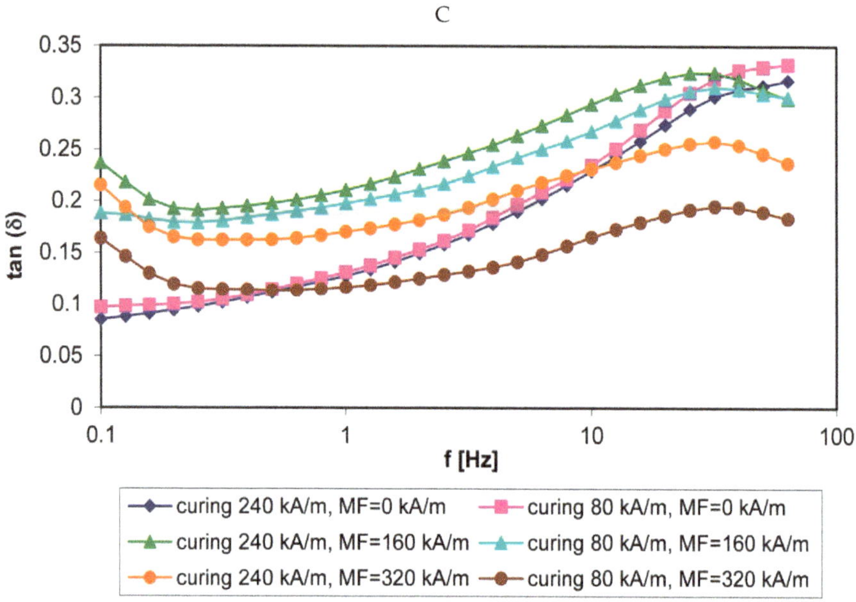

Figure 15. The change of rheological properties in the magnetic field as a function of frequency at a constant angular deformation of 0.1% for the MRE with PU matrix containing 11.5 vol.% Fe particles, cured in the 80 and 240 kA/m: A - storage modulus G', B - the loss modulus G", C - damping factor tan(δ).

MREs cured at different magnetic field strengths, tested without magnetic field have a similar storage (elasticity) modulus, loss (viscosity) modulus and the loss angle, which increases with increasing oscillation frequency of deformation. This pattern is characteristic for viscoelastic materials for which the higher the frequency of oscillation, the higher the value of the modules G' and G" [44]. The sharp increase in tan(δ) as a function of oscillation frequency shows an increase in the share of the viscous properties of MREs.

The increase in the magnetic field causes an increase of both modules G' and G", much higher values were obtained in 160 kA/m for the MRE with a higher magnetic and structural anisotropy (cured in the 240 kA/m). For larger magnetic field the difference is not so evident. In a magnetic field damping is characterized by smaller changes in the function of frequency, but its value clearly depends on the magnetic field strength (tan (δ) increases with increasing field) and to a lesser extent, but also depends on the microstructure formed during curing. MREs cured in the field of higher intensity are characterized by higher values of tan(δ). With increasing magnetic field, the share of the viscous properties is greater, and increases with the structural and magnetic anisotropy. The material absorbs more energy which is dissipated as heat.

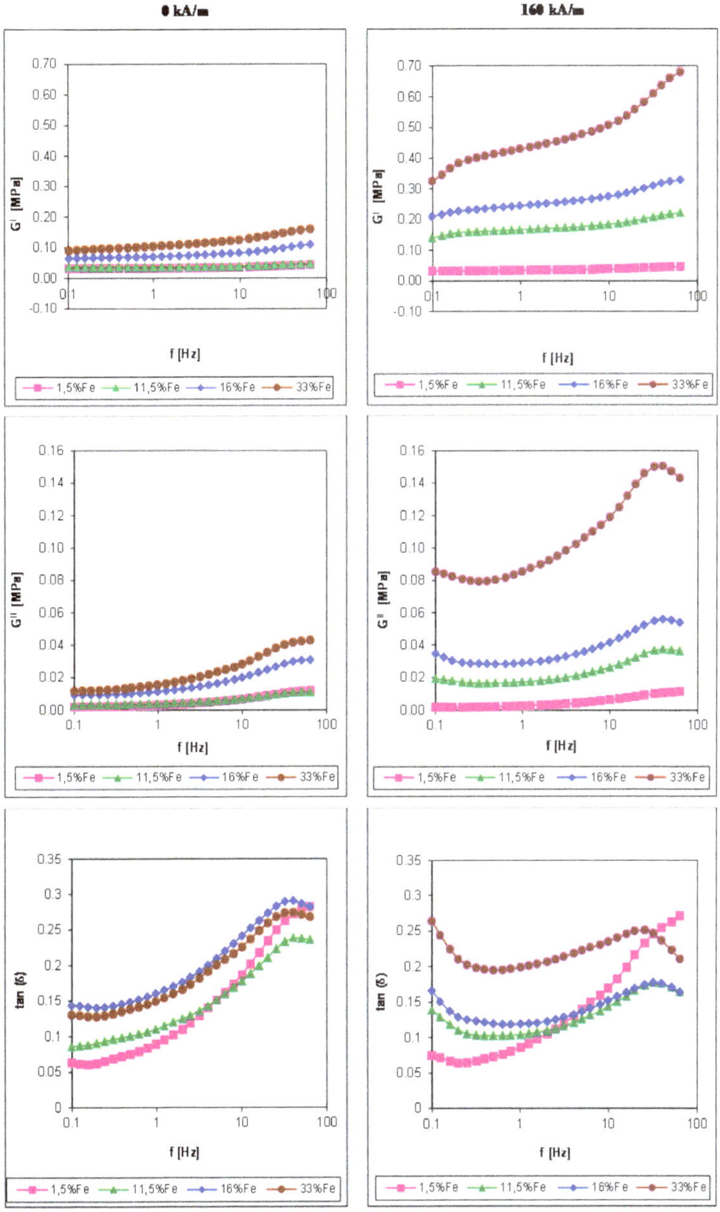

Figure 16. Changes in modules G' and G'' and damping factor tan(δ) as a function of oscillation frequency (at constant strain of 0.1%) for the MREs with PU matrix and different content of Fe particles without and in the field 160 kA/m. Particle chains parallel to the MF in rheometer.

Selected plots of rheological properties as a function of frequency of oscillation, described by the storage modulus G', loss modulus G'' and loss angle tan(δ) without field and in a magnetic field of 160 kA/m, for MREs with a polyurethane matrix and different content of carbonyl iron particles, are shown in Fig 16. The samples were cured in the magnetic field of 240 kA/m, and the particle chains were oriented parallel to the direction of the MF applied in rheometer.

It can be concluded that the increase in the particle content causes a slight increase in both modules in the initial state (without magnetic field) and they are dependent on the frequency of oscillation. Modules (storage and loss) in the initial state increase with increasing frequency of oscillation. Loss angle also depends on the content of particles and is greater with the higher particle content. The higher the value of tan (δ), the greater the ability to absorb energy. However, note that the values obtained for the studied MREs are well below 1, which means a significant advantage of the elastic properties over the viscous.

In the magnetic field storage and loss modulus increase, and this increase depends on the content of the particles. The more particles the greater the value of the modules, and both increase with increasing frequency. For high frequencies, a decrease in loss modulus is observed, inversely proportional to the particle content.

Increase of loss angle as a function of frequency in the magnetic field is much smaller, except for the MREs with very low amount of particles 1.5 vol.%, for which the tan(δ) in the MF is the same as without magnetic field. The values of tan(δ) clearly increase for the low oscillation frequency, and rise with the rising particle content. The higher the value of tan(δ), the greater the share of energy dissipated, and the material exhibits better damping properties.

The influence of the size of particles on the microstructure and rheological properties of MREs has also been investigated. The particle size does not affect the possibility of obtaining anisotropic structure consisting of chains of particles. In case of irregularly shaped particles (PYRON iron) during curing in magnetic field, they turn so their long axis is parallel to the direction of the magnetic field, which is consistent with the direction of easy magnetization and the tendency to minimize the energy associated with the shape anisotropy.

Rheological parameters were measured for a selected frequency of 10 Hz, the parameters G', G'', and their absolute (ΔG', ΔG'') and relative ($\Delta G'/G'_0$, $\Delta G''/G''_0$) changes describing absolute and relative MR effect are shown in Figure 18 as a function of magnetic field strength. Compared MREs have parallel arrangement of particles chains to the magnetic field in rheometer, the same content of particles but different sizes.

Storage modulus, loss modulus, and their absolute and relative changes increase with increasing magnetic field to the value of 320 kA/m, above the increase is weaker, which is associated with reaching a saturation magnetization of particles. In case of loss modulus G'' and the absolute and relative changes, significantly higher values as a function of field were obtained for the largest particles, with an average size of ~ 74 microns and irregular shapes.

Storage modulus G' values were clearly lower for MRE with particles of small diameter (~ 1.4 μm), while for other types of particles are similar. Noticeable difference in favor of larger particles occurs in the field of 320 kA/m, which comes from the fact that particle size affects the strength of magnetic interactions between them. The bigger they are the stronger the interactions, the particles are attracted to each other in a chain with higher force, which ultimately leads to a greater MR effect. The maximum relative MR effect for the MREs with 11.5 vol.% of particles is 760% of the relative change in storage modulus, and 770% for the loss modulus. For this field strength these values are much higher than the values found in the literature.

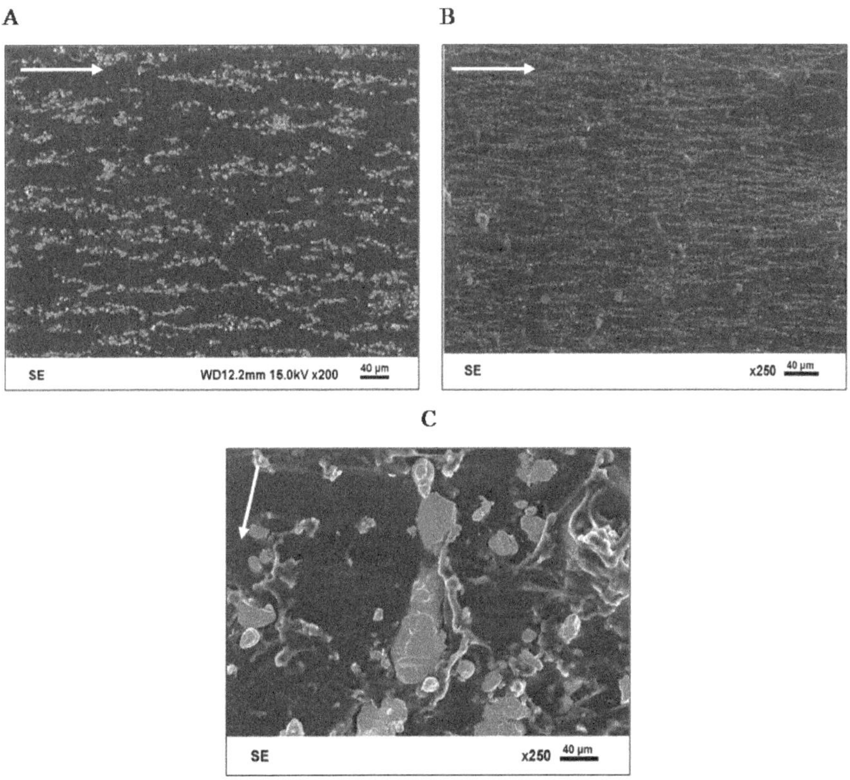

Figure 17. Microstructure ofs MRE with 11.5 vol.% of particles of different sizes: A - Fe ~ 9 μm, B - Fe HQ ~ 1.4 μm, C - Fe PYRON ~ 74 μm. Arrows shows the direction of the magnetic field during curing.

The alignment of particle chains within the matrix may have a significant effect on the performance of the MREs. In order to investigate this effect, samples with different alignment of particles were produced and tested. A schematic representation of the samples, which contained 11.5 vol.% of Fe particles (Fluka, 9 μm), is shown in Fig. 19.

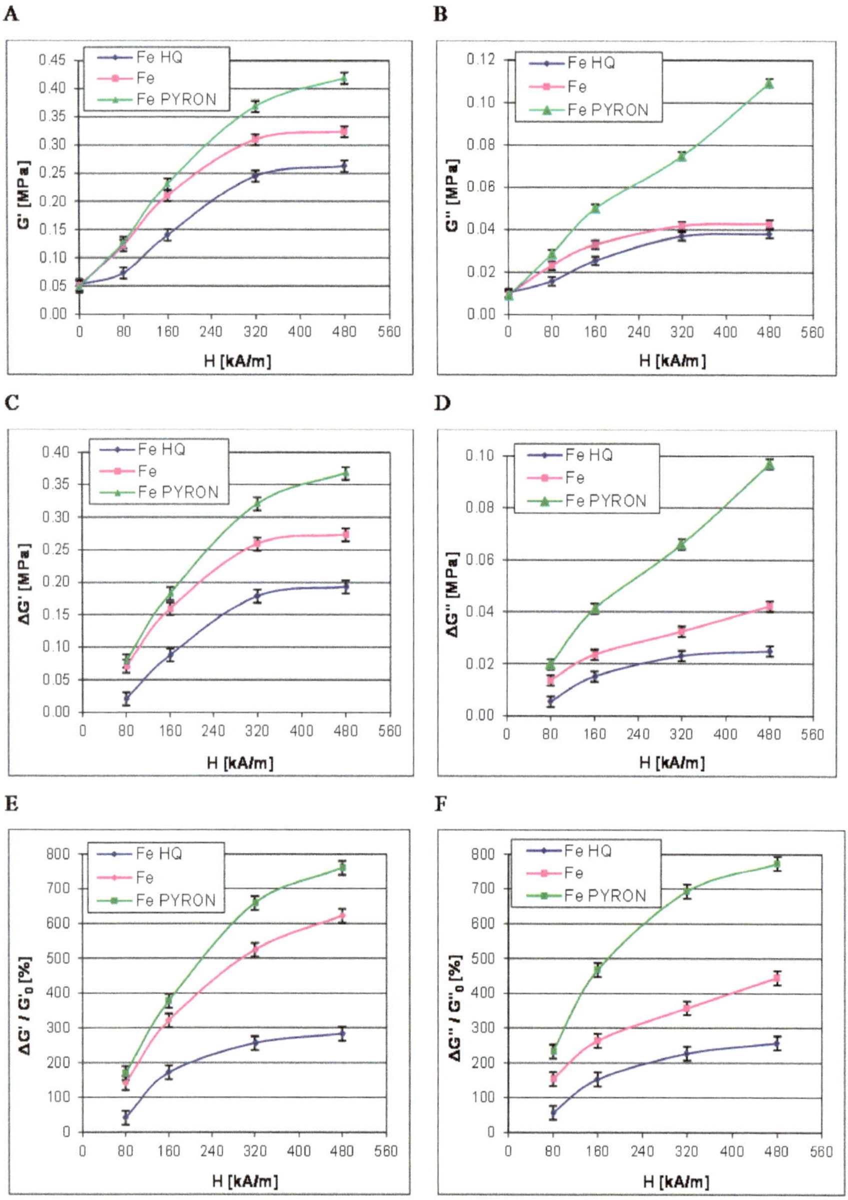

Figure 18. Changes of the storage modulus G' (A), loss modulus G'' (B) and the absolute (C, D) and relative (E, F) magnetorheological effects as a function of magnetic field for the MRE with polyurethane matrix and content of 11.5 vol.% of particles. Sizes of particles: Fe ~ 9 μm, Fe HQ ~ 1.4 μm, Fe PYRON ~ 74 μm. Magnetic field in rheometer parallel to the chains of particles.

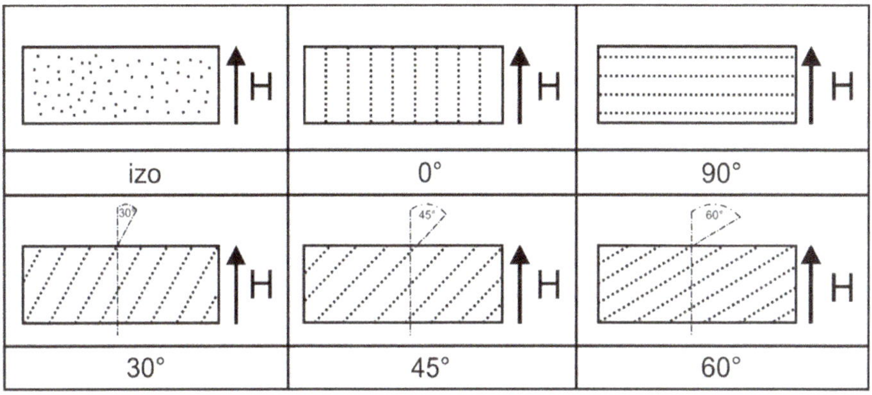

Figure 19. Schematic representation of the particle distribution with respect to the magnetic field direction.

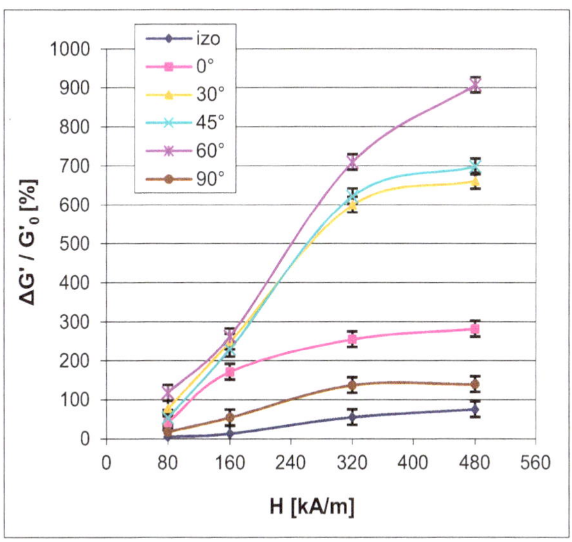

Figure 20. The effects of different orientation of carbonyl iron particles in elastomer matrix, changes of relative MR effect as a function of the magnetic field.

It can be concluded that the values of storage and loss modules, and their absolute and relative changes strongly depend on the orientation of the particles chains. The largest absolute and relative MR effects were obtained for the arrangement of particles at an angle of 60° to the direction of the magnetic field. Under the influence of a magnetic field the direction of easy magnetization comes along the chains of ferromagnetic particles. To minimize the Zeeman energy the chains tend to set to the direction of the magnetic field, which involves the displacement of particles. Because they are embedded in the elastomer matrix and restrained by interactions at the interface, their displacements will introduce additional shear stress into the matrix, causing increased stress concentration in the matrix between the particles [71]. This does not take place when the particles chains are perpendicular to the direction of the field, and the resulting MR effect is much higher than for the isotropic structure. The lowest MR effect as a function of the field, both absolute and relative has been obtained in isotropic MREs.

The properties of the elastomeric matrix have also a great influence on MREs rheological properties. Rheological properties measured for the MREs fabricated from elastomers with different stiffness and hardness are shown in Fig. 21. Elastomer matrix with higher stiffness and hardness (EPU 2.5) leads to the higher storage and loss modulus of the MREs measured without magnetic field, while under the magnetic field the values are significantly lower in comparison to MREs obtained from soft elastomer (PU 70/30). As a result the absolute and relative MR effect is much higher in MREs obtained from the soft elastomer.

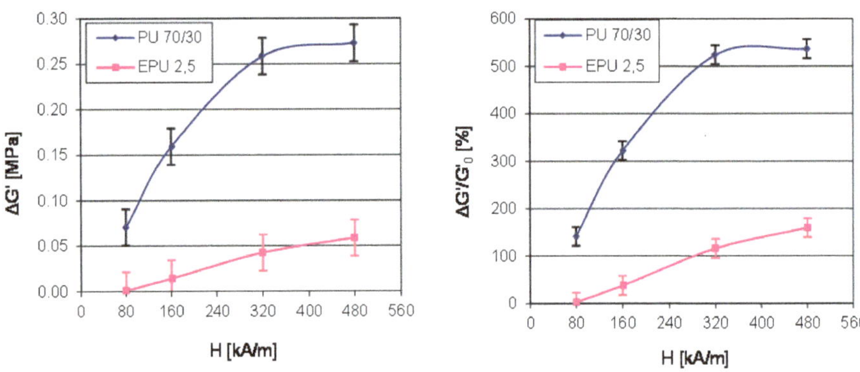

Figure 21. Absolute and relative MR effect of the MREs based on PU 70/30 and EPU 2.5 elastomers vs. the magnetic field strength (H).

MREs with the stiffer matrix (EPU 2.5) exhibit significant lower MR effect because the stiffer matrix makes impossible the particles to displace when they are subjected to the magnetic field [72].

Developed magnetorheological composites belong to a group of smart materials, and under the influence of the magnetic field changes of the properties should be fully reversible. It is a complete reversibility of change in modules G' and G", as shown in Fig. 22 for the MRE containing 11.5 vol.% Fe particles oriented at an angle of 45 degrees in the increasing magnetic field, which was alternating turned on and off. Each time after switching off the magnetic field the modules immediately returned to initial state.

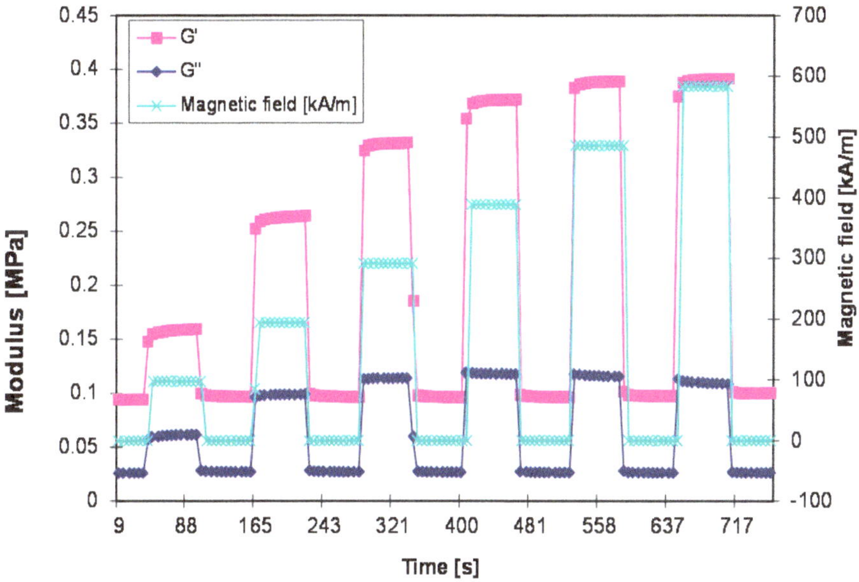

Figure 22. The changes of rheological properties of MRE in the magnetic field alternating switched on and off.

5. Conclusions

Studies on fabrication of MREs were carried out using different elastomers as a matrix, ferromagnetic particles with various shape and size. Samples with isotropic and anisotropic particles arrangement, were examined. Particles were oriented into chains under external magnetic field. Special attention was put on fabrication of samples with different orientation of chains to the magnetic field direction.

It was found that the microstructure of the MREs depends on the amount of ferrous particles and manufacturing conditions. The orientation of the iron particles into aligned chains is possible for lower volume content of the ferromagnetic fillers. High carbonyl-iron volume content in the matrix leads to the formation of more complex three-dimensional lattices. Also the magnetic measurements confirmed the existence of the microstructure anisotropy for the lower volume content of the iron particles. The structural and magnetic anisotropy has not been found in the MREs with 33 vol. % of particles. To evaluate the effect of the external magnetic field on the magnetorheological properties compressive strength, storage and loss modulus, as well as loss factor were measured. Both, the content of particles and their arrangement have significant effect on the properties of magnetorheological urethane elastomers. Compression test results showed that under external magnetic field samples are distinguished by a higher compressive strength.

Rheological properties of magnetorheological urethane elastomers also depend on the content of particles and their arrangement. Application of an external magnetic field leads to a significant increase in elastic modulus. Absolute and relative changes of storage modulus, calculated from obtained curves, show that the microstructure of the samples has a significant effect on their magnetorheological effect. Magnetorheological effect, expressed by relative change of storage modulus under magnetic field, is the highest for the sample with the highest magnetic and structural anisotropy.

The aligned particle network structure has a significant influence on the elastic properties of the composite material. Inside the chains the effective filler content is higher than the average filler content. By optimizing the content of particles and their alignment, either the stiffness or the damping properties of MREs can be increased by applying a magnetic field.

As a result of the studies it was also found, that the MREs having anisotropic microstructure exhibit, for the same particles content, much higher magnetorheological effect in comparison to the isotropic ones. The non-linear change of the rheological properties versus particles fraction was found. It is due to structural and magnetic anisotropy of the MREs, which has the greatest influence on the changes of the properties under the magnetic field, i.e. magnetorheological effect. Moreover, it was found that the magnetorheological effect can be controlled by the particles alignment to the magnetic field lines. It means that it is possible to obtain high magnetorheological effect not by increasing of the particles volume fraction, but by the formation of appropriate microstructure. It can be achieved for lower particles volume fraction, what advantageously decreases weight of devices based on the MREs. As a result of this work the new MREs, with application capabilities, characterized by extremely high magnetorheological effect (900% in MF of 480 kA/m), were elaborated.

Author details

Anna Boczkowska and Stefan Awietjan
Warsaw University of Technology, Faculty of Materials Science and Engineering, Warsaw, Poland

6. References

[1] Encyclopedia of Smart Materials vol. 1, editor Mel Schwartz, Wiley, New York 2002
[2] Encyclopedia of Smart Materials vol. 2, editor. Mel Schwartz, Wiley, New York 2002
[3] Garg D.P, Anderson G.L, Research in active composite materials and structures: an overview, Proc. SPIE 3992 (2000) 2-12
[4] Takagi T., Present State and Future of the Intelligent Materials and Systems in Japan, J. Intelligent Mater. Syst. Struct. 10 (1999) 575
[5] Zrinyi M., Barsi L., Buki A., Ferrogel: a new magneto-controlled elastic medium, Polymer Gels and Networks 5 (1997) 415-427

[6] Ginder J.M., Rheology controlled by magnetic fields, in: Encyclopedia of Applied Physics, vol. 16, VCH Publisher Inc., New York (1996)
[7] Bednarek S., Elastomery magnetoreologiczne – właściwości, technologia, zastosowania, Inżynieria Materiałowa 1 (2003) 39-44
[8] Zhou G.Y., Shear properties of magnetorheological elastomer, Smart Materials and Structures 12 (2003) 139-146
[9] Banks H.T., Pinter G.A., Potter L.K., Gaitens M.J., Yanyo L.C., Modeling of Nonlinear Hysteresis in Elastomer under Uniaxial Tension, Journal of Intelligent Material Systems and Structures, vol. 10-February 1999
[10] Carlson J.D., Jolly M.R., MR fluid, foam and elastomer devices, Mechatronics 10 (2000), pp. 555-569
[11] Farshad M., Le Roux M., Compression properties of magnetostrictive polymer composite gels, Polymer Testing 24 (2005) 163-168
[12] Kankanala S.V., Triantafyllidis N., On finitely strained magnetorheological elastomers, Journal of the Mechanics and Physics of Solids 52 (2004) 2869-2908
[13] Wang Y., Hu Y., Chen L., Gong X., Jiang W., Zhang P., Chen Z., Effects of rubber/magnetic particle interactions on the performance of magnetorheological elastomers, Polymer Testing 25 (2006) 262–267
[14] Guan X., Dong X., Ou J., Magnetostrictive effect of magnetorheological elastomer, Journal of Magnetism and Magnetic Materials 320 (2008) 158–163
[15] Lokander M., Reitberger T., Stenberg B., Oxidation of natural rubber-based magnetorheological elastomers; Polymer Degradation and Stability 86 (2004) 467-471
[16] Jolly M. R., Carlson J.D., Munoz B.C, Bullions T.A., The Magnetoviscoelastic Response of Elastomer Composite Consisting of Ferrous Particles Embedded in a Polymer Matrix; Journal of Intelligent Material Systems and Structures, vol. 7-November 1996 613-622
[17] Vicente J., Bossis G., Lacis S., Guyot M., Permability measurements in cobalt ferrite and carbonyl iron powders and suspensions; Journal of Magnetism and Magnetic Materials 251 (2002) 100-108
[18] An Y., Shaw M.T., Actuating properties of soft gels with ordered iron particles: basis for a shear actuator; Smart Materials and Structures 12 (2003) 157-163
[19] Kallio M., Aalto S., Lindroos T. i inni, Preliminary test on a MRE device, AMAS Workshop on Smart Materials and Structures SMART'03, Jadwisin, 2-5.09. 2003, conference proceedings pp. 353-360
[20] Lokander M., Stenberg B., Improving the magnetorheological effect in isotropic magnetorheological rubber materials, Polymer Testing 22 (2003) 677-680
[21] Dishovsky N., Ruskova K., Radulov I., "In situ" magnetic modification of polar elastomers, Materials Research Bulletin 36 (2001) 35–45
[22] Hansaka M., Preparation and Properties of Magnetic Vibration Damper, Proc. Inter-noise 93 (1993) 859-862
[23] Hansaka M., Mifune N., Damping properties of Magnetic Vibration Damper (MVD), Proc. Inter-noise 94 (1994) 693-696

[24] Wang D., Chen J-S., Sun L., Homogenization of magnetostrictive particle-filled elastomers using an interface-enriched reproducing kernel particle method, Finite elements in Analysis and Design 39 (2003) 765-782
[25] Yina H.M., Suna L.Z., Chen J.S., Magneto-elastic modeling of composites containing chain-structured magnetostrictive particles, Journal of the Mechanics and Physics of Solids 54 (2006) 975–1003
[26] Scheerbaum N., Hinz D., Gutfleisch O., Muller K.H., Schultz L., Textured polymer bonded composites with Ni-Mn-Ga magnetic shape memory particles, Acta Materialia 55 (2007) 2707-2713
[27] Scheerbaum N., Hinz D., Gutfleisch O., Muller K.H., Schultz L., Textured polymer bonded composites with Ni-Mn-Ga magnetic shape memory particles, Acta Materialia 55 (2007) 2707-2713
[28] Feuchtwanger J., Griffin K., Huang J.K., Bono D., O'Handley R.C., Allen S.M., J Magn Mat. 2038 (2004) 272-276
[29] Bednarek S., The coupling of magnetostriction and magnetoresistance in elastic ferromagnetic composites with conducting matrix, Materials Science and Engineering B63 (1999) 228–233
[30] Ulański J., Krzyszewski M., Polymers, electrical and electronic properties, in: Encyclopedia of Applied Physics, vol. 14, VCH Publisher Inc., New York 1996
[31] Roth S., Survey of industrial applications of conducting polymers, Acta Physica Polonica 87 (1995) 4-5
[32] Farshad M., Benine A., Magnetoactive elastomer composites, Polymer Testing 23 (2004) 347-353
[33] Coquelle E., Bossis G., Mullins effect in elastomers filled with particles aligned by a magnetic field, International Journal of Solids and Structures 43 (2006) 7659–7672
[34] Chen L., Gong X.L., Li W.H., Microstructures and viscoelastic properties of anisotropic magnetorheological elastomers, Smart. Mater. Struct. 16 (2007) 2645-2650
[35] Boczkowska A., Awietjan S.F, Wróblewski R., Microstructure – property relationships of urethane magnetorheological elastomers, Smart Materials and Structures 16 (2007) 1924-1930
[36] Boczkowska A., Awietjan S.F., The influence of mictrostructural anisotropy on the magnetorheological effect in elastomer-based composites with iron particles, Composites 8: 3 (2008) 327
[37] Varga, Z. Filipcsei G., Zrinyi M., Smart composites with controlled anisotropy, Polymer 46 (2005) 7779–7787
[38] Gong X.L., Zhang X.Z., Zhang P.Q., Fabrication and characterization of isotropic magnetorheological elastomers, Polymer Testing 24 (2005) 669–676
[39] Hua Y., Wang Y.L., Gong X.L., Gong X.Q., Zhang X.Z., Jiang W.Q., Zhang P.Q., Chen Z.Y., New magnetorheological elastomers based on polyurethane/Si-rubber hybryd, Polymer Testing 24 (2005) 324–329
[40] Wang Y., Hu Y., Deng H., Gong X., Zhang P., Jiang W., Chen Z., Magnetorheological Elastomers Based on Isobutylene–Isoprene Rubber, Polymer Engineering And Science DOI 10.1002 (2006) 204-208

[41] Gasperowicz A., Kaleta J., Zając P., Ziętek G., Badanie i modelowanie izotropowych elastomerów magnetoreologicznych, Prace Naukowe Politechniki Wrocławskiej, Mechanika (2007) 25-30
[42] Wang Y., Hu Y., Chen L., Gong X., Jiang W., Zhang P., Chen Z., Effects of rubber/magnetic particle interactions on the performance of magnetorheological elastomers, Polymer Testing 25 (2006) 262–267
[43] Hayashi S., Kondo S., Kapadia P., Ushioda E., Room temperature shape memory polymers, Plast. Eng. 2 (1995) 29
[44] Goupta R. K., Polymer and Composite Rheology, CRC Press, 2000
[45] Wilczyński K., Reologia w przetwórstwie tworzyw sztucznych, WNT, Warszawa 2001
[46] Rheology – theory and applications course, TA Instruments, USA 2004
[47] Li W.H., Du H., Chen,G., Yeo S.H.. Viscoelastic properties of MR fluids under oscillatory shear. Smart Structures and Materials 2001: Damping and Isolation. Inman, D.J. (Ed.). Proceedings of SPIE, 4331 (2001) 333-342
[48] Ferry J.D., Viscoelastic properties of polymers, John Wiley & Sons, Inc. 1980
[49] Kelar K., Ciesielska D., Fizykochemia polimerów. Wybrane zagadnienia, Wydawnictwo Politechniki Poznańskiej, Poznań 1997
[50] Lokander M., Reitberger T., Stenberg B., Oxidation of natural rubber-based magnetorheological elastomers; Polymer Degradation and Stability 86 (2004) 467-471
[51] Borcea, L., Bruno, O., On the magneto-elastic properties of elastomer-ferromagnet composites. J. Mech. Phys. Solids 49 (2001) 2877–2919
[52] Demchuk S.A., Kuzmin V.A., Viscoelastic properties of magnetorheological elastomers in the regime of dynamic deformation. Journal of Engineering Physics and Thermophysics, Vol. 75, No. 2, (2002) 396-400
[53] Yalcintas M., Dai H., Vibration suppression capabilities of magnetorheological materials based adaptive structures. Smart Mater. Struct. 13 (2004) 1-11
[54] Zhou, G.Y., Jiang, Z.J., Deformation in magnetorheological elastomer and elastomer-ferromagnet composite driven by a magnetic field. Smart Mater. Struct. 13 (2004) 309-316
[55] Dorfmann A., Ogden R.W. Magnetoelastic modeling of elastomers, European Journal of Mechanics A/ Solids 22 (2003) 497-507
[56] Böse H., Röder R., Magnetoreological elastomers with high variability of their mechanical properties, 11th Conference on Electrorheological Fluids and Magnetorheological Suspensions, Journal of Physics: Conference Series 149 (2009) 1-6
[57] Jolly, M.R., Carlson, J.D., Munoz, B.C., A model of the behaviour of magnetorheological materials, Smart Mater. Struct. 5, (1996) 607-614
[58] Ginder J.M., Clark S.M., Schlotter W.F., Nichols, M.E. Magnetostrictive phenomena in magnetorheological elastomers, Int. J. Modern Phys. B, Vol. 16, Nos. 17-18, (2002) 2412-2418
[59] Ray S., Shanmugharaj M., Bhowmick K., A new parameter for interpretation of polymer-filler and filler-filler interactions in rubber vulcanizates, J. Mater. Sci. Letters 21 (2002) 1097-1100

[60] Ginder J.M., Schlotter W.F., Nichols M.E., Magnetorheological elastomers in tunable vibration absorbers. Smart Structures and Materials 2001: Damping and Isolation, Inman, D.J (Ed.). Proceedings of SPIE, Vol. 4331 (2001) 103-110

[61] Lokander M., Stenberg B., Performance of isotropic magnetorheological rubber materials, Polymer Testing 22 (2003) 245-251

[62] Banks H.T., Pinter G.A., Potter L.K., Gaitens M.J., Yanyo L.C., Modeling of Nonlinear Hysteresis in Elastomer under Uniaxial Tension, Journal of Intelligent Material Systems and Structures, vol. 10-February 1999

[63] Zhou G.Y., Shear properties of magnetorheological elastomer; Smart Materials and Structures 12 (2003) 139-146

[64] Zhou G.Y., Li J.R., Dynamic behavior of a magnetorheological elastomer under uniaxial deformation: I. Experiment, Smart Mater. Struct.12 (2003) 859-872

[65] Bose H., Viscoelastic properties of silicon based magnetorheological elastomers, International Journal of Modern Physics B, vol. 21, no 28-29 (2007) 4790-4797

[66] Gong X.L., Chen L., Li J.F., Study of utilizable magnetorheological elastomers, International Journal of Modern Physics B, vol. 21, nos. 28 & 29 (2007) 4875-4882

[67] Bellan C., Bossis G., Field dependence of viscoelastic properties of MR elastomers. Int. J. Modern Phys. B, Vol. 16, No. 17 i 18, (2002) 2447-2453

[68] Jiang W.G., Yao J.J., Gong X.L., Chen L., Enhancement in Magnetorheological Effect of Magnetorheological Elastomers by Surface Modification of Iron Particles, Chin. J. Chem. Phys. 21 (2008) 87-92

[69] Zhang X.Z., Gong X.L., Zhang P.Q, Li W.H., Existence of Bound-Rubber in Magnetorheological Elastomers and Its Infuence on Material Properties, CHINESE JOURNAL OF CHEMICAL PHYSICS VOLUME 20, NUMBER 2 (2007), 173-179

[70] Li J., Gong X., Zhu H., Jiang W., Influence of particle coating on dynamic mechanical behaviors of megnetorheological elastomers, Polymer Testing 28 (2009) 331-337

[71] W. Szymczyk, A. Boczkowska, T. Niezgoda, K. Zubko; Experimental Validation of Numerical Methods of MRE Simulations, Solid State Phenomena, 154 (2009) 113-120

[72] A. Boczkowska, S.F. Awietjan, Effect of the elastomer stiffness and coupling agents on rheological properties of magnetorheological elastomers, Materials Characterisation V, WIT Transactions on Engineering Sciences, Vol 72, (2011) 263-274

Chapter 7

Thermo-Shrinkable Elastomers

Magdalena Maciejewska and Alicja Krzywania-Kaliszewska

Additional information is available at the end of the chapter

1. Introduction

The ability of polymeric materials to sense and respond to external stimuli has great scientific and technological significance. It enables these materials to change their properties, such as shape, colour and electrical conductivity, as a result of pH, temperature, chemicals, light, or stimulation by an electric or magnetic field. Materials that respond dynamically to external stimuli are called intelligent or smart materials, and it is important that their response should be repeatable and controllable (Landlein, 2010). One of the most important classes of smart materials is shape memory polymers (SMPs), which can change their shape in a predetermined way upon the application of an external stimulus. The shape memory effect in polymers depends primarily on the existence of separated phases that are related to the coiled structure, crosslinks (covalent bonds), hydrogen or ionic bonding or physical intermolecular interactions of the polymer (Hu, 2007). Covalent crosslinks are formed during suitable crosslinking of the polymer, whereas physical crosslinks are obtained when the polymer morphology consists of segregated domains, such as crystalline and amorphous phases or hard and soft segments (Landlein, 2010) (e.g., linear block copolymers). In multiphase polymers, the hard segments act as the frozen phase, which is usually semi-crystalline or physically crosslinked and provides stiffness and reinforcement to the material (Hu, 2007), while the soft segments are responsible for the thermo-elastic behaviour of polymers and act as the reversible phase. In this case, the shape memory effect is produced by the reversible phase transformation of the soft segments (Hu, 2007). Thermo-shrinkable polymers, which change their shape as the temperature changes, are a unique class of SMP materials with interesting properties and many potential applications. The shape transition temperature (T_{trans}) for this type of SMP can be the melting point (T_m) or glass transition temperature (T_g) of the soft phase (Ratna & Karger-Kocsis, 2008). Melting points are preferred because the transition is sharper than the glass transition; therefore, the temperature of the shape recovery can be better determined. Heating the SMP above the T_m or T_g of the hard segment enables its processing. This original (permanent) shape can be memorised by cooling the material below the T_m or T_g of the hard phase. Cooling the SMP below the T_m or T_g of the soft segment while the shape is deformed allows a temporary shape to be fixed. The permanent shape of the SMP is recovered by heating it above the T_m or T_g of the soft phase (Hu, 2007). Another way to fix the temporary shape of a SMP is to deform it at a temperature lower than the T_m or T_g of the soft phase, which causes stress and strain absorption by the soft segments. When the material is heated above the T_m or T_g of the soft segments, the stresses and strains are relieved, causing the material to return to its original shape (Leng & Du, 2010). The thermally induced shape memory effect in polymers is schematically presented in Figure 1. SMPs that use T_m as the shape transition temperature are represented by polymers such as polyurethanes, block copolymers of polyethyleneterephthalate and polyethyleneoxide, and copolymers consisting of polystyrene and poly(1,4-butadiene) (Wang et al., 1998). T_g is the shape transition temperature for thermoset SMPs; e.g., styrene-based resins (Ivens et al., 2011). SMPs can also be single phase materials with a certain number of crosslinks between their polymer chains. In this case, the crosslinks are the net points that enable fixing and storage of the permanent shape of the polymer, whereas the free polymer chains between the crosslinks act as switching segments that possess increased mobility above T_g (Ratna & Karger-Kocsis, 2008). Stretching the polymer chain segments in a certain direction reduces their entropy. At the same time, the net points deform elastically. The overall result of this process is an increase in the SMP enthalpy. The loss of polymer chain segment mobility in the switching segments stabilises the temporary shape of the SMP when the material is allowed to cool in the deformed state (Ivens et al., 2011).

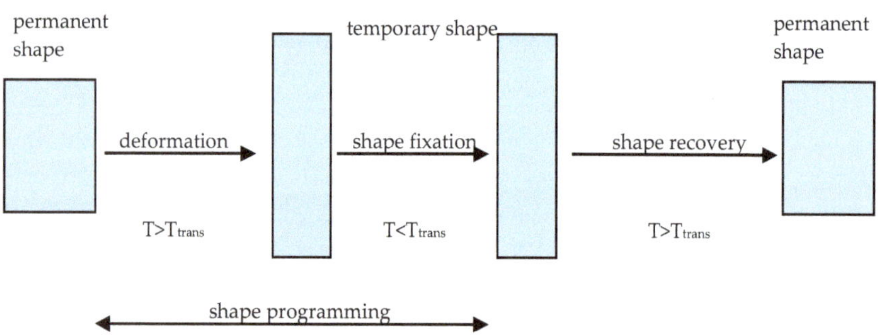

Figure 1. Thermally induced shape memory effect in polymers

Recently, research activities related to the development of elastomeric composites that exhibit a thermally induced shape memory effect have intensified. Thermoplastic elastomers (TPEs) are an interesting class of materials. TPEs behave like cured elastomers at room temperature, and they can be processed as plastics at higher temperatures because of their special multiphase morphology. TPEs consist of plastic phases embedded in a continuous elastomer phase, and they form physical crosslinks. Among the TPEs, segmented polyurethane elastomers and ionomers or blends of elastomers with thermoplastic polymers have many potential applications. In shape memory polyurethanes, the hard segment phases with the highest thermal transition temperatures (T_{perm}) acts as the physical crosslink points and control the permanent shape of the polymer. When the polymer is heated above T_{perm}, the physical crosslinks between the hard segments are destroyed. The molecular chains melt, and the polymer can be processed to a permanent shape like a thermoplastic material. When the shape memory polyurethane is cooled below T_{perm} and above T_g or T_m of its soft segments, the polymer becomes relatively soft, but it cannot flow because of the physical crosslinks. Consequently, it can be easily deformed to a temporary shape by stretching or compression. Reheating the polymer above T_g or T_m of its soft segments but below T_{perm} induces shape recovery (Hu, 2007). In the case of polyolefin/elastomer blends, the crosslinked elastomeric phase causes an enhancement of the blend shrinkability upon heating. Crosslinked points in the elastomer network are believed to serve as memory points, increasing the heat shrinkability (Patra & Das, 1997). Weiss et al. designed a new type of SMP based on blends of an elastomeric ionomer and low molar mass fatty acids or their salts (Weiss et al., 2008). Nanophase separation of the ionomer was used to develop the permanent network, and the fatty acids or salts were used to produce a secondary network. The role of the ionomer was to provide a strong intermolecular bond between crystals of the fatty acids or salts and the polymer, acting as a physical crosslink below the fatty acids or salt T_m and allowing reshaping of the material above T_m. The polymer films were heated to 100°C and stretched to 47% strain, then cooled to room temperature to fix an elongated temporary shape. When the samples were reheated to 100°C, the film recovered to the permanent shape with a length recovery of approximately 92%. A two-way temperature-induced shape memory effect was observed for polymer laminates. SMP-laminated composites were prepared with SMP polyurethane films and elastic polymers films. Their two-way shape memory behaviour was produced by bending upon heating and reverse bending upon cooling. The shape memory mechanism was ascribed to the release of the elastic strain of the SMP layer upon heating and the recovery of the elastic strain induced by the bending force of the substrate layer upon cooling (Chen et al., 2008). Mishra et al. have studied the heat shrinkability behaviour of grafted low-density polyethylene/polyurethane elastomers. They have suggested that the interchain crosslinking between the grafted polyethylene and the elastomer improves their shrinkability (Mishra et al., 2004).

Zhang et al. reported a novel type of shape memory polymer blend that consisted of two immiscible components, an elastomer and a switch polymer. The elastomer could be a rubber or thermoplastic elastomer, and the switch polymer could be an amorphous or crystalline polymer. Styrene–butadiene–styrene tri-block copolymer (SBS) was chosen as the elastomer, and poly(ε-caprolactone) (PCL) was used as the switch polymer (Zhang et al., 2009). The SBS/PCL blends demonstrated good shape recovery performance, with a shape recovery ratio of approximately 100%. The shape memory effect was also observed in elastomer networks containing reversibly associating side-groups. The supramolecular shape-memory elastomer consisted of a lightly crosslinked polymer network that was covalently bonded to reversibly associating side-groups. These elastomers exhibited shape-memory effects arising from reversible hydrogen bond association, and the shape memory recovery rate was strongly dependent on the temperature. Hydrogen-bonding interactions could stabilise mechanically strained states in these elastomers, and the thermo-mechanical cycling produced a strain fixity of approximately 90% and a strain recovery of approximately 100% (Li et al., 2007).

Thermal shrinkability was also reported for elastomer blends containing carboxylated groups, which could form ionic labile bonds during crosslinking, and low-density polyethylene (Mishra et al., 2000) or poly(ethylene-vinyl-acetate) (Raychowdhury et al., 2000).

In our previous studies, we proved that the thermally induced shape memory effect was also exhibited by a carboxylated nitrile elastomer cured with zinc oxide because it contained labile ionic crosslinks, which were able to rearrange upon external deformation (Przybyszewska & Zaborski, 2009).

In this work, the thermo-shrinkable properties of carboxylated acrylonitrile – butadiene elastomer (XNBR) and hydrogenated acrylonitrile – butadiene elastomer (HNBR) containing ionic crosslinks were studied. The XNBR was vulcanised with nanosized calcium, magnesium oxide or zinc oxide to ensure the formation of ionic crosslinks during vulcanisation. In the case of HNBR, nanosized calcium and magnesium oxides were coated with unsaturated carboxylic acids (itaconic, 2,4-pentadienoic, oleic, linoleic and linolenic acids) and applied as coagents in the peroxide vulcanisation of the elastomer. The application of these coagents led to formation of ionic crosslinks in the elastomer network.

Heat-shrinkable polymers are widely used in packaging and in the cable industry; therefore, the shrinkability of XNBR and HNBR vulcanisates is technologically important.

2. Experimental section

2.1. Materials

Carboxylated nitrile elastomer XNBR (Krynac X7.50) containing 27 wt % acrylonitrile and 6.7 wt % carboxylic groups was obtained from Bayer C.O. The Mooney viscosity was (ML1+4 (100°C):47). Nanosized calcium oxide CaO (Aldrich), magnesium oxide MgO (Nanostructured & Amorphous Materials Inc., Houston, USA) and zinc oxide ZnO (Nanostructured & Amorphous Materials Inc., Houston, USA) were used as crosslinking agents. Hydrogenated acrylonitrile-butadiene elastomer HNBR (Therban 3407) containing 34 wt % acrylonitrile and 0.9 wt % of residual double bonds after hydrogenation was obtained from Bayer C.O. The Mooney viscosity was (ML1+4 (100°C):70). It was vulcanised with dicumyl peroxide DCP (Aldrich). Nanosized calcium oxide (Aldrich), magnesium oxide (Nanostructured & Amorphous Materials Inc., Houston, USA), itaconic acid IA (Fluka), 2-4-pentadienoic acid (Aldrich), oleic acid, linoleic acid and linolenic acid (Aldrich) were applied as coagents.

2.2. Preparation of coagents

The nanosized metal oxides were mixed with a solution of modifying agent (unsaturated carboxylic acid) in acetone for 30 minutes during ultrasonic treatment (BANDELIN DT 255) at a frequency of 35 kHz. The mixture was left for 24 hours. Then, the solvent (acetone) was evaporated using a vacuum evaporator at 50°C. The coagents obtained were dried in a vacuum drier at 70°C for 96 hours.

2.3. Preparation and characterisation of rubber compounds

Rubber compounds with the formulations given in Table 1 were prepared using a laboratory two-roll mill. The samples were cured at 160°C until they developed a 90% increase in torque, as measured by an oscillating disc rheometer.

The crosslink densities (ν_T) of the vulcanisates were determined by their equilibrium swelling in toluene, based on the Flory-Rehner equation (Flory & Rehner, 1943). The Huggins parameter of the XNBR-solvent interaction (χ) was calculated from the equation $\chi = 0.487 + 0.228V_r$ (Equation 1) (Przybyszewska & Zaborski, 2008), where V_r is the volume fraction of elastomer in the swollen gel, and $\chi = 0.501 + 0.273V_r$ for HNBR-solvent interaction (Equation 2) (Przybyszewska & Zaborski, 2009). To determine the content of ionic crosslinks in the elastomer network, samples were swollen in toluene in a dessicator with saturated ammonia vapour (25% aqueous solution). The ionic crosslink content ($\Delta\nu$) was calculated from Equation 3, where ν_A is the crosslink density determined for samples treated with ammonia vapour.

$$\Delta\nu = \frac{\nu_T - \nu_A}{\nu_T} \bullet 100\% \quad (3)$$

The tensile properties of the vulcanisates were determined according to ISO-37 with a ZWICK 1435 universal machine.

XNBR		HNBR	
Elastomer	100	Elastomer	100
Nanosized metal oxide	4	DCP	2
		Coagent	7

Table 1. Composition of the XNBR and HNBR-based rubber compounds [phr]

2.4. Dynamic-mechanical analysis

Dynamic - mechanical measurements were carried out in the tension mode using a DMA/SDTA861^e analyser (Mettler Toledo). Measurements of the dynamic moduli were performed over the temperature range (-60 - 120°C) for XNBR and (-80 - 100°C) for HNBR with a heating rate of 2°C/min, a frequency of 1 Hz and a strain amplitude of 4 μm. The temperature of the elastomer glass transition was determined from the maximum of $\tan\delta = f(T)$, where $\tan\delta$ is the loss factor and T is the measurement temperature.

2.5. Shrinkability measurements

To measure the shrinkability of the XNBR and HNBR vulcanisates, the samples were stretched above their T_g at a temperature of 100°C until they reached an elongation of 200%

and left in the stretched form for 48 h. They were then stabilised in the stretched form for 4 h at (-7°C). Finally, the stretched samples were allowed to shrink above the ionic transition temperature at 70°C for 48 h. The length of the samples at each state of study was measured using the digital callipers (Preisser) with the measurement error 1 mm. The lengthwise shrinkage was calculated according to Equation 4, in which S_h is the percentage of shrinkability, L_{str} is the length of the sample after stretching, and L_{shr} is the length of the shrunk sample. The maximum shrinkage, S_{hmax}, was calculated from the length of the sample before stretching according to Equation 5, in which L_0 is the original length of the sample before stretching. The physical properties of the vulcanisates were studied before and after the thermal treatment.

$$S_h\% = \frac{L_{str} - L_{shr}}{L_{str}} \bullet 100 \tag{4}$$

$$S_{h\max}\% = \frac{L_{str} - L_0}{L_{str}} \bullet 100 \tag{5}$$

The continuous increase in temperature causes the recovery of the stretched sample deformation, which reflects the memory effect of the vulcanisate. The percentage recovery R is the ratio of the lengthwise shrinkage to the maximum shrinkage (Equation 6) (Khonakdar et al., 2007).

$$R\% = \frac{S_h}{S_{h\max}} \bullet 100 \tag{6}$$

2.6. Scanning Electron Microscopy (SEM)

The morphology of the metal oxide particles and their dispersion in the elastomer matrix were estimated using scanning electron microscopy with a LEO 1530 SEM. The vulcanisates were broken down in liquid nitrogen, and the surfaces of the vulcanisate fractures were examined. Prior to the measurements, the samples were coated with carbon.

3. Results and discussion

3.1. Thermo-shrinkable XNBR vulcanisates

3.1.1. Crosslink density and ionic crosslink content of XNBR vulcanisates

The carboxylated nitrile elastomer XNBR reacts with metal oxides to form carboxylic salts, which behave as ionic crosslinks. These salts are able to associate, forming multiplets and clusters. This association is caused by the electrostatic interactions between multiplets, and it is impaired by the retractive elastic forces of the backbone chains. The restricted elastomer chain mobility in the proximity of the ionic clusters produces a hard phase surrounded with the soft elastomer matrix (Mishra et al., 2000). The biphasic structure of the XNBR crosslinked with metal oxide and the presence of labile ionic crosslinks provide possible routes to obtain thermo-shrinkable composites.

Because the ionic crosslinks play a crucial role in the return of the sample to the original shape and serve as memory points, the crosslink density and ionic crosslinks content in the elastomer network were determined in the first stage of the study. These results are presented in Figs. 2 and 3.

The nanosized calcium, magnesium and zinc oxides exhibited high crosslinking activity in XNBR. Their application led to the formation of ionic crosslinks in the elastomer network. The content of ionic crosslinks was in the range of 55% to 76%. Nanosized zinc oxide, for which the highest vulcanisate crosslink density and ionic crosslink content was observed (approximately 76%), appeared to be the most active, whereas the lowest activity was exhibited by CaO (with an ionic crosslink content of approximately 55%). Therefore, it could be supposed that the highest shape recovery would be obtained for vulcanisates crosslinked with nanosized ZnO.

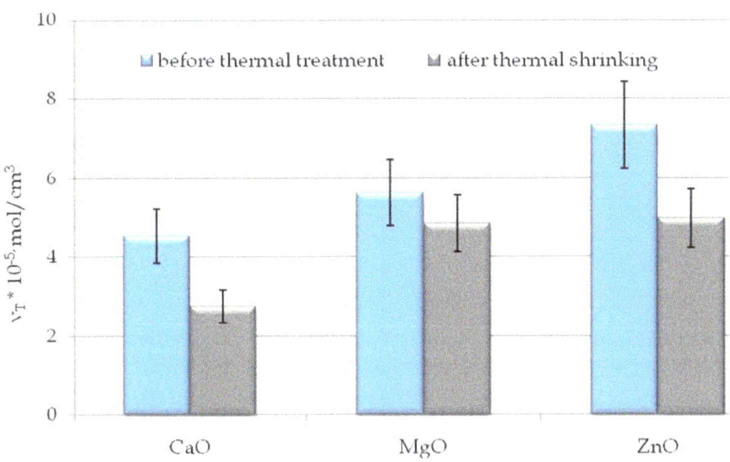

Figure 2. Crosslink density of XNBR vulcanisates

The stretched samples were allowed to shrink above the ionic transition temperature at 70°C to initiate their return to the original shape. The influence of this thermal treatment on the ionic crosslink content and crosslink density of the vulcanisates was studied. From the data presented in Fig. 3, it follows that heating the samples to 70°C caused decomposition of the ionic crosslinks. The number of ionic crosslinks in the elastomer network was reduced by 21-34% compared to the vulcanisates before thermal treatment. The greatest number of ionic crosslinks was decomposed in the case of vulcanisates containing nanosized ZnO.

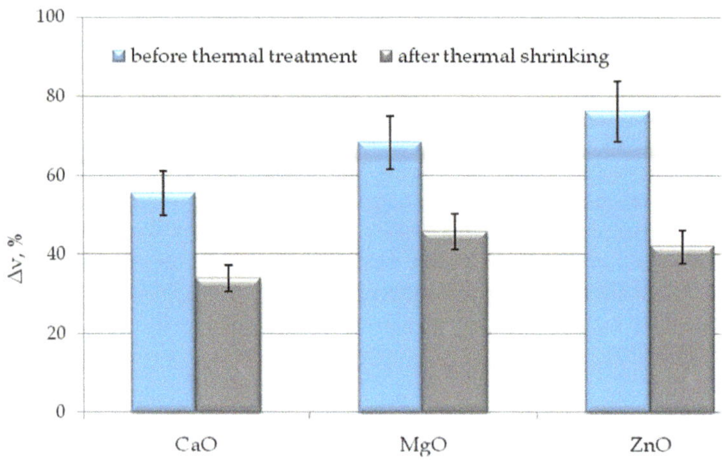

Figure 3. Ionic crosslink content in XNBR vulcanisates

3.1.2. Thermo-shrinkability of XNBR vulcanisates

Having determined the number of ionic crosslinks in the elastomer network and confirmed that these crosslinks could decompose during thermal treatment, we then examined the thermo-shrinkability of the vulcanisates. In Fig. 4, the percentage of thermo-shrinkability and the percentage shape recovery of the vulcanisates are presented.

The vulcanisates of XNBR crosslinked with nanosized calcium, magnesium and zinc oxides exhibited heat shrinkability. The greatest shrinkage upon heating (50%) was achieved for vulcanisates containing ZnO nanoparticles. The lower shrinkability of the vulcanisates with calcium and magnesium oxides (42% and 41%, respectively) was produced by the lower crosslink density and ionic crosslink content in their elastomer networks. Because the crosslinked points in the elastomer network serve as shape memory sites, a greater crosslink density improves the shrinkability of the XNBR. The vulcanisates demonstrated good shape recovery performance. The percentage shape recovery was in the range of 90-100% (Fig. 4). The highest percentage recovery was observed for vulcanisates containing ZnO, which had the greatest ionic crosslink content. The stretched XNBR samples shrunk upon heating above the temperature of the ionic transition because of the occurrence of ionic clusters in the elastomer network, which could rearrange or decompose. The results described in the previous section confirm that the decomposition of the ionic crosslinks is one of the reasons for the heat shrinkability of the XNBR vulcanisates containing nanosized CaO, MgO and ZnO.

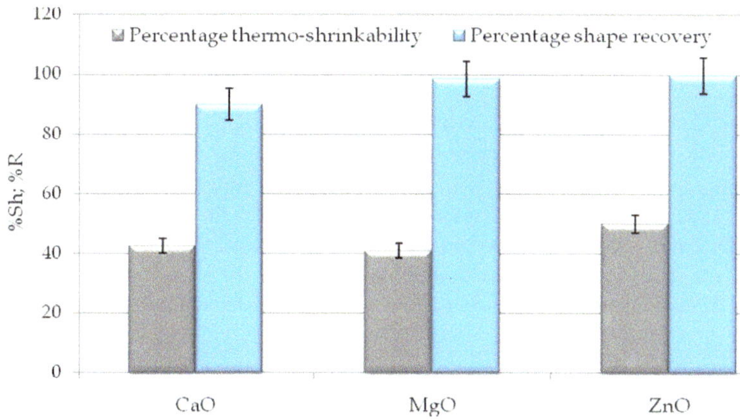

Figure 4. Percentage thermo-shrinkability and shape recovery of XNBR vulcanisates

3.1.3. Dynamic mechanical properties of XNBR vulcanisates

Dynamic - mechanical analysis was performed to confirm the biphasic structure of XNBR crosslinked with nanosized metal oxides, as well as the existence of ionic clusters in the elastomer network. The values of the glass transition temperature (T_g) are given in Table 2. The loss factor, tanδ, is presented in Fig. 5 as a function of temperature for the vulcanisates before thermal treatment.

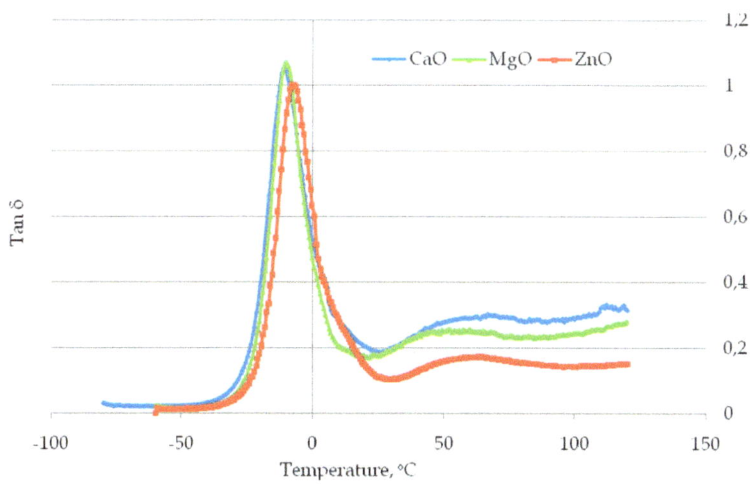

Figure 5. Tan δ versus temperature for XNBR vulcanisates before thermal treatment

The results of the DMA analysis confirm the biphasic structure of the XNBR crosslinked with nanosized CaO, MgO and ZnO. The existence of two phase transitions was observed. The first transition is the glass transition of the XNBR at low temperatures, the maximum of which represents T_g. The determined glass transition temperatures for the vulcanisates with MgO and ZnO were approximately (-11.1°C) and (-10.5°C), respectively, whereas the T_g value was (-13.0°C) for the vulcanisate with CaO nanoparticles. The CaO vulcanisate most likely exhibits the lowest T_g because it contains the lowest crosslink density of the vulcanisates. The second peak, which is fuzzy and has a low-intensity, was observed in the temperature range of (50-100°C). This peak corresponds to the ionic transition at high temperature that is caused by the occurrence of a hard phase arising from ionic associations (ionic clusters or aggregates). These transitions were observed for all XNBR vulcanisates. Therefore, the existence of a biphasic structure in XNBR crosslinked with metal oxides was confirmed.

Vulcanisate	T_g before thermal treatment, °C	T_g after thermal shrinking, °C
CaO	-13.0	-15.0
MgO	-11.1	-13.5
ZnO	-10.5	-16.0

Table 2. Glass transition temperature of XNBR vulcanisates

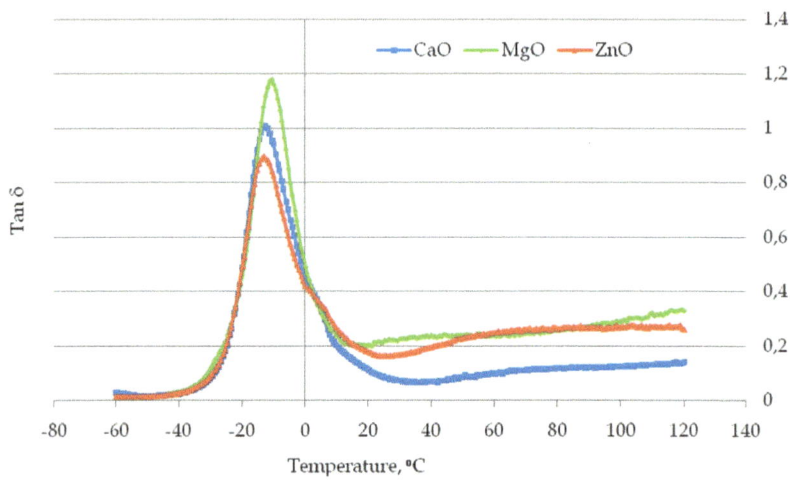

Figure 6. Tan δ versus temperature for XNBR vulcanisates after thermal shrinking

DMA measurements were also performed for the vulcanisates after the thermal shrinking process. These results are presented in Fig. 6. Heating the samples to the ionic transition temperature decreased the glass transition temperatures of the vulcanisates. This reduction was most likely caused by the decomposition of the ionic crosslinks, leading to a reduction of the crosslink density of the vulcanisates. The highest decrease in the T_g value was observed for the vulcanisate containing nanosized ZnO, for which the greatest amount of ionic crosslinks decomposed during the thermal treatment. The second fuzzy peak corresponding to the ionic transition disappeared, which suggests that the return of the vulcanisates to their original shape was caused by the decomposition of ionic crosslinks, along with changes in or the disappearance of the biphasic structure of the crosslinked elastomer.

3.1.4. Mechanical properties of XNBR vulcanisates

The mechanical properties of the vulcanisates (especially their tensile strength) are technologically important. Therefore, the effect of the thermal shrinking process on the tensile strength and elongation at break of the vulcanisates was studied. These results are presented in Table 3.

Vulcanisate	TS_0, MPa	TS_{shr}, MPa	EB_0, %	EB_{shr}, %
CaO	17.6	12.0	803	734
MgO	37.0	34.4	704	544
ZnO	28.1	20.7	634	525

Table 3. Tensile strength and elongation at break of XNBR vulcanisates before thermal treatment (TS_0, EB_0) and after thermal shrinking (TS_{shr}, EB_{shr})

Regarding the properties of the vulcanisates before the thermal treatment, the greatest tensile strength was exhibited by the vulcanisate with nanosized MgO, whereas the lowest was produced by CaO. One possible reason for the different activities of the metal oxides is their tendency to agglomerate in the elastomers (see Fig. 7).

The CaO nanoparticles were poorly dispersed in the elastomer matrix (Fig. 7a). They created microsized agglomerates with complex structures, which displayed poor adhesion to the elastomer. These agglomerates acted as centres for stress concentration in the vulcanisates during the deformation and initiate breakage of the sample under external stress. As a result, the tensile strength of the vulcanisates decreased. The ZnO nanoparticles also created microsized agglomerates that were smaller than the CaO particles and surrounded by an elastomer film (Fig. 7c). The wetting of the ZnO agglomerates with the elastomer probably produced the better mechanical properties of the ZnO vulcanisates, despite the heterogeneous dispersion of the nanoparticles. The MgO nanoparticles revealed the weakest ability to agglomerate in the XNBR, creating clusters of approximately 3 μm in size that were tightly bound to the elastomer matrix (Fig. 7b). The highest tensile strength was observed for the vulcanisate with MgO. The elongation at break was the lowest for the vulcanisate containing ZnO nanoparticles, and this result was correlated with the crosslink density of the examined vulcanisates.

The stretching of the samples at high temperature, their stabilisation in the stretched state and the shrinkage above the ionic transition temperature reduced the tensile strength and elongation at break of the vulcanisates by decomposing the ionic crosslinks and reducing the crosslink density of the vulcanisates. However, these mechanical properties remained

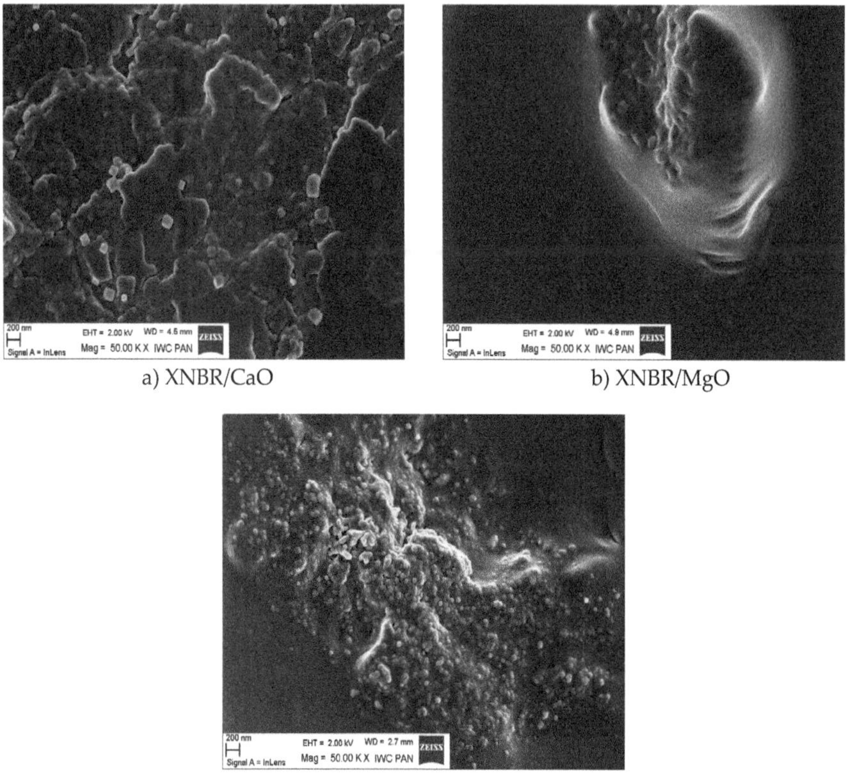

a) XNBR/CaO b) XNBR/MgO

c) XNBR/ZnO

Figure 7. SEM images of XNBR vulcanisates

satisfactory, especially in the case of the vulcanisates containing MgO and ZnO (TS_{shr}, EB_{shr} in Table 3).

3.2. Thermo-shrinkable HNBR vulcanisates

3.2.1. Crosslink density and ionic crosslink content for HNBR vulcanisates

Studies performed on the XNBR elastomer crosslinked with nanosized metal oxides confirmed the existence of a biphasic structure and the presence of ionic crosslinks in the elastomer network, which are able to rearrange or decompose upon heating to produce thermo-shrinkable vulcanisates. Therefore, to obtain thermo-shrinkable vulcanisates from the HNBR elastomer, multifunctional crosslinking coagents based on nanosized calcium and magnesium oxides were used in combination with unsaturated carboxylic acids (UCAs). UCAs containing easily abstractable hydrogen atoms and readily accessible double bonds were grafted onto the powder surface during the modification process. Because of their multifunctionality, this type of coagent based on zinc oxide was proven to be able to react directly with the elastomer and form effective covalent bonds. Moreover, they can form ionic crosslinks that increase the vulcanisate crosslink density and improve tensile strength (Przybyszewska & Zaborski, 2009). Because the presence of ionic crosslinks is crucial for the thermo-shrinkability of vulcanisates, the effect of these coagents on the crosslink density of the vulcanisates and their ionic crosslink content was studied. These results are given in Figs. 8 and 9.

The application of coagents based on nanosized CaO and MgO grafted with UCA caused the formation of ionic crosslinks in the elastomer network. The greatest content of ionic crosslinks was obtained for vulcanisates containing nanosized CaO in combination with oleic and linoleic acid, as well as for MgO with itaconic and linoleic acid. The results of these studies confirmed that these coagents contributed to the increased vulcanisation efficiency. The ionic crosslink content should affect the thermo-shrinkability of the vulcanisates considerably. The crosslinked points in the elastomer network are believed to serve as memory points, enhancing the heat shrinkability.

To confirm that the decomposition and/or rearrangement of the ionic crosslink aggregates is one of the causes of heat shrinkability in the examined HNBR vulcanisates, the crosslink density and ionic crosslink content were determined for the vulcanisates after thermal shrinking and compared with the values of the vulcanisates before thermal treatment.

The data presented in Figs. 8 and 9 suggest that the ionic crosslinks that decomposed during the shrinking of vulcanisates at 70°C allowed the samples to return to their original shape. As a result of the ionic crosslink decomposition, the crosslink density of the vulcanisates decreased. The most considerable reduction of the ionic crosslink number after shrinkage above the ionic transition temperature was observed for the vulcanisates containing CaO-OA, CaO-LA, MgO-IA and MgO-OA as coagents, which were characterised by the highest ionic crosslink content before the heat treatment. These vulcanisates may be expected to show the greatest thermal shrinkage and shape recovery.

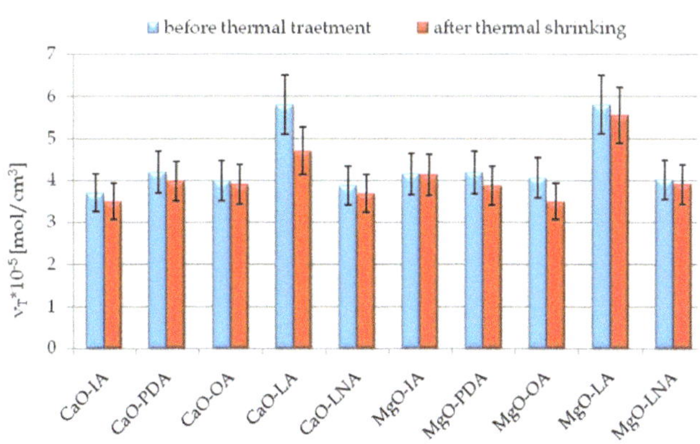

Figure 8. Crosslink density of HNBR vulcanisates

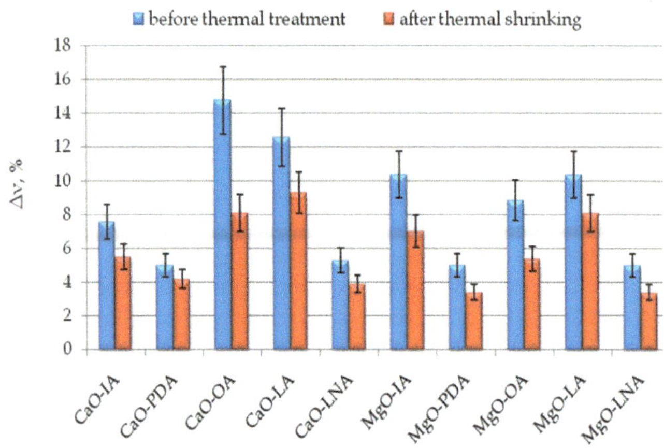

Figure 9. Ionic crosslink content of HNBR vulcanisates

3.2.2. Thermo-shrinkability of HNBR vulcanisates

As mentioned before, the shrinkage of polymer is caused by an internal rearrangement of the structural elements within the stretched sample (Mishra et al., 2000). In contrast to the covalent crosslinks formed during conventional vulcanisation with peroxides, ionic crosslinks are multifunctional and labile. Ionic crosslinks group together, forming clusters that are immersed in the elastomer matrix. Moreover, ionic clusters can rearrange in the elastomer matrix upon external deformation or temperature change.

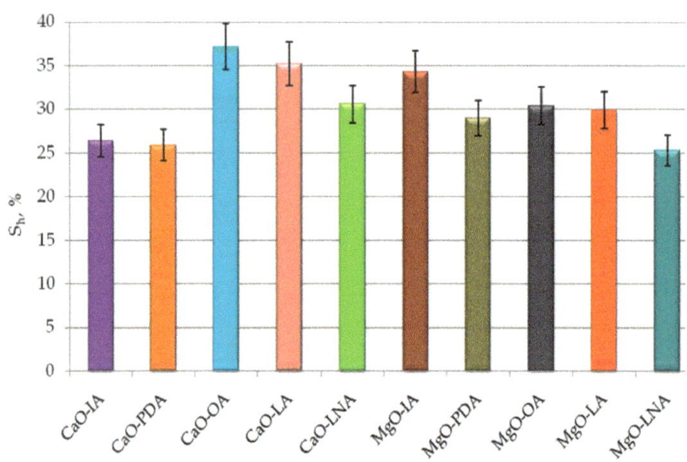

Figure 10. Percentage thermo-shrinkability of HNBR vulcanisates

The crosslinked samples were heated and stretched at 100°C (a temperature above the glass transition temperature). Cooling these samples in this stretched state stabilised the temporary shape. Finally, the stretched samples were allowed to shrink above the ionic transition temperature at 70°C. The heat shrinkability values and percentage recovery of the HNBR samples containing 7 phr of the coagents are presented in Figs. 10 and 11.

The vulcanisates of the HNBR elastomers crosslinked in the presence of coagents based on nanosized CaO and MgO in combination with unsaturated carboxylic acids exhibited heat shrinkability. The greatest shrinkage upon heating (approximately 35%) was achieved for vulcanisates containing the CaO-OA, CaO-LA and MgO-IA coagents (Fig. 10). The lower shrinkability of the vulcanisates with other coagents is a result of the lower crosslink density and ionic crosslink number in their elastomer networks (see Figs. 8 and 9). Because the crosslinked points in the elastomer network serve as shape memory sites, a higher crosslink density improves the shrinkability of the vulcanisate.

The HNBR-based vulcanisates demonstrated good shape recovery performance, with a shape recovery percentage in the range of 91-100% (Fig. 11). The greatest percentage recovery was obtained for the vulcanisates with the highest ionic crosslink content. Moreover, the most considerable reduction of the ionic crosslink number during the thermal shrinking process was observed for these vulcanisates. This result confirms the assumption that the decomposition of ionic crosslinks is the most important reason for the shrinkability and shape recovery of HNBR vulcanisates. It can be concluded that the ionic crosslinks formed in the elastomer by the coagents serve as memory points in the elastomer network.

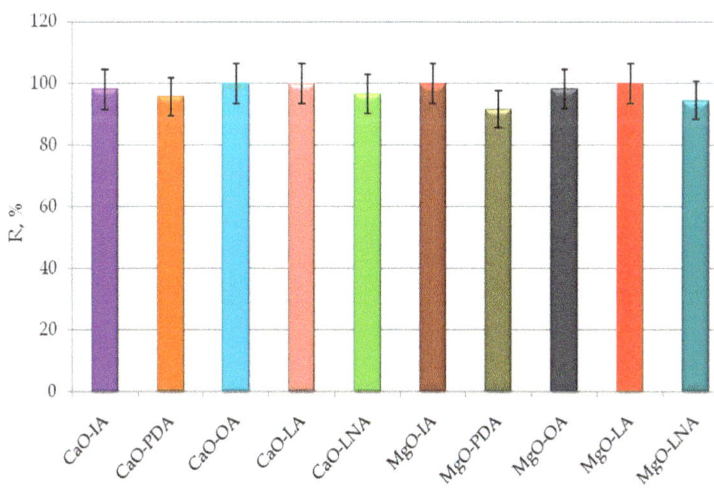

Figure 11. Percentage recovery of HNBR vulcanisates

3.2.3. Dynamic mechanical properties of HNBR vulcanisates

Dynamic - mechanical analysis was performed to confirm the existence of ionic clusters in the elastomer network. The loss factor (tanδ) of the HNBR vulcanisates with coagents is presented as a function of temperature in Figs. 12 and 13 as an example. The values of the glass transition temperature (T_g) are given in Table 4.

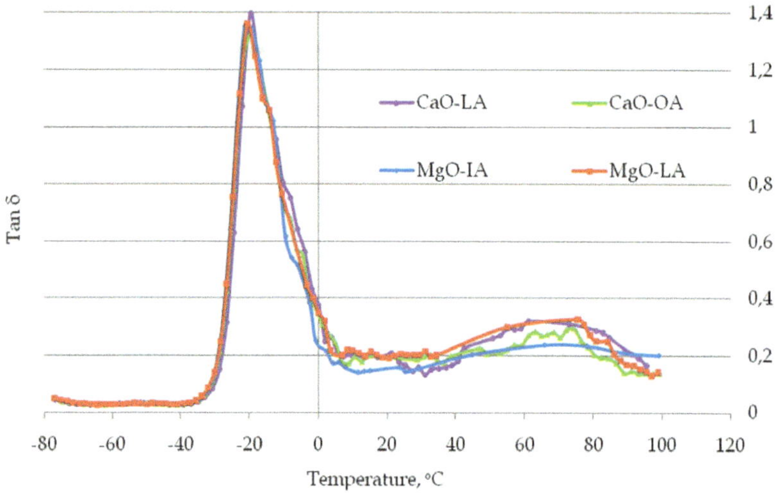

Figure 12. Tan δ versus temperature for HNBR vulcanisates before thermal treatment

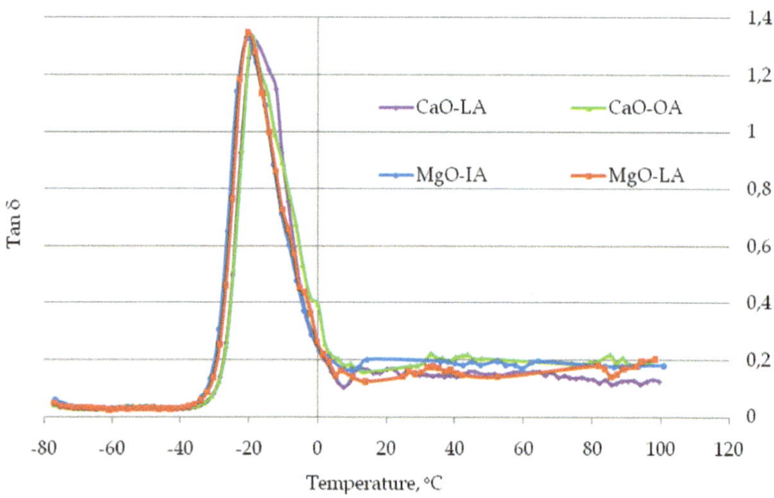

Figure 13. Tan δ versus temperature for HNBR vulcanisates after thermal shrinking

Vulcanisate	T_g before thermal treatment, °C	T_g after thermal shrinking, °C
CaO-IA	-21.2	-22.5
CaO-PDA	-20.7	-21.3
CaO-OA	-20.4	-22.7
CaO-LA	-19.1	-22.0
CaO-LNA	-19.6	-20.8
MgO-IA	-19.4	-22.9
MgO-PDA	-20.3	-21.5
MgO-OA	-20.6	-23.1
MgO-LA	-17.8	-20.4
MgO-LNA	-19.3	-21.0

Table 4. Glass transition temperature of HNBR vulcanisates

The DMA analysis revealed the biphasic structure of the HNBR crosslinked in the presence of coagents based on nanosized CaO and MgO grafted with UCA. Two phase transitions were observed. The glass transition of the HNBR, which occurs at T_g, was observed in the range from (-21.2°C) to (-19.1°C) for vulcanisates with CaO and from (-20,6°C) to (-17.8°C) for vulcanisates containing MgO. The determined T_g values were correlated with the crosslink density of the vulcanisates. A fuzzy low-intensity peak was observed in the temperature range of (40-100°C) (corresponding to the ionic transition) because of the occurrence of a hard phase arising from the existence of the ionic aggregates.

Heating to the ionic transition temperature decreased the glass transition temperature of the vulcanisates by decomposing ionic crosslinks, reducing the crosslink density of the vulcanisates. The greatest decrease of the T_g value occurred in the case of the vulcanisates for which the greatest number of ionic crosslinks decomposed during heating (CaO-OA, CaO-LA, MgO-IA, MgO-OA). The fuzzy peak corresponding to the ionic transition disappeared; consequently, it can be concluded that the shape recovery of the HNBR vulcanisates was caused by the decomposition of the ionic crosslinks and changes in the biphasic structure of the crosslinked elastomer, similarly to XNBR.

3.2.4. Mechanical properties of HNBR vulcanisates

Having established the ability of the HNBR vulcanisates to shrink upon exposure to the ionic degradation temperature and return to their original shape, we then examined their mechanical properties.

It is known that the formation of coagent bridges, which are labile ionic crosslinks inside the elastomer network formed during vulcanisation, improved the tensile properties of the vulcanisates (Przybyszewska & Zaborski 2008, 2009). However, it is reasonable to investigate the effect of the heat treatment of the vulcanisates on their mechanical properties.

As could be supposed, the thermal treatment of the vulcanisates during their shrinkage deteriorated their tensile strength as a result of the decomposition of their ionic crosslinks.

Their tensile strength was reduced by 5% to 30% in comparison with that of the vulcanisates before thermal treatment (Fig. 14). The vulcanisates subjected to shrinking at elevated temperature also exhibited a lower elongation at break (Fig. 15).

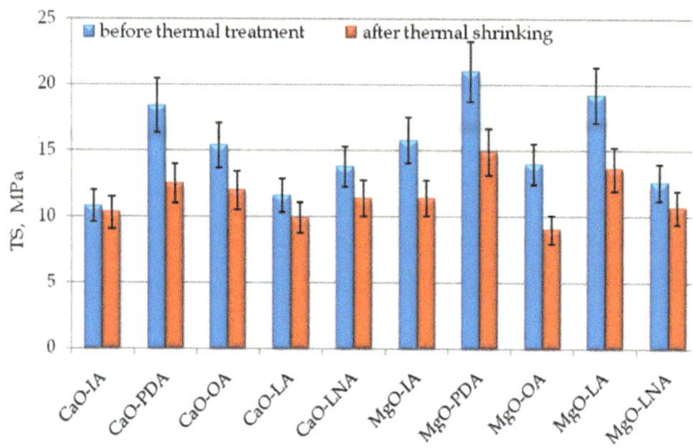

Figure 14. Tensile strength of HNBR vulcanisates

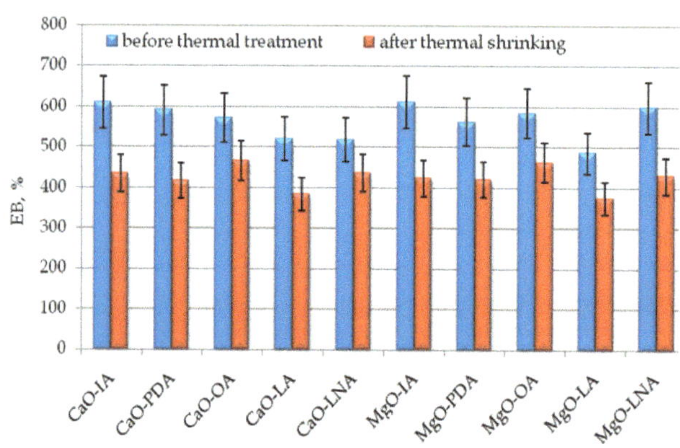

Figure 15. Elongation at break of HNBR vulcanisates

4. Conclusions

The thermo-shrinkable properties of carboxylated (XNBR) and hydrogenated (HNBR) acrylonitrile butadiene elastomer were studied. XNBR was crosslinked with nanosized calcium, magnesium and zinc oxide to ensure the formation of ionic crosslinks in the elastomer matrix, which can serve as memory points and enhance its heat shrinkability. Similarly, HNBR was crosslinked in the presence of coagents based on nanosized calcium and magnesium oxides grafted with unsaturated carboxylic acids to achieve ionic crosslinks. The examined samples were allowed to shrink at a temperature above the ionic transition temperature. The XNBR vulcanisates revealed thermo-shrinkability in the range of 41% for MgO to 50% for ZnO, whereas the percentage of shape recovery was in the range of 90-100%. Good shape recovery performance was also observed for the HNBR vulcanisates. The greatest shrinkage upon heating (approximately 35%) was achieved for the vulcanisates containing CaO-OA, CaO-LA and MgO-IA coagents. The percentage of shape recovery of these vulcanisates was in the range of 91-100%.

The thermo-shrinkability value and shape recovery ratio were strongly correlated with the ionic crosslink content of the elastomer network, and the number of ionic crosslinks was reduced by heating the samples above their ionic transition temperature. Greater ionic crosslink contents and their more significant decomposition corresponded to vulcanisates with greater thermo-shrinkability. DMA measurements confirmed the presence of ionic crosslinks and the existence of a biphasic structure in both the XNBR and HNBR elastomers. A fuzzy low-intensity peak was observed in the (tan δ) curve, in the temperature range of (50-100°C), which corresponds to the ionic transition that occurs at high temperatures as a result of the occurrence of a hard phase arising from ionic aggregates. This peak disappeared when the samples were heated above their ionic transition temperature. Therefore, it could be concluded that the thermo-shrinkability of the XNBR and HNBR vulcanisates and their shape recovery was a result of the decomposition of ionic crosslinks and changes in or the disappearance of the biphasic structure of the crosslinked elastomer.

Author details

Magdalena Maciejewska and Alicja Krzywania-Kaliszewska
Technical University of Lodz; Institute of Polymer and Dye Technology, Poland

Acknowledgement

The authors wish to acknowledge the Polish Ministry of Science and Higher Education as well as the National Centre of Research and Development for supporting this study.

5. References

Chen, S., Hu, J., Zhuo, H., Zhu, Y. (2008). Two-way shape memory effect in polymer laminates. *Materials Letters*, Vol. 62, No. 25, pp. (4088-4090), ISSN 0167-577X

Flory, P.J., Rehner, J. (1943). Statistical mechanics of cross-linked polymer networks. II. Swelling. *Journal of Chemical Physics*, Vol. 11, No. 11, pp. (521-526), ISSN 1089-7690

Hu, J. (2007). *Shape memory polymers and textiles* (1), Woodhead Publishing Limited, ISBN 978-1-84569-047-2, Cambridge

Ivens, J., Urbanus, M., De Smet, C. (2011). Shape recovery in a thermoset shape memomy polymer and its fabric-reinforced composites. *eXPRESS Polymer Letters*, Vol. 5, No. 3, pp. (254-261), ISSN 1788-618X

Khonakdar, H.A., Jafari, S.H., Rasouli, S., Morshedian, J., Abedini, H. (2007). Investigation and modeling of temperature dependence recovery behavior of shape-memory crosslinked polyethylene. *Macromolecular Theory and Simulation*, Vol. 16, No. 1, pp. (43-52), ISSN 1521-3919

Landlein, A (Ed(s).). (2010). *Shape-Memory Polymers*, Springer, ISBN 978-3-642-12358-0, New York

Leng, J., Du, S. (Ed(s).). (2010). *Shape-Memory Polymers and Multifunctional Composites*, Taylor and Francis Group, ISBN 978-1-4200-9019-2, Boca Raton

Li, J., Viveros, J.A., Wrue, M.H., Anthamatthen, M. (2007). Shape-memory effects in polymer networks containing reversibly associating side-groups. *Advanced Materials*, Vol. 19, No. 19, pp. (2851-2855), ISSN 1521-4095

Mishra, J.K., Kim, I., Chang-Sik, H. (2004). Heat shrinkable behavior and mechanical response of a low-density polyethylene/millable polyurethane/organoclay ternary nanocomposite. *Macromolecular Rapid Communications*, Vol. 25, No. 21, pp. (1851-1855). ISSN 1521-3927

Mishra, J.K., Raychowdhury, S., Das, C.K. (2000). Effect of interchain crosslinking on the shrinkability of the blends consisting of grafted low-density polyethylene and carboxylated nitrile rubber. *Materials Letters*, Vol. 46, No. 4, pp. (212-218), ISSN 0167-577X

Patra, P.K., Das, C.K. (1997). Blends of polyolefins and chlorosulphonated polyethylene (CSM) with special reference to their shrinkability and flame retardancy. *International Journal of Polymeric Materials*, Vol. 35, No. 1-4, pp. (103-118), ISSN 1563-535X

Przybyszewska, M., Zaborski, M. (2008). The effect of zinc oxide nanoparticle morphology on activity in crosslinking of carboxylated nitrile elastomer. *eXPRESS Polymer Letters*, Vol. 3, No. 9, pp. (542-552), ISSN 1788-618X

Przybyszewska, M., Zaborski, M. (2009). New coagents in crosslinking of hydrogenated butadiene-acrylonitrile elastomer based on nanostructured zinc oxide. *Composite Interfaces*, Vol. 16, No. 2-3, pp. (131-141), ISSN 0927-6440

Ratna, D., Karger-Kocsis, J. (2008). Recent advances in shape memory polymers and composites: A review. *Journal of Materials Science*, Vol. 43, No. 1, pp. (254-269), ISSN 1573-4803

Raychowdhury, S., Mishra, J.K., Das, C.K. (2000). Structure, shrinkability and thermal property correlations of ethylene vinyl acetate (EVA)/carboxylated nitrile rubber (XNBR) polymer blends. *Polymer Degradation and Stability*, Vol. 70, No. 2, pp. (199-204), ISSN 0141-3910

Wang, M., Luo, X., Ma, D. (1998). Dynamic mechanical behavior in the ethylene terephthalate-ethylene oxide copolymer with long soft segment as a shape memory material. *European Polymer Journal*, Vol. 34, No. 1, pp. (1-5), ISSN 0014-3057

Weiss, A., Izzo, E., Mandelbaum, S. (2008). New design of shape memory polymers: mixtures of an elastomeric ionomer and low molar mass fatty acids and their salts. *Macromolecules*, Vol. 41, No. 8, pp. (2978-2980), ISSN 1520-5835

Zhang, H., Wang, H., Zhong, W., Du, Q. (2009). A novel type of shape memory polymer blend and the shape memory mechanism. *Polymer*, Vol. 50, No. 6, pp. (1596-1601), ISSN 0032-3861

Section 5

Elastomers in Microelectronics and Microfluidics

Chapter 8

Elastomeric Electronics: A Microfluidic Approach

Shi Cheng

Additional information is available at the end of the chapter

1. Introduction

Thousands of millions of electronic units have been produced since the first invention of electronics. These devices and systems, in which stiff, solid metals, insulators, and semiconductors have dominated for ages, are exclusively made in the rigid format, and usually retain a static shape once fabricated. Things have however changed recently. Researchers are currently attempting to shape a soft and rubbery future for electronics, where elastic materials like elastomers would be widely used to replace conventional rigid materials. The reason is rather simple. We as human beings are soft, just like most other living creatures in nature, but existing electronics are not. Shouldn't we create some kind of new electronics that resemble ourselves?

Extensive research activities on developing elastomeric electronics that may withstand severe bending, twisting, and straining, with maintained electronic functions, have been recently carried out. The mechanical characteristics of gold thin films on polydimethylsiloxane (PDMS) surfaces: wrinkles and cracks, have been investigated by Whitesides and Hutchinson for the first time in 1998 [1]. Subsequently, wrinkled electrodes on pre-stretched elastic substrates were employed to implement electroactive polymer actuators [2], and studies on wrinkled gold electrodes on PDMS substrates were conducted in the Wagner's and Suo's groups at Princeton [3],[4]. Micromechanics and manufacturing processes based on relaxed and pre-stressed PDMS slabs were developed in-depth. The most significant contributions were made by the Rogers's group at University of Illinois, Urbana-Champaign, where stretchable and foldable silicon integrated circuits (ICs) on "wavy" silicon ribbons (from Silicon On Insulator (SOI)) in PDMS were proposed for fully integrated stretchable electronics [5],[6]. Dozens of appealing devices based on stretchable silicon ICs, including electronic eye sensor [7], smart gloves/skins [8], implanted medical devices [9], and wearable ergonomic biomedical sensors [10], have been demonstrated. An alternative approach utilizing anisotropic etching of bulk wafers was also introduced [11],[12]. Later, thin meandered stretchable interconnects encased in silicone rubber substrates, operating at different frequencies, were presented [13]-[15]. New electrical nanocomposite materials containing sliver nano-particles or carbon nanotubes/graphene that can flex and strain showed promising results as well [16],[17].

Microfluidics based stretchable electronics were initiated by Whitesides, who first studied low temperature metling solder filled microstructured elastomeric channels [18],[19]. Recently, this concept was extended to implement elastic direct current (DC) circuits, by incorporating eutectic gallium and indium alloy (EGaln) into microfluidic channels in thin sheets of elastomers [20]-[22]. However, all these previously mentioned studies deal with either low-frequency ICs or relatively simple interconnects. Stretchable electronics at radio frequencies remained an unexploited field until early 2009, at which time Cheng demonstrated the first stretchable fluidic antennas, for enabling wireless communication and remote sensing [23],[24]. Similar work was reported afterwards using EGaln alloy as conductors for the antennas instead of Galinstan [25]. Significantly enhanced elasticity of the resulting antennas could be achieved by introducing a different type of siloxane [26], and mechanically reconfigurable antennas could also be realized [27]. In 2010, Cheng proposed a hybrid integration strategy for the first demonstration of active microfluidic stretchable RF electronics (μFSRFEs), an integrated RF radiation sensor, *cf.* Fig. 1.a. [28]. Cheng further developed the μFSRFEs to the multi-layer configurations, and reported an integrated stretchable large-area wireless strain sensor, as seen in Fig. 1.b. [29].

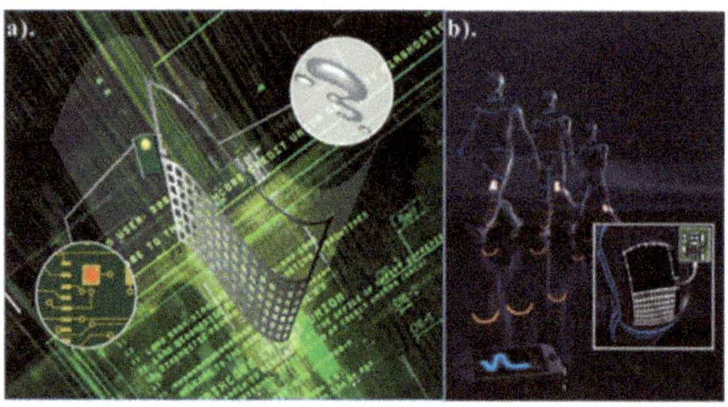

Figure 1. Foldable, flexible, stretchable elastomeric electronic devices: a). Microfluidic stretchable radiation sensor. b). Elastomeric, reversibly stretchable, large-area wireless strain sensor.

The following sections in this chapter address the recent progresses in the emerging field of microfluidics based elastomeric electronics. Fabrication processes, integration techniques, application examples, as well as future perspectives are presented and discussed.

2. Fabrication of single-layer microfluidics based passive elastomeric electronic devices

In brief, the manufacturing process steps can be summarized as follows: master construction, molding, liquid-alloy injection or filling, and encapsulation, *cf.* Fig. 2.

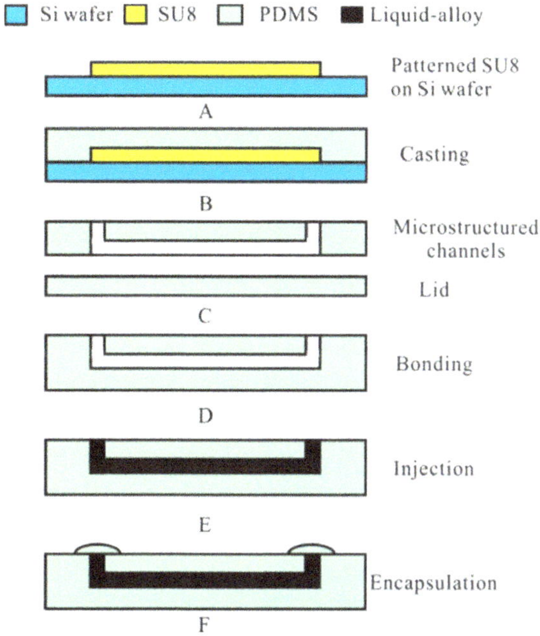

Figure 2. Manufacturing process steps of single-layer passive microfluidics based elastomeric electronics.

First of all, the design patterns are transferred to a 100 μm thick SU-8 100 (MicroChem, Newton, MA) layer on a silicon wafer with a standard soft lithography, and then developed and thermally stabilized at 150°C for 30 min to enhance the adhesion between SU-8 layer and silicon substrate. Subsequently, the PDMS prepolymer and cross linker (Elastosil RT601A and B, Wacker Chemie, Munich, Germany) are thoroughly mixed at a ratio of 9:1 (wt:wt) and poured onto the SU-8 master. After degassing, the PDMS mixture is cured at 70°C in an oven for 30 min. Then the cured thin PDMS replica is peeled off and a couple of holes are punched out for liquid alloy injection or filling. Meanwhile a thin blank PDMS lid is prepared using blank silicon wafer. Later, the PDMS replica and blank lid are bonded, using corona discharging (ETP, Chicago, IL, USA) activation. Through the punched holes in the PDMS, the liquid metal alloy (Galinstan, 68.5 % Ga, 21.5 % In, 10 % Sn, σ=3.46·106 S/m) is manually injected into the channels. This alloy remains in liquid state from -19°C to 1300°C. The ventilation outlets are encapsulated with uncured PDMS mixture as mentioned above.

3. Liquid metal, elastic, microfluidic unbalanced loop antenna

The fabrication process described in the previous section has been successfully employed to implement a single-layer, liquid metal, stretchable, unbalanced loop antenna, which consists of a radiating element and a semi-circular ground plane, as depicted in Fig. 3 [23]. Several cylindrical reservoirs are aligned along the upper tube of the antenna prototype to ensure good electrical connections while being flexed. The presence of these reservoirs introduces slight resonance frequency decrease of the antenna. The lower semi-circular area acts as a ground plane for the antenna. A number of posts are aligned to space the top and bottom PDMS membranes in the ground area, which has negligible influence on the electrical characteristics of the antenna.

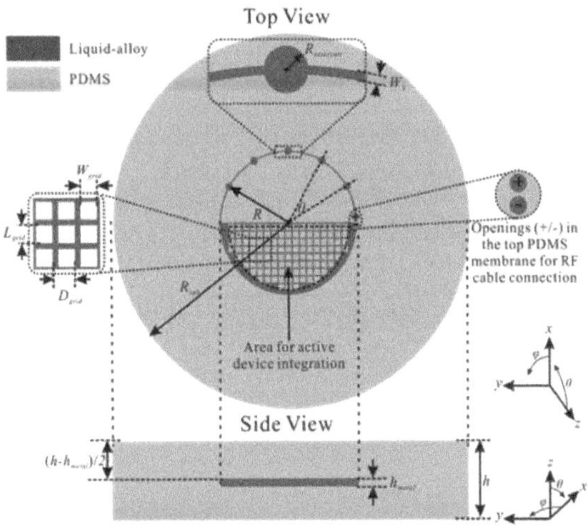

Figure 3. Geometrical schematic of stretchable unbalanced loop antenna: R=18.1 mm, R_{sub}=47.5 mm, $R_{reservior}$=1 mm, W_S=400 μm, L_{grid}=1.5 mm, W_{grid}=1.5 mm, D_{grid}=1.9 mm, h=1.0 mm, h_{metal}=100 μm, and θ=30°.

Mechanical properties of the antenna prototype were characterized experimentally. Fig. 4 shows that the resulting antenna is reversibly foldable, flexible, and stretchable. Severe bending and twisting in the tests did not cause any mechanical damages. In addition, the antenna can withstand extreme levels of strains of up to 40% along multiple axis. As a result of the uneven thickness of the PDMS substrate and the heterogeneous pattern of the antenna, slight mechanical inhomogeneity was observed while stressing along different orientations, *cf.* Figs. 4. b and c. In principle, elasticity of up to 100% should be in reach. Nevertheless, the openings in the membranes for electrical tests limit the mechanical deformability, reliability, and robustness, as the antenna can be easily torn from these openings.

Electrical properties of the antenna in its original state were obtained from numerical simulations. Experiments on port impedance and radiation characteristics of the antenna in both relaxed and flexed states were performed. Fig. 5 presents the simulated and measured reflection coefficients (S_{11}). The non-strained antenna features good impedance matching at 2.4 GHz, with an input impedance of approximately 75+15j Ω, Fig. 5.a. Applying strains on the antenna introduces an increase of the length of the upper radiating loop so that leads to a decrease on its resonance frequency, Fig. 5.b. In the case of stressing along y-axis, the antenna input resistance decreases due to the stronger coupling between the upper antenna arm and the lower ground plane, whereas this resistances goes up while stretching along x-axis.

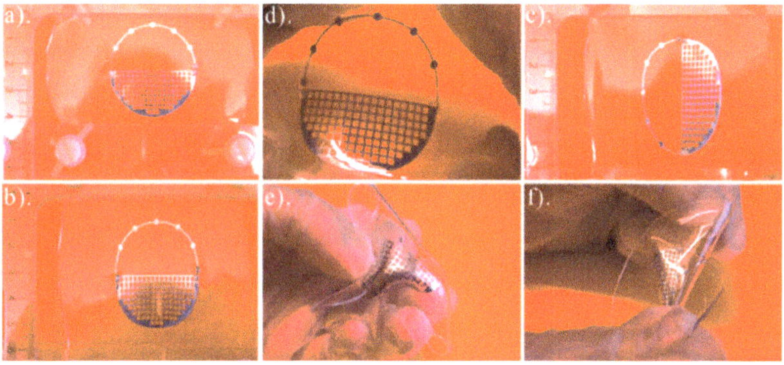

Figure 4. Photographs of the elastomeric fluidic unbalanced loop antenna: a). relaxed state, b). with 40% x-axis elongation, c). with 40% y-axis elongation, d). with biaxial elongation, e). bent state, and f). twisted state.

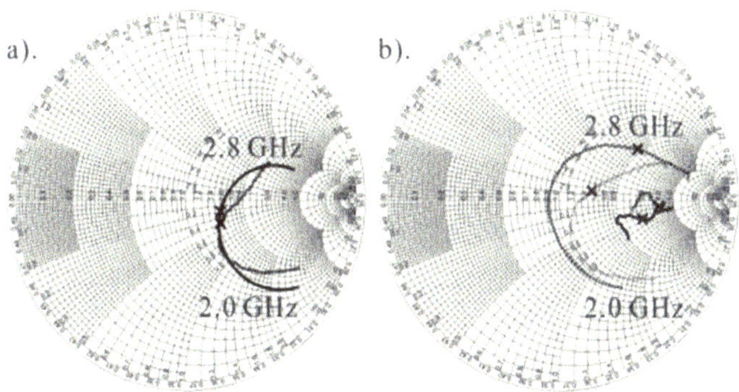

Figure 5. Simulated and measured port impedance of the elastic unbalanced loop antenna operating between 2.0 and 2.8 GHz: a). simulated (solid) and measured (dashed) S_{11} of the relaxed antenna; b). measured S_{11} of the flexed antenna with 20% x-axis (solid), 40% x-axis (dashed) and 20% y-axis (dotted), 40% y-axis (dashdot) elongation. The S_{11} at 2.4 GHz is marked by a cross sign on each curve.

Fig. 6 shows simulated and measured radiation patterns of the antenna prototype at 2.4 GHz.

As expected, the relaxed antenna resembles conventional unbalanced loops that exhibit broad beam coverage, especially in the yz-plane where nearly perfect omnidirectionality is seen. The maximum antenna gain is around 2.7 dBi, and the measured cross-polarization (G_ϕ in the xz-plane and G_θ in the yz-plane) is approximately 15 dB lower than the corresponding co-polarization. Variations of the radiation patterns of the strained antenna can be found, *cf.* Fig. 6.b. The omnidirectionality of the yz-plane radiation patterns degrades, particularly while the antenna is stressed along y-axis. The increasing cable influence caused by the y-axis strain explains the ripples within the angle of 45°-135° on the measured xz-plane radiation pattern. Also, the level of the measured cross-polarization of the stretched antenna slightly increases compared to that of the relaxed one.

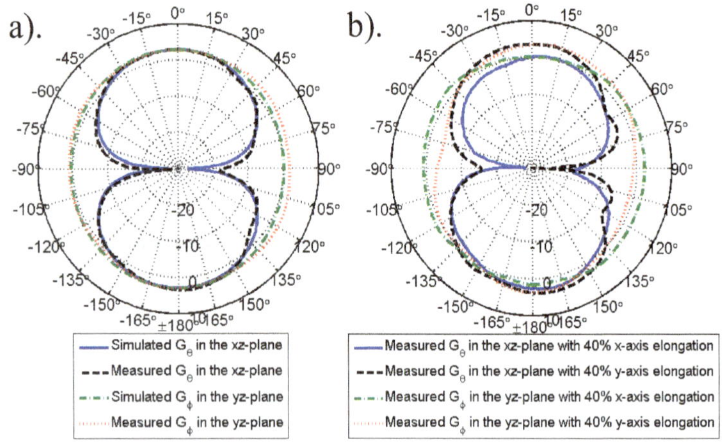

Figure 6. Simulated and measured radiation patterns of the a). relaxed and b). flexed antenna at 2.4 GHz. The corresponding coordinate system is depicted in Fig. 3. The antenna gain along ϕ and θ is defined as G_ϕ and G_θ, respectively.

High conductivity of Galinstan and large cross dimensions of the microfluidic channels attribute to low conductive losses and high radiation efficiency of the antenna. The measured data indicate a resonance frequency decrease of up to 18% while stretching, but the radiation efficiency at 2.4 GHz remains greater than 80%. In the experiments, no significant radiation efficiency decrease can be seen. This fact implies that galvanic connections in the micro-structured elastomeric channels are not interrupted by straining. Good electrical continuity of the liquid alloy owes to its excellent wettability on PDMS substrate surfaces. Though the presented unbalanced loop antenna achieves good radiation properties, its resonance frequency detuning introduced by stretching leads to relatively poor total efficiency at 2.4 GHz. Other antenna concepts with more robust port impedance and radiation characteristics in response to deformation would be feasible alternative solutions.

4. Bendable, stretchable, fluidic UWB antenna

The concept of microfluidics based elastomeric electronics is then extended to realize a planar inverted cone antenna (PICA) for the ultrawideband (UWB) frequency range of 3.1-10.6 GHz [24]. The reason for choosing the PICA is that its impedance matching and radiation characteristics are expected to be insensitive to its mechanical deformation. Also, its uniplanar configuration makes it a suitable antenna type for a design that should be bendable and stretchable. Fig. 7 shows the design schematically, with a leaf-shaped radiation and a large ground plane.

Resembling to the previously demonstrated elastic unbalanced loop, the implemented PICA prototype features excellent reversible deformability in the mechanical tests, as shown in Fig. 8. Extreme levels of straining of up to 40% along either *x*- or *y*-axis do not cause any mechanical failures, *cf.* Figs. 8.b and d. Moreover, the resulting prototype also withstands severe mechanical folding and twisting. After removal of applied mechanical forces, the antenna returns to its original state.

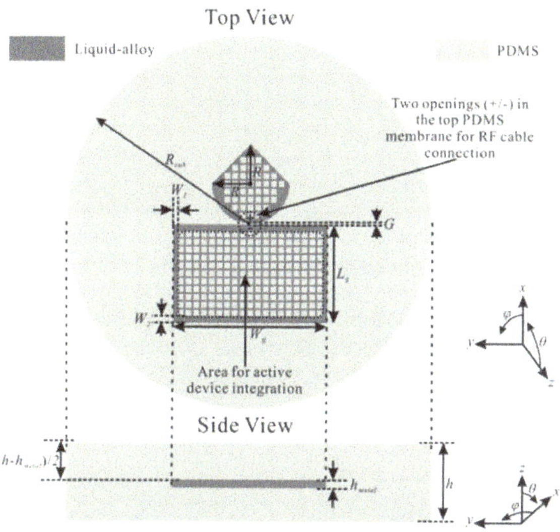

Figure 7. Geometry of the 2-D stretchable PICA. Dimensions are: R=10 mm, R_{sub}=47.5 mm, L_g=25 mm, W_g=40 mm, W_1=1.25 mm, W_2=1.75 mm, G=300 μm, h=1 mm, and h_{metal}=100 μm.

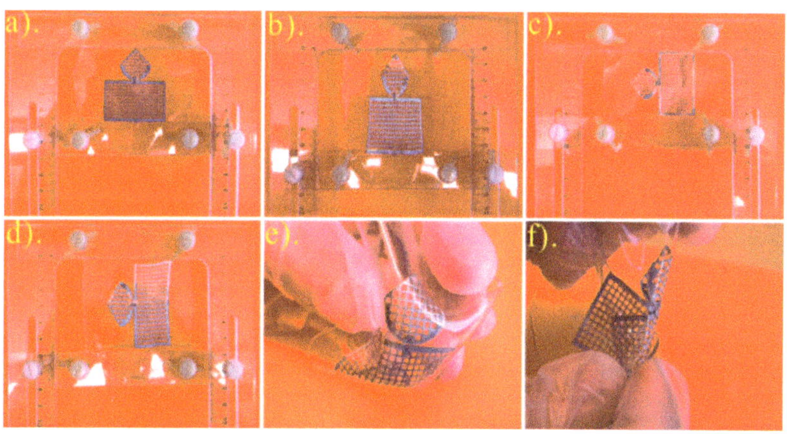

Figure 8. Photographs of the bendable, flexible, stretchable PICA: a). and c). relaxed state, b). strained antenna with 40% x-axis elongation, d). strained antenna with 40% y-axis elongation, e). folded antenna, and f). twisted antenna. The corresponding coordinate system is presented in Fig. 7.

Figs. 9 and 10 present simulated and measured reflection coefficients of the elastic, microfluidic PICA. The relaxed antenna shows good impedance match (S_{11} < -10 dB), within 3-11 GHz, both in numerical simulations and experiments.

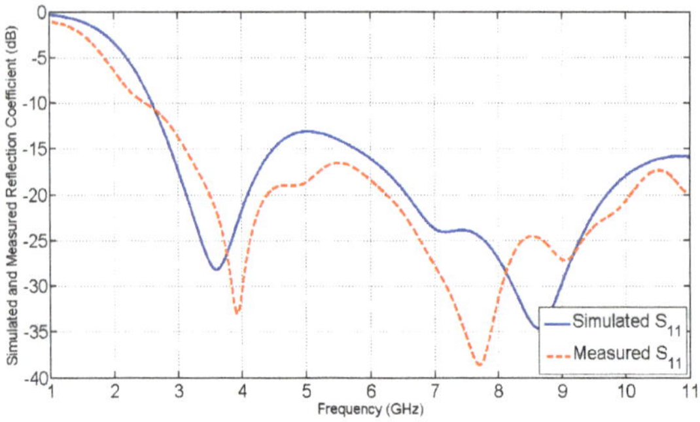

Figure 9. Simulated and measured S_{11} of the non-stretched antenna.

Applying strains along the *x*-axis results in the increased height of the antenna radiator so that its first resonance frequency decreases, *cf.* Fig. 10. Port impedance of the antenna is somewhat sensitive to its geometry, and consequently the antenna exhibits slightly varying impedance matching while stressed. However, good impedance match maintains at the entire UWB frequency band even if the antenna is strained to 40%.

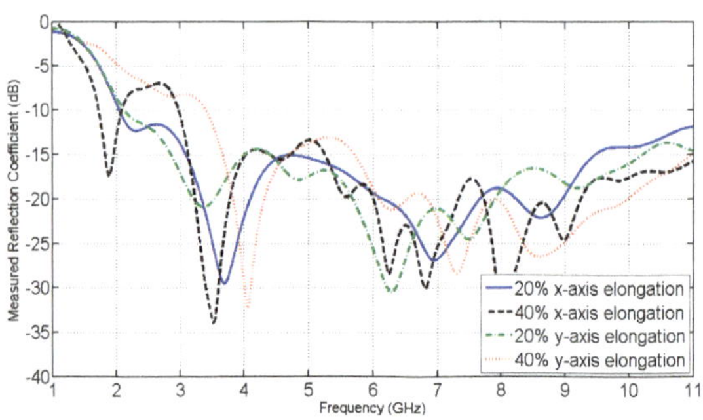

Figure 10. Measured reflection coefficients of the flexed antenna.

Measured radiation patterns at 2.5 GHz of the antenna in its relaxed and flexed states are displayed in Figs. 11 and 12. Similar to conventional fat monopole antennas, the non-strained antenna features broad coverage, especially in the *yz*-plane. The maximum antenna gain at 2.5 GHz was measured to be 2.2 dBi. The cross-polarization discrimination is very good. Numerical simulations are in line with the corresponding experimental data. Stretching the antenna along either *x*- or *y*-axis up to 40% introduces slight variations in the measured radiation patterns at 2.5 GHz, but without any significant gain reduction.

Similar simulations and experiments are also performed at 5 GHz, where ripples and slight asymmetry occur in the radiation patterns caused by disturbance from feed cable. The presence of higher order modes at 5 GHz together with the increasing cable influence degrades the cross-polarization discrimination. Compared to the experimental data at 2.5 GHz, larger variations on the measured radiation patterns at 5 GHz are observed while the antenna is in its strained states, particularly in the *yz*-plane.

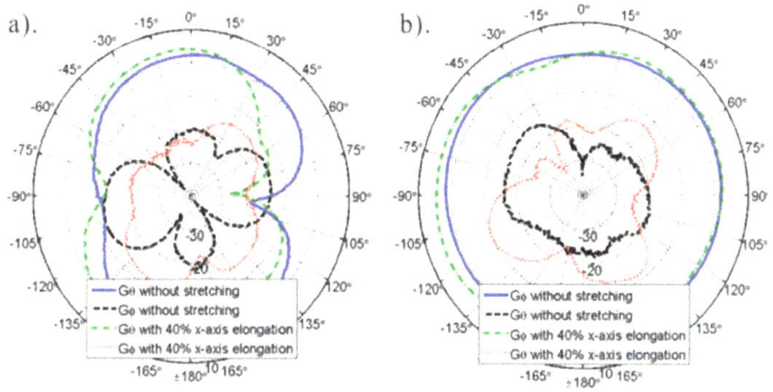

Figure 11. Measured a). xz- and b). yz-plane (according to the coordinate system in Fig. 7) radiation patterns at 2.5 GHz of the non-stretched and stretched antenna with 40% x-axis elongation.

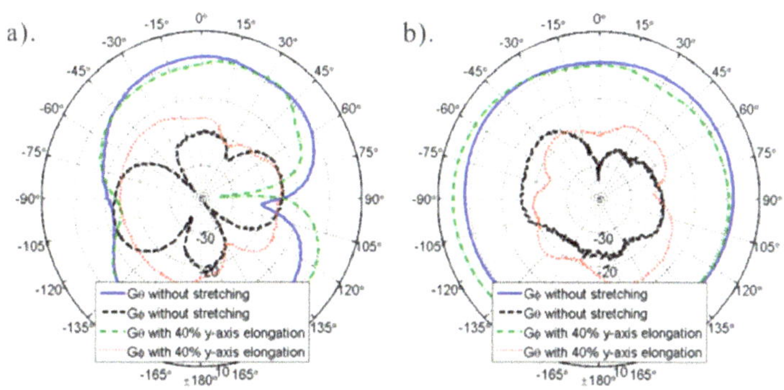

Figure 12. Measured a). xz- and b). yz-plane (according to the coordinate system in Fig. 7) radiation patterns at 2.5 GHz of the non-stretched and stretched antenna with 40% y-axis elongation.

The measured radiation efficiency within the frequency range of 3-10 GHz is shown in Fig. 13. Although the radiation efficiency at the lower end of the frequency range decreases when the antenna is stretched, it is still greater than 70%. This, together with the experimental data on port impedance, indicates that the PICA prototype achieves good total antenna efficiency regardless of stretching.

Much work however remains. First of all, comprehensive environmental reliability and durability tests, e.g. vibration, temperature cycling, and aging, are needed, since it is indeed a new way of fabricating antennas, with new materials. Electrical characteristics of the antenna at extreme temperature condition, e.g. below the melting point of the liquid alloy, should be evaluated when a special measurement setup is established. Studies on antenna radio interfaces are also necessary. Development of fully integrated stretchable wireless electronic systems consisting of flexible thin embedded active chips, stretchable interconnects, and highly efficient stretchable antennas are the end objective, with a need for much research efforts.

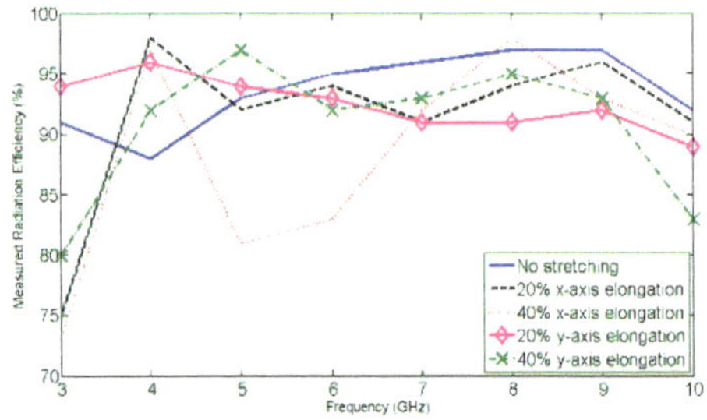

Figure 13. Experimental results on the radiation efficiency of the relaxed and flexed antennas.

5. Single-layer stretchable elastomeric integrated active RF electronics

The concept of microfluidics based passive stretchable elastomeric electronics has been further developed to the integrated device level utilizing localized stiff cells (LSCs), as illustrated in Fig. 14 [28]. A microfluidic, elastic, large-area antenna is realized in the same manner as the previous stretchable antennas, by incorporating liquid alloy into microstructured elastomeric channels. Established IC chips associated with passive components are assembled on small flexible laminates. Subsequently, a few tin-plated contact pins resembling cantilevers are soldered to the flexible circuits, and a semi-spherical solder ball is then mounted on the bottom surface of each contact pin on the other end to improve galvanic connection to the liquid fluid, *cf.* Fig. 14.e. Whereafter, the flexible circuits are embedded into the antenna substrate, with each contact pin immersing in the liquid alloy. In the end, uncured PDMS mixture droplets are deposited on top of the flexible circuits to locally enhance the stiffness of the elastomeric substrate and encapsulate the flexible circuits as well as the flex-to-stretch interfaces. The LSCs with enhanced stiffness than surrounding areas ensures that nearly zero stress and displacement between the rigid and the elastic structures would arise inside the LSCs when applying strains to the heterogeneously integrated device. The overall elasticity of the hybrid device is degraded compared to the standalone stretchable antennas due to the presence of the LSCs. But the reliability of the integrated device is greatly improved. Also, good wetting of the liquid metal alloy on tin-plated pins and solder balls ensures reliable electrical connections between contact pins and the antenna.

The proposed hybrid integration strategy was employed to implement a 900 MHz microfluidic stretchable RF radiation sensor, as seen in Fig. 15. The resulting sensor prototype is capable of performing real-time monitoring on the human exposure level to electromagnetic fields (EMF). Once the exposure level to EMFs exceeds the threshold power, the integrated light emission diode (LED) will be switched on. This sensor is of importance to human health as more and more EMFs are generated by today's telecom networks and mobile terminals, and might cause health issues. The integrated radiation sensor contains three sub-modules fully encased in a large-area elastomeric substrate, including a stretchable unbalanced loop antenna for capturing RF radiation from ambient environments, an RF power detection unit for converting received RF energy to the corresponding DC voltages, and a LED indicator for visualizing.

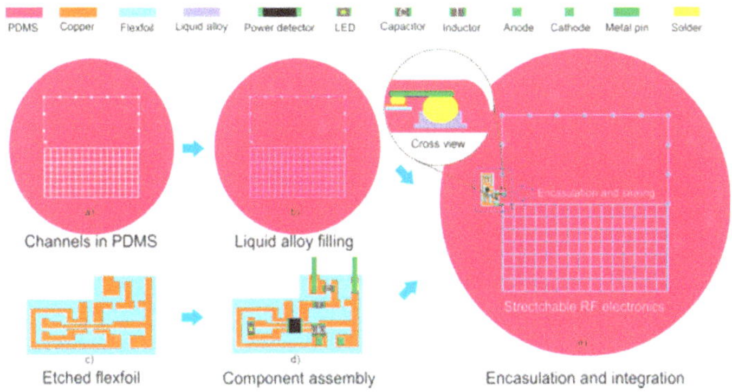

Figure 14. Schematic of the hybrid integration of the single-layer stretchable elastomeric RF electronics.

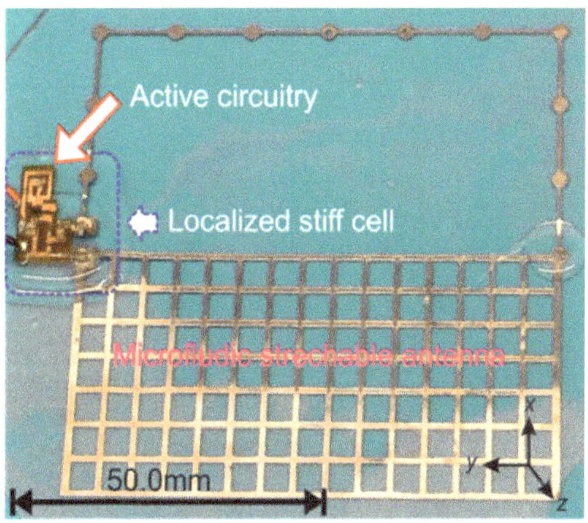

Figure 15. Photograph of the stretchable elastomeric RF radiation sensor prototype.

Prior to the hybrid integration, experimental evaluations on electrical characteristics of the standalone antenna and the RF power detection sub-module assembled on a flex foil are conducted. The implemented stretchable fluidic antenna exhibits similar mechanical and electrical performance as the previously presented unbalanced loop, but operates at a lower frequency of approximately 900 MHz. The power conversion sub-module in the integrated RF radiation sensor involves an off-the-shelf power detector IC chipset (Linear Technology, LT 5534), two decoupling and one coupling capacitors, an inductor for impedance matching at the RF input, and a green LED indicator, all assembled on a small flex foil with a size of 10 mm × 18 mm, as seen in Fig. 15. When the input RF energy at the detector exceeds the threshold power of the integrated device, the LED indicator is switched on, and vice versa. The entire integrated power detection unit can be powered by four serially connected AA rechargeable batteries with a DC supply voltage of 5.23 V. The power conversion behavior of the integrated device is first characterized using a signal generator and a digital multimeter. Fig. 16 presents the measured DC voltages at the output of the power detection unit in response to varying input RF power at 900 MHz. A dynamic range of -60 dBm to 0 dBm can be achieved, and the threshold power for turning on/off the LED indicator is found to be slightly higher than -30 dBm.

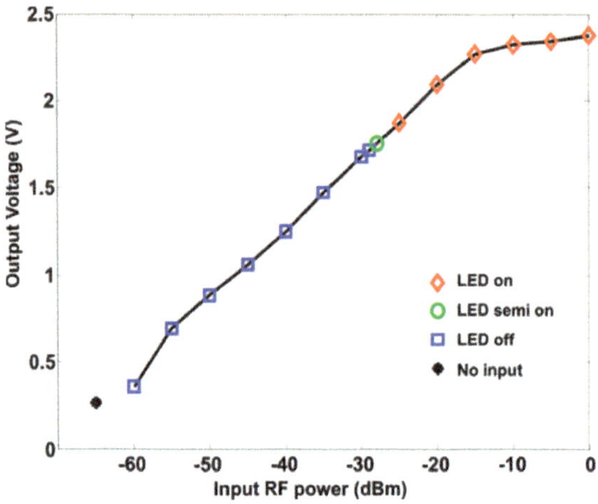

Figure 16. Measured output DC voltages of the RF power detection sub-module and the LED on/off states versus varying input RF power.

The integrated elastic radiation sensor device is tested in a demonstration setup in Fig. 17. The RF radiation source consists of an RF signal generator and a horn antenna placed 5 m away from the sensor device in the line-of-sight.

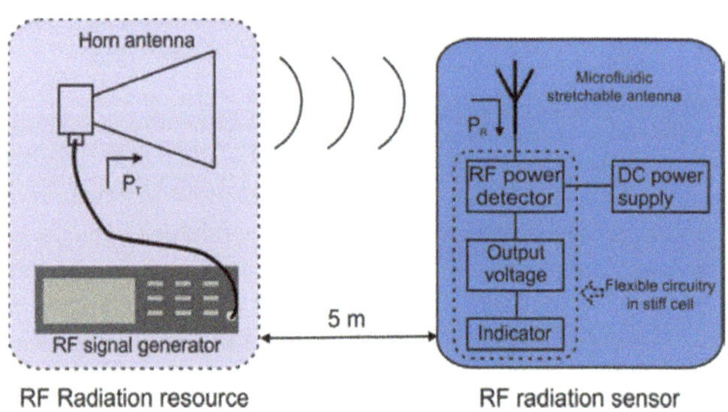

Figure 17. Schematic illustration of the RF radiation sensing demonstration setup.

Experiments presented in Fig. 18 verify that the resulting microfluidics based elastomeric integrated device maintains its radiation sensing capabilities even if it is strained along multiple axis. Extreme levels of twisting do not cause any failures in its operation either.

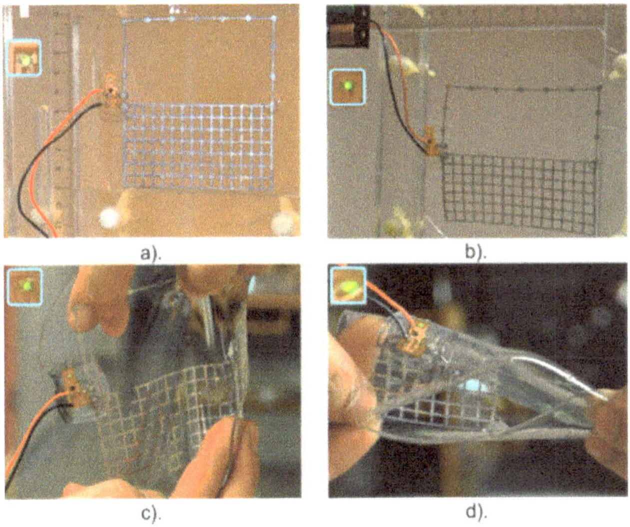

Figure 18. Photographs of the elastomeric RF radiation sensor device operating in ordinary office environment: a). relaxed state, b). with 15% strain applied along y-axis, e). with manually applied strain along multiple directions, and f). with severe twisting. The sensor was directly illuminated by a radiation source 5 m away. The coordinate system is shown in Fig. 15.

6. Multi-layer stretchable elastomeric integrated active RF electronics

Very recently, the emerging field of microfluidics based elastomeric electronics has been further advanced to multi-layer μFSRFEs, with the demonstration of a microfluidic, reversibly stretchable, large-area wireless strain sensor [29]. The manufacturing process for multi-layer elastomeric passive components resembles the single-layer fabrication, but with a few minor modifications and process steps added, *cf.* Fig. 19.

In summary, the upper and lower microfluidic channels are respectively constructed in the top and the bottom PDMS sheets, using standard soft lithography techniques. Then a few inlets and outlets are punched. In addition, the blank middle PDMS slab is prepared. The microstructured top elastomer layer is bonded to the blank silicone rubber slab using plasma boning, and addition inlets are punched on the bonded PDMS sheet. The bottom elastomer substrate is bonded to the previously bonded PDMS layers afterwards, using corona discharging activation. Prior to filling the upper microfluidic channels with galinstan fluid metal, the inlet 3 is sealed with a piece of Scotch® tape. Later, the ventilation outlets in the top silicone rubber sheet are encapsulated using PDMS prepolymer. Whereafter, both in the inlets 1 and 3 are taped, and galinstan alloy is injected into the lower microfluidic channels from the bottom side. All remaining ventilation outlets and the inlet 2 are then encapsulated, and the inlets 1 and 3 are reserved for connecting active circuitry in a heterogeneously integrated device.

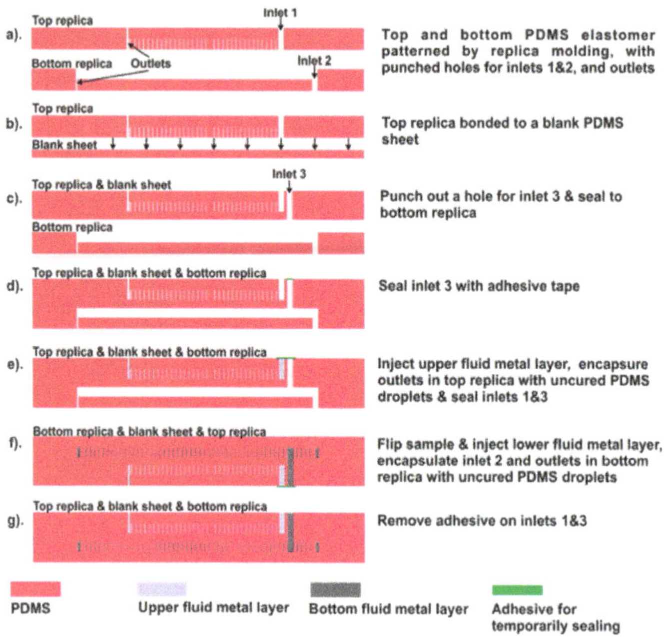

Figure 19. Fabrication process of a multi-layer microfluidics based stretchable elastomeric passive electronic device.

The subsequent fabrication process steps are active circuit assembly and hybrid device integration, which resemble the manufacturing and the integrated processes presented in the previous section. Schematic illustrations describing the entire assembly and integration procedure are displayed in Fig. 20, and not discussed in detail in this section.

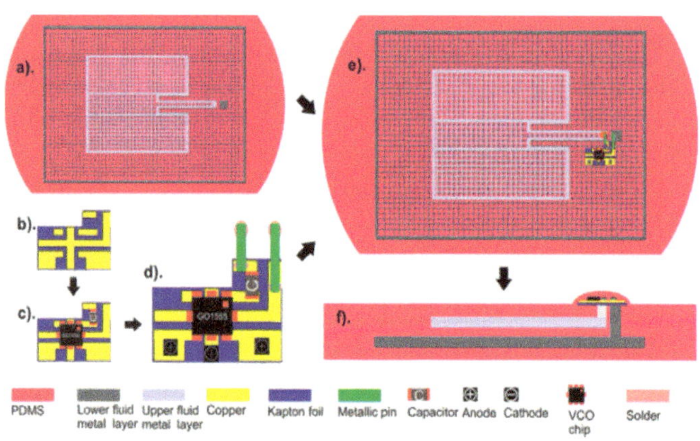

Figure 20. Schematic drawings of the integration procedure for a multi-layer stretchable, microfluidic integrated electronic device.

The first resulting integrated device based on multi-layer μFSRFEs is a self-contained wireless strain sensor, fully encased in a thin large-area silicone elastomer, as shown in Fig. 21. A stretchable liquid metal microstrip patch antenna comprising an upper rectangular meshed patch and a lower ground plane constructed in the same manner, takes up the major area of the hybrid device. It is worthy mentioning that this antenna not only serves as a radiator like any other conventional antennas, but also acts as a sensing device owing to its varying electrical characteristics in response to mechanical deformation. The entire integrated sensor device is slightly larger than its meshed fluid metal ground with a size of 100 mm × 80 mm, and approximately four times as big as the antenna patch. Though the liquid metal patch is the actual strain sensing element, the large ground plane can also serve an effective sensing area, and extend the strain sensing functionality to almost the entire device. Apart from the self-contained wireless strain sensor, a custom-designed personal computer (PC)-assisted radio receiver for remotely collecting, processing, and storing the measured data wirelessly transmitted from the integrated sensor device is also implemented. It removes the need for costly RF measurement facilities, and significantly reduces the cost for building up a complete system.

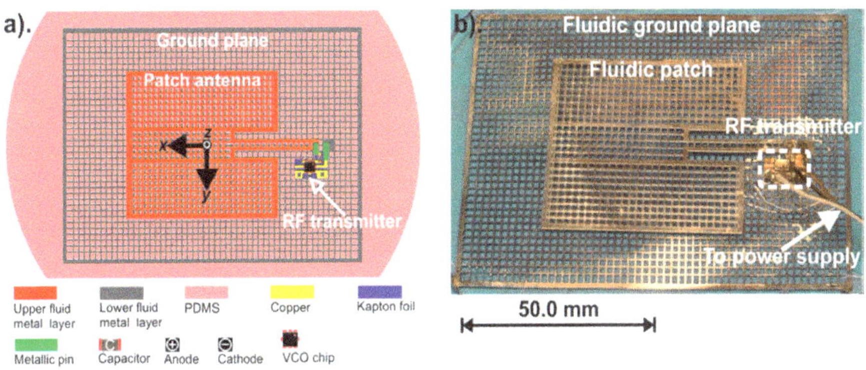

Figure 21. Demonstrated microfluidic reversibly stretchable wireless strain sensor: a). schematic illustration, and b). optical photograph of the resulting device.

Port impedance and radiation characteristics of the standalone, mechanically reconfigurable, elastomeric patch antenna in its relaxed and flexed states are measured. Excellent port impedance matching around 1.46 GHz along with very good correlation between the simulated and measured reflection coefficients of the non-stretched antenna are seen in Fig. 22.a. Applying increasing strain of up to 15% along its x-axis introduces persistent downshift of its resonant frequency of the patch antenna. The lowest resonant frequency of approximately 1.33 GHz is achieved at the maximum x-axis elongation of 15%.The measured resonant frequency of the relaxed antenna is slightly less than the minimum operational frequency of the integrated transmitter circuit. This negative offset is crucial to strain sensing, since persistently rising antenna mismatch losses at the operational frequencies in response to increasing stressing along the x-axis of the hybrid device.

Furthermore, this frequency offset should be as little as possible to avoid too high mismatch losses so that reasonably long remote sensing ranges as well as sufficient sensing sensitivity can be attained. Placing the antenna original resonant frequency above the highest operational frequency of the transmitter is not an option, as inconsistent variations of mismatch losses would occur if the antenna based strain sensor is stressed from its relaxed state to a high tensile strain.

Measured radiation patterns, including mismatch losses, at 1.46 GHz, of the relaxed and flexed antennas are shown in Figs. 22.b and c. The realized peak gain is measured along the +z-axis (according to the coordinate system depicted in Fig. 21.a.) of the mechanically reconfigurable antenna, and greatly degraded from 2.0 dBi to -10.7 dBi while applying an increasing strain from 0% to 15% along its *x*-axis. The meshed ground plane directs the radiation forward, with a front-to-back ratio as high as 10.0 dB, regardless of stretching. Moreover, the cross-polarization discrimination is very good in both relaxed and strained cases.

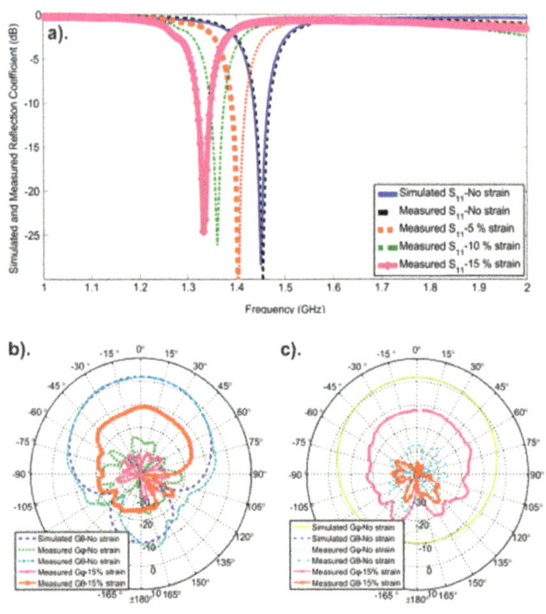

Figure 22. a). Simulated and measured reflection coefficients of the elastomeric fluid metal patch antenna in its relaxed state, and with different strains along the *x*-axis (according to the coordinate system in Fig. 21). b). *xz*- and c). *yz*-plane radiation patterns (including mismatch losses) at the original resonant frequency of the elastic antenna in its relaxed state, and with 15% elongation along its *x*-axis. The antenna gain along ϕ and θ orientations is defined as G_ϕ and G_θ, respectively. The antenna co- and cross-polarization along +z-axis are in parallel to the *x*- and *y*-axis.

Mechanical properties of the elastomeric patch antenna are evaluated after characterizations on its electrical performance. The implemented antenna prototype is more than twice as thick as the previously demonstrated stretchable RF electronic devices, and thus features degraded mechanical deformability. Yet moderate twisting or folding do not cause any mechanical damages to the fluidic patch antenna during experiments.

The measured total efficiency of the standalone stretchable patch antenna including mismatch losses is 36.9% in its relaxed state at 1.46 GHz. Straining the patch to 15% along its x-axis leads to considerable total efficiency drop of 33.8%. In the intermediate states with the strains between 5% and 10%, the total efficiency of 11.5% and 5.8% is achieved according to the experiments.

The system demonstration setup is first calibrated by characterizing the output DC voltages of the RF power detector with different static stresses applied to the integrated wireless strain sensor, as presented in Fig. 23. A fairly linear decline of the measured DC voltages versus increasing mechanical elongation along the x-axis of the self-contained sensor device is seen, and can be explained by decreased resonant frequencies and increased mismatch losses of the transmitting, elastomeric antenna as a result of incremental stretching. As co-polarization components dominate in both microfluidic stretchable patch antenna in the hybrid sensor and the receiving horn in the custom-designed RF receiver, the measured output voltages with respect to the cross-polarization are considerably lower than that of the co-polarization, and also exhibit smaller variations versus different strains. The steepest voltage decline in response to increasing stretch is found in the case of the co-polarization measured in an anechoic chamber. When it comes to ordinary office environment, the presence of reflections and scatterings limits the range of voltage variations.

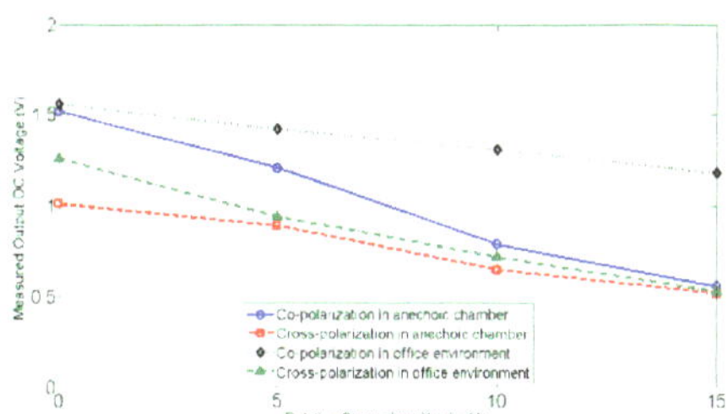

Figure 23. Measured output DC voltages at the receiver versus varying mechanical stress along the x-axis (according to the coordinate system in Fig. 21) applied to the integrated strain sensor. Its co- and cross-polarizations along the +z-axis were in parallel to the x- and y-axis, and the receiving standard horn was horizontally polarized.

Fig. 24 shows the demonstration of remote sensing of repeated body motion in a corridor of ordinary office environment, using the resulting wireless strain sensor prototype. Periodically repeated dynamic strains of up to 15% along its x-axis, with a period of 10s and a duty cycle of 50%, are manually applied to the two shorter edges of the integrated, elastic, strain sensor. The output DC voltages in the PC-assisted RF receiver varying with mechanical strains of the integrated, body-worn, sensor device, are continuously monitored and recorded, *cf.* Fig. 24.c. The measured data of six cycles, well correlating with the applied varying tensile strains, are presented in the subplot in Fig. 24.c, in which six fairly uniform, quasi-rectangular waves with varying amplitude between 1.28V and 1.55V are recorded in an overall period of 60s. This experimental data also verifies that the implemented wireless strain sensor can rapidly return to its relaxed state without any hysteresis, once removing the applied stress. This great feature reflects reversible deformability as well as high degree of elasticity of the multi-layer μFSRFEs based sensor device.

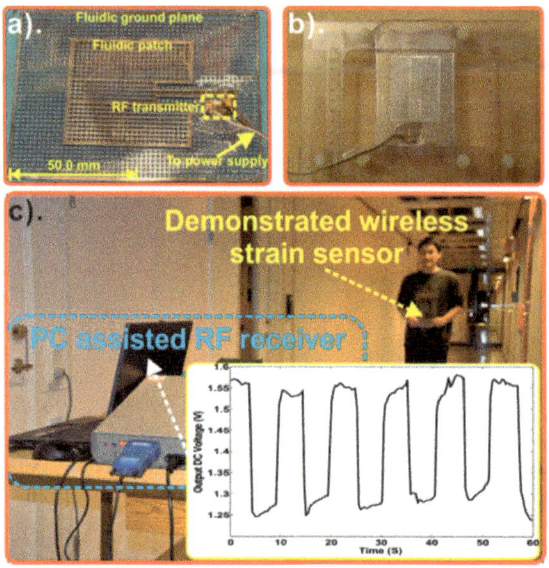

Figure 24. Integrated elastomeric strain sensor a). in its original state and b). with 15% vertical strain. c). system demonstration (periodical manual straining). Real-time recorded data at the PC-assisted RF receiver 5m away is presented in the subplot.

7. Conclusion

Recent advances in the emerging field of elastomeric electronics that are able to be conformed into complex curvilinear shapes, or be compressed, twisted, and stressed to extreme degrees, have been briefly reviewed. Various techniques and strategies for realizing bendable, flexible, stretchable elastomeric electronic devices and system have been discussed. As the main focus, elastomeric electronics based on microfluidic approaches have been addressed in detail. Fabrication processes, hybrid integration techniques, as well as appealing application examples involving single- and multi-layer μFSRFEs have also been presented.

Microfluidics based elastic electronics together with other members in the family of elastomeric electronics are introducing a revolution to the world of electronics, and shaping the future for electronics so as to change our tomorrow's daily life and contribute to our networked society. It is anticipated that more than 50 billion devices will be wirelessly connected by 2020, which would involve units as intelligent as smartphones/tablets, and as soft as elastomeric electronics.

Acknowledgement

The author greatly appreciates the Editor Board and the publisher.

Author details

Shi Cheng
Ericsson AB, Stockholm, Sweden

8. References

[1] Bowden N, Brittain S, Evans AG, Hutchinson JW, Whitesides GM (1998) Spontaneous Formation of Ordered Structures in Thin Films of Metals Supported on an Elastomeric Polymer. Nature 393: 146-149.
[2] Watanabe M, Shirai H, Hirai T (2002) Wrinkled Polypyrrole Electrode for Electroactive Polymer Actuators. J. Appl. Phys. 92: 4631.
[3] Lacour SP, Wagner S, Huang ZY, Suo ZG (2003) Stretchable Gold Conductors on Elastomeric Substrates. Appl. Phys. Lett. 82: 2404.
[4] Lacour SP, Jones J, Wagner S, Li T, Suo ZG (2005) Stretchable Interconnects for Elastic Electronic Surfaces. Proc. IEEE 93: 1459-1467.
[5] Kim DH, Ahn JH, Choi WM, Kim HS, Kim TH, Song JZ, Huang Y, Liu ZJ, Lu C, Rogers JA (2008) Stretchable and Foldable Silicon Integrated Circuits. Science 320: 507-511.
[6] Kim DH, Song JZ, Choi WM, Kim HS, Kim RH, Liu ZJ, Huang Y, Hwang KC, Zhang YW, Rogers JA (2008) Materials and Noncoplanar Mesh Designs for Integrated Circuits With Linear Elastic Responses to Extreme Mechanical Deformations. Proc. Nat. Acad. Sci. USA 105: 18675-18680.
[7] Ko HC, Stoykovich MP, Song J, Malyarchuk JV, Choi WM, Yu CJ, Geddes JB, Xiao J, Wang S, Huang Y, Rogers JA (2008) A Hemispherical Electronic Eye Camera Based on Compressible Silicon Optoelectronics. Nature 454: 748-753.
[8] Kim DH, Kim YS, Wu J, Liu Z, Song J, Kim HS, Huang Y, Hwang KC, Rogers JA (2009) Ultrathin Silicon Circuits With Strain-Isolation Layers and Mesh Layouts for High-Performance Electronics on Fabric, Vinyl, Leather, and Paper. Adv. Mater. 21: 3703–3707.
[9] Kim DH, Lu N, Ghaffari R, Kim YS, Lee SP, Xu L, Wu J, Kim RH, Song J, Liu Z, Viventi J, de Graff B, Elolampi B, Mansour M, Slepian MJ, Hwang S, Moss JD, Won SM, Huang Y, Litt B, Rogers JA (2011) Materials for Multifunctional Balloon Catheters With Capabilities in Cardiac Electrophysiological Mapping and Ablation Therapy. Nat. Mater. 10: 316-323.
[10] Kim DH, Lu N, Ma R, Kim YS, Kim RH, Wang S, Wu J, Won SM, Tao H, Islam A, Yu KJ, Kim TI, Chowdhury R, Ying M, Xu L, Li M, Chung HJ, Keum H, McCormick M, Liu P,VZhang YW, Omenetto FG, Huang Y, Coleman T, Rogers JA (2011) Epidermal Electronics Science 333: 838-843.

[11] Baca AJ, Meitl MA, Ko HC, Mack S, Kim HS, Dong J, Ferreira PM, Rogers JA (2007) Printable Single-Crystal Silicon Micro/Nanoscale Ribbons, Platelets and Bars Generated from Bulk Wafers. Adv. Func. Mater. 17: 3051–3062.
[12] Mack S, Meitl MA, Baca AJ, Zhu ZT, Rogers JA (2006) Mechanically Flexible Thin-Film Transistors That Use Ultrathin Ribbons of Silicon Derived from Bulk Wafers. Appl. Phys. Lett. 88: 213101.
[13] Brosteaux D, Axisa F, Gonzalez M, Vanfleteren J (2007) Design and Fabrication of Elastic Interconnections for Stretchable Electronic Circuits. IEEE Electron Device Lett. 28: 552-524.
[14] Huyghe B, Rogier H, Vanfleteren J, Axisa F (2008) Design and Manufacturing of Stretchable High-Frequency Interconnects. IEEE Trans. Adv. Packaging. 31: 802-808.
[15] Carta R, Jouranda P, Hermansa B, Thonéa J, Brosteauxb D, Vervustb T, Bossuytb F, Axisab F, Vanfleteren J, Puersa R (2009) Design and Implementation of Advanced Systems in A Flexible-Stretchable Technology for Biomedical Applications. Sensors Actuat. A-Phys. 156: 79-87.
[16] Niu X, Peng S, Liu L, Wen W, Sheng P (2007) Characterizing and Patterning of PDMS-Based Conducting Composites. Adv. Mater. 19: 2682.
[17] Sekitani T, Noguchi Y, Hata K, Fukushima T, Aida T, Someya T (2008) A Rubberlike Stretchable Active Matrix Using Elastic Conductors. Science 321: 1468-1472.
[18] Siegel AC, Shevkoplyas SS, Weibel DB, Bruzewicz DA, Martinez AW, Whitesides GM (2006) Cofabrication of Electromagnets and Microfluidic Systems in Poly(dimethylsiloxane). Angew. Chem. Int. Edit. 45: 6877-6882.
[19] Siegel AC, Bruzewicz DA, Weibel DB, Whitesides GM (2007) Microsolidics: Fabrication of Three-Dimensional Metallic Microstructures in Poly(dimethylsiloxane). Adv. Mater. 19: 727-733.
[20] Hu H, Shaikh K, Liu C (2007) Super Flexible Sensor Skin Using Liquid Metal As Interconnect. Proceeding of IEEE Sensors 1:815-817.
[21] Kim HJ, Son C, Ziaie B (2008) A Multiaxial Stretchable Interconnect Using Liquid-Alloy-Filled Elastomeric Microchannels. Appl. Phys. Lett. 92: 011904.
[22] Kim HJ, Maleki T, Wei P, Ziaie B (2009) A Biaxial Stretchable Interconnect With Liquid-Alloy-Covered Joints on Elastomeric Substrate. J. Microelectromech. S. 18: 138-146.
[23] Cheng S, Rydberg A, Hjort K, Wu ZG (2009) Liquid Metal Stretchable Unbalanced Loop Antenna. Appl. Phys. Lett. 94: 144103.
[24] Cheng S, Wu ZG, Hallbjörner P, Hjort K, Rydberg A (2009) Foldable and Stretchable Liquid Metal Planar Inverted Cone Antenna. IEEE Trans. Antennas Propag. 57: 3765–3771.
[25] So JH, Thelen J, Qusba A, Hayes GJ , Lazzi G, Dickey MD (2009) Reversibly Deformable and Mechanically Tunable Fluidic Antennas. Adv. Func. Mater. 19: 3632-3637.
[26] Kubo M, Li X, Kim C, Hashimoto M, Wiley BJ, Ham D, Whitesides GM (2010) Stretchable Microfluidic Radiofrequency Antennas. Adv Mater. 22: 2749 - 2752.
[27] Khan MR, Hayes GJ, So JH, Lazzi G, Dickey MD (2011) A Frequency Shifting Liquid Metal Antenna With Pressure Responsiveness. Appl. Phys. Lett. 99: 013501.
[28] Cheng S, Wu ZG (2010) Microfluidic Stretchable RF Electronics. Lab Chip 10: 3227 - 3234.
[29] Cheng S, Wu ZG (2011) A Microfluidic, Reversibly Stretchable, Large-Area Wireless Strain Sensor. Adv. Func. Mater. 21: 2282-2290.

Chapter 9

Elastomer Application in Microsystem and Microfluidics

Shuang (Jake) Yang and Kunqiang Jiang

Additional information is available at the end of the chapter

1. Introduction

Elastomer has been widely used in academic and industry since it was invented in the nineteen century [1]. By definition, elastomer is a polymer which is viscoelastic and able to regain its shape after deformation. Compared to other materials, elastomer has notably low Young's modulus and high yield strain. In general, amorphous elastomer is made of carbon, hydrogen, oxygen and/or silicon whose glass transition temperature is well below the application temperature, e.g., room temperature. As a result, it has excellent properties in elasticity [2], transparency [3], permeability [4], and insulation [5].

One of the most commonly used elastomers is silicone elastomer such as poly(dimethylsiloxane) (PDMS), which has been widely used in electronics and microfluidics due to its elasticity, optical transparency, UV transmission, permeability, biocompatibility, and availability. PDMS generally consists of repeat unit dimethylsiloxane. The repeat methyl is modifiable using functional groups such as urea or bis-urea for development of high mechanical strength of elastomer. The modified PDMS presents both a high mechanical strength and elasticity at room temperature due to functionality of the cross-linking domains and intrinsic reversibility of the supramolecular chains. PDMS is optically transparent and is usually inert, non-toxic and non-flammable [6]. PDMS is viscoelastic and is normally measured using dynamic mechanical analysis (DMA), in which a specialized instrument is used to determine the material's flow characteristics over a wide range of temperature, flow rates, and deformations. In general, the shear modulus of PDMS is in the range of 100 kPa to 3 MPa, varied with the preparation conditions.

Characterization of elastomer by dynamic-mechanical testing includes temperature sweep, frequency sweep, strain sweep, damping and friction properties. The dynamic mechanical measurement is particularly useful to characterize conductive particles in elastomer for understanding the particles effect on overall elastomer mechanical properties. It is very common to measure damping factor (tan δ) change with temperature for understanding the elastomer degree of crosslinking [7]. The dynamic mechanical testing provides information on both uncrosslinked and crosslinked phases of the elastomer, as well as viscoelasticity. To this end, nanoindentation plays a significant role on viscoelastic characterization of elastomers for their creep and relaxation properties [8, 9].

Elastomer is an excellent matrix which can mix with other materials such as conductive particles and thermoplastics to enhance mechanical properties. For example, silver particles are added into silicone elastomer matrix to formation of conductive elastomer for high-density electrical contacts in microelectronics. Polypropylene and ethylene-propylene copolymer elastomer significantly improves the dynamic mechanical, tensile and impact properties and even changes the modes of failure [10]. The size, shape and spatial packing of elastomer inclusions are the most significant factors in controlling the mechanical behavior of blends. Since PDMS can be cured with different ratio of dimethylsiloxane and curing reagent, it is very useful to fabricate microfluidic devices for a variety of applications, including cell culture [11], ELISA assay [12], capillary electrophoresis [13], DNA sequencing [14].

In this chapter, we briefly discussed elastomer elastic and viscoelastic properties, as well as permeability. Elastomer is widely used in microelectronics and microfluidics due to all or part of these characteristics. Then we discussed several applications in microelectronics and microfluidics using elastomer as substrate. Finally, we introduce potential development of elastomer.

2. Elastomer property and its applications

Elasticity is the ability of a material to return to its original shape and size after being stretched, compressed, twisted or bent [15]. Elastic deformation (change of shape or size) lasts only as long as a deforming force is applied to the object, and disappears once the applied force is removed. The elasticity of elastomer makes it extremely useful in microelectronics and microfluidics. For example, the Intel CPU uses land grid array sockets (LGA), in which the socket has over thousands of elastomer contacts that consist of silver particles in elastomer matrix [16, 17]. When socket, the CPU and printed circuit board (PCB) are assembled by compressing force, the silver-embedded elastomer is compressed to increase contact area between socket and the CPU, and between socket and the PCB. The excellent elastic deformation of elastomer can keep the electrically tight contact for minimum contact resistance. Elasticity of PDMS minimizes its stress relaxation and creep effect, which is severe in other polymer with low elasticity. The excellent elasticity of PDMS allows to release from microfluidic masters without damaging PDMS device [18, 19] and significant deflection to occur when a pressure difference develops [20]. Therefore, it is important to characterize PDMS elasticity and understand how to improve such property for various applications.

The most common elasticity experiment for many decades involved mechanically blending a reinforcing filler such as carbon black into natural rubber, crosslinking the elastomer with sulfur, and characterizing it network structure by stress-strain measurements in simple extension, or by extents of equilibrium swelling [21]. However, in most instances, such as thin-film elastomer, land grid array elastomer contact, and PDMS microfluidics, these methods may not be applicable due to effects of supporting materials. To characterize elastomer in those applications, nanoindentation has been an emerging technology for elastomer characterization and is an indispensable tool in many instances where other techniques are inapplicable. Nanoindentation is such tool for characterization of polymer on its viscoelasticity. Indentation technique was first introduced as advent of indentation fracture mechanics [22] in 1975 and ever since it has been developed for characterization of elastic-plastic materials using the Oliver-Pharr's method [23]. No valid method was reported for polymer nanoindentation analysis based on stress-strain experiments until the viscoelastic model was proposed in 2004 and other emerging studies that followed [8, 24, 25]. Because of viscoelasticity, elastic-plastic model cannot be directly used to derive modulus and hardness. Also these measurements are dependent on the experimental parameters, such as loading rate, maximum load, holding duration, unloading rate, and minimum unloading force [26]. To minimize the effects by those experimental parameters during indentation, the experimental conditions were proposed in such a way that fast loading was applied for minimizing viscoelastic deformation, long dwell time for full viscoelastic deformation, and quick unloading rate. The fast loading and unloading showed that only elastic-plastic deformation was observed, while viscoelastic deformation was observed in the dwell duration. Also, it was observed that the viscoelastic deformation can be measured by keeping a minimum force on the indenter so that the instrument can record the deformation progress with time. The modulus and viscosity coefficient of elastomer can be derived using flat-ended punch-tip [8], the elastic-viscoelastic-viscous deformation is described by following equation,

$$h = \frac{P_0 h_{in}}{E_0 A_0} + \sum_{1}^{n} \frac{P_0 h_{in}}{E_i A_0}\left(1 - e^{-E_i t/\eta_i}\right) + \frac{h_{in}}{\eta_o} t$$

where P represents load, h is deformation and subscript e is elastic, η coefficient of viscosity, E modulus, t time. Elastic modulus is thus derived by $P_0 h_{in}/A_0 h_e$. This model provides a means for characterization of elastomer, such as investigation of effect of crosslinking, work of adhesion and fluid environment on elastic modulus [24]. Nanoindentation on PDMS indicated that it is capable of measuring several order of magnitudes of elastic modulus of PDMS, with expected increase of the elastic modulus at higher crosslinked PDMS (2.80 MPa at 10:1 ratio, and 0.88 MPa at 30:1 ratio; monomer:crosslinker, wt%) [24, 27].

Elastomer is permeable to solvents, gases, moisture or even some small molecules [4, 28-30]. Permeability has been employed to fabricate non-contact pump for the manipulation of aqueous solutions within PDMS microfluidic devices [20] or supply oxygen for large-scale culture of hepatocytes [31]. On the other hand, PDMS permeability limits its application to microfluidic technology, in which coating such as glass-like layer using sol-gel chemistry is developed on PDMS to minimize permeability [32]. The permeability of a PDMS film or membrane to a penetrant (i.e., gas or solvent) is defined as [28, 33],

$$P_A = \frac{N_A l}{p_{2A} - p_{1A}}$$

Where P_A is the permeability coefficient, N_A is the steady-state penetrant flux through the film, p_{1A} and p_{2A} are the downstream and upstream partial pressures of component A, and l is the membrane or film thickness. In general, PDMS permeability depends strongly on the feeding gas/solvent composition and temperature: an increase in vapor concentration or liquid pressure leads to increased chain mobility at constant feeding pressure and temperature, resulting in high diffusivity and permeability. The increase of permeability with pressure is attributed to the swelling of PDMS.

Due to the permeability nature of PDMS, many researchers synthesized co-polymer which contained PDMS as a permeable medium for gas/solvent. For example, the 'smart' tri-component amphiphilic membranes consisting of poly(ethylene glycol) (PEG), PDMS and polypentamethylcyclopentasiloxane (PD5) domains were synthesized for biological applications, including immunoisolation of cells [29, 30]. The high oxygen permeability of PDMS makes it feasible for the application in microdevices in cell culture [34]. Compared to large-scale system / platform, microfluidics provides suitable environments for sample interaction, such as cell culture, with the significant increase of surface-to-volume ratio while the fluidic behavior is similar to the environments in vivo [29]. Integration of PDMS into microfluidic platform can implement studies on cell-to-cell interactions and understanding of cell behaviors in vitro, emulation of situations observed in vivo. Several studies have demonstrated that microfluidic devices consisted of two PDMS layers were capable for cell culture [35]. They further fabricated a complex microfluidics which consisted of stacking ten PDMS layers, in which four cell culture chambers and one-oxygen chamber inserts between.

3. Elastomer as electrical contact matrix in microelectronics

Insulation elastomer in microelectronics plays a significant role to prevent a short-circuit [36]. Electrical insulating polymers have been introduced nearly one hundred years ago and proved to be excellent long-term physical and electrical properties, including weatherability, moisture-sealing, corrosion resistant and durability. The polymer insulating materials are used in a wide range of applications such as surge arresters, insulators, insulation enhancement, and bushings [5]. The initial polymers used in insulating materials were mainly on radiation-crosslinked and semicrystalline polyolefin copolymers. With development of manufacturing and materials technologies, elastomer has been increasingly designed for insulating applications [5, 37].

The elastomer formulations have been optimized so that an exceptional electrical and weathering performance is achieved, which are equal or above that of polyolefin copolymer materials. The silicone elastomer has many advantages such as hydrophobicity, hydrophobic recovery, weathering resistance, processability, and elastomeric mechanical properties. These advantages make silicone elastomer suitable for outdoor insulating applications. For example, PDMS has been used for high-voltage (HV) outdoor insulation in the real world as a replacement over conventional porcelain and glass insulation. As one silicone elastomer, PDMS as HV insulating materials is light weight, has superior vandal resistance, and contains better contamination performance. However, many research works have been conducted on molecular modification in order to improve PDMS properties, including degradation under exposure to discharge and arcing, mechanical strength etc. It has been demonstrated that hydrophobicity of PDMS can be rendered by exposure to discharges, which can be recovered when exposure ceases. Their studies showed that the lowly crosslinked PDMS can readily reorient between hydrophilic and hydrophobic states, therefore improving its weathering capabilities [38].

Recently, PDMS has been micropatterned using a photoresist lift-off technique for selective electrical insulation in microelectrode array applications [39]. The micropatterning technique is able to manufacture PDMS patterns with feature as small as 15 μm and various thicknesses down to 6 μm. With micropatterned PDMS insulation layer, the electrical resistance between adjacent electrodes is within the specification. Additionally, the micropatterned PDMS insulation can be applied to biosensing microdevices such as in an extensive neuronal network. Other researchers used PDMS to develop a replaceable insulator for a single-use planar microelectrode array (MEA) in the study of electrogenic tissues [40]. They demonstrated applications using microstencils for rejuvenation of an old MEA and the fabrication of a single-use MEA.

Conductive materials are used to produce shielding gaskets for military, aerospace, electronics, and communications. The conductive elastomers are designed to balance requirements for electrical conductivity, thermal management, and cost performance [41]. Conductive elastomer materials are ideal for customer applications requiring both excellent electromagnetic interference shielding and environmental sealing across a wide range of temperature. The conductive elastomers could be blended polyacetylene with different types of thermoplastic elastomers: styrene-butadiene-styrene, styrene-isoprene-styrene, and styrene-ethylenebutylene-styrene tri-block copolymers [42]. The polyacetylene is a highly conjugated polymer which contains high concentration of unsaturated sites. Those sites are easily attacked by ozone / UV light and reacted with elastomer, formation of conductive elastomer composites. Or insulation elastomers are coated with compliant electrode material on both sides of the elastomer film [43]. The dielectric elastomers were actuated by means of electrostatic forces applied via compliant electrodes. The performance of conductive elastomers is dependent on many factors, e.g., elastomer crosslinking method and type of monomer. In an electrically conductive heterogeneous binary polymer blends, consisting of ethylene-propylene-diene-monomer (EPDM) and polyaniline (PAni), the performance was significantly affected by the crosslinking method [44]. In the blend, PAni underwent doping upon exposure to a protonic acid and became electrically conductive. Several methods have been studied for monomer crosslinking, such as the use of a phenolic resin for unsaturated rubbers and high service temperature products, inter-chain C-C bonds formation after the reticulation reaction, electron beam induced crosslinking for 3-D network. Results indicated that electron beam irradiation crosslinking method was not affected by the presence of the acid necessary for doping the conductive elastomers.

Insulation elastomer has been used as a matrix to hold conductive particles in land grid array socket since it significantly improves contact resistance and increases interconnect density. The need for high density electronics has led to the development of semiconductor array packages, such as ball grid array (BGA) and land grid array (LGA) packages. However, soldering a package to a printed circuit board (PCB) is challenging for input/output (I/O) counts greater than a thousand, and solder joint defects can induce reliability problems in the application. These problems can be solved by using an LGA socket, which eliminates the soldering process, provides a separable interconnect between the component and the PCB, and also enables easy product rework, repair, and upgrade. The conductive particles in elastomer matrix has been introduced into electronics in 1980's [16, 45]. One such conductive elastomer composite is called Z-direction anisotropically conductive materials which are made of the magnetic alignment of conductive particles in elastomers [46]. The composites contain many vertically aligned but laterally isolated chains of ferromagnetic metal spheres, and the ends of which protrude from the surfaces for better electrical contact. These materials exhibit Z-direction only electrical conduction, in combination with the compliant nature of the materials, can be exploited advantageously for a variety of electronic applications, including fine-pitch, area-array, circuit interconnects, circuit-testing, heat sink interfacing, and sensor devices [47, 48]. The schematic diagram of this design structure is shown in Figure 1. This design has been used in a tactile shear sensor for applications such as robotic skins or grippers, touch sensitive actuators, and finger operated controls [49].

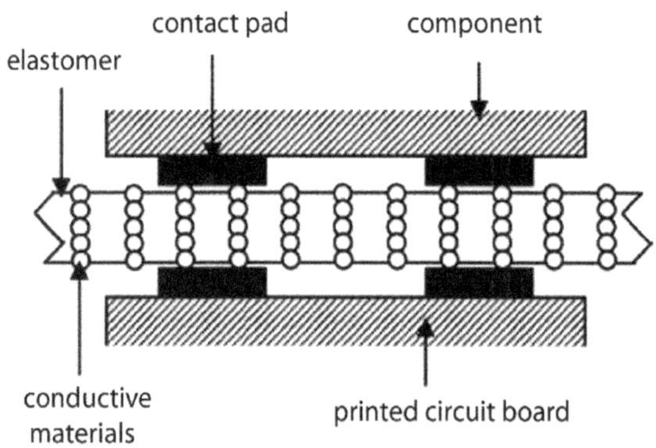

Figure 1. Schematic diagram of a Z-direction only electrical conduction, in which the component is electrical connected to the printed circuit board through vertically aligned and laterally isolated conductive materials. The contact is made by external contact force. Reprint and permission from [16]

An LGA socket assembly consists of component, socket, and board as shown in Figure 2. The contact force is applied through a fixture [16, 45]. The socket design can vary in terms of target component, contact, and housing. One class of LGA sockets incorporates conductive particles in elastomer (polymer) as electrical interconnects for LGA. This class of socket has comparatively low cost, ease of assembly, and short signal path; the elastomer has a sealing effect. However, elastomer-based interconnection is susceptible to failure due to creep and stress relaxation, whenever a force or deformation is provided. A decrease in the contact force due to stress relaxation may lead to the degradation of contact resistance. Increase of elasticity can effectively mitigate creep and stress relaxation, in that crosslinked elastomer has instantaneous recovery after removal of stress.

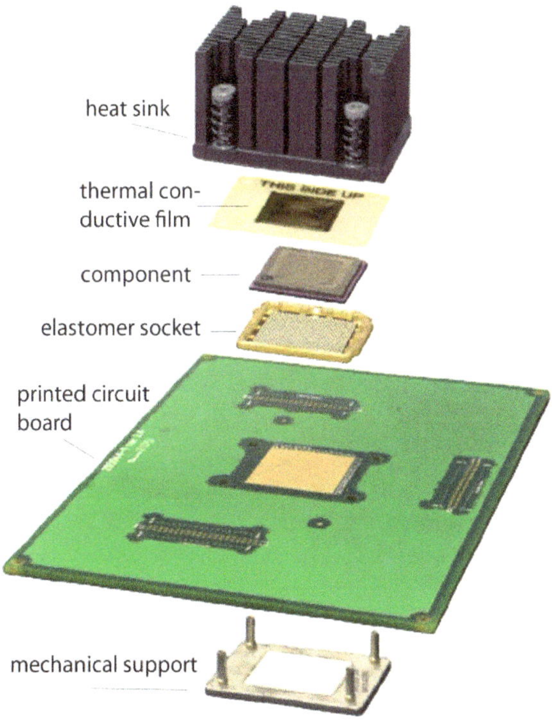

Figure 2. Schematic diagram of elastomer land grid socket assembly. The interconnection part, elastomer socket, can be replaced with different type of sockets, including metalized particle interconnect (MPI), PariPoser film, and conductive rubber sockets (CRS). Reprint and permission from [16]

Several mechanisms can lead to the elastomer failure, such as electrochemical migration [50-52], stress relaxation and creep [45, 53] and corrosion [16]. This section will extensively emphasize these mechanisms and use respective models to predict the lifecycle of the components. The metal electrochemical migration was first identified in the short-circuit failure of Sun Microsystems Sun Fire 6800. Investigation on the failed parts did not observe any dendrite to bridge adjacent components and/or contacts. In general, dendrites were considered to short the electrical circuit. This finding indicated that even without dendrite formation a short-circuit can happen. The design of the experiment, which mimicked the device environment with accelerated conditions, was conducted to identify the failure mechanism and electrochemical migration processes based on the failure products [50-52]. The new failure mechanism in the metal-in-elastomer socket was proposed on the facts that surface insulation resistance has catastrophically decreased and unknown materials were formed on surface after a certain period of time incubation in the temperature-humidity-bias conditions. The metal-in-elastomer socket consists of silver particles in elastomer matrix. Previous studies showed that silver can migrate in the humidity-temperature-bias conditions [54]. However, those studies added the external conductive ions, e.g., contaminants or silver ions, into the system to initialize the migration process [55]. In recent studies, several substantial results exploited different electrochemical migration, including silver electrochemical migration (ECM) processes, migration products, and failure criterion on surface insulation resistance (10^5 ohm).

The silver ECM could occur when bias was applied to silver-in-elastomer socket in the humidity-temperature. Initially, the water layer absorbed on the surface is electrolyzed to hydrogen and hydroxide ions, which further migrate to cathode and anode respectively. Silver oxides to silver ions by accepting electrons through electrolyzed water solution under bias, and silver ions are accumulated in absorbed water layer. The silver ions are also migrated to cathode through water layer. Overall, the surface insulation resistance (SIR) decreases when ions are generated in the absorbed water layer, or formation of conductive electrolyte. The short-circuit failure occurs when silver ion concentration increases to the specific value when the surface insulation resistance decreases to 10^5 ohm.

Both film and dendrites were observed on the film after permanent failure when the surface insulation resistance of the silver-in-elastomer was less than 1000 ohm. Studies on this regard showed that the film growth was developed in deionized water, while the dendrite growth was developed in high conductivity water. It was found that the film grew on the surface without distinctive direction while dendrites grew from cathode to anode. The analysis of film and dendrites from electron scanning spectrometry and X-ray photoelectron microscopy showed that the film consists of silver oxide and dendrites are silver only. The process of the silver electrochemical migration is progressed as follow: silver oxidation at anode to formation of silver ion; silver ion migration from anode to cathode through absorbed water layer (6-7 water monolayers); partial silver ion reduction to silver oxide with hydroxide from water electrolysis (film); other silver ion reduction to formation of silver at the cathode (dendrites) [50-52].

New failure criterion on surface insulation resistance was proposed. In the IPC standard (IPC-TR-476A), the failure was defined as the surface insulation resistance below 10^6 ohm. New experimental findings showed that the component didn't fail when resistance below the resistance, no electrochemical migration related products detected either [50-52]. In addition, the surface insulation resistance immediately decreased to the order of 10^6 ohm when the metal-in-elastomer was conditioned in the highly accelerated stress testing (130°C/85% RH/bias). Until SIR was below 10^5 ohm, the silver dendrites were observed. The resistance ultimately was within hundreds of ohms when the dendrites bridged the adjacent contacts.

The new ECM process was proposed as: presence of moisture, anodic metal dissolution or ion generation and ion migration to cathode (electrochemical reaction), ion accumulation, and metal dendritic growth. The ion accumulation was found to be the rate controlling step in which the surface insulation resistance degraded significantly when the ion concentration accumulated to a specific value. This added step explained the failure occurrence without dendritic formation/bridging and perfected the process. Time-to-failure (TTF) model considered these factors to estimate the failure time, including water adsorption, ion generation, ion migration, and ion accumulation. The dendritic growth was excluded from the model based on the experimental findings. Furthermore, incorporation of dendritic growth model with TTF model can be used to estimate time-to-permanent component failure, since the dendritic bridge creates a permanent conductive path between conductors.

The ion-accumulation based model assumed that the silver discharged at the anode was directly proportional to the amount of electrical charge that passed through the electrode, or Faraday's Law, the total mass of metal dissolution equaled the total mass of metal ions discharged into the electrolyte. Unlike the bulk resistance of conductive particles in elastomer [56], by use of Ohm's law and Brunauer-Emmett-Teller (BET) model [57], the time-to-failure model was given by,

$$TTF = nF \times \frac{m_0}{M} \times \beta \times \frac{1}{V} \times \frac{(1-RH)[1+(c-1)RH]}{cRH} \times e^{(\frac{E_\sigma}{RT})}$$

Where σ_s is the surface insulation conductivity, R is the resistance, m_0 is the critical mass loss required to dissolve to reach the critical ion concentration when resistance decreases to value of failure criterion, V is potential, RH is relative humidity, n is chemical valence, M is the molecular weight, F and R are constant, β, c and E_σ are coefficient relevant to relative humidity, temperature and materials properties. Using this model, the time-to-failure is estimated for elastomer sockets. For silver in elastomer socket, the TTF can be simplified as,

$$TTF = 1.11 \times 10^{-5} \times \frac{1}{V} \times \frac{(1-RH)}{RH} \times e^{(\frac{7846}{T})}$$

Based on this model, the silver-in-elastomer socket fails in 150 hours after incubation in 90°C/90%/20 V conditions; while it can resist up to 27000 h in service at conditions of 25°C/50%/110 V.

Another failure mechanism is called creep and stress relaxation due to viscoelastic elastomer as matrix in socket. As shown in Figure 2, the mechanical support fixed elastomer deformation as constant. As a result, the stress applied to the socket is expected to decrease gradually due to the stress relaxation of elastomer. To understand failure mechanism and its effect on contact resistance, the dynamic mechanical analyzer (DMA) was used to characterize elastomer stress relaxation. The chamber of DMA can control temperature by air gas and liquid nitrogen. Creep is performed by constant stress and stress relaxation by constant strain. The creep process is often categorized into three stages: primary creep (a stage of decreasing creep rate), secondary creep (a stage of constant creep rate), and tertiary creep (a stage of increasing creep rate). When the load is initially applied, there is an instantaneous elongation, a primary stage of a transient nature during which slip and work hardening occur in the most favorably oriented grains. Then there is a secondary stage of steady-state creep during which the deformation continues at an approximately constant rate. The third stage (tertiary stage) takes place when the stress is high enough that the creep rate accelerates until fracture occurs [58]. The results for the creep of the silicone elastomer are shown in Figure 3 (a). Both primary and secondary creeps are observed. The results show that the creep deformation increases with temperature. The constant creep rate during the secondary stage was also a function of temperature.

Stress relaxation refers to the time-dependent contact force decrease under a constant strain. Figure 3 (b) plotted behavior of stress relaxation of elastomer under various temperatures. The significance of stress relaxation on elastomer socket is that contact resistance increases at a decreased contact force. It is thus important to derive the relationship time-dependent contact resistance due to stress relaxation of elastomer. Combination of stress relaxation and contact resistance, the time-dependent contact resistance model is derived [45],

$$R_c = \frac{\rho}{2}\sqrt{\frac{\pi H}{F_0[1 - Bln\left(\frac{t}{C}+1\right)]} + \frac{\sigma_f H}{F_0[1 - Bln\left(\frac{t}{C}+1\right)]}}$$

Where the R_c is contact resistance, ρ is electrical resistivity (ohm.m), H is contact hardness, F_0 is initial contact force, t is time, and B, C are temperature dependent parameters, given by,

$$B = 1.20 \times 10^{-4}T, \text{ and } C = 6.62 \times 10^{-7}Te^{\frac{1130}{T}} \text{ (for silicon elastomer)}$$

The $Rc - t$ can be used for estimation of contact resistance degradation of metal-in-elastomer socket for scheduled socket replacement.

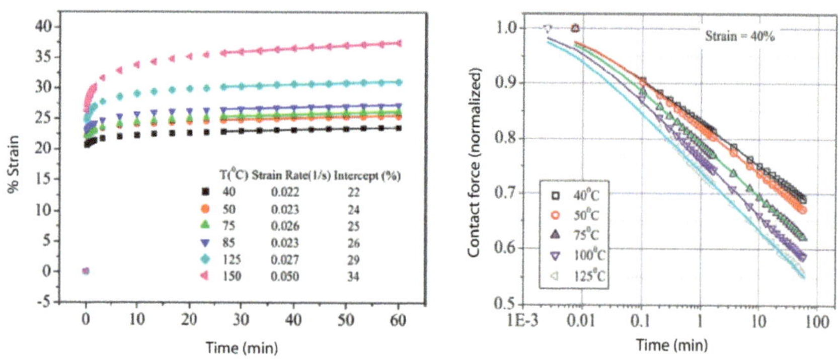

(a) elastomer creep (b) elastomer stress relaxation

Figure 3. Elastomer creep (a) and stress relaxation (b) behavior at various temperatures. Dots: experimental data; solid line: fitted curve. Reprint and permission from [45]

4. Elastomer as substrate in microfluidics

Elastomer is also one of popular materials for microfluidics application as substrate, micro-valves, pumps [59-61]. Elastomer can be easily patterned by curing on a micromachining mold as an alternative microfabrication technique. The soft lithography technique, which generates micropatterns by contact printing and microstructures by embossing and replica molding, has been used to manufacture blazed grating optics, stamps for chemical patterning, and microfluidics devices [19, 62, 63]. Different from photolithography technique [64], the soft lithography is not subject to the limitations set by optical diffraction and optical transparency, with feature size down to 10-100 nm. Soft lithography techniques is termed as 'multilayer soft lithography' and developed by Quake et al. [59] in 2000. Briefly, multilayer structures were formed by bonding PDMS layers from separated cast in a micromachining mold. Since each layer has excess of one of the components, the reactive molecules remained at the interface form permanent bonding after further curing. They used such technique for development of microfluidic large-scale integration chips that contain plumbing networks with thousands of micromechanical PDMS valves and hundreds of individually addressable chambers. Their chips can be used to construct the microfluidic analog of a comparator array and a microfluidic memory storage device similar to the behavior of random-access memory.

PDMS microfluidics have great potential to be biomedical applications which only require small amounts of sample, routine operation by untrained personnel and low cost [65]. One of such examples was developed on a simple, inexpensive PDMS microfluidic diagnostic device, which performs sandwich immunoassays for medicine and biological studies [12, 63]. Screws are used in the device as virtual valves to on and off fluids. The low-cost on soft lithography microfabrication and raw materials makes this device great potential for portable healthcare delivery and monitoring. PDMS microfluidics can be used for reagent mixing in microscale channel [66]. The challenge for flow mixing in a small dimension is mainly due to the difficult liquid turbulence inside chip. By use of PDMS as microfluidics substrate, a topological structure can be fabricated to exploit the laminarity of the flow to repeatedly fold the flow and double the lateral concentration gradient in a compact chip. Two different solutions are contained in a T-junction, and fluid flow is repeatedly split, rotated, and merged [67]. The chip design was obtained by a rapid-prototyping master mold, and results showed that effective mixing can be achieved on short length scales.

PDMS microfluidics has applied to a variety of applications such as cell sorting, DNA sizing and sorting. DNA sequence is successfully performed in PDMS devices including sample preparation and electrophoresis analysis [14]. PDMS has substantial advantages over glass or silicon dioxide substrates, since these latter materials were adhesive for the cells of interest. In addition, Chou et al. [68] developed a PDMS device for the sizing and sorting of restriction fragments of DNA based on single DNA molecule detection. The device sizes DNA on the basis of fluorescence intensity in which the DNA was labeled with fluorescence dye.

PDMS device can be easily fabricated by non-covalent bonding or plasma treated permanent bonding after replica by molding. The PDMS microfluidics for capillary electrophoresis was developed by Effenhauser et al. [69] in 1997. They used replica molding to obtain PDMS device and bonded reversibly the device with a flat piece of PDMS. The performance of PDMS CE is not ideal on unmodified surface, especially for proteins/peptides analysis, due to hydrophobicity of the unmodified PDMS surface. Later, Whitesides et al. developed PDMS microfluidics device by a rapid prototyping and irreversible bonding through plasma oxidation[19]. Different PDMS devices were developed to demonstrate separation performance of fluorescently labeled amino acids and proteins using capillary zone electrophoresis (CZE) and gel electrophoresis (GE).

One-dimensional sodium dodecyl sulfate (SDS) capillary gel electrophoresis (CGE) can be also performed in PDMS microfluidics due to its versatile surface chemistry. PDMS surface is treated by UV-Ozone so that it could immobilize other hydrophilic molecules to prevent electroosmotic flow and protein/peptide non-specific binding. For complex cell lysate analysis, isoelectric focusing (IEF) can be interfaced with SDS-CGE as the first dimension separation. What makes PDMS special for IEF-SDS/CGE is because of conventional procedures for fabricating 3-D microfluidic channels [70], and the facility with which PDMS-based devices can be readily assembled and disassembled. For example, after IEF separation, the PDMS was disassembled and connected with a 3-D channel for SDS/CGE. Recent development on 2-D IEF-SDS/CGE showed that other thermoplastic would be better substrate of choice for integrated 2-D complex cell lysate analysis [71, 72].

The design of elastomer microfluidics can be made based on these studies [71-73]. The electrokinetic injection of defined sample plugs within single dimension microfluidic systems has been extensively studied, with the most widely used injector configurations including the cross, double-T, and triple-T topologies [74, 75], and extended configurations employing continuous sample injection leveraging flow switching techniques. Floating injection and pinched-injection methods [76, 77] have been employed in simple cross-injectors for the controlled definition of small sample plugs, while double- and triple-T designs are generally employed for the introduction of larger sample volumes. Sample leakage [78] was observed in a 2-D device during sample injection which was the central issue for separation performance as shown in Figure 4. After the desired sample plug has been transferred to the separation channel, additional samples enter the injection region due to diffusion and fringing of the electrical field during the electrokinetic transfer process. The excess sample tailing degrades separation performance. To solve this issue, backbiasing is a commonly used to eliminate sample leakage. Practically, bias voltages can be applied to sample inlet and waste reservoirs, electrokinetically pulling back excess sample from the injection zone. The effective of this approach depended on the accurate selection of backbiasing bias. This requires characterization of device geometry, medium in device, and buffer resistivity. A few groups [49, 72, 79] have proposed 2-D chip designs consisting of an array of second dimension separation channels aligned with an identical number of injection channels on the opposite side of the first dimension microchannel, a more efficient design for spatially multiplexed 2-D separation chips is shown in Figure 4 (a). The design consists of a single first dimension separation channel intersected by multiple second dimension separation channels on one side, and sample injection channels on the opposite side, with the injection channels staggered with respect to the separation channels to ensure complete and simultaneous sampling of species separated in the first dimension channel. In this staggered design, the parallel second dimension microchannels may be regarded as an array of double-T injectors operating in parallel. However, unlike the single-channel case, the double-T injectors are not electrically isolated in the multidimensional system.

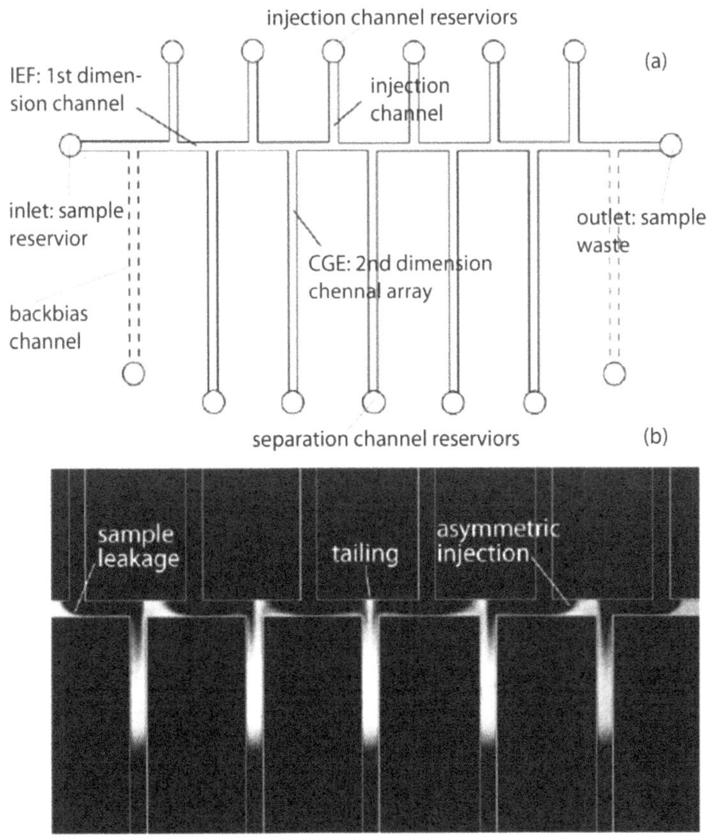

Figure 4. Schematic diagram of two-dimensional capillary electrophoresis microfluidics design with (a) simplified device of a spatially-multiplexed separation platform with five second dimension microchannels, and (b) image during the sample transfer process using a fabricated chip without backbias channels. Reprint and permission from [71]

This interconnected design can result in significant variations in performance between the different injectors, as depicted in Figure 4 (b). In this image, the sample within the first dimension channel is being electrokinetically transferred into the second dimension channel array by applying a uniform bias voltage in the injection channel reservoirs while grounding the second dimension reservoirs. Three main features are evident in this injection process. The substantial tailing of sample occurs at the head of each second dimension channel, resulting in sample dispersion during transfer; in all but the center second dimension channel, the injection is highly asymmetric, reflecting a non-uniform electric field distribution within the first dimension channel; sample from the outermost regions of the first dimension channel continually leaks into the second dimension array due to a combination of diffusion and electric field fringing, leading to additional tailing which can continue long after sample from the center region of the first dimension channel has been fully transferred separation devices .

The solution for unsymmetrical injection and tailing was proposed by adding backbias channel s parallel to the separation channels as shown in Figure 4 (b) [70]. Analytical calculation indicated that two-fold of resistance in the backbias channel generated uniform current and potential distribution in the 2-D channel network, not only in 1-D channel, but also injection and separation channels. Additionally, the angled 1-D channel can level the electric field distribution in the local region of T intersection. As a result, the angled 1-D channel substantially reduced sample tailing with negligible effect on the first dimension separation. These design perspectives have been used for complex biological sample analysis.

Use of PDMS for liquid chromatography has been limited largely due to its low bonding strength [80] and high gas/liquid permeability [81]. It was reported to fabricate column in a device using PDMS [82-84]. Since PDMS is inexpensive and thus it can be fabricated as a disposable column, it can be used as a stationary phase in liquid chromatography columns. The siloxane monomer was added into a molding to formation of desired features. The featured PDMS was reversibly sealed with cover PDMS by placing the molded chip in contact with cover PDMS, or irreversibly sealed by plasma oxidation. The PDMS chip was treated with silanes to minimize sample non-specific absorption. In general, PDMS is not the good materials of choice for high performance liquid chromatography. In such instance, other thermoplastics such as polyimide (PI) [85], fuse silica capillary [75], cyclic olefin copolymer (CoC) [86], are recommended.

Droplets formed by mixing one fluid in another or emulsions are very useful in a wide range of applications including personal care products, foods, microbiology, cell biology, and drug delivery vehicles [87, 88]. It has been recognized that a successful droplet device for applications is able to control over size and distribution in microscale and nanoscale emulsions [88]. Droplet formation can be obtained by either turbulence to break apart an immiscible mixture or continuous flow. PDMS devices for micro-droplet formation are fabricated by soft lithography in a molding. The native PDMS surface is usually hydrophobic and excellent surface for water-in-oil emulsions. The hydrophobic PDMS surface ensures that the oil droplets encapsulation of water phase inside without being contacted the PDMS channel walls [89, 90]; while hydrophilic PDMS surface can be used for oil-in-water droplets and biological sample analysis [91, 92]. Several surface modifications have been reported by the sequential layer-by-layer deposition of polyelectrolytes yielding hydrophilic microchannels. Others used oxygen-plasma and oxygen-C2F6 to modify PDMS surface in order to enhance hydrophilicity and reduce channel electroosmotic flow (EOF). Alternatively, use of chemical modification, e.g., 2-hydroxyethyl methacrylate (HEMA), can permanently enhance PDMS hydrophilicity [91]. In layer-by-layer (LbL) surface modification approach, segments of sodium chloride, poly(allylamine hydrochloride), and poly(sodium 4-styrenesulfonate) solutions are separated by air plugs and loaded into a piece of tubing and sequentially flushed through the channel at a constant flow rate. This LbL modification provides long storage time up to 5 months without noticeable behavior change of droplet formation. The LbL modified PDMS device can be used for water-oil-water double emulsion system for applications such as pharmaceutical compound delivery.

An interesting PDMS device for static microdroplet arrays is fabricated to droplet trapping, incubation, and release of enzymatic and cell-based assays [93]. They simply used a single-layered PDMS microfluidic structure. The aqueous droplets are trapped en masse and optically monitored for extended periods of time. The array droplet approach is used to characterize droplet shrinkage, aggregation of encapsulated biological cells and enzymatic reactions. On the other hand, to optimize the droplet formation, a design which use feedback control loop is developed [88]. They successfully demonstrated that the closed loop control system with negative feedback can produce the required droplet size with minimal noise and no systematic steady sate error. This close-loop design is utilized for a number of applications, including micro-chemistry and emulsion production.

5. Outlook of elastomer application

The application of elastomer is being continuously exploited. The future of elastomer is closely related to the specific requirement on applications. One of the hot topics is development of nano- and microfluidics for assisted reproductive technologies [94]. For example, the studies have demonstrated that thickness of PDMS layers at 100 μm in chips produced significantly fewer embryos due to drastically shift in osmolality of embryo-culture media. It could be the solution by development of a hybrid membrane consisting of PDMS-parylene-PDMS for maintaining a stable cell growth environment. Therefore, investigation of new type of PDMS materials or hybrid PDMS can provide means for such applications.

Due to the low bonding strength, PDMS has seldom used for high performance liquid chromatography application (HPLC). The HPLC can be directly interfaced with detection system such as electrospray ionization or mass spectrometry. As a result, it becomes research of interest to synthesize PDMS which has strong bonding strength (via covalent bond) and low permeability, while PDMS optical transparency and elasticity are largely remained. In summary, PDMS is becoming more and more popular as the materials of choice for biological analysis due to its good physical, mechanical, and chemical properties, as well as its excellent biocompatibility.

A thermoplastic rubber material has recently been developed that can completely mend itself when the fracture interfaces are rejoined and left to heal for a moderate time [95]. This "smart" rubber is easy to synthesize and displays excellent mechanical properties. This material could be used for microelectronics, microfluidics and biology system.

Author details

Shuang (Jake) Yang[*]
Department of Pathology, Johns Hopkins University, Baltimore, USA

Shuang (Jake) Yang and Kunqiang Jiang
Department of Mechanical Engineering, University of Maryland, USA
Department of Chemistry and Biochemistry, University of Maryland, USA

[*] Corresponding Author

6. References

[1] Legge, N.R. Thermoplastic elastomers. 60, 83 (1987).

[2] Eisenbach, C., Baumgartner, M. & Gunter, C. Advances in Elastomers and Rubber Elasticity. 51 (1986).

[3] Markham, R.L. & Mueller, W.J. (Google Patents, 1986).

[4] Massey, L.K. Permeability Properties of Plastics and Elastomers: A Guide to Packaging and Barrier Materials, Edn. 2nd. (Plastics Design Library / William Andrew Publishing, Norwich, NY; 2002).

[5] Simmons, S., Shah, M., Mackevich, J. & Chang, R.J. Polymer outdoo insulating materials Part Ill-silicone elastomer considerations. 13, 8 (19978).

[6] Lin, Y.H., Kang, S.W. & Wu, T.Y. Fabrication of polydimethylsiloxane (PDMS) pulsating heat pipe. 29, 573-580 (2009).

[7] Jafari, S.H. & Gupta, A.K. Impact strength and dynamic mechanical properties correlation in elastomer-modified polypropylene. *J Appl Polym Sci* 78, 962-971 (2000).

[8] Yang, S., Zhang, Y.W. & Zeng, K.Y. Analysis of nanoindentation creep for polymeric materials. *J Appl Phys* 95, 3655-3666 (2004).

[9] White, C.C., Vanlandingham, M.R., Drzal, P.L., Chang, N.K. & Chang, S.H. Viscoelastic characterization of polymers using instrumented indentation. II. Dynamic testing. *J Polym Sci Pol Phys* 43, 1812-1824 (2005).

[10] Jancar, J., Dianselmo, A., Dibenedetto, A.T. & Kucera, J. Failure Mechanics in Elastomer Toughened Polypropylene. *Polymer* 34, 1684-1694 (1993).

[11] Toepke, M.W. & Beebe, D.J. PDMS absorption of small molecules and consequences in microfluidic applications. 6, 1484-1486 (2006).

[12] Eteshola, E. & Leckband, D. Development and characterization of an ELISA assay in PDMS microfluidic channels. 72, 129-133 (2001).

[13] Martin, R.S., Gawron, A.J., Lunte, S.M. & Henry, C.S. Dual-electrode electrochemical detection for poly (dimethylsiloxane)-fabricated capillary electrophoresis microchips. 72, 3196-3202 (2000).

[14] Paegel, B.M., Blazej, R.G. & Mathies, R.A. Microfluidic devices for DNA sequencing: sample preparation and electrophoretic analysis. 14, 42-50 (2003).

[15] Scott, N. An area modulus of elasticity: definition and properties. 58, 269-275 (2000).

[16] Yang, S., Wu, J. & Pecht, M.G. Reliability Assessment of Land Grid Array Sockets Subjected to Mixed Flowing Gas Environment. *Ieee T Reliab* 58, 634-640 (2009).

[17] Liu, W. & Pecht, M. IC component sockets. (Wiley-Blackwell, 2004).

[18] Lee, J.N., Park, C. & Whitesides, G.M. Solvent compatibility of poly (dimethylsiloxane)-based microfluidic devices. 75, 6544-6554 (2003).

[19] McDonald, J.C. et al. Fabrication of microfluidic systems in poly(dimethylsiloxane). *Electrophoresis* 21, 27-40 (2000).

[20] Eddings, M.A. & Gale, B.K. A PDMS-based gas permeation pump for on-chip fluid handling in microfluidic devices. 16, 2396 (2006).

[21] Mark, J.E. Some unusual elastomers and experiments on rubberlike elasticity. *Prog Polym Sci* 28, 1205-1221 (2003).

[22] Lawn, B. & Wilshaw, R. Indentation Fracture - Principles and Applications. *J Mater Sci* 10, 1049-1081 (1975).
[23] Oliver, W.C. & Pharr, G.M. An Improved Technique for Determining Hardness and Elastic-Modulus Using Load and Displacement Sensing Indentation Experiments. *J Mater Res* 7, 1564-1583 (1992).
[24] Carrillo, F. et al. Nanoindentation of polydimethylsiloxane elastomers: Effect of crosslinking, work of adhesion, and fluid environment on elastic modulus. *J Mater Res* 20, 2820-2830 (2005).
[25] Bao, T., Morrison, P.W. & Woyczynski, W. AFM nanoindentation as a method to determine microhardness of hard thin films. *Mater Res Soc Symp P* 517, 395-400 (1998).
[26] Oyen, M.L. Sensitivity of polymer nanoindentation creep measurements to experimental variables. 55, 3633-3639 (2007).
[27] Mata, A., Fleischman, A.J. & Roy, S. Characterization of polydimethylsiloxane (PDMS) properties for biomedical micro/nanosystems. 7, 281-293 (2005).
[28] Singh, A., Freeman, B.D. & Pinnau, I. Pure and mixed gas acetone/nitrogen permeation properties of polydimethylsiloxane [PDMS]. *J Polym Sci Pol Phys* 36, 289-301 (1998).
[29] Kurian, P. et al. Synthesis, permeability and biocompatibility of tricomponent membranes containing polyethylene glycol, polydimethylsiloxane and polypentamethylcyclopentasiloxane domains. *Biomaterials* 24, 3493-3503 (2003).
[30] Pinnau, I. & He, Z.J. Pure- and mixed-gas permeation properties of polydimethylsiloxane for hydrocarbon/methane and hydrocarbon/hydrogen separation. *J Membrane Sci* 244, 227-233 (2004).
[31] Leclerc, E., Sakai, Y. & Fujii, T. Microfluidic PDMS (polydimethylsiloxane) bioreactor for large-scale culture of hepatocytes. 20, 750-755 (2004).
[32] Abate, A.R., Lee, D., Do, T., Holtze, C. & Weitz, D.A. Glass coating for PDMS microfluidic channels by sol–gel methods. 8, 516-518 (2008).
[33] Ghosal, K. & Freeman, B.D. Gas separation using polymer membranes: an overview. 5, 673-697 (1994).
[34] Shiku, H. et al. Oxygen permeability of surface-modified poly(dimethylsiloxane) characterized by scanning electrochemical microscopy. *Chem Lett* 35, 234-235 (2006).
[35] Leclerc, E., Sakai, Y. & Fujii, T. Cell culture in 3-dimensional microfluidic structure of PDMS (polydimethylsiloxane). *Biomed Microdevices* 5, 109-114 (2003).
[36] Tummala, R.R., Rymaszewski, E.J. & Klopfenstein, A.G. Microelectronics Packaging Handbook. (Kluwer Academic Publishers Group, Norwell, MA; 1997).
[37] Kim, J., Chaudhury, M.K. & Owen, M.J. Hydrophobicity loss and recovery of silicone HV insulation. *Ieee T Dielect El In* 6, 695-702 (1999).
[38] Liu, H.P., Cash, G., Birtwhistle, D. & George, G. Characterization of a severely degraded silicone elastomer HV insulator - an aid to development of lifetime assessment techniques. *Ieee T Dielect El In* 12, 478-486 (2005).
[39] Park, J., Kim, H.S. & Han, A. Micropatterning of poly(dimethylsiloxane) using a photoresist lift-off technique for selective electrical insulation of microelectrode arrays. *J Micromech Microeng* 19 (2009).

[40] Nam, Y., Musick, K. & Wheeler, B.C. Application of a PDMS microstencil as a replaceable insulator toward a single-use planar microelectrode array. *Biomed Microdevices* 8, 375-381 (2006).

[41] Tong, X.C. Advanced Materials for Thermal Management of Electronic Packaging, Edn. 1st. (Springer Science & Business Media, LLC, New York; 2010).

[42] Lee, K.I. & Jopson, H. Electrically Conductive Thermoplastic Elastomer Polyacetylene Blends. *Polym Bull* 10, 105-108 (1983).

[43] Pelrine, R., Kornbluh, R., Pei, Q.B. & Joseph, J. High-speed electrically actuated elastomers with strain greater than 100%. *Science* 287, 836-839 (2000).

[44] Faez, R., Schuster, R.H. & De Paoli, M.A. A conductive elastomer based on EPDM and polyaniline. II. Effect of the crosslinking method. *Eur Polym J* 38, 2459-2463 (2002).

[45] Yang, S., Wu, J., Tsai, D. & Pecht, M.G. Contact resistance estimation for time-dependent silicone elastomer matrix of land grid array socket. *Ieee T Compon Pack T* 30, 81-85 (2007).

[46] Jin, S. et al. New, Z-Direction Anisotropically Conductive Composites. *J Appl Phys* 64, 6008-6010 (1988).

[47] Jin, S., Tiefel, T.H., Wolfe, R., Sherwood, R.C. & Mottine, J.J. Optically Transparent, Electrically Conductive Composite Medium. *Science* 255, 446-448 (1992).

[48] Ausanio, G. et al. Giant resistivity change induced by strain in a composite of conducting particles in an elastomer matrix. *Sensor Actuat a-Phys* 127, 56-62 (2006).

[49] Chen, L.H., Jin, S. & Tiefel, T.H. Tactile Shear Sensing Using Anisotropically Conductive Polymer. *Appl Phys Lett* 62, 2440-2442 (1993).

[50] Yang, S., Wu, J. & Pecht, M. Electrochemical migration of land grid array sockets under highly accelerated stress conditions. 238-244 (2005).

[51] Yang, S., Wu, J. & Christou, A. Initial stage of silver electrochemical migration degradation. *Microelectron Reliab* 46, 1915-1921 (2006).

[52] Yang, S.A. & Christou, A. Failure model for silver electrochemical migration. *Ieee T Device Mat Re* 7, 188-196 (2007).

[53] Liu, W.F., Pecht, M.G. & Xie, J.S. Fundamental reliability issues associated with a commercial particle-in-elastomer interconnection system. *Ieee T Compon Pack T* 24, 520-525 (2001).

[54] Krumbein, S.J. Metallic Electromigration Phenomena. *Ieee T Compon Hybr* 11, 5-15 (1988).

[55] Zamanzadeh, M., Meilink, S.L., Warren, G.W., Wynblatt, P. & Yan, B. Electrochemical Examination of Dendritic Growth on Electronic Devices in Hcl Electrolytes. *Corrosion* 46, 665-671 (1990).

[56] Ruschau, G.R., Yoshikawa, S. & Newnham, R.E. Resistivities of Conductive Composites. *J Appl Phys* 72, 953-959 (1992).

[57] Brunauer, S., Emmett, P.H. & Teller, E. Adsorption of gases in multimolecular layers. *J Am Chem Soc* 60, 309-319 (1938).

[58] Findley, W.N., Lai, J.S. & Onaran, K. Creep and Relaxation of Nonlinear Viscoelastic Materials. (Dover, New York; 1975).

[59] Unger, M.A., Chou, H.P., Thorsen, T., Scherer, A. & Quake, S.R. Monolithic microfabricated valves and pumps by multilayer soft lithography. *Science* 288, 113-116 (2000).

[60] Melin, J. & Quake, S.R. Microfluidic large-scale integration: The evolution of design rules for biological automation. *Annu Rev Bioph Biom* 36, 213-231 (2007).

[61] Erickson, D. & Li, D.Q. Integrated microfluidic devices. *Anal Chim Acta* 507, 11-26 (2004).

[62] de Mello, A. On-chip chromatography: the last twenty years. *Lab Chip* 2, 48n-54n (2002).

[63] Xia, Y.N. & Whitesides, G.M. Soft lithography. *Angew Chem Int Edit* 37, 551-575 (1998).

[64] Levenson, M.D., Viswanathan, N. & Simpson, R.A. Improving resolution in photolithography with a phase-shifting mask. 29, 1828-1836 (1982).

[65] Paguirigan, A.L. & Beebe, D.J. Microfluidics meet cell biology: bridging the gap by validation and application of microscale techniques for cell biological assays. 30, 811-821 (2008).

[66] Song, H. & Ismagilov, R.F. Millisecond kinetics on a microfluidic chip using nanoliters of reagents. 125, 14613-14619 (2003).

[67] Chen, H. & Meiners, J.C. Topologic mixing on a microfluidic chip. 84, 2193 (2004).

[68] Chou, H.P., Spence, C., Scherer, A. & Quake, S. A microfabricated device for sizing and sorting DNA molecules. 96, 11 (1999).

[69] Effenhauser, C.S., Bruin, G.J.M., Paulus, A. & Ehrat, M. Integrated capillary electrophoresis on flexible silicone microdevices: analysis of DNA restriction fragments and detection of single DNA molecules on microchips. 69, 3451-3457 (1997).

[70] Jo, B.H., Van Lerberghe, L.M., Motsegood, K.M. & Beebe, D.J. Three-dimensional micro-channel fabrication in polydimethylsiloxane (PDMS) elastomer. 9, 76-81 (2000).

[71] Yang, S., Liu, J. & DeVoe, D.L. Optimization of sample transfer in two-dimensional microfluidic separation systems. *Lab Chip* 8, 1145-1152 (2008).

[72] Yang, S., Liu, J.K., Lee, C.S. & Devoe, D.L. Microfluidic 2-D PAGE using multifunctional in situ polyacrylamide gels and discontinuous buffers. *Lab Chip* 9, 592-599 (2009).

[73] Liu, J.K., Yang, S., Lee, C.S. & Devoe, D.L. Polyacrylamide gel plugs enabling 2-D microfluidic protein separations via isoelectric focusing and multiplexed sodium dodecyl sulfate gel electrophoresis. *Electrophoresis* 29, 2241-2250 (2008).

[74] Fu, L.M., Yang, R.J., Lee, G.B. & Liu, H.H. Electrokinetic injection techniques in microfluidic chips. 74, 5084-5091 (2002).

[75] Blas, M., Delaunay, N. & Rocca, J.L. Electrokinetic-based injection modes for separative microsystems. 29, 20-32 (2008).

[76] Hirokawa, T., Takayama, Y., Arai, A. & Xu, Z. Study of a novel sample injection method (floating electrokinetic supercharging) for high-performance microchip electrophoresis of DNA fragments. 29, 1829-1835 (2008).

[77] Zhang, L., Yin, X. & Fang, Z. Negative pressure pinched sample injection for microchip-based electrophoresis. 6, 258-264 (2006).

[78] Duffy, D.C., McDonald, J.C., Schueller, O.J.A. & Whitesides, G.M. Rapid prototyping of microfluidic systems in poly (dimethylsiloxane). 70, 4974-4984 (1998).

[79] Li, Y., Buch, J.S., Rosenberger, F., DeVoe, D.L. & Lee, C.S. Integration of isoelectric focusing with parallel sodium dodecyl sulfate gel electrophoresis for multidimensional protein separations in a plastic microfludic network. 76, 742-748 (2004).
[80] Gent, A. & Tobias, R. Effect of interfacial bonding on the strength of adhesion of elastomers. III. Interlinking by molecular entanglements. 22, 1483-1490 (1984).
[81] Kuncová-Kallio, J. & Kallio, P.J. 2486-2489 (IEEE, 2006).
[82] Weng, X., Chon, C.H., Jiang, H. & Li, D. Rapid detection of formaldehyde concentration in food on a polydimethylsiloxane (PDMS) microfluidic chip. 114, 1079-1082 (2009).
[83] Slentz, B.E., Penner, N.A., Lugowska, E. & Regnier, F. Nanoliter capillary electrochromatography columns based on collocated monolithic support structures molded in poly (dimethyl siloxane). 22, 3736-3743 (2001).
[84] Ocvirk, G. et al. Electrokinetic control of fluid flow in native poly (dimethylsiloxane) capillary electrophoresis devices. 21, 107-115 (2000).
[85] Levkin, P.A. et al. Monolithic porous polymer stationary phases in polyimide chips for the fast high-performance liquid chromatography separation of proteins and peptides. 1200, 55-61 (2008).
[86] Mair, D.A., Geiger, E., Pisano, A.P., Fréchet, J.M.J. & Svec, F. Injection molded microfluidic chips featuring integrated interconnects. 6, 1346-1354 (2006).
[87] Anna, S.L., Bontoux, N. & Stone, H.A. Formation of dispersions using "flow focusing" in microchannels. 82, 364 (2003).
[88] Miller, E., Rotea, M. & Rothstein, J.P. Microfluidic device incorporating closed loop feedback control for uniform and tunable production of micro-droplets. 10, 1293-1301 (2010).
[89] Link, D.R. et al. Electric control of droplets in microfluidic devices. 45, 2556-2560 (2006).
[90] Fujii, T. PDMS-based microfluidic devices for biomedical applications. 61, 907-914 (2002).
[91] Bodas, D. & Khan-Malek, C. Formation of more stable hydrophilic surfaces of PDMS by plasma and chemical treatments. 83, 1277-1279 (2006).
[92] Bauer, W.A.C., Fischlechner, M., Abell, C. & Huck, W.T.S. Hydrophilic PDMS microchannels for high-throughput formation of oil-in-water microdroplets and water-in-oil-in-water double emulsions. 10, 1814-1819 (2010).
[93] Huebner, A. et al. Static microdroplet arrays: a microfluidic device for droplet trapping, incubation and release for enzymatic and cell-based assays. 9, 692-698 (2009).
[94] Beebe, D., Wheeler, M., Zeringue, H., Walters, E. & Raty, S. Microfluidic technology for assisted reproduction. 57, 125-135 (2002).
[95] Wietor, J.L. & Sijbesma, R.P. A Self-Healing Elastomer. 47, 8161-8163 (2008).

Chapter 10

Use of Silicone Elastomer-Based Microfluidic Devices and Systems in Reproductive Technologies

Koji Matsuura and Keiji Naruse

Additional information is available at the end of the chapter

1. Introduction

Reproductive technologies (RT) play important roles in the conservation of biodiversity, the production of domestic animals such as pigs and cows, and in human infertility treatments [1-3]. Artificial insemination (AI) is a very useful tool to introduce superior genes with a minimal risk of disease as compared with natural mating [4]. Improvements in cryopreservation of semen as well as storage in its liquid form have made AI more accessible [5, 6]. *In vitro* manipulation of gametes for the production of embryos was first successfully applied in mice in 1958 and in rabbits in 1959 [7]. In initial experiments, matured oocytes retrieved from a female were fertilized with sperm in the laboratory and then transferred to a recipient mother. In 1981, the first live calf was produced by *in vitro* fertilization (IVF). Currently, the development and refinement of AI and IVF are concentrating on gamete and embryo collection, sorting and preservation of sperm, gametes, and embryos, and *in vitro* production of embryos, culture, and embryo transfer. These developments are facilitated by modern equipment for ultrasonography, microscopy, cryopreservation, endoscopy, and flow cytometry, as well as micromanipulation and centrifugation [1]. Microfluidic technologies have also contributed to recent developments in RT, especially over the past decade.

Microfluidic devices are powerful tools for handling reproductive cells, because the sizes of these cells are in the micrometer range. The length of a mammalian sperm is 50 μm, and the diameter of the mammalian oocyte and early embryo is approximately 100 μm; these are similar in all mammalian species. For cellular-level observations, transparent materials must be used. More than 10 years ago, microdevices for manipulation of reproductive cells were hand-made and sometimes reproducing these devices was laborious work. For example, to settle embryos in microwells, heated polished steel rods were pressed by hand to the bottom of the well of a polystyrene dish to prepare the microwells [8]. The current most popular technology for the fabrication of microfluidic devices for cell biological application is based on the soft-lithography of poly-dimethylsiloxane (PDMS). PDMS is a silicone elastomer, it is transparent, not detrimental to cells, possesses high permeability to gas, and is easy to fabricate; therefore, fabricated devices can be used for observing cells. Through simple molding procedures, the material can be made into microfluidic devices in short periods of time (typically, <1 day), and multiple copies of a device can be generated in several hours [9]. Its widespread use as a material of choice is because of its mechanical properties, which are amenable to integration of fluidic valves, essential elements for major microfluidic applications [10]. For biological use, reported applications of PDMS microfluidic devices are analyzing cells [11, 12], applying forces [13, 14], substrate patterning [15, 16], and creating chemical microenvironments [17-19]. Because of laminar flow in microfluids, experiments using these microfluidic devices for chemical gradient production and cellular manipulations are easily reproduced. These elastomer technologies and applications can be transferred to the development of RT.

This chapter introduces the usability of silicone elastomer-based microfluidic devices in RT. Since most readers are not likely to be familiar with RT, we will cover the following points: (1) the background to the use of elastomers in RT, (2) details of current research on sorting and analyzing motile sperm, (3) silicone elastomer-based microfluidic devices for creating static and dynamic mammalian embryo culture systems that can mimic the motion of the oviduct, and (4) the conclusion. This chapter introduces the current research areas for developing improvements in RT and suggests the possibility of using elastomers in human ART.

1.1. Background to the use of elastomers in reproductive technologies

RT refers to procedures that include *in vitro* handling of cattle or human oocytes, sperm, or embryos to induce pregnancy [20, 21]. To improve fertilization rate in pigs and cattle, AI using washed sperm is the most widely used technique. Because the outcome of AI largely depends on the quality of semen and the insemination procedure, sperm washing and sperm motility analyses are part of conventional AI procedures [4]. Sperm washing by differential gradient centrifugation and swim-up of motile sperm can reduce the amount of virus present in the sample [22]. However, harmful effects of washing and centrifugation have been attributed to the generation of reactive oxygen species, which can irreversibly damage the sperm and impair their fertilizing ability [23-25]. This damage may be severe for cryopreserved sperm that have already been injured during freezing [25]. Some sperm motility analyses provide reliable estimation of the fertilizing ability of mammalian sperm such as computer-assisted sperm analysis (CASA) [26, 27]. A significant positive correlation with fertility was found for the linear-motility parameter [28]. Furthermore, the generally reduced fertility of cryopreserved semen is considered to be due to the short life-span of frozen–thawed sperms [29, 30]. Before AI, the likelihood of successful fertilization can be predicted by motility analyses.

The most common procedure performed to assist reproduction is *in vitro* fertilization-embryo transfer (IVF-ET). Conventional IVF treatment requires that the ovaries be stimulated with gonadotropins, which contain follicle-stimulating hormone (FSH) and luteinizing hormone (LH) in order to increase the number of mature oocytes retrieved, and thus the number of embryos available for transfer [31]. After washing the sperm to be used, the sperm-containing medium is dispensed in microdrops under paraffin oil or more simply in wells without oil [32]. At the same time, matured oocytes are introduced into the fertilization medium and co-incubated with sperm for 18–24 h at 37–39°C in 5% CO_2. The fertilization rate can be assessed indirectly by examination of the cleavage rate 40–42 h post insemination. The basic procedures of IVF in cattle and humans are similar [33]. Another *in vitro* procedure is intracytoplasmic sperm injection (ICSI), where a selected single sperm is directly injected into the oocyte. For the ICSI procedure, the oocyte is immobilized using a holding pipette; an injection pipette with an internal diameter of 6 μm is used to aspirate a single spermatozoon. Before aspiration, the sperm is immobilized in polyvinylpyrrolidone. A morphologically normal sperm is aspirated into the injection needle, tail first. Immobilization of the sperm can also be achieved by crushing the tail with the injection pipette. The injection pipette is passed through the zona pellucida and the membrane of the oocyte into the cytoplasm in a position sufficiently distant from the first polar body. After ICSI using ejaculated sperm, more than two-thirds of the injected oocytes became normally fertilized [34]. This procedure is prominent in assisted human reproduction. ICSI is feasible in cattle, even with freeze-dried sperm; however, it is not yet widely applied [20, 35].

The fertilized embryos are cultured in microdrops under paraffin oil. First cleavage occurs 23–29 h after fertilization, and the cleaved embryo is called a 2-cell embryo [36]. The number of blastomeres in an embryo increases to 4 and 8 at the second and third cleavage, which occur at days 2 and 3, respectively. Compaction of the cleaving embryo begins on day 3 due to formation of tight inter-cellular junctions. The embryos secrete factors that sustain their development, and for this reason they grow better in groups than alone [37]. During blastocyst formation, two clearly distinguishable cell lines are formed, the inner cell mass (ICM) and the trophectoderm (TE) [36]. A full human blastocyst at day 5 of development should consist of more than 60 cells and should at least double its cell number on day 6 [36]. Embryos of excellent and good quality at the compact morula to blastocyst developmental stages yield the highest pregnancy rates [38]. Recently, non-surgical embryo transfer techniques involving the use of specialized embryo transfer pipettes have been developed [38]. Because the most important factor influencing the rate of multiple births is the number of embryos transferred, in human assisted RT (ART), single embryo transfer is considered to be appropriate [39, 40]. IVF-ET is the most important process in animal and human RT. The scheme of human ART is summarized in Figure 1.

Sperm washing and selection are important processes for improving the fertilization rates in IVF and ICSI, and these procedures are the same as those used in AI. The quality of the embryo formed in *in vitro* embryo culture is related to the fertilization rate. Therefore, these processes need to be studied further in order to improve the success rate of RT. *In vitro* culture systems can and must be improved to obtain embryos of normal quality, equal to those obtained by development in the oviduct. A better understanding of the many factors regulating embryonic development and interaction with the female reproductive tract might provide some insights [41]. In conventional IVF-ET procedures, sperm are separated by non-physiological centrifugation and embryos are cultured under static conditions. The far from physiological conditions of these methods could be one of the reasons for the reduced pregnancy rate following the use of RT. Researchers in this field have to reconsider these current methods in order to further improve the success rate of RTs.

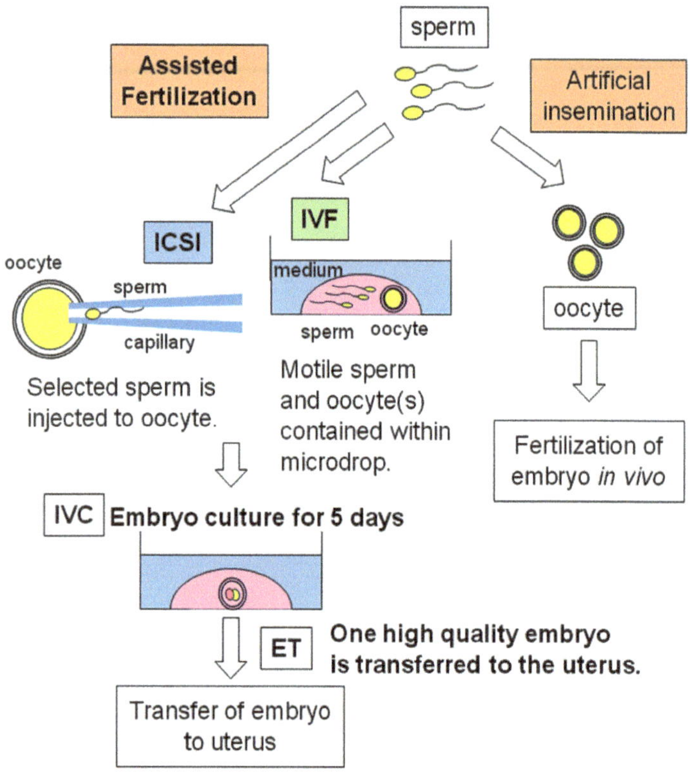

Figure 1. Schematic representation of human ART

The mammalian oviduct and uterus under physiological conditions have been studied to aid the design of microfluidic systems for RT. The oviduct consists of three segments, each with different functions: the uterotubal junction, the isthmus, and the ampulla [42]. The uterotubal junction provides a barrier to infectious microbes that might enter the oviduct from the uterus. The isthmus serves as a sperm storage organ and the ampulla provides an environment conducive to fertilization and early embryonic development. Figures 2A and B show the dynamic structural changes in the diameter of the oviduct induced by the peristaltic movement of the rat oviduct stained with fluorescent dyes. The stained microstructure in the isthmus has moved. The embryos' motion is caused by this peristaltic movement, and non-motile sperm are washed out from the oviduct.

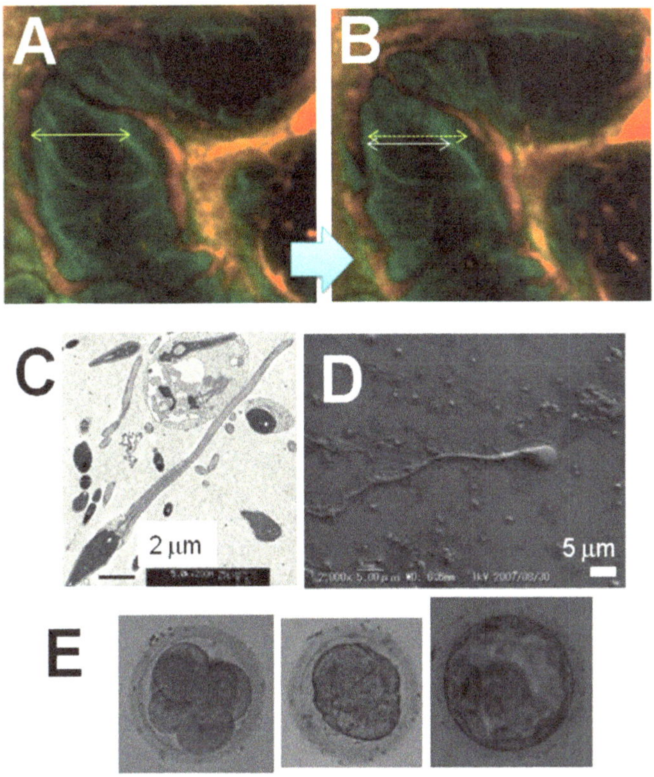

Figure 2. (A) and (B) The rat oviductal structure. Fluorescence image (B) was recorded 10 s after recording (A). The yellow arrow in (A) and the yellow dotted arrow in (B) are of the same length. The distance of the inner wall of the oviduct decreased in (B) due to peristaltic movements. (C) TEM and (D) SEM images of mammalian sperm. (E) Human embryos: (left) Cleaving embryo, (center) Morula, (right) Blastocyst.

Figures 2C and D show electron microscopic images of human sperms, indicating that the length is approximately 50 μm. The sperm head contains the DNA, the midpiece contains mitochondria, and the tail assists with swimming towards the oocyte [43]. Ejaculated sperm in the oviduct undergo capacitation, including motility hyperactivation, as the time of ovulation approaches. Capacitating sperm shed proteins that bind them to the mucosal epithelium, while hyperactivation assists the sperm in pulling away from the epithelium and escaping out of mucosal pockets. The process of sperm release is gradual, reducing the chances of polyspermic fertilization that reduces embryo quality. Released sperm may be guided towards the oocyte by secretions of the oviduct, cumulus cells, or oocyte. Hyperactivation likely assists sperm in penetrating the cumulus matrix and is absolutely required for penetrating the zona pellucida and achieving fertilization. Mammalian oocytes take up only a small area of the lumen [42]. After fertilization in the ampulla, the cleaving embryo (shown in Figure 2E) develops for 5–7 days *in vivo*. In rats, mice, and humans, once the blastocyst reaches the uterine cavity, it orientates its embryonic pole towards the uterine epithelium and bursts out of the zona pellucida, a process known as hatching [44]. The epithelium is invaded from the lateral side of the blastocyst and the fragmented epithelial cells are phagocytosed by TE [44-46]. The blastocyst develops from a spherical to a tubular form and then elongates to a filamentous conceptus. In domestic ruminants (sheep, cattle, and goats), the elongation of the blastocyst marks the beginning of implantation, which involves apposition, followed by transient attachment and finally firm adhesion [47]. On the basis of these physiological events in the female reproductive tract, we aim to improve RT by mimicking the dynamic environment *in vitro* through development of elastomer-based microfluidic devices.

1.2. Microfluidic channels to handle sperm and embryos

Microfluidic devices which come into contact with reproductive cells can be fabricated using mechanical drilling or molding. Mechanical drilling is feasible for rapid prototyping because of the small volume to be removed; however, tool marks in the fabricated microfluidic channel caused by the drilling are undesirable for use in cell observation. For prototyping of the devices, molding methods using elastomers to create plastic devices are based on replication and are faster than those used on glass and harder plastics [48]. A combination of molding and use of elastomers is applicable to RT to mimic the physiological environment. The hydrophobic silicone elastomer PDMS, having a contact angle of 110°, is a key material capable of extending device applications for RT because it is nontoxic, transparent, inexpensive, and easy to handle [9]. The Young's modulus of PDMS is 100 kPa [49]. The softness of the material enables easy fabrication, soft mechanism, and appropriate combinations with actuation systems. We classified the characteristics of the material and found it to be hydrophobic, transparent, elastic, and easy to pattern. PDMS microfluidic devices prepared by molding the microstructure and bonding the cured structure with a cover glass or glass slide can be used for manipulation and culture of cells to investigate their physiological functions. Within the last decade, studies using PDMS microfluidic channels or funnels have suggested novel solutions for oocyte manipulation, sperm sorting, and embryo culture [50-74]. Microfluidic systems that mimic oviductal structures and functions for use in RT are divided into those used for sperm motility control or monitoring [50-60], regulation of chemical gradients for *in vitro* fertilization and embryo culture [61-69], and applying mechanical stimuli to the developing embryo [70-74]. Although the unique characteristics of elastomers have been exploited for microstructure fabrication and/or micro-pumping, mechanical deformation of the elastomer membrane has not been used to produce new devices for RT. Applications developed by our group based on this elastomeric characteristic are introduced in this chapter.

2. Sorting and analyzing motile sperm

2.1. Elastomer chip device for sperm sorting

Ejaculated motile sperm is selected for successful fertilization in the oviduct. In ART, selection and sorting of motile sperm are routine processes. Some procedures may take up to 2 h for semen processing by conventional protocols, such as density gradient centrifugation and subsequent swim-up [75]. To reduce treatment times and physical damage induced by centrifugation, microfluidic sperm sorting (MFSS) chip devices have been developed for selecting motile sperm for use in ART [50-55]. As shown in Figure 3A, two gravity-driven laminar flows within the microfluidic channel are important for sperm selection. The fluids flowing through the semen inlet (A) and the medium inlet (B) should move parallel to each other and then exit through their respective outlets (A→D and B→C). Sperm are sorted on the basis of their ability to swim across the streamline into the medium stream, and hence only motile sperm are recovered in outlet C. Using an MFSS device, embryologists can perform a 1-step sorting protocol without centrifugation and complete processing within 30 min [76]. Reducing the treatment time and eliminating the centrifugation step minimizes the exposure of sperm to concentrated reactive oxygen species (ROS) and prevents DNA fragmentation [77]. Schulte *et al.* previously reported that DNA fragmentation was significantly decreased in MFSS-treated sperm [53]. Technologies that facilitate the identification and selection of sperm with high motility should increase the ART success rate, and the linear velocities (LV) of sperm separated using an MFSS device have been shown to increase compared with those of untreated sperm [54, 55]. On the basis of these results, an MFSS device can be used in clinical semen processing protocols for efficient ICSI and IVF. In the first report, the device was fabricated using PDMS bonded to a glass slide as shown in Figure 3B. Currently, disposable plastic devices for human ART are commercially available, as shown in Figure 3C. This device is a successful example of facilitated prototyping using elastomers.

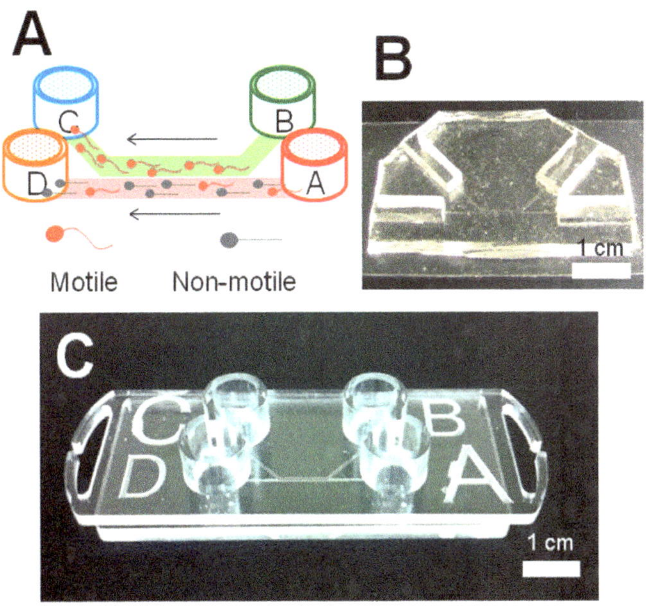

Figure 3. (A) Principle of sperm sorting using an MFSS device. Arrows show the direction of the laminar flow. (B) A glass-bottomed PDMS-MFSS. (C) A cycloolefin polymer-based MFSS device for human ART (Menicon Co., Ltd.).

2.2. Elastomer chip device for analyzing motility of motile sperm

Sperm tracking and motility analyses are usually performed before and after sorting in ART. Optical microscopes have been used to image sperm for CASA and for manual identification of sperm motility for ART. A minimal image platform is desired to enable compactness, ease-of-use, minimized footprint, and portability to monitor sperm motility by a preclinical assay at home. Some researchers have developed a microfluidic chip that can be used by an individual himself at home on convenient moments to check his semen quality [56-59]. For example, if a person finds that his semen quality is low or that there are no motile sperm in his semen, he can opt for ICSI for his fertility treatment. McCormack *et al.* developed a fluorescent labeling method for simultaneously analyzing motile sperm concentration and motility by monitoring fluorescent intensity of a region where motile sperm were collected [56]. After a certain time period, the intensity of fluorescence was measured to quantify motile sperm concentration. Both motile sperm concentration and motility were taken into account by this design. However, fluorescent labeling might not be suitable for routine use in hospitals or at home. The microdevice produces a flow field for sperm to swim against, and sperm that overcome the flow within a specified time are propelled along in a separate channel and counted by the resistive pulse technique [57]. The microdevice could be used for simple self-assessment of sperm quality at home. Recently, an on-chip device for concentration of sperms using electrical impedance measurements was developed by Segerink *et al.* [58]. Using this device, the concentration of sperm and other cells, such as HL-60 cells, in suspension can be determined. A lensless charge-coupled device (CCD) was integrated with a microfluidic chip to create a wide field of view combined with automatic recording as the sperm move inside a microfluidic channel [59]. These devices can evaluate sperm concentration and motility and the microfluidic channels can be fabricated using glass or poly(methyl methacrylate) (PMMA). However, for effective working of these devices, the physical characteristics of silicone elastomer such as hydrophobicity and elasticity are not required.

2.3. Elastomeric chambers for recording trajectories of motile sperm

Sperm motility analysis is not only used in human ART, but is also one of the key procedures used to analyze cattle and porcine sperm, especially frozen or transferred sperm used for AI. As discussed above, diluted semen is usually sandwiched between hydrophilic glass slides or glass- or plastic-bottomed microfluidic channels for observation of motile sperm. The trajectories of human and bovine sperm can be recorded using glass equipment; however, it is difficult to record the trajectory of motile porcine sperm using such a device because they adsorb to the surface of glass and hydrophilic plastics such as PMMA. Using a hydrophobic PDMS preparation (as shown in Figure 5), it would be possible to record the trajectories of motile sperm without problems associated with adsorption, making it possible to compare sperm motility parameters [60]. Because of the elasticity of PDMS, we propose using a preparation device such as that shown in Figure 5 for conventional CASA to reduce the overlap of motile sperm images (Figure 5D), which are artifacts of CASA. This device could thus be used to check for the quality of adhesive motile sperm and to investigate porcine sperm motility and AI success rate. The quality of porcine semen affects the nutrition or transfer of semen. Before AI, a technician needs to check the sperm quality. A device such as this could also be used to record adhesive human motile sperm adsorbed to glass in human ART.

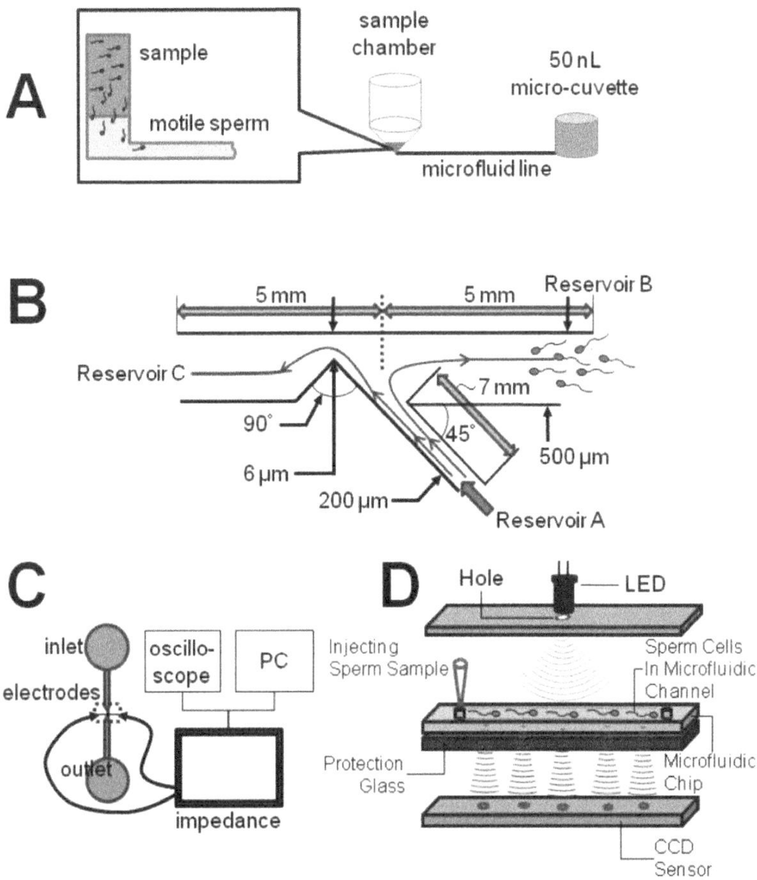

Figure 4. Sperm-analyzing microfluidic devices (A) A microfluidic device for monitoring fluorescent intensity of motile sperm through a microfluid line in the 50 nL micro-cuvette [56]. (B) A home-use device that allows rapid and quantitative sperm quality analysis using voltage measurements [57]. After a set time period, sperm arrive at the junction and the sperms swimming against the flow from A to B are flushed out through the aperture to reservoir C by the fluid flow from A to C. (C) On-chip device for sperm concentration using electrical impedance measurements [58]. (D) A lensless charge-coupled device (CCD) integrated with a microfluidic chip [59].

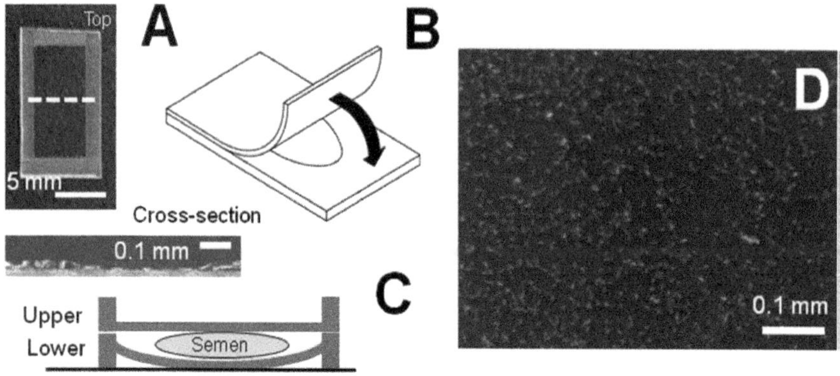

Figure 5. Use of a PDMS membrane for sperm motility analysis. (A) A PDMS membrane before use, with an area of 0.5 × 1 mm^2. (B) Method of sandwiching semen between the two membranes with a thickness of 0.1 mm. (C) Cross-sectional image of the preparation. A water droplet is sandwiched between two PDMS membranes. (D) Sperm in this preparation is displayed by CASA. No overlap of motile sperm is observed in this frame.

3. Static and dynamic mammalian embryo culture systems

3.1. Embryo culture systems using PDMS chambers or channels

Fertilized oocytes can be cultured in several microliters of optimized medium for culture for longer than 5 days. In various animal models, increased embryo density has been shown to improve development, possibly through secretion of autocrine/paracrine factors. These growth factors may influence embryo development [78-80]. Based largely on data from animal models, several novel culture approaches utilizing reduced volumes to culture embryos appear to offer potential benefits for subsequent embryo development [79, 80]. However, using a low volume of medium may render the embryos susceptible to detrimental changes in conditions such as osmolality or pH [81]. To control beneficial and detrimental chemical environments during *in vitro* embryo culture, various systems including PDMS-based microfluidic devices have been developed. These are known as microwells or microfluidic channels.

The well-of-the-well system (WOW) has been used successfully with embryos from a variety of species including mouse, pig, cow, and human and entails using small impressions, or microwells, of varying sizes and arrangements created in the bottom of a vessel/dish [8, 61] (Figure 6A). An alternate approach that permits a commercial means of utilizing microwells in conjunction with existing dishware in the lab involves using microwell-inserts consisting of several rows of tiny culture wells composed of PDMS [38]. Within PDMS and glass-bottomed microchannels under static media conditions, 2-cell mouse embryos can be cultured to the blastocyst stage [65, 66]. However, researchers found that culture in microchannels resulted in significantly greater blastocyst formation and hatched blastocyst development at 72 h and 96 h, respectively. These static culture systems could therefore be useful in investigating the impact of autocrine/paracrine compounds versus embryo spacing.

A weak point of the use of PDMS for embryo development was evaporation of the medium through the PDMS due to the porous structure of the cured devices. This evaporation through the PDMS is significant and could result in marked shifts in medium osmolality and a resulting reduction in embryo quality. This drawback of PDMS was circumvented by design and use of a sandwich membrane that remained flexible, yet protected against evaporation [70]. A nonporous PDMS microwell culture system supported the developmental competence of bovine and human embryos cultured individually [64]. An important process in the fabrication of nonporous PDMS microwells is curing under low pressure (−0.08 mPa). As a result, reports suggest that evaporation and osmolality changes can be reduced by appropriate fabrication techniques [64, 70].

3.2. Elastomer channels and devices for dynamic embryo culture systems

In vivo the preimplantation embryo is continually moving due to muscle contractions and movement of epithelial cell cilia [80]. This movement also disrupts cell-surface gradients that can form around embryos in static culture. It has been experimentally demonstrated that gradients exist in culture medium under static culture conditions because of embryo secretions or depletion of medium components. Gradients of potassium, calcium, and oxygen have been measured around mouse embryos [82, 83] and dynamic culture systems (DCSs) may disrupt these gradients, providing a more homogenous environment that more closely recapitulates the *in vivo* growth environment. These cilia and oviduct movements (shown in Figures 2A and B) also result in mechanical effects such as shear stress (SS), compression, and friction in the fallopian tube, which may influence developing embryos before nidation. DCSs may apply MS to the cultured embryo and thus replicate the environment it experiences under physiological conditions.

When using DCSs, there is a suggested threshold of SS that causes detrimental effects, such as apoptosis, on the cultured embryo. Xie *et al.* reported that SS in excess of 1.2 dyn/cm^2 can cause damage to blastomeres, up-regulation of stress signaling pathway constituents, and embryo degeneration [84]. Therefore, the SS in the vicinity of a cultured embryo induced by motions of DCSs should be kept below the threshold. In a study using a microfunnel as a DCS (Figure 6B), embryos were placed into the funnel reservoir, while medium was added and removed via a microfluidic channel connected to the bottom of the funnel via actions of a Braille actuator [71]. Mouse embryos were cultured in microdrop-static control, microfunnel-static control, or microfunnel-dynamic conditions with microfluidics. Progression to the blastocyst developmental stage was significantly enhanced under dynamic microfunnel culture conditions, as evidenced by an increased percentage of hatching or hatched blastocysts and a significantly higher average number of cells per blastocyst. Kim *et al.* developed a DCS using a combination of a PDMS microfluidic channel and a tilting machine, as shown in Figure 6C [72]. Bovine embryos were loaded and incubated by simply placing them on a tilting machine to provide embryo movement *via* gravity. The proportion of embryos that progressed to the eight-cell stage was superior in the constricted channel to that in the straight channel. Positive effects using tilting culture systems were also found in a microdroplet environment. We developed a tilting embryo culture system (TECS) to move mouse, porcine, and human embryos by fluid motion of the microdroplet, and we have shown that blastocyst development rates of these embryos using TECS improved over those of static controls [85-87]. Gentle vibration of porcine and human embryos also appeared to yield improved blastocyst formation rates [88-90]. These DCSs can be combined with microwells and microfluidic channels for mammalian embryo cultures.

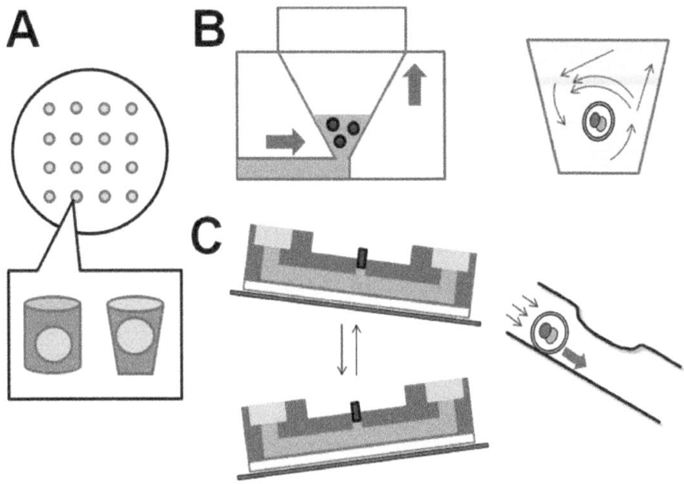

Figure 6. PDMS-based microfluidic devices for embryo culture. (A) The WOW system, (B) A microfunnel (C) A microfluidic IVC channel. In (B) and (C), the left and right figures show reservoir or microfluidic channels of the system and embryo and fluid motions, respectively.

3.3. Peristaltic movement of an elastomer membrane to create a dynamic embryo culture system

MS may be generated by fluid dynamics and compression of embryos due to interactions with the wall of the oviduct. We developed an air actuating system with microfluidic channels to apply MS by deforming a 0.1-mm-thick PDMS membrane and evaluated the MS applied to mouse embryos inside the microfluidic channel. Using an air actuating system as shown in Figures 7A and B, we applied compression to mouse embryos inside the medium channel and estimated SS based on the velocity of the embryos' motion [74, 75]. It has been demonstrated that this culture system can be employed to investigate the relationship between MS and molecular mechanisms. Because this culture system could be directly positioned on the stage of a fluorescence microscope, fluorescence images of embryos in the microfluidic channel could be recorded, indicating the distribution of intracellular calcium concentration ($[Ca^{2+}]_i$). Both MS and $[Ca^{2+}]_i$ were quantified based on time-resolved confocal microscopy images. When blastocysts were compressed, FI increased in response to the applied MS, as shown in Figure 7C. Molecular mechanosensing systems such as mechanosensitive ion channels could play an important role in responses to these MS. Using this device to investigate the applied MS and resultant molecular response, we can investigate the functions of the embryonic sensor proteins. The ultimate goal is to use this mechanism as an artificial oviduct for clinical use based on the soft mechanism of elastomers.

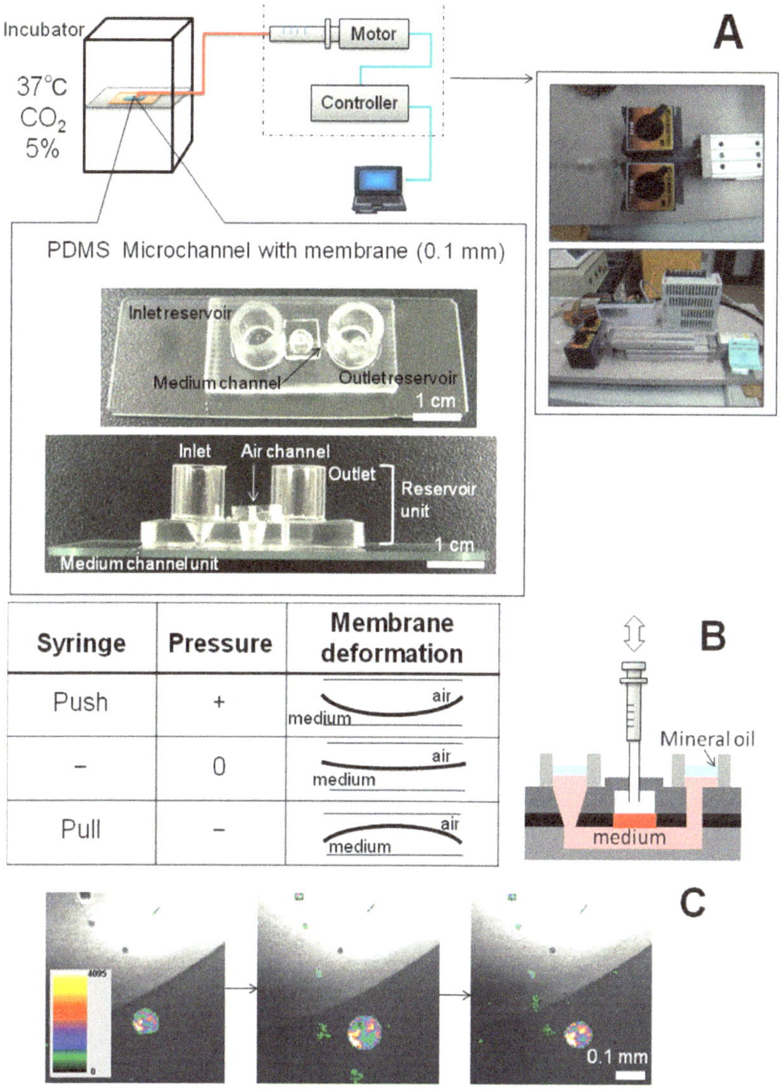

Figure 7. A pneumatic microfluidic actuation system for dynamic embryo culture. (A) Mechanical drive system and glass-bottomed PDMS microfluidic channel. (B) Schematic view of PDMS membrane deformation. The thick line represents the PDMS membrane. (C) $[Ca^{2+}]_i$ changes in a compressed mouse blastocyst in the microfluidic channel induced by membrane deformation. The center image shows embryo compression.

4. Conclusion

We developed microfluidic devices and systems for use in motile sperm sorting, sperm motility checking, and static and dynamic embryo culture systems based on the physical and chemical properties of silicone elastomers. Utilizing these properties, methods for RT can be improved, and novel strategies can be developed to implement more physiological treatments for human ART. However, most commercial elastomers are not recommended for medical use. To apply microfluidic channels in dynamic human embryo culture for ART, it is necessary to select a soft material approved for clinical use to prepare the microfluidic channels. Silicone hydrogel, which is used in the manufacture of contact lenses, is a candidate material for this application; therefore, we developed silicone hydrogel microfluidic channels for embryo culture. Once commercial elastomers have been approved for clinical use, elastomer-based medical microdevices are likely to become widespread and routinely used. The material development and approval of the developed materials for medical use are bottlenecks for the medical application of elastomeric devices and systems. This problem also applies to implantation and regenerative medicine. Over the last decade, great progress has been made in RT and human ART, and these improvements can contribute to other fields of medicine utilizing microfluidic systems.

Author details

Koji Matsuura*
Research Core for Interdisciplinary Sciences, Okayama University, Okayama, Japan

Keiji Naruse
Cardiovascular Physiology, Graduate School of Medicine, Dentistry and Pharmaceutical Sciences, Okayama University, Okayama, Japan

Acknowledgement

This study was partly supported by a grant-in-aid for Scientific Research for Young Scientists (B) and (A), and Challenging Exploratory Research (Nos. 20700380, 22680036, 23650262 to K. M.) and Special Coordination Funds for Promoting Sciences and Technology from the Ministry of Education Science Sports and Culture, Japan (K. M.). K. M. thanks to Mieko Kodama (Okayama University) for her assistance with figure preparation and formatting. The authors would like to thank Enago for the English language review.

5. References

[1] Okere C, Nelson L (2002) Novel Reproductive Techniques in Swine Production–A Review. Asian-Aust. J. Anim. Sci. 15: 445-452.

[2] William VH, Amanda RP (1999) Role of reproductive technologies and genetic resource banks in animal conservation. Rev. Reprod. 4: 143-150.

* Corresponding Author

[3] Gardner DK (2006) In vitro fertilization–A practical approach. Informa health care
[4] Maes D, Nauwynck H, Rijsselaere T, Mateusen B, Vyt P, Kruif AD, Soom AV (2008) AI transmitted diseases in swine: an overview. Theriogenology. 70: 1337-1345.
[5] Vishwanath R (2003) Artificial insemination: the state of the art. Theriogeneology. 59: 571-584.
[6] Thibier M, Wagner HG (2000) World statistics for artificial insemination in cattle. Proc. 14th Int. Congr. Anim. Reprod. 22: 76.
[7] Moore K, Thatcher WW (2006) Major Advances Associated with Reproduction in Dairy Cattle. J. Dairy Sci. 89: 1254-1266.
[8] Vajta G, Peura TT, Holm P, Paldi A, Greve T, Trounson AO, Callesen H (2000) New method for culture of zona-included or zona-free embryos: the Well of the Well (WOW) system. Mol. Reprod. Dev. 55: 256-264.
[9] Douglas BW, George MW (2006) Applications of microfluidics in chemical biology. Cur. Opin. Chem. Biol. 10: 584-591.
[10] Velve-Casquillas G, Berre ML, Piel M, Tran PT (2010) Microfluidic tools for cell biological research. Nano Today. 5: 28-47.
[11] Holmes D, Pettigrew D, Reccius CH, Gwyer JD, Berkel CV, Holloway J, Davies DE, Morgan H (2009) Leukocyte analysis and differentiation using high speed microfluidic single cell impedance cytometry. Lab Chip. 9: 2881-2889.
[12] Wang J, Heo J, Hua SZ (2010) Spatially resolved shear distribution in microfluidic chip for studying force transduction mechanisms in cells. Lab Chip. 10: 235-239.
[13] Tan JL, Tien J, Pirone DM, Gray DS, Bhadriraju K, Chen CS (2003) Cells lying on a bed of microneedles: An approach to isolate mechanical force. Proc. Natl. Acad. Sci. U.S.A. 100: 1484-1489.
[14] Roure OD, Saez A, Buguin A, Austin RH, Chavrier P, Silberzan P, Ladoux B (2005) Force mapping in epithelial cell migration. Proc. Natl. Acad. Sci. U.S.A. 102: 2390-2395.
[15] Vanapalli SA, Duits MHG, Mugele F (2009) Microfluidics as a functional tool for cell mechanics. Biomicrofluidics. 3: 012006.
[16] Chiu DT, Jeon NL, Huang S, Kane RS, Wargo CJ, Choi IS, Ingber DE, Whitesides GM (2000) Patterned deposition of cells and proteins onto surfaces by using three-dimensional microfluidic systems. Proc. Natl. Acad. Sci. U.S.A. 97: 2408-2413.
[17] Hung PJ, Lee PJ, Sabounchi P, Lin R, Lee LP (2005) Continuous perfusion microfluidic cell culture array for high-throughput cell-based assays. Biotechnol. Bioeng. 89: 1-8.
[18] Koyama S, Amarie D, Soini HA, Novotny MV, Jacobson SC (2006) Chemotaxis assays of mouse sperm on microfluidic devices. Anal. Chem. 78: 3354-3359.
[19] Shamloo A, Ma N, Poo M, Sohn LL, Heilshorn SC (2008) Endothelial cell polarization and chemotaxis in a microfluidic device. Lab Chip. 8: 1292-1299.
[20] Mapletoft RJ, Hasler JF (2005) Assisted reproductive technologies in cattle: a review. Rev. sci. tech. Off. int. Epiz. 24: 393-403.
[21] Chan WS, Dixon ME (2008) The "RT" of thromboembolism : A review of assisted reproductive technology and thromboembolic complications. Thromb. Res. 121: 713-726.

[22] Kim LU, Johnson MR, Barton S, Nelson MR, Sontag G, Smith JR, Gotch FM, Gilmour JW (1999) Evaluation of sperm washing as a potential method of reducing HIV transmission in HIV-discordant couples wishing to have children. AIDS. 13: 645-651.

[23] Aitken RJ, Clarkson JS (1988) Significance of reactive oxygen species and antioxidants in defining the efficacy of sperm preparation techniques. J. Androl. 9: 367-376.

[24] Alvarez1 JG, Lasso JL, Blasco L, Nuñez RC, Heyner S, Caballero PP, Storey BT (1993) Centrifugation of human spermatozoa induces sublethal damage; separation of human spermatozoa from seminal plasma by a dextran swim-up procedure without centrifugation extends their motile lifetime. Hum. Reprod. 8: 1087-1092.

[25] Sharma RK, Vemulapalli S, Kohn S, Agarwal A (1997) Effect of Centrifuge Speed, Refrigeration Medium, and Sperm Washing Medium on Cryopreserved Sperm Quality After Thawing. Sys. Biol. Reprod. Med. 39: 33-38.

[26] Hirano Y, Shibahara H, Obara H, Suzuki T, Takamizawa S, Yamaguchi C, Tsunoda H, Sato I (2001) Relationships Between Sperm Motility Characteristics Assessed by the Computer-Aided Sperm Analysis (CASA) and Fertilization Rates In Vitro. J. Assist. Reprod. Genet. 18: 215-220.

[27] Agarwal A, Sharma RK (2007) Automation is the key to standardized semen analysis using the automated SQA-V sperm quality analyzer. Fertil. Steril. 87: 156-162.

[28] Zhang BR, Larsson B, Lundeheim N, Rodriguez-Martinez H (1998) Sperm characteristics and zona pellucida binding in relation to field fertility of frozen–thawed semen from dairy AI bulls. Int. J. Androl. 21: 207-216.

[29] England GCW (1992) The Cryopreservation of Canine Semen. Thesis, University of London, UK.

[30] Rotaa A, Penã AI, Linde-Forsberg C, Rodriguez-Martinez H (1999) In vitro capacitation of fresh, chilled and frozen–thawed dog spermatozoa assessed by the chlortetracycline assay and changes in motility. Anim. Reprod. Sci. 57: 199-215.

[31] Holzer HE, Chain RC, Demirtas E, Ba-Akdah H, Tan SL (2006) In vitro maturation of oocytes. In: Gardner DK editor, In vitro fertilization–A practical approach. Informa health care. pp. 127-145.

[32] Galli C, Lazzari G (1996) Practical aspects of IVM/IVF in cattle. Anim. Reprod. Sci. 42: 371-379.

[33] Hoover L, Baker A, Check JH, Lurie D, Summers D (1997) Clinical outcome of cryopreserved human pronuclear stage embryos resulting from intracytoplasmic sperm injection. Fertil. Steril. 67: 621-624.

[34] Devroey P, Steirteghem AV (2004) A review of ten years experience of ICSI. Hum. Reprod. Update. 10: 19-28.

[35] Keskintepe L, Pacholczyk G, Machnicka A, Norris K, Curuk MA, Khan I, Brackett BG (2002) Bovine blastocyst development from oocytes injected with freezedried spermatozoa. Biol. Reprod. 67: 409-415.

[36] Ebner T (2006) Embryo development and assessment of viability. In: Gardner DK, editor. In vitro fertilization–A practical approach. Informa health care, pp 199-220.

[37] Gardner DK, Lane M, Spitzer A, Batt PA (1994) Enhanced rates of cleavage and development for sheep zygotes cultured to the blastocyst stage in vitro in the absent of

serum and somatic cells-amino acids, vitamins, and cuhuring embryos in groups stimulate development. Biol. Reprod. 50: 390-400.
[38] Mapletoft RJ (2006) Bovine Embryo Transfer, IVIS Reviews in Veterinary Medicine, I.V.I.S. (Ed.). International Veterinary Information Service, Ithaca NY
[39] Strandell A, Bergh C, Lundin K (2000) Selection of patients suitable for one-embryo transfer reduces the rate of multiple births by half without impairment of overall birth rates. Hum. Reprod. 15: 2520-2525.
[40] Bergh C (2005) Single embryo transfer: a mini-review. Hum. Reprod. 20: 323–327.
[41] Gandolfi F (1994) Autocrine, paractine and environmental factors influencing embryonic development from zygote to blastocyst. Theriogenology. 41: 95-100.
[42] Suarez SS (2008) Regulation of sperm storage and movement in the mammalian oviduct, Int. J. Dev. Biol. 52: 455-462.
[43] Suarez SS (2008) Control of hyperactivation in sperm, Hum. Reprod. Update. 14: 647-657.
[44] Yoshinaga K (2010) Research on Blastocyst Implantation Essential Factors (BIEFs). Am. J. Reprod. Immun. 63: 413-424.
[45] Enders AC, Schlafke S (1967) A morphological analysis of the early implantation stages in the rat. Am. J. Anat. 120: 185-226.
[46] Bischof1 P, Campana A (1996) A model for implantation of the human blastocyst and early placentation. Hum. Reprod. Update. 2: 262-270.
[47] Spencer TE, Johnson GA, Bazer FW, Burghardt RC (2004) Implantation mechanisms: insights from the sheep. Reproduction. 128: 657-668.
[48] Duffy DC, McDonald JC, Schueller OJA, Whitesides GM (1998) Rapid Prototyping of Microfluidic Systems in Poly(dimethylsiloxane). Anal. Chem. 70: 4974-4984.
[49] Gray DS, Tien J, Chen CS (2003) Repositioning of cells by mechanotaxis on surfaces with micropatterned Young's modulus. J. Biomed. Mater. Res. 66A: 605-614.
[50] Cho BS, Schuster TG, Zhu X, Chang D, Smith GD, Takayama S (2003) Passively Driven Integrated Microfluidic System for Separation of Motile Sperm. Anal. Chem. 75: 1671-1675.
[51] Schuster TG, Cho B, Keller LM, Takayama S, Smith GD (2003) Isolation of motile spermatozoa from semen samples using microfluidics. Reprod. Biomed. Online. 7: 75-81.
[52] Wu JM, Chung Y, Belford KJ, Smith GD, Takayama S, Lahann J (2006) A surface-modified sperm sorting device with long-term stability. Biomed. Microdev. 8: 99-107.
[53] Schulte RT, Chung YK, Ohl DA, Takayama S, Smith GD (2007) Microfluidic sperm sorting device provides a novel method for selecting motile sperm with higher DNA integrity. Fertil. Steril. 88: S76.
[54] Shibata D, Ando H, Iwase A, Harata T, Kikkawa F, Naruse K (2007) Analysis of sperm motility and fertilization rates after the separation by microfluidic sperm sorter made of quartz. Fertil. Steril. 88: S110.
[55] Matsuura K, Takenami M, Kuroda Y, Hyakutake T, Yanase S, Naruse K (2012) Screening of Sperm Velocity by Fluid Mechanical Characteristics of a Cycloolefin Polymer Microfluidic Sperm Sorting Device. Reprod. Biomed. Online. 24: 109-115.

[56] McCormack MC, McCallum S, Behr B (2006) A novel microfluidic device for male subfertility screening. J. Urol. 175: 2223-2227.
[57] Chen YA, Huang ZW, Tsai FS, Chen CY, Lin CM, Wo AM (2011) Analysis of sperm concentration and motility in a microfluidic device. Microfluid. Nanofluid. 10: 59-67.
[58] Segerink LI, Sprenkels AJ, Braak PMT, Vermes I, Berg AVD (2011) On-chip determination of spermatozoa concentration using electrical impedance measurements. Lab Chip. 10: 1018-1024.
[59] Zhang X, Khimji I, Gurkan UA, Safaee H, Catalano PN, Keles HO, Kayaalp E, Demirci U (2011) Lensless imaging for simultaneous microfluidic sperm monitoring and sorting. Lab Chip. 11: 2535-2540.
[60] Matsuura K, Kuroda Y, Yamashita K, Funahashi H (2011) Hydrophobic Silicone Elastmer Chamber for Recording Trajectories of Porcine Motile Sperms without Adsorption. J. Reprod. Dev. 57: 163-167.
[61] Vajta G, Korosi T, Du Y, Nakata K, Ieda S, Kuwayama M, Nagy ZP (2008) The Well-of-the-Well system: an efficient approach to improve embryo development. Reprod. Biomed. Online. 17: 73-81.
[62] Sugimura S, Akai T, Somfai T, Hirayama M, Aikawa Y, Ohtake M, Hattori H, Kobayashi S, Hashiyada Y, Konishi K, Imai K (2010) Time-lapse cinematography-compatible polystyrene-based microwell culture system: a novel tool for tracking the development of individual bovine embryos. Biol. Reprod. 83: 970-978.
[63] Cheng CM, Matsuura K, Wang IJ, Kuroda Y, Leduc PR, Naruse K (2009) Fabricating small-scale, curved, polymeric structures with convex and concave menisci through interfacial free energy equilibrium. Lab Chip. 9: 3306-3309.
[64] Hashimoto S, Kato N, Saeki K, Morimoto Y, (2012) Selection of high-potential embryos by culture in poly(dimethylsiloxane) microwells and time-lapse imaging. Fertil Steril. 97: 332-337.
[65] Raty S, Davis J, Beebe D, Rodriguez-Zas S, Wheeler M (2001) Culture in microchannels enhances in vitro embryonic development of preimplantation mouse embryos. Theriogenology. 55: 241.
[66] Raty S, Walters EM, Davis J, Zeringue H, Beebe DJ, Rodriguez-Zas SL, Wheeler MB (2004) Embryonic development in the mouse is enhanced via microchannel culture. Lab Chip. 4: 186-190.
[67] Zeringue HC, Rutledge JJ, Beebe DJ (2005) Early mammalian embryo development depends on cumulus removal technique. Lab Chip. 5: 86-90.
[68] Beebe D, Wheeler M, Zeringue H, Walters E, Raty S (2002) Microfluidic technology for assisted reproduction. Theriogenology. 57: 125-135.
[69] Krisher RL, Wheeler MB (2010) Towards the use of microfluidics for individual embryo culture. Reprod. Fertil. Dev. 22: 32-39.
[70] Heo YS, Cabrera LM, Song JW, Futai N, Tung YC, Smith GD, Takayama S (2007) Characterization and resolution of evaporation-mediated osmolality shifts that constrain microfluidic cell culture in poly(dimethylsiloxane) devices. Anal. Chem. 79: 1126-1134.

[71] Heo YS, Cabrera LM, Bormann CL, Shah CT, Takayama S, Smith GD (2010) Dynamic microfunnel culture enhances mouse embryo development and pregnancy rates. Hum. Reprod. 25: 613-622.
[72] Kim MS, Bae CY, Wee G, Han YM, Park JK (2009) A microfluidic in vitro cultivation system for mechanical stimulation of bovine embryos. Electrophoresis. 30: 3276-3282.
[73] Li JC, Matsuura K, Kuroda Y, Funahashi H, Naruse K (2010) Application of mechanical stimuli using a microfluidic air actuating system to cultured mammalian embryos. MHS 2010 abstract: 29-34.
[74] Matsuura K, Watanabe K, Kodama M, Kuroda Y, Naruse K (2011) Development of observation system to investigate both intracellular calcium concentration and mechanical stimuli to mammalian embryos. MHS 2011 abstract: 99-104.
[75] Gautam N, Allahbadia R, Basuray D, Rubina M (2003) The Art and Science of Assisted Reproductive Techniques. Taylor & Francis: 213-221. Chapter 30.
[76] Hughes CM, Lewis SE, McKelvey-Martin VJ, Thompson W (1998) The effects of antioxidant supplementation during Percoll preparation on human sperm DNA integrity. Hum. Reprod. 13: 1240-1247.
[77] Mortimer D (1991) Sperm preparation techniques and iatrogenic failures of in-vitro fertilization. Hum. Reprod. 6: 173-176.
[78] Richter KS (2008) The importance of growth factors for preimplantation embryo development and in-vitro culture. Curr. Opin. Obstet. Gynecol. 20: 292-304.
[79] Swain JE, Smith GD (2011) Advances in embryo culture platforms: novel approaches to improve preimplantation embryo development through modifications of the microenvironment. Hum. Reprod. Update. 17: 541-557.
[80] Smith GD, Takayama S, Swain JE (2012) Rethinking In Vitro Embryo Culture: New Developments in Culture Platforms and Potential to Improve Assisted Reproductive Technologies. Biol. Reprod. 86: 62.
[81] Swain JE (2010) Environmental factors and manual manipulations during preparation influence embryo culture media osmolality. Fertil. Steril. 94(Suppl.): S32.
[82] Trimarchi JR, Liu L, Smith PJ, Keefe DL (2000) Noninvasive measurement of potassium efflux as an early indicator of cell death in mouse embryos. Biol. Reprod. 63: 851-857.
[83] Trimarchi JR, Liu L, Porterfield DM, Smith PJ, Keefe DL (2000) Oxidative phosphorylationdependent and -independent oxygen consumption by individual preimplantation mouse embryos. Biol. Reprod. 62: 1866-1874.
[84] Xie Y, Wang F, Zhong W, Puscheck E, Shen H, Rappolee DA (2006) Shear stress induces preimplantation embryo death that is delayed by the zona pellucida and associated with stressactivated protein kinase-mediated apoptosis. Biol. Reprod. 75: 45-55.
[85] Matsuura K, Hayashi N, Kuroda Y, Takiue C, Hirata R, Takenami M, Aoi Y, Yoshioka N, Habara T, Mukaida T, Naruse K (2010) Improved Development of Mouse and Human Embryos by Tilting Embryo Culture System. Reprod. BioMed. Online. 20: 358-364.
[86] Koike T, Matsuura K, Naruse K, Funahashi H (2010) In-vitro Culture with a tilting device in chemically defined media during meiotic maturation and early development

improves the quality of blastocysts derived from in-vitro matured and fertilized porcine oocytes. J. Reprod. Dev. 56: 552-557.

[87] Hara T, Naruse K, Matsuura K, Kodama T, Sato K, Tateaki Y (2010) A novel tilting embryo culture system (TECS) improves human blastocyst quality clinically. ESHRE2010.

[88] Mizobe Y, Yoshida M, Miyoshi K (2010) Enhancement of cytoplasmic maturation of in vitro-matured pig oocytes by mechanical vibration. J. Reprod. Dev. 56: 285-290.

[89] Isachenko E, Maettner R, Isachenko V, Roth S, Kreienberg R, Sterzik K (2010) Mechanical agitation during the in vitro culture of human pre-implantation embryos drastically increases the pregnancy rate. Clin. Lab. 56: 569-76.

[90] Isachenko V, Maettner R, Sterzik K, Strehler E, Kreinberg R, Hancke K, Roth S, Isachenko E (2011) In-vitro culture of human embryos with mechanical micro-vibration increases implantation rates. Reprod. BioMed. Online. 22: 536-544.

Section 6

Elastomers in Cement Concrete and Mortar

Chapter 11

Technology, Properties and Application of NRL Elastomers

Bala Muhammad

Additional information is available at the end of the chapter

1. Introduction

1.1. General appraisal

Polymer latexes are being increasingly used in the construction industry as modifiers, especially in hydraulic cement concrete and mortar. Figure 1 presents polymeric latexes used as cement modifiers. Among the different presentations of polymeric latexes, elastomers are the most widely used (Ohama, 1995). In practice, there are two basic elastomers applicable to cement mixes and these are natural rubber latex (NRL) and synthetic latexes. Figure 2 shows some prominent derivatives of these elastomers. Indeed, elastomers are mainly added into hydraulic cement concrete and mortar in order to improve their performance properties.

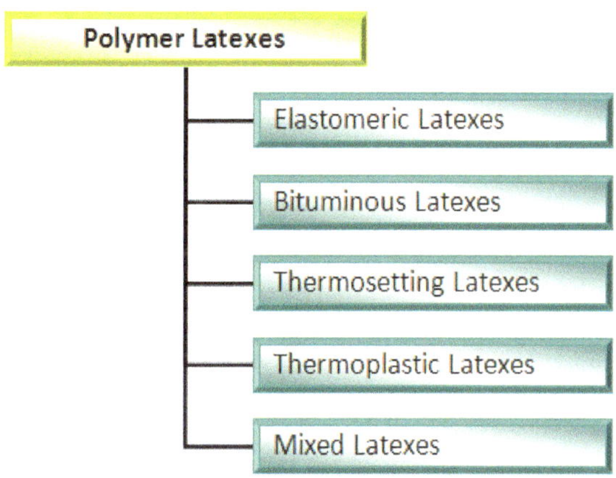

Figure 1. Polymer latexes for concrete and mortar modifications (Ohama 1995)

Even though, concrete is the most preferred construction materials on earth, but it has some limitations which inevitably affects its quality and general performance. These limitations include; delayed hardening, inherent brittleness, weak tensile capacity, low flexural strength, small failure strain, large drying shrinkage, susceptibility to frost damage, high moisture absorption and most critically low resistance to chemicals. To improve on these deficiencies, elastomers are added as modifiers (Ohama, 1995; Joao & Marcos, 2004; Rajni, Asthana & Anupam 2006; Bala *et al.*, 2011a).

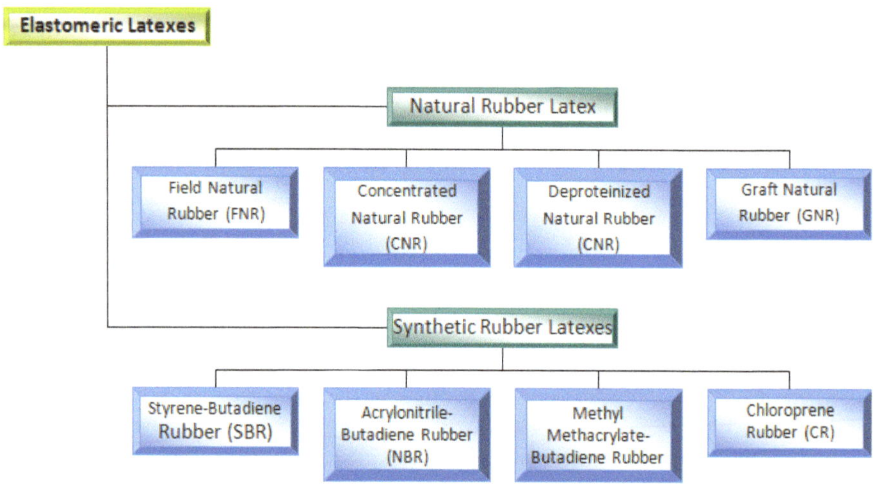

Figure 2. Elastomeric latexes for concrete and mortar modifications

1.2. Performance of elastomers in cement mixes

Depending on the type of elastomer used, modified concrete (MC) and mortar exhibit excellent performance in both mechanical and durability functions. Normally, addition of appropriate quantity of elastomer in cement-based mix improves; workability, through ball bearing actions; cement hydration, through water retention qualities; compressive, tensile and flexural strengths, through latex-film formation; and water tightness, through effective void and capillary fillings (Ramakrishnan, 1992).

Application of elastomers into cement-based infrastructures has been successful in the last few decades perhaps due to prominent qualities usually exhibited by these substances. The most desired of these qualities include high amorphousness and low particle size which makes it possible for the elastomers to effectively fill cavities and capillaries of hardened cement paste, and ability to coalesce upon withdrawal of the surrounding moisture in the cement-elastomer co-matrix systems (Bala, Yussuf & Mohammad, 2009). Admixing rather than impregnation is mostly applied in both laboratories and on construction sites. Comparatively, admixing entails simple operations and it ensures overall homogeneity between cement and elastomer. Compatibility of elastomer in the voids of hardened cement paste provides protection against ingress of moisture which normally serves as the main transporter to agents of chemical attack. In addition, the ability of an elastomer to coalesce in the co-matrix usually leads to latex-film formation which forms the basis for improvements in mechanical properties.

Although, inclusion of elastomers in hydraulic cement mixes is tremendously becoming popular worldwide, elastomers are basically rubbers, and rubbery substance softens when exposed to high temperatures. Thus, in the event of elevated temperatures, it is feared that the comparatively low softening point temperature of elastomers may cause failure in MC members or promote durability threats much earlier than the recommended design limits. Therefore, as much as elastomers are desired in the construction industry for their excellent performance, a clear margin of performance at elevated temperatures is necessary, so that both mechanical integrity and durability functions are ensured. Indeed, absence of clear margin of performance in cement-elastomer phase could render structures made with cement-elastomer co-matrix unsafe particularly at high temperatures. This creates a major point of concern to professionals in the technology sector since structures are often exposed to temperatures beyond the normal ambient limits (Bala et al., 2011b). This issue has been addressed in the experimentation which follows. In fact, a section which presents experimental findings regarding behavior of elastomeric concrete under elevated temperatures has been included in the analysis and discussion of results.

Considering the general performance of elastomers in the hydraulic cement concrete and mortar particularly their positive and negative roles in the cement-elastomer phase, the following advantages and disadvantages could be enumerated.

Advantages of elastomers as modifiers in cement concrete and mortar

1. Improvement of workability which normally provides additional ease to complete mixing, proper placement and adequate compaction.
2. Enhancement of cement hydration with consequent increase in strength due to water retention capacities of elastomers.
3. Greater mechanical properties especially tensile and flexural strengths. This is mainly the result of reduced brittleness in the modified phase.
4. Improvement of water tightness, resistance to chemical aggression and freeze-thaw. Ability of elastomers to fill capillaries and voids of the hardened phase normally contributes to these achievements.

Disadvantages of elastomers as modifiers in cement concrete and mortar

1. Some elastomers impair certain qualities of concrete. For example, polyvinyl acetate and chloroprene rubber were observed to increase the drying shrinkage of concrete.
2. Inclusion of high dosage of elastomer such as NRL into concrete mix could render hardened phase susceptible to strength weakening especially at elevated temperatures.
3. In some cases, allergic reactions from exposure to proteins found in NRL could be experienced. Therefore, the use of deproteinized NRL is most recommended.

1.3. Natural Rubber Latex

NRL is a whitish to off-white milky fluid usually obtained by tapping the bark of Para tree (*Hevea Brasiliensis*). Figure 3 portrays some of the principal operations involved in the production of NRL. In its fresh state, NRL comprises "30%" – "40%" rubber hydrocarbon particles (C_5H_8) suspended in a serum together with about "6%" non-rubber substances (Ong, 1998; Jitladda, 2006). The non-rubber substances include proteins, lipids, carbohydrates, sugars and traces of some metals such as zinc, magnesium, copper and iron. Natural rubber is a high molecular weight polymer of isoprene (cis-1,4-polyisoprene). It has a particle diameter of 0.1 – 4.0 μm and a chemical structure as shown in Figure 4 (Rattana, 2003; David & Richard, 2002).

Figure 3. Production of natural rubber latex

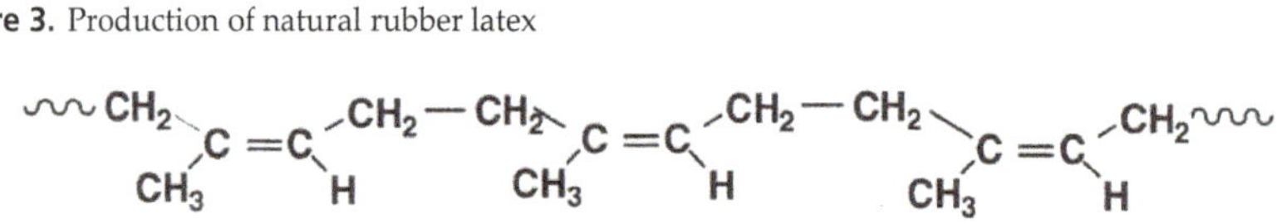

Figure 4. Chemical structure of cis -1,4-polyisoprene (David & Richard, 2002)

Most of the properties of NRL are determined during the process of natural polymerization rather than controlled as normally is the case with emulsion polymerization. This forms the basis for the presence of non-rubber substances. Bacterial activities and coagulation are known to exist in NRL also. An in-depth analysis on the non-rubber contents (NRC) present in NRL is given in this chapter under properties of NRL. In order to combat bacterial growth as well as coagulation, NRL is usually preserved with ammonia when harvested from the tree and again after concentration (Esah & Paul, 2002). Preservation is generally achieved through high or low ammonia-tetramethythiuram disulfide/zinc oxide (HA-TZ or LA-TZ). However, LA-TZ ensures good color, chemical stability and low toxicity (Bala, 2009). Medium of dispersion in the NRL is greatly reduced after concentration so that density of the rubber hydrocarbons is increased to about "60%" (David & Richard, 2002).

While both NRL and synthetic latexes are used for the purpose of academic research works, the later is mostly applied in practice. Meanwhile, due to the prevailing increase in global awareness of environmental issues, a high level of interest in NRL and its derivatives has been triggered. The reason for this is that NRL is a renewable resource, whereas its synthetic counterparts are mostly manufactured from non-renewable oil-based resources. Thus, by virtue of its renewable origin in addition to its higher sticky quality, superior building tack, extremely high resilience, and excellent mechanical characteristics (John, 1987; Schneider, 1997; Kondou, 2006), NRL is chosen for the experimentations that follow in this chapter.

2. Aims of the chapter

This chapter aims at expounding properties, application and performance of NRL as a modifier in cement concrete and mortar. Properties of NRL influencing performance in mechanical and durability functions are dealt with in the first part of the discussion section. Furthermore, influence of high temperature on these two important qualities of concrete; strength and durability was investigated in the subsequent sections. Recent trends in research activities and challenges facing applications of elastomers are provided in the concluding parts of the chapter.

Regarding properties of the NRL relevant to concrete and mortar applications, the clonal lattices involved in the experiments were analyzed for sixteen compositional properties each. In addition, scanning electron microscope (SEM) captions of microstructural units were observed. The objective of these rigorous chemical analyses was to evaluate nature and contents of each substance present in the latexes so that factors affecting performance of each of the latexes when added to the cement-mixes are identified. The SEM on the other hand provides additional information on the structure of the individual phases as well as that of the integrated phase. Eventually, the much desired properties for optimum performance of the latexes in cement-based mixes and developments in the physical structure of the co-matrix systems were noted and concisely reported.

In order to address one of the most important problems affecting elastomerically modified cement concrete and mortar, an in-depth study on the mechanical integrity and durability characteristics of cement-elastomer phase due to changes in temperature was conducted and observations were discussed. In particular, loss in compressive strength and crack formations under high temperatures were assessed. While details on the factors surrounding the excellent roles of elastomers in cement-elastomer co-matrix systems was based on performance assessments, intermolecular softening believed to be caused by the presence of elastomers particularly at elevated temperatures was evaluated through a comparative study between compressive strength test and thermogravimetric analysis (TGA).

Finally, critical limit of performance at elevated temperatures for cement-elastomer systems has been proposed. Parameters considered for the critical limit of performance include content of elastomer in the cement-elastomer co-matrix, temperature margin, degree of degradation entertained and loss in the compressive strength of the hardened phase.

3. Experimentation

3.1. Materials

OPC conforming to BS 12: 1989 was used throughout. Chemical composition and physical properties of the cement are shown in Table 1. Naturally occurring river-washed quartz sand passing through ASTM sieve No. 4 and crushed granite stones with nominal maximum sizes 10 mm are used as fine and coarse aggregate respectively. NRL derivatives which include six clonal latexes and concentrated latexes were involved. Stabilization was achieved through the use of LA-TZ. Latexes were supplied by Rubber Research Institute and Sime Darby Research Center, Malaysia.

Chemical Composition	("%")
Silicon dioxide (SiO2)	20.1
Aluminum oxide (Al2O3)	4.9
Ferrous oxide (Fe2O3)	2.4
Calcium oxide (CaO)	65.0
Sulphur oxide (SO3)	2.3
Magnesium oxide (MgO)	3.1
Insoluble residue	1.9
Loss of ignition	1.0
Lime saturated factor	0.85
Physical Property	**Value**
Surface area (Blaine)	3.14 m^2/kg
Initial setting time	105 min
Final setting time	190 min
Soundness	8.7 mm

Table 1. Chemical composition and physical properties of cement

3.2. Methods

Principal operations involved in this section are preparation of specimen for various property assessments, mode of curing regimes and performance tests. Relevant standards are employed for each of the operations and these are spelt out accordingly. Details on mix-design, mixing, casting, demoulding and curing are included.

Batching and mixing were conducted in accordance with BS 1881-125: 1986. Mixing involves careful dispersion of latex into the mixing water followed by overall mixing in a pan mixer conforming to BS 1881-125:1986. Casting was conducted in accordance with BS 1881-108:1983. Standard steel cube moulds; 100 mm were employed for all specimens. However, 75 mm diameter cored specimens are used as water absorption samples. In fact, coring was diligently carried out in order to avoid visible cracks.

In the case of specimens for chemical attack, two curing mediums containing "5%" sulfuric acid (H_2SO_4) and "2.5%" (Na_2SO_4) each are used as aggressive environments. In order to understand what really happens when concrete is 'cast-in-situ'; where concrete is introduced to the service area during its initial age, specimens for H_2SO_4 were placed into the aggressive curing environment immediately after removal from moulds. However, specimens for Na_2SO_4 were given one month treatment in ordinary water at 23 ^{o}C and "80 ± 5%" RH before subjection to the aggressive medium. In fact, in the case of specimens for Na_2SO_4, 72 h air drying at 20 ± 3 ^{o}C and "80 ± 5%" RH was entertained before immersion into the diluted curing medium.

Although, air-curing in addition to the moist-curing is necessary for latex-film to develop in the modified specimens, these were given similar moist-curing treatment with the NC. Uniform curing treatment was considered important so that MC do not absorb higher content of the simulated aggressive moisture when immersed into the curing medium. This could obviously render results from the two categories of concrete non comparative.

In the case of morphological observations, samples were cured for 6 months under laboratory atmosphere; 20 ± 3 ^{o}C and "80 ± 5%" RH. Latex-film was obtained by drying few drops of latex in an oven at 85 ^{o}C. The latex-film was also stored under similar laboratory atmosphere until day for testing. Conditioning specimen for SEM observations is necessary in order to expel moisture so that a clear realization of microstructural matrix is achieved.

3.3. Testing program

3.3.1. Chemical analysis

The clonal latexes were analyzed for physical properties and chemical compositional contents. The chemical parameters and respective standards employed in each case are shown in Table 2. However, polydispersity index (PDI) is calculated as the weight average molecular weight divided by the number average molecular weight.

3.3.2. Scanning electron microscopy

JEOL Scanning Electron Microscope JSM 6390 LV was used. Morphologies were obtained at a current and working distance of 15 Kv and 9 mm respectively. Specimens were coated with 10 nm platinum in an 'Auto Fine Coater' before positioning against electron gun. Platinum coating was carried out at 20 mA for about 60 sec.

Parameter	Standard
Total solid content (TSC)	BS ISO 124:2008
Dry rubber content (DRC)	BS ISO 126:2005
Volatile fatty acid (VFA)	ISO 506:1992
Sludge content (SC)	ISO BS 2005:1992
Protein content	MS 1392:1998
Zinc	MS 1449: Part 1:1999
Copper	MS 1449: Part 3:1999
Manganese	MS 1449: Part 4:1999
Iron	MS 1449: Part 5:1999
Magnesium	MS 1449: Part 6:1999
Molecular weight (M_w)	BS ISO 16564:2004
Alkalinity (as NH_3)	ISO BS 125:2003
pH	ISO 976:1996

Table 2. Chemical analysis and standards

3.3.3. Mechanical properties

Compressive and flexural strength tests were conducted. In addition, compressive strength test was also conducted on specimens after chemical attack and firing. While compressive strength test was in conformity with BS EN 12390-3:2002, flexural strength test was conducted in accordance with BS EN 12390-5:2000. At the end of every curing period, an average of three cubes was taken as the strength of each particular batch.

3.3.4. Water absorption test

Measurement of water absorption was conducted in accordance with BS 1881–122: 1983; 'Method for Determination of Water Absorption'. The cores were kept in an oven for 72 h at 105 ± 5 °C followed by subsequent cooling for 24 h in dry airtight vessels. At the end of drying and cooling processes as described in the standard specimens were immersed in water for 30 ± 0.5 min at 20 ± 1 °C and then weighed. Average water absorption of three cored specimens expressed as a percentage of dry samples is considered as the water absorbed in each particular batch.

3.3.5. Fire endurance

Heating was performed in an automatic furnace with an average temperature gradient of 15.6 °C/min, evaluated from the ratio $(T_f - T_o)/M_t$. Where T_o and T_f are the initial and final temperatures respectively, and M_t represents minutes taken to raise the temperature from 27 °C to 800 °C. However, further heating beyond 800 °C was conducted mainly for the purpose of monitoring disintegration at failure. Heating is terminated once the desired temperature is attained. Specimens are cooled to room temperature (27 °C) in the furnace and then taken out for testing.

3.3.6. Thermogravimetric analysis

Thermogravimetry analyzer, Mettler Toledo TGA/STDA 851e was used for the thermal degradation assessments. Observations were made by raising applied temperature from 25 - 900 °C, with heating rate and flow of nitrogen at 10 °C/min and 10 ml/min respectively.

4. Results and discussions

4.1. Properties of NRL

Properties of NRL influencing cement mixes were obtained through chemical analysis. Table 3 presents results of the six clonal latexes. It is important to note that DRC represents the main hydrocarbon substance sought in civil engineering applications for the purpose of void-filling and latex-film formation in the modified matrix.

The main importance of these results is to identify other properties in addition to DRC which influence performance of the NRL in cement concrete and mortar. This issue was appropriately dealt with after compressive strength assessments on the performance of each of the latexes. Thus, observations were fully discussed in section 4.2.

Parameter	*Clonal Latexes*					
	KT 3935	*RRIM 2015*	*RRIM 937*	*RRIM 926*	*PB 350*	*PB 260*
TSC ("%")	30.34	33.47	35.43	45.54	33.56	27.20
DRC ("%")	27.75	31.03	32.64	41.66	31.09	25.39
Mol. Wt. (u)	3390292	3146156	3141284	3230285	3059752	2863935
PDI	1.7176	1.6397	1.7535	1.6564	1.6009	1.7485
VFA ("%")	0.045	0.041	0.285	0.029	0.041	0.038
Protein ("%")	4.94	3.69	4.19	3.25	4.31	3.81
Zinc (ppm)	170.8	694.3	702	659.8	266.9	273.7
Magnesium (ppm)	139.9	101	102.1	100	9.53	9.64
Copper (ppm)	5.48	4.93	4.62	4.30	8.71	8.10
Iron (ppm)	3.38	1.97	2.29	1.92	0.97	1.14
Manganese (ppm)	0.82	0.40	0.09	0.25	0.56	0.51
Sludge ("%")	0.3368	0.3548	0.1048	0.1032	0.1397	0.1223
NH_3 ("%")	1.77	1.92	0.36	0.35	0.37	0.37
pH	11.27	11.45	10.01	10.08	10.15	10.24

Table 3. Properties of NRL-clonal latexes

4.2. Influence of compositional properties on compressive strength of concrete

Figure 5 presents results of the compressive strength for concretes modified with the six clonal latexes each. The results were obtained through an addition of "2%" latex/water ratio. However, the compressive strength of the control concrete is 34.87 N/mm^2. Concretes modified with RRIM 2015 and PB 260 exceeded control concrete by "4%" and "2%"

respectively. On the other hand, concretes modified with RRIM 926 and RRIM 937 yielded lowest compressive strengths values. In fact, RRIM 937 depicted strength reduction of up to "12.4%" when compared with that of control concrete. Differences in the amounts of the compositional substances present in the latexes are believed to be responsible for the variations in the compressive strength.

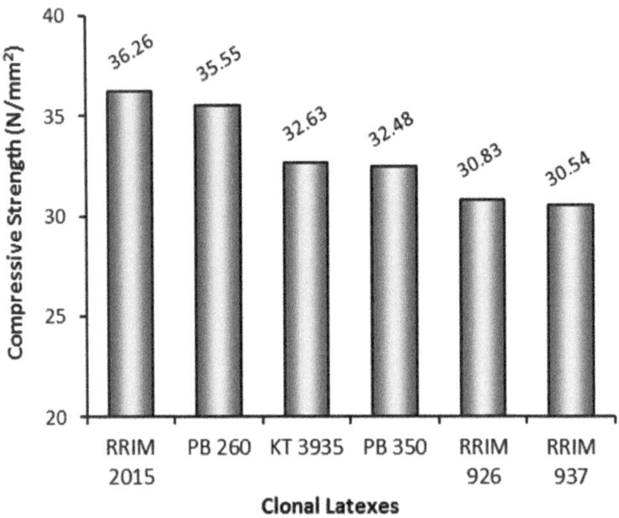

Figure 5. Compressive strength of NRL concrete due to clonal latexes

Even though, RRIM 937, RRIM 926 and PB 350 contain the highest amounts of DRC but these three latexes gave rise to the least three strength values; 30.54 N/mm^2, 30.83 N/mm^2 and 32.48 N/mm^2 respectively. On the other hand, specimens modified with latexes associated with comparatively low DRC yielded the best three compressive strength results; 32.63 N/mm^2, 35.55 N/mm^2 and 36.26 N/mm^2. Therefore, it could be suggested that what really influences strength is not the actual quantity of DRC, rather, it is to do with other compositional substances.

Close observations on the properties of clonal latexes given in Table 3 in relation with the compressive strength assessment revealed that parameters with strong potentialities of positive impact on the strength are sludge and pH. Conversely, those with high negative tendencies of impairing the strength are VFA and zinc. The rest of the parameters appeared to have little or no impact on the strength evaluations. This could be a result of their insignificant quantities in the NRL.

From the results, RRIM 2015 contains highest sludge content and the concrete modified with this latex yielded the best compressive strength. In addition, KT 3935 and PB 260 with second and third highest sludge contents gave rise to the next two best compressive strength values. On the other hand, RRIM 937 and RRIM 926 which contains least quantities of sludge were related to the lowest compressive strengths. Thus, higher sludge content seems to favor strength development.

By virtue of the fact that sludge normally consists of fine particles of heavier characteristics especially when compared with hydrocarbon particles, this could lead to void and micro-structural fillings as these fine particles occupy spaces between larger aggregate particles in concrete phase. Obviously, this action might result into higher material compaction with eventual increase in the compressive strength of the concrete. However, where sludge is relatively high in quantity its presence in between aggregate particles may cause aggregate particle displacements with eventual decrease in compressive strength. For instance, the use of ready-mixed concrete plant sludge water in making concrete was observed to cause "4-8%" reduction in the compressive strength of control concrete (Chatveera & Lertwattanaruk, 2009).

Considering the pH results, RRIM 2015 and RRIM 937 were observed to have highest and lowest values respectively. Concretes modified with these two clonal latexes depicted first and last compressive strengths accordingly. These outcomes therefore suggest that compressive strength concrete modified with NRL is favored by relatively higher pH values. This could be a result of contributions due to precipitation of metals particularly zinc. In fact, according to Boardman (1999) zinc is amphoteric and the optimum pH it will precipitate is about 9.45. Above this pH, hydroxide precipitates would re-dissolve in an excess of hydroxide ions to form soluble zincates (ZnO_2) and the solubility increases with increase in pH. Therefore, the higher pH values in RRIM 2015, KT 3935 and PB 260 might have aided the solubility of zinc, hence, less precipitation in the form of hydroxides. Thus, the retardation effect on cement hydration due to metal precipitation is minimized under the higher pH conditions.

RRIM 937 with its outstanding quantity of VFA stands a good test for the impact of fatty acids on the concrete strength. Indeed, concrete modified with RRIM 937 yielded the least compressive strength, which indicates strong possibilities that VFA have contributed to its poor performance. Statistically, RRIM 937 suffers "12.4%" and "15.8%" losses in compressive strength when compared with control concrete and RRIM 2015 respectively. In fact, acidic substances are known to harm cement products (Bala et al., 2012; Ali et al., 2005; Bertrona, Escadeillasa, & Duchesn 2004).

Zinc content seems to play a role in the strength decrease observed among the modified phases. For instance, while RRIM 937 and RRIM 926 which contain high zinc contents yielded concretes with the lowest compressive strength values, PB 260 and KT 3935 with low metal contents gave rise to concretes with relatively high compressive strengths. In fact, RRIM 937 depicted "14.1%" loss in compressive strength when compared with PB 260. Thus, presence of metals in NRL appeared to impair compressive strength of the modified concrete. Indeed, in a study conducted on the hydration of tricalcium silicate, zinc seems to be the most severe hydration retardant among the metals considered in the study (Chen et al., 2007).

Even though, protein which is popularly known to exist in NRL basically consists of large molecules of one or more chains of amino acids, but correlating protein contents in the individual latexes with the compressive strengths results, it appears that protein has little or no impact on the strength. Similarly, the trend in compressive strength of the modified concretes does not reflect that of molecular weight distribution among the latexes. Indeed, modified-concretes associated with strong or weak strengths values do not show any preference to high or low molecular weights. However, according to Ramakrishnan (1992), concrete modified with latex of lower molecular weight particles may not have the same strength as that modified with latex of higher molecular weight.

4.3. Performance of NRL on flexural strength of concrete

Figure 6 presents results of flexural strength test. The optimum latex/water ratio was "4%". The corresponding increase percent of this optimum value over control strength was "4.7%".

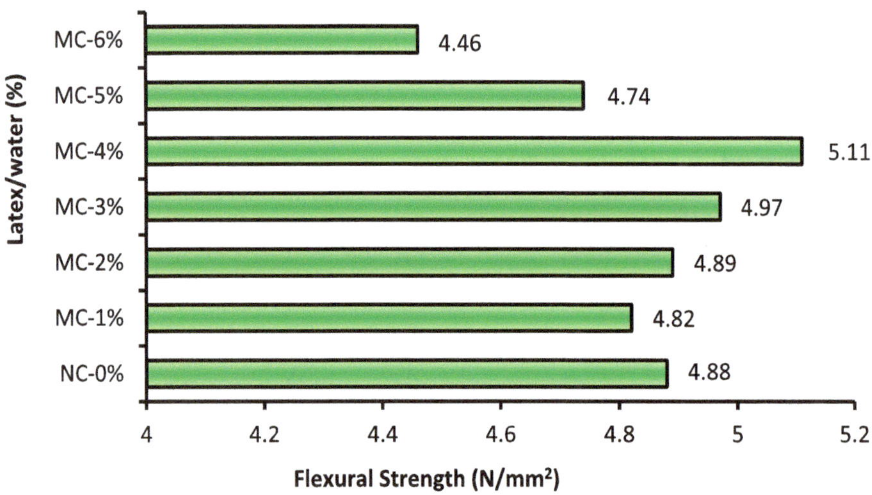

Figure 6. Effect of latex content on the flexural strength of NRL concrete

Maximum flexural strength was observed at higher latex content when compared with that of the compressive strength. This could be related to the nature of the applied force during test. The applied bending stresses in flexure are invariably causing elongations most especially in the lowermost layers of the section, and this act synonymous to a direct test on the elasticity of the latex-films. Thus the higher amount of the latex became an advantage as it will provide additional elastic capabilities. In fact, strength development in flexural parameters of latex modified concrete as observed by (Ramakrishnan, 1992) is the combined actions of cement hydrate-aggregate bond and elastic strength of latex-films. Furthermore, NRL has been credited for this specialty property, where according (David & Richard, 2002), NRL still dominates rigid elasticity applications such as; production of dipped goods, extruded threads and water based adhesives for its high strength characteristics. However, increase in latex beyond "4%" resulted in significant decrease in the strength, an indication of elastic performance limit by the latex-films.

Another interesting observation arising from these results is in relation to artificial latexes; unlike most of the synthetic latexes where optimum strength can be attained only by adding "10%" or more of latex-cement ratios (Ramakrishnan 1992; Ohama, 1995; Barluenga, 2004; Abdullah, 2008), in this case it can be seen that peak value is readily possible by adding a smaller amount. However, optimum values in mechanical capacities are generally greater when synthetic latexes are used as modifiers. The reason for this is to do with the polymerized contents in synthetic latexes which are usually predetermined as against natural polymerization in NRL.

4.4. Performance of NRL on water exclusion of concrete

Based on the fact that durability relates to the ability of concrete to withstand environmental factors of destruction while serving its purpose, the main gateway to aggressive attack was investigated and the findings were displayed in this section. Moisture being the chief transporter to destructive agents was evaluated in both unmodified and modified phases. Meanwhile, Figure 7 presents SEM captions for NRL, unmodified and modified phases of mortar mixes. These captions revealed the microstructural nature of the material orientation in all the three phases.

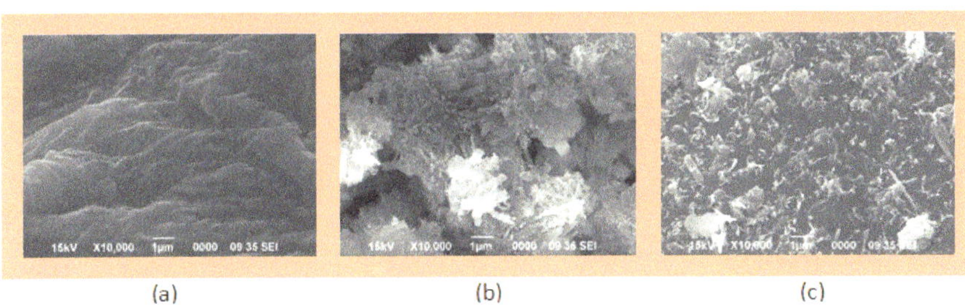

Figure 7. Microstructural units: (a) NRL, (b) normal mortar, (c) NRL modified mortar

Microstructural details of the NRL portrayed impervious characteristics which is quite suitable for moisture exclusion. On the other hand, microstructural unit of the normal mortar phase describes the porous nature of the matrix. In fact, hardened cement paste alone is known to have capillaries and pores initially occupied by mixing water. Introduction of aggregates therefore give rise to additional voids through gaps between the larger particles and microcracks. However, inclusion of NRL into the normal mix yielded the fairly cementitious phase shown in Figure 7 (c). Fragments of latex-films are seen scattered on the caption and this shows how blended the mortar was with the NRL.

Indeed, the modified mortar caption suggested a much denser matrix and since presence of voids in the unmodified mix renders the product more vulnerable to aggressive attack, better moisture exclusion capacity is expected from the modified phase. Figure 8 presents results of moisture absorption of normal concrete (NC) and concrete modified with three clonal latexes. The assessment was based on modification with "5%" latex/water ratio. This percentage was observed as the optimum content of NRL to be added into concrete in order to yield maximum exclusion of moisture (Bala et al., 2012).

In a similar manner to assessments on mechanical properties, modification with clonal latexes also resulted in different levels of water absorption. Furthermore, concrete modified with PB 260 excelled the other two modifications by marked reductions. Even though, this clone has the least DRC, its presence in the concrete however yielded the most effective blockage to the water passage. One obvious advantage which PB 260 has over the other two clones is higher sludge content and this has been manifested in Table 3. Possible effect of sludge in cement aggregate conglomerate has been explained in section 4.2. An important role could be that of void and micro-structural fillings as sludge occupy spaces between larger aggregate particles in concrete phase with eventual increase in water tightness.

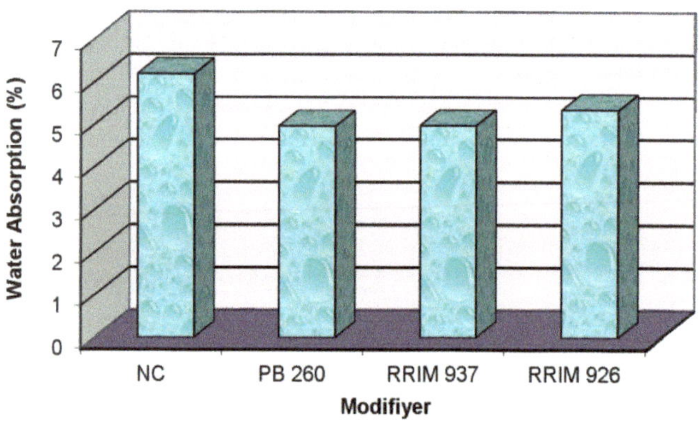

Figure 8. Effect of clonal latexes on water absorption of NRL concrete

4.5. Performance of NRL on chemical resistance of concrete

Physical appearance of specimens subjected to H_2SO_4 is presented in Figure 9. The originally gray colour of the specimen was observed to turn white with eroded fragments especially at edges. However, Figure 10 shows compressive strength results of concrete modified with different percentages of NRL after the aggressive attack. From the results, low content of latex has shown mechanical superiority over higher contents. Indeed, modification with "2.5%" latex/water not only exceeded higher modifications in quality but also the NC.

The entire compressive strength values were observed to be low. This was understood to be the consequence of early subjection of specimens into the aggressive curing medium. For instance, at 28 days old, the NC should have reached the designed concrete strength, but the impact of attack coming from the sulfuric acid clearly hindered this normalcy. Instead, strength reduction was registered not only in NC but also among all modifications. In fact, this is similar to real situations where concrete is cast in-situ particularly aggressive environments.

Figure 9. Effect of sulfuric acid on the surface of NRL modified concrete

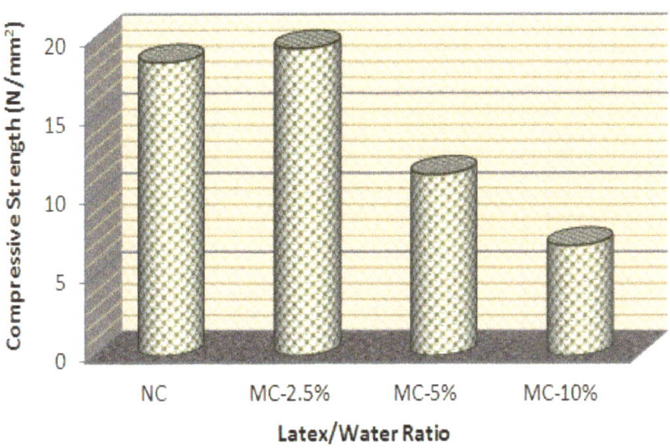

Figure 10. Compressive strength of NRL-MC due to sulfuric acid attack; 84 days

The improvement associated with "MC-2.5%" is an indication of the impact of NRL at that particular amount. Higher amount appeared to attract severe attack through destruction of latex by the acid. Indeed, physical observations in the event of carrying out experimental processes revealed a high volume change of relatively more than "100%" in the affected latex. Previous findings regarding acidic substances coming into contact with polymeric materials have indicated possible harm on the polymeric Si-O-Al bonds with consequence strength weakening. This is probably the main reason behind the greater strength reductions witnessed in the higher modifications, "5 – 10%" (Allahverdi, 2005).

However, it worth noting that while a positive volume change accompanied the attack on the latex, a negative volume change was entertained in the concrete matrix. Close inspections on the attacked specimens discloses leaching of the cement, gradual disintegration of the fine aggregate particles and protrusions of the coarse aggregates. This indeed correlates the findings of Eglinton (1987), where the author pointed out that the action of acids on Portland cements is to leach calcium hydroxide from the cement paste. Thus, when the acid first comes into contact with the concrete it reacts with the calcium hydroxide of the cement to form calcium sulfate, or gypsum which then attacks the hydrated calcium aluminates to give ettringite. Meanwhile, in the present case, measurement on weight change was generally obscured as the expanded latex contains moisture which compensates for the lost in cement and fine aggregate particles.

Figure 11 shows the compressive strength results of concrete subjected to sodium sulfate environment. Unlike in the sulfuric acid test, the compressive strength values of all specimens were comparatively higher. In this case, higher strength values were observed due to mainly the initial normal curing of 28 days given to specimens before immersion into the simulated sodium sulfate environment. Also, contrary to the foregone assessment, all modifications were observed to yield greater strength values than the NC. But, just like the previous case, lower latex content ("1.5%") proves to be the best modification. Usually, the optimum content for maximum compressive strength is less than "2%" depending on the concentration of the NRL (Bala, 2009).

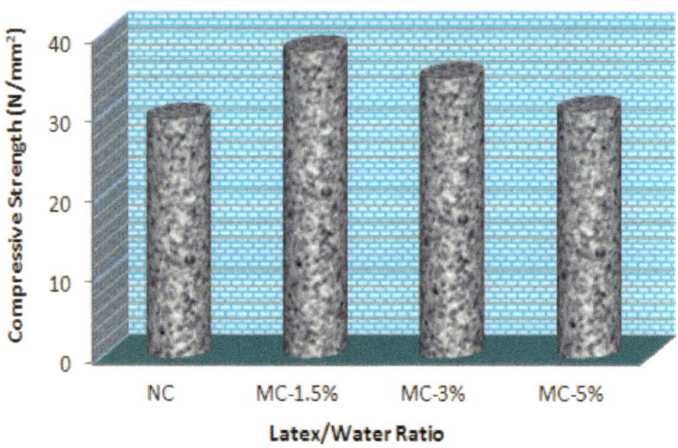

Figure 11. Compressive strength of NRL-MC due to sodium sulfate attack; 84 days

Superior performance observed in NRL modified concretes was a result of the inclusion of the elastomer. Thus, as the latex coats the aggregate and blocks the passage of the Na_2SO_4 particles into the hardened paste, development of expansive forces from high volume increase which normally evolves when calcium sulfate reacts with tricalcium aluminate was discouraged (Vedalakshmi et al., 2005). In fact, while sign of attack in the form of leaching was manifested on the surfaces of specimens subjected to acidic environment, similar sign has not been witnessed on the surfaces of specimens introduced into the sulfated environment. In other words, surface erosion does not occur in the later. Henceforth, variations in weights of specimens at the end of the curing exercise were also noticed to be insignificant.

4.6. Performance of NRL concrete at elevated temperatures

Physical appearances of specimens during the entire heating range (27 – 1300 °C) are shown in four stages in Figure 12. Prior to heating, the specimens are perfectly cubic in shape with smooth surfaces and straight edges. However, during the process of heating many physical changes took place. These include colour changes, crack formations, distortions and spallings. Indeed at 1300 °C, the concrete is not only so week to be tested for compressive strength but also not strong enough to resist pressure from forefinger and thumb as indicated in the disintegration stage.

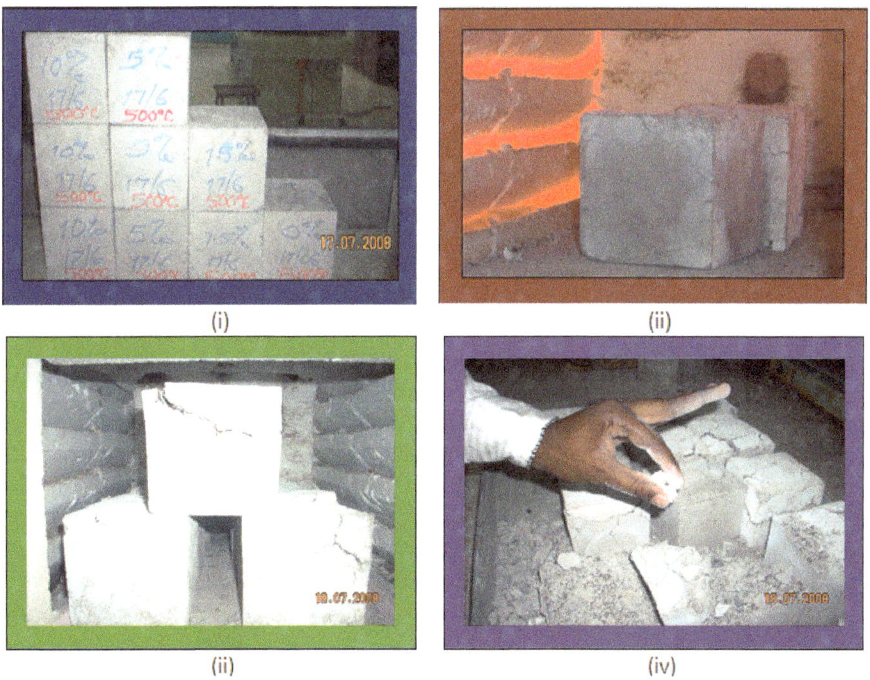

Figure 12. States of concrete in the process of firing; (i) before firing, (ii) during firing, (iii) after firing, and (iv) disintegration

Compressive strength test results after heating at various temperature levels particularly within 27 – 800 °C are presented in Figure 13. Continuous increase in temperature is seen to have caused corresponding decrease in compressive strength to both categories of concrete. However, least drop in the compressive strength was entertained by NC and the highest was associated with "MC-10%". In fact, the more the latex percent added into the concrete the higher the strength loss. From this, it shows that inclusion of latex into concrete has a negative impact on the compressive strength of concrete at elevated temperatures and the impact increases with increasing latex content.

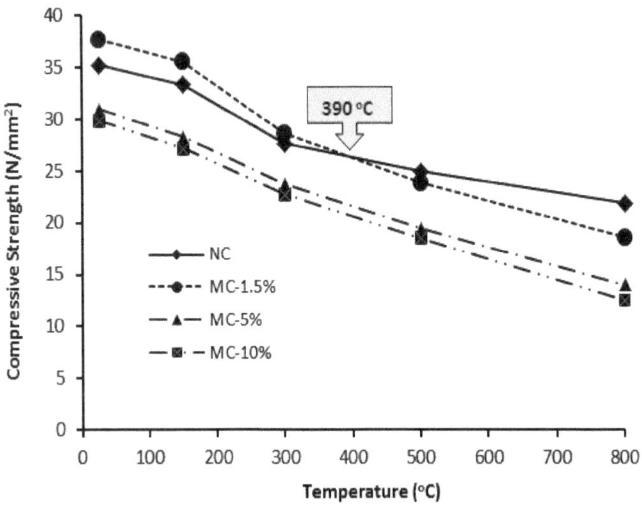

Figure 13. Effect of high temperature on compressive strength of NRL modified concrete

"MC-1.5%" surpassed NC by "6.8%" at the initial stage and this superiority was maintained up to 300 - 500 °C where the MC dropped below NC. The temperature at the intersection point between NC and "MC-1.5%" was about 390 °C. Thus, the critical limit of superiority of NRL-MC over NC is 390 °C. Nevertheless, up to 500 °C the difference in strength between these two categories is only "4.4%", which may be considered as insignificant.

Categorically, NRL-MC suffered more strength loss than NC. The main reason behind higher strength depreciations in the NRL-MC could be the early degradation of the latex-films which gives rise to undue formation of unacceptable cracks during the compression test. To this regard, the temperature limit at which these cracks were formed was postulated to be related to the superiority limit of "MC-1.5%" over NC as indicated in Figure 13. Eventually, the temperature limit was realized to be about 390 °C. However, this hypothesis was investigated further using TGA.

Thermal degradations which accompanied temperature rise in latex-film and cement-latex blend are presented in Figure 14. While the latex-film in its entity suffers a single major weight loss, the cement-latex blend entertained relatively two minor and one major weight losses as witnessed in its TGA.

Latex-film lost more than "95%" of its weight within 340-460 °C. This marked the softening point and the main degradation limit of NRL. The remaining decomposed content afterwards represents the filler. Previous works have also indicated similar weight losses in polymeric substances at temperatures above 300 °C. For instance, more than "95%" weight loss was observed on polyethylene between 380 and 510 °C (Zhengzhou et al. 2003).

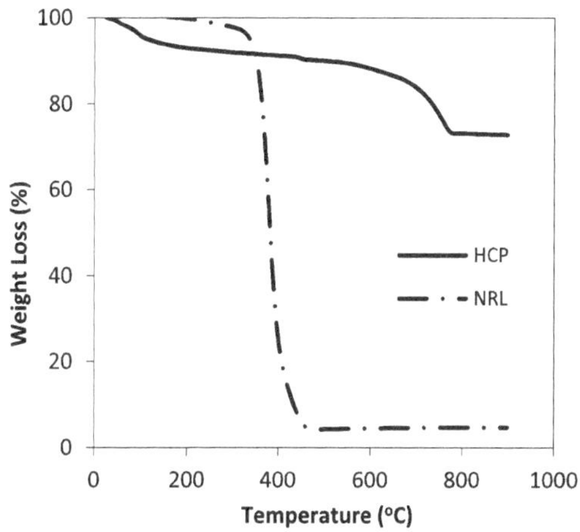

Figure 14. Thermal degradations of HCP and NRL-film

On the other hand, the TGA presented by cement-latex blend also entertained its major weight loss within the range 350-430 °C. This forms an interesting part of the result since it indicates the impact of the blended latex. Indeed, out of the overall weight loss of about "39%" as witnessed in the blend, this degradation was observed to be about "29%", which amounts to "74%" of the overall degradation. This amount of degradation could possibly be the main factor responsible for the significant fall in the compressive strength of MC at temperatures within 300-500 °C.

This observation was further strengthened by the fact that the mean of the temperature at which the sudden weight loss was registered in the cement-latex blend (350-430 °C) is 390 °C, a temperature value which corresponds to the superiority performance limit of MC over NC. In other words, cement-latex blend lost much of its strength at the softening point of the NRL. Therefore, the critical limit of performance of elastomerically modified cement mix should be expected at the softening point temperature of the latex involved.

4.7. Recent trend in research activities and challenges facing applications of elastomers

The act of directing great sources of power in nature for the benefit of man has been described as 'Civil Engineering' (Navin, 2000). Since the beginning of this profession several efforts have been made by research individuals, groups, private organizations and statutory parastatals in order to identify materials of potential qualities. Consequently, numerous materials have so far been classified as civil engineering materials, particularly in the last three decades during which the profession experiences dramatic changes in the way of thinking about process engineering, modification technologies and innovative material evolutions. Thus, construction industry is already in a new era where not only there exist interest in the expansion of mankind's frontiers enabling new technologies to be developed, but also the need for a harmonious co-existence with the eco-sphere through optimization of material technology and infrastructural creativity, so that a healthy life in harmony with nature is fully enforced.

Indeed, the cement industry, in its effort towards evolving befitting products for the new era has taken a positive step by developing 'eco-cement', a product that helps to clean the environment by utilizing municipal solid wastes in its manufacture and is even said to have the potential of absorbing carbon dioxide (CO_2) from the environment as it hydrates (Ampadu & Toni, 2001). However, even before the concept of eco-cement is matured enough to make impact in the present world of advancing technology, global worries over the existing environmental threats, have triggered a tremendous force towards the search for eco-friendly materials in all respects. Thus, frantic efforts towards inclusion of eco-friendly materials into OPC mixes originated from the rising concern and worries over environmental degradations such as; global warming atmospheric pollutions and waste disposals. These coupled with the recent worldwide economic recession have indeed strengthened the inspiration for the continuous efforts towards the search for durable and sustainable infrastructures [Walter, et al., 2004; Dionys, 2007].

At present, the act of directing great sources of power in nature has dramatically turn to redirecting waste materials as well as agricultural byproducts to construction materials. Even though, polymers have contributed immensely on the performance of cement mixes but the rush towards consuming waste materials and agro-based substances may open an entirely new challenge to the world of polymer concrete. Meanwhile, pozzolans from agricultural waste are receiving more attention since their uses generally not only improve the properties of the blended cement concrete but also reduce the environmental problems.

Therefore, sooner or later, research activities in the world of polymer concrete may face compounding challenges. According to Lech (2006) the search for new research simulations and applications of polymeric substances is still essential in order to make their improvements more efficient and reliable. Like other advanced materials, polymeric substances need reliable theories based on firm scientific findings. In fact, elastomers which currently dominate construction applications are no exception in the dear need for scientific knowhow on their performance. This, together with the possible chemical metamorphism which could erupt when cement, elastomer and agricultural byproduct intermingled together in a single co-matrix system needs intensive investigations.

5. Conclusions

In this chapter, properties, application and performance of elastomers in cement mixes have been investigated. In particular, the roles of NRL in the two major properties of concrete; strength and durability have been examined. Based on the experimental framework adopted in this investigation, the following conclusions were drawn:

- Inclusion of NRL in concrete enhances its performance especially durability functions such as water tightness and chemical resistance. The improvement was achieved mainly through void and capillary fillings of the cis-1,4-polyisoprene particles.
- The dry rubber content of NRL is not the major factor responsible for strength development in modified concrete. Instead, sludge content and pH value are the chief factors contributing to strength development in the cement concrete.
- NRL consists of some non-rubber substances which are likely to cause damage to the mechanical properties of concrete. The chief substances causing strength impairment are volatile fatty acids (VFA) and metals particularly zinc.
- SEM captions expose the microstructural details of latex film, hardened cement paste and cement-latex co-matrix. Presence of latex in the hardened cement paste transforms the initially porous phase to a more cementitious matrix.
- Deterioration of concrete due to chemical attack, particularly those from acidic and sulfated environments is discouraged by the inclusion of NRL. However, efficiency of the latex was observed to be higher in sulfated environment rather than acidic.
- The critical limit of performance of NRL concrete at high temperatures was observed to be about 390 °C. Meanwhile, TGA result has shown that NRL softens and degrades to about "5%" at similar temperature. Therefore, as the NRL softens at this temperature significant loss in the compressive strength of the NRL modified concrete was observed.

Author details

Bala Muhammad
Bayero University Kano, Nigeria

6. References

Abdullah, S. (2008). Characteristics and performance of concrete modified with prepackaged polymer-modified mortar. PhD Thesis. Universiti Teknologi Malaysia.

Afridi, M. U. K., Ohama, Y., Demura, K. & Iqbal, M. Z. (2003). Development of polymer films by the coalescence of polymer particles in powdered and aqueous polymer-modified mortar. Cement and Concrete Research, 33:1715-21.

Ali, T. A., Muzaffer, Y., Ali, M. G. & Ismet, G. (2005). Investigation of the effects of fatty acids on the compressive strength of the concrete and the grindability of the cement. Cement and Concrete Research, 35 (2):400-4

Allahverdi, A., and Skvara, F. (2005). Sulfuric acid attack on hardened paste of geopolymer cements Part 1: Mechanism of corrosion at relatively high concentrations. Ceramics-Silikaty, 49, 225-229.

Ampadu, K. O., & Toni, K. (2001). Characterization of ecocement pastes and mortars produced from incinerated ashes. Cement Concrete Research, 31, 431-436.

Bala, M., A. A. Yussuf & Mohammad, I. (2009). Making Void-Free Cement-Latex Blend using Morphology and Thermal Degradation Analysis. Indian Concrete Journal (ICJ). 83(11):32-39.

Bala, M., Mohammad, I., A. A. Yussuf & M. A. R. Bhutta. (2011[a]) Elastomeric Influence of Natural Rubber Latex on Cement Mortar at High Temperatures using Thermal Degradation Analysis. Construction & Building Materials 25: 2223-2227.

Bala, M., Mohammad, I., Muhammad, A. R. B. & Zaiton, A. (2012). Influence of non-hydrocarbon substances on the compressive strength of natural rubber latex-modified concrete. Construction & Building Materials 27: 241-247.

Bala, M., Mohammad, .I, Zaiton, H. & Abdirahman, A. Y. (2011[a]) Elastomeric Effect of Natural Rubber Latex on Compressive Strength of Concrete at High Temperatures. Journal of Materials in Civil Engineering (JMCE-ASCE). Online Production, 5; 2011[b]

Bala, M. & Mohammad, I. (2012). Performance of natural rubber latex modified concrete in acidic and sulfated environments. Construction and Building Materials 31 129–134

Bala, M. (2009), Impact of natural rubber latex on engineering properties of concrete. PhD thesis. Universiti Teknologi Malaysia.

Barluenga, G., & Hemandez-Olivares, F. (2004). SBR Latex Modified Mortar Rheology and Mechanical Behavior. Cement and Concrete Research, 34(3), 527-535.

Bertrona, A., Escadeillasa, G. & Duchesn, J. (2004) Cement pastes alteration by liquid manure organic acids - chemical and mineralogical characterization. Cement and Concrete Research, 34:1823–35

Boardman, D. I. (1999). Lime stabilization: Clay-metal-lime-interactions, PhD Thesis. Loughborough University United Kingdom.

BS EN 12390:2002. Testing hardened concrete – part 3. Compressive strength of test specimens. London: BSI Publications.

Chatveera, B. & Lertwattanaruk, P. (2009). Use of ready-mixed concrete plant sludge water in concrete containing additive or admixture. Journal of Environmental Management, 90:1901-8.

Chen, Q. Y., Hills, C. D., Tyrer, M., Slipper, I., Shen, H. G. & Brough, A. (2007). Characterization of products of tricalcium silicate hydration in the presence of heavy metals. Journal of Hazardous Materials, 147:817-25

David, J. D. & Richard, H. D. (2002). Natural and Synthetic Latex Polymers. UK: Rapra Technology Limited.

Dionys, V. G. (2007). Cement-Concrete and Concrete-Polymer Composites: Two Merging Worlds 12th International Congress on Polymers in Concrete (ICPIC). Chuncheon, Korea. September 27-28, pp.3-15

Eglinton, M.S. (1987). *Concrete and its Chemical Behaviour*. London, United Kingdom: Thomas Telford.

Esah, Y. & Paul, C. (2002). The Manufacture of Gloves from Natural Rubber Latex. Malaysian Rubber Export Promotion Council, 110(2), S3-S14.

Illston, J. M., Dinwoodie, J. M., & Smith, A. A. (1979). Concrete Timber and Metals - the Nature and Behavior of Structural Materials. Van Nostrand, Berkshire, England: Reinhold Company.

Jitladda, T. S. (2006). Structural characterization of natural rubber based on recent evidence from selective enzymatic treatments. Journal of Bioscience and Bioengineering, 103, 287-292.

Joao, A.R. & Marcos, V. A. (2004). Durability of polymer-modified lightweight aggregate concrete. Cement & Concrete Composites 26:375–380.

John, S. D. (1987). Compounding Materials for the Polymer Industries, A Concise Guide to Polymers, Rubbers, Adhesives and Coatings. New Jersey USA, Noyes Publications.

Kondou, H. (2006). Natural rubber latex, natural rubber, rubber compositions containing the same and tire: Washington DC U. S. Patent, No. 537698.

Lech, C. (2006). Concrete-polymer composite: Overview. 5th Asian Symposium on Polymers in Concrete (ASPIC) September, 11-12. Chennai, India. Allied Publishers Pvt. Ltd., 87-104.

Navin, F. P. D. (2000). A Descriptive Risk Model for Civil Engineering Schemes, why some succeed and others crash and burn. Proceedings of the 2000 Annual Canadian Conference, June 7-10. Montreal Canadian Society for Civil Engineering.

Ohama, Y. (1995). Handbook of polymer-modified concrete and mortars. New Jersey (USA): Noyes Publications.

Ong, E. L. (1998). Latex Protein Allergy and Your Gloves. Kuala Lumpur, Malaysia: Malaysian Rubber Board.

Rajni, L., Asthana, K. K. & Anupam, S. (2006). Polymer modified cementitious (PMC) systems for building applications. 5th Asian Symposium on Polymers in Concrete (ASPIC), September 11-12, 2006. Chennai, India. Allied Publishers Pvt. Ltd., 235-242.

Ramakrishnan, V. (1992). Synthetic of highway practice 179 – latex modified concrete and mortars. Washington (DC): NCHRP.

Rattana, T. (2003). Reinforcement of Natural Rubber Latex by Nanosize Montmorillonite Clay. PhD Thesis. The Pennsylvania State University.

Schneider, M., Pith, T. & Lambla, M. (1997). Toughening of polystyrene by natural rubber-based composite particles. Part I: Impact reinforcement by PMMA and PS grafted core-shell particles. Journal of Materials Science, 32 6331-6342.

Vedalakshmi, R., Rajf, A.R., Srinivasan, S., & Babu, K.G. (2005). Effect of magnesium ions on the sulphate resistance of blended cements in low and medium strength concrete. Advances in Cement Research, 17(2), 47-55.

Walter, O. O., Kunitomo, S. & Eiichi, W. (2004). Flexural response of polymer concrete filled steel beams. Construction and Building Materials 18(6), 367-376

Zhengzhou, W., Yuan, H., Zhou, G. & Ruowen, Z. (2003). Halogen-free flame retardation and silane crosslinking of polyethylenes. Polym Test 22:533–8.

Section 7

Experimental Testing and Modelling of Elastomer Properties

Chapter 12

Transmittance and Reflectance of an Imprinted Cholesteric Elastomer During a Segregation Process

P. Castro-Garay and Jesus Manzanares-Martinez

Additional information is available at the end of the chapter

1. Introduction

The interest in the separation of enantiomers of organic compounds has been recently increased in the food, medicine and cosmetic industries. As it is well know, the chirality is extremely important for living systems, because most of the biomolecules are chiral ones. For example, glucose, a sugar exists in one conformation; while the opposite enantiomer cannot be used as a food source. Biochemistry uses only one hand and not the other however the reason for this has not been understood, so far. Over half of the organic compounds in drugs are chiral whose different enantiomers have different effects. For instance, Ritalin prevents hyperactivity in children in one enantiomer, whereas it has not effect when the opposite enantiomer is used. Another example is thalidomide: one enantiomer of thailomide can cure morning sickness of pregnant women, whereas the other causes birth defects. The sense of smell is also sensitive to the chirality of the odor molecules. For instance, limonene has an orange smell whereas the opposite enantiomer smells like lemon [1].This exemplifies the fact that separation of molecules by chirality is an important process in biochemistry and in the pharmaceutical industry. In this paper we suggest a method to optically monitoring the chiral sorting of molecules performed by an elastomer liquid crystal.

Liquid crystal elastomers combine the rubbery elastic properties with the liquid crystals (LC) anisotropy [2]. When these stiff molecules are joined to a large and flexible polymer chain, they leave their muddy liquid macroscopic form and acquire the appearance of a rubbery network. Particularly, imprinted cholesteric elastomers acquire its chirality from an induced chiral structure instead of chiral molecules. Upon cross-linking nematic polymers in a chiral solvent to form a gel, the solvent, which causes a natural twist to the nematic is thus removed but nonetheless leaves behind an imprinted cholesteric elastomer [3]. A spectacular property of imprinted networks is their capacity to preferentially absorb and retain right or left molecules from a racemic solvent which renders them chiral pumps. Moreover the elastomers are photonic materials in the sense that exhibit an optical band structure just as the one shown by electrons in metals and semiconductors. Elastomers provide a large spectrum of applications that exploit the possibility of externally handle a band structure by various types of agents like mechanical stress, electric fields chemical solvents for designing devices such as optical sensors, mirrors, laser and filters [2][4].

The purpose of this chapter is to propose an optical method to determine the volume fraction of preferentially absorbed molecules by measuring the transmittance and reflectance spectra during a preferentially absorbing process. Specifically, we are interested in calculate the reflectance and transmittance of circularly polarized light for oblique incidence due to a cholesteric elastomer, assuming arbitrary values of chiral order parameter, α. We have thoroughly discussed the solution of the boundary value problem for cholesteric elastomers in sections 4 and have determined the optical spectra of the system in Sec. 5.

2. Elastic formulation

A cholesteric elastomer has locally a nematic structure. Consequently, the director vector $\hat{n} = (\cos\phi(z), \sin\phi(z), 0)$ is at $x-y$ plane and $\phi(z)$ is the angle between director vector and x axis (see Fig.1). The director vector has a continuous rotation as function of z, which describes a periodic helical structure characterized by a pitch p (or equivalent by wave number $q_0 = 2\pi/p$) [5][6].

The chiral order parameter $\alpha = \xi q_0$ is related with the imprinting efficiency of the cholesteric phase, where $\xi = \sqrt{\frac{K_2}{D_1}}$ is the nematic penetration depth in rubbery networks, K_2 is the twist elastic constant and D_1 is local anchoring of the director to the rubbery network. Variations in the chiral order parameter imply changes in the robustness of chiral imprinting, which can be controlled by varying the density of cross-links in the network (affecting D_1) [7].

The energy for an elastomer formed under the presence of a chiral solvent which is subsequently replaced with an achiral one is given by [8],

$$F = \frac{1}{2}\int \left[K_2 \left(\frac{d\phi}{dz}\right)^2 + D_1 \sin^2(\phi - q_0 z) \right] dz. \tag{1}$$

The Frank free energy to twist the material is given by

$$F_f = \frac{1}{2} K \left(\hat{n}\cdot\nabla\times\hat{n}\right)^2 \tag{2}$$

associated with the first term of the eq. (1)- tend to unwind the cholesteric helix but it finds resistance to rotate in the anchoring of the rubbery network- associate with the second term of eq. (1). D_1 is proportional to rubbery elasticity modulus and the rubbery elasticity modulus is related with cross-linked density [7]. The scale length or nematic penetration depth rubbery at which the two energy contributions are comparable is usually small $\xi \simeq 10^{-8}$m for a typical $K_2 \simeq 10^{-11}$ J/m , $D_1 \simeq \mu(r-1)^2/r$, where $r \sim 0.9$ is the anisotropy chain number

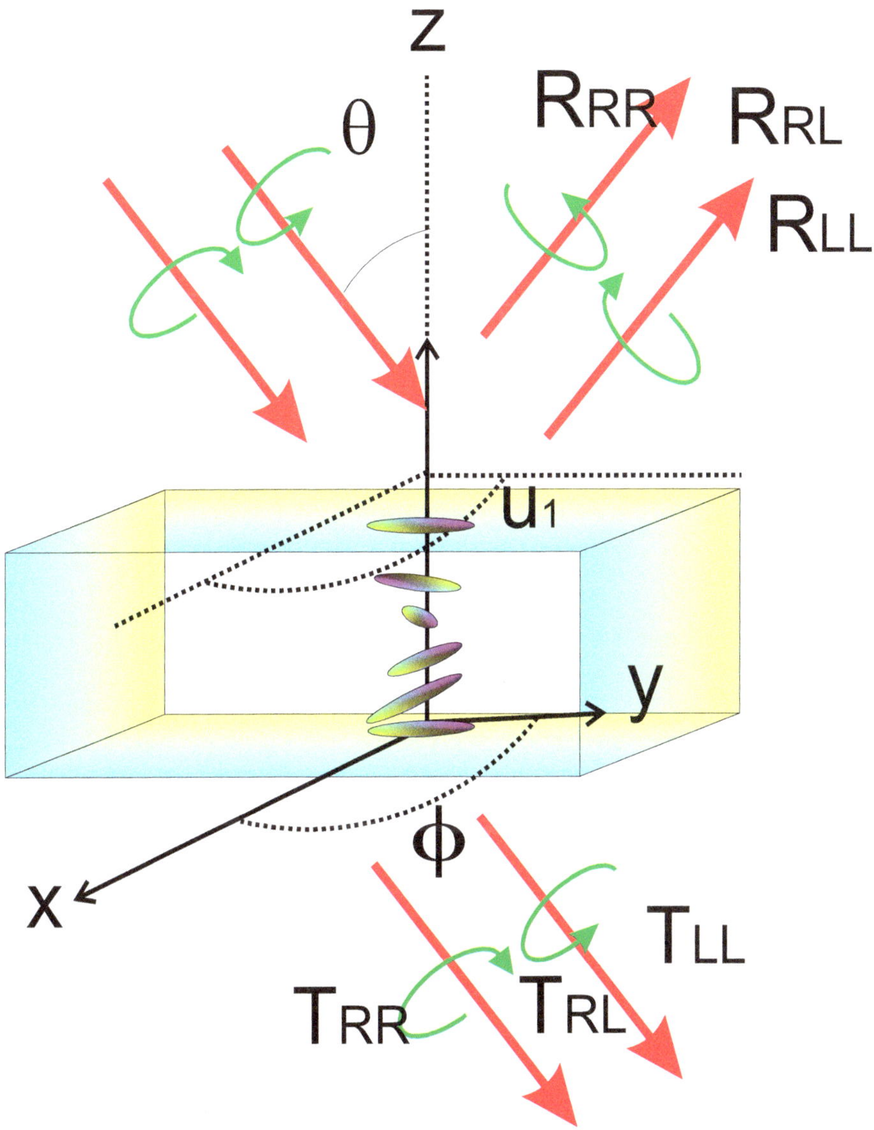

Figure 1. Schematic representation of a liquid crystal cholesteric elastomer. The circular components of the obliquely incident beam as well as the components of the transmitted an reflected parts are shown.

or cross-linked number and $\mu \simeq 10^5 \mathrm{J/m^3}$ is the elasticity rubber modulus. The pitch is 10^3 bigger than nematic penetration depth ξ, then

$$\xi \ll p = \frac{\pi}{q_0}. \tag{3}$$

There are two forms of modifying this relation, i) by increasing the nematic penetration depth which amounts to have a weaker gel or one with smaller anisotropy, ii) by enlarging the Frank free energy. The chiral imprinting will be lost if D_1 diminish, i.e. so that the director vector anchoring to rubbery will be almost null. Similarly, if the wave number q_0 or twist elastic constant K_2 are bigger the chiral imprinting will be also lost. The imprinting efficiency is reached for values of chiral parameter order $\alpha_c = \xi q_o$ lower than $2/\pi$.

Eq. (1) can be expressed in terms of chiral parameter order as,

$$F = \frac{1}{2}\int \left[\frac{\alpha^2}{q_0^2}\left(\frac{d\phi}{dz}\right)^2 + \sin^2(\phi - q_0 z)\right] \tag{4}$$

The equilibrium configuration can be obtained by minimizing F with respect to the angle, then we have got the Euler-Lagrange equations,

$$-\partial_z\left(\frac{\partial F}{\partial_z \phi}\right) = 0, \tag{5}$$

which in this case turns out to be

$$\frac{\alpha^2}{q_0^2}\frac{d^2\phi}{dz^2} + \sin 2(q_0 z - \phi) = 0. \tag{6}$$

Notice that if $\alpha = 0$ the solution of eq. (6) is $\phi_0(z) = q_0 z$ which corresponds to an ideal or undistorted cholesteric helix. By introducing the variable $w = q_0 z - \phi + \pi/2$, Eq.(6) transforms in the expression

$$\frac{d^2 w}{dz^2} + \frac{q_0^2}{\alpha^2}\sin 2w = 0. \tag{7}$$

which is the simple pendulum equation. Hence, for large values of α, small harmonic oscillations of frequency q_0/α around $\phi = q_0 z + \pi/2$ are expected. Nevertheless, for smaller values of α anarmonic oscillations take place until w abandons its oscillatory dependence and grows without limit following the dependence corresponding to a rotatory motion in a simple pendulum. This behavior can be inferred by using the general solution of eq.(6) which is given by

$$\phi(z) = q_0 z - Am(cz/\xi, 1/c^2) + \pi/2, \tag{8}$$

where $Am(z,m)$ is the Jacobian amplitude and c is a constant related with the reduced elastic energy [8] $g = 2\xi F/TD_1$. In the Fig. 2 is shown the director vector angle $\phi(z)$ against position for different values of the parameter c. In this case, the director vector changes the amplitude of its oscillations around the ideal case solution $\phi_0(z)$, while the spatial period is maintained constant to increase the values of parameter c from zero until to reach the critical value in $c = 1$. Schematic representation of the director vector is shown in the the Fig. 3. In the Fig. 4 is shown the director vector angle $\phi(z)$ against position for different values of chiral order parameter α. It is interesting to note that by increasing ξ from zero, $\phi(z)$ presents oscillations without changing its amplitude, around the undistorted solution $\phi_0(z)$ whose spatial period increases, until ξ reaches the critical values: $\xi_c = 3$. After this value $\phi(z)$ deviate considerably from $\phi_0(z)$ and enlarges its effective q-value for $\xi \geq \xi_c$ which is equivalent to have an untwisted helix than that of $\phi_0(z)$ as is shown in the Fig. 5.

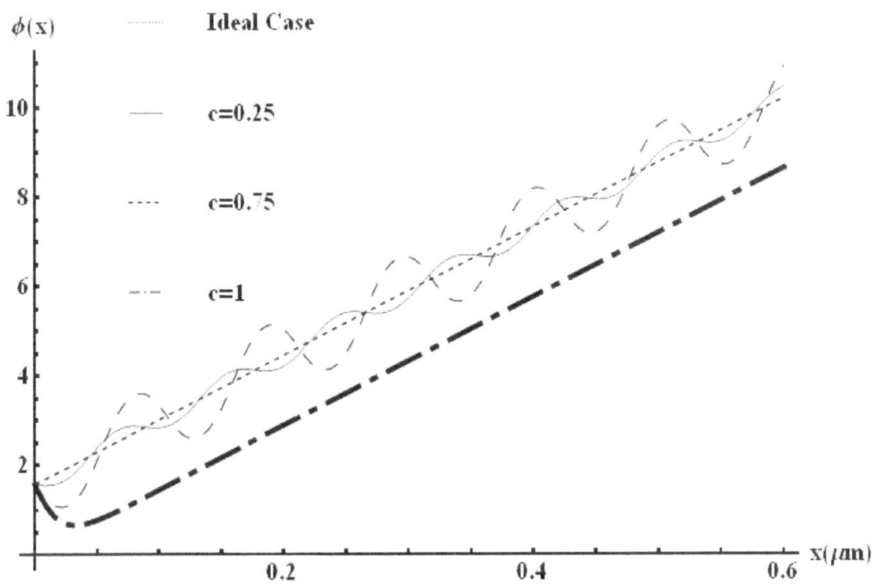

Figure 2. Director vector angle as function of the position for different values of the parameter,c.

3. Optical description

As any chiral material whose axis of chiral nonhomogeneity is along the $z-$axis, the optical relative permittivity matrix of the chiral elastomer may be stated as

$$\bar{\epsilon}(z) = \begin{pmatrix} 0 & 0 & 0 \\ \epsilon_{\perp}^{(0)} + \epsilon_a \cos^2\phi & \epsilon_a \sin\phi\cos\phi & 0 \\ \epsilon_{\perp} \sin\phi\cos\phi & \epsilon_{\perp}^{(0)} + \epsilon_a \sin^2\phi & \epsilon_{\perp}^{(0)} \end{pmatrix} \tag{9}$$

where $\epsilon_a = \epsilon_{\parallel} - \epsilon_{\perp}$ is the dielectric anisotropy with $\epsilon_{\parallel}$ and $\epsilon_{\perp}$ the dielectric constants in the optical regime parallel and perpendicular to the director $\hat{\mathbf{n}}$, respectively.

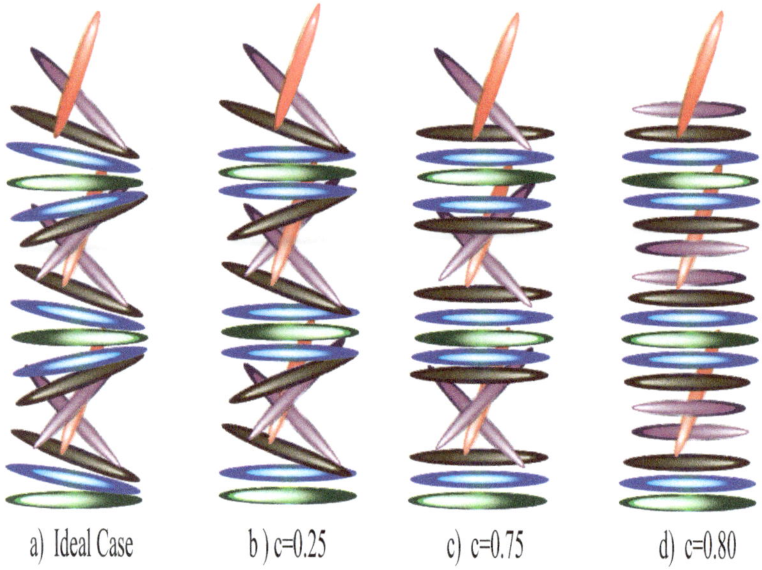

Figure 3. Schematic representation of director vector for different values of parameter, c, a) ideal case, b) $c = 0.25$ and c) $c = 0.75$.

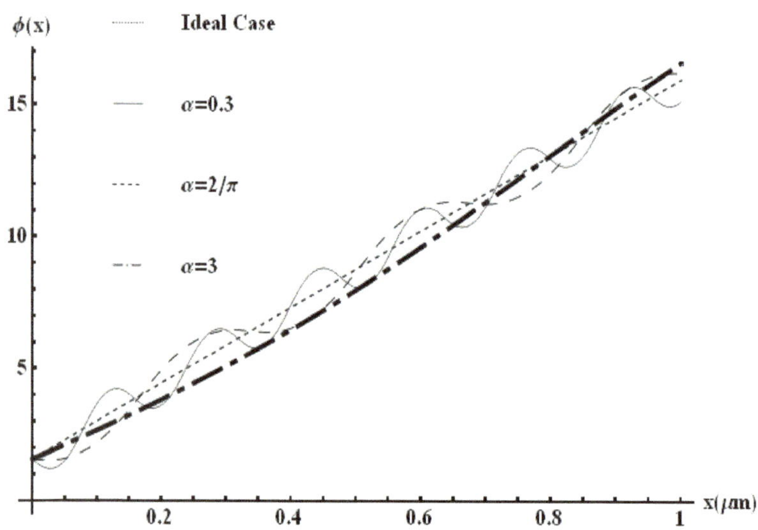

Figure 4. Director vector angle as function of the position for different values of the chiral order parameter, α.

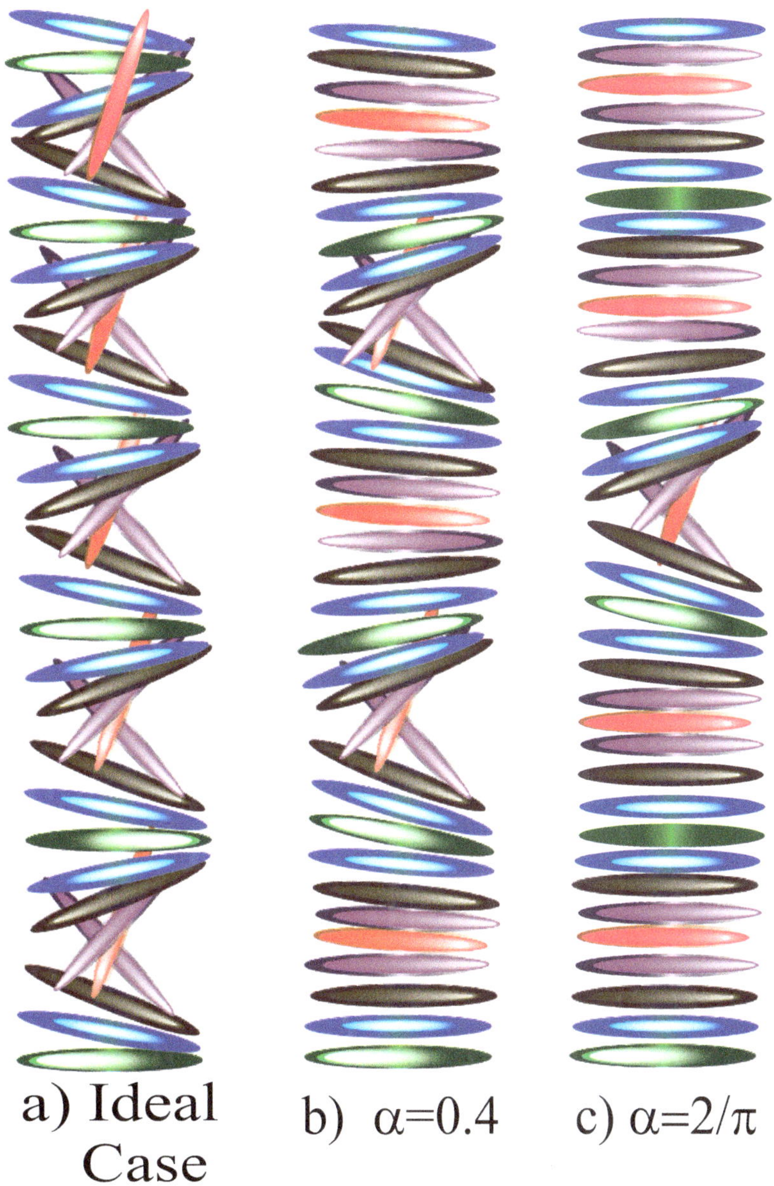

Figure 5. Schematic representation of director vector for different values of chiral order parameter, α, a) ideal case, b) $\alpha = 0.3$ and c) $\alpha = 2/\pi$.

Maxwell curl equations (in MKS units) for media whose dielectric tensor is only $z-$dependent, are given by

$$\left.\begin{array}{l}\nabla \times \mathbf{E}(x,y,z) = i\omega\mu_o \mathbf{H}(x,y,z) \\ \nabla \times \mathbf{H}(x,y,z) = -i\omega\epsilon_o \bar{\epsilon}(z) \cdot \mathbf{E}(x,y,z)\end{array}\right\}, \qquad 0 < z < L, \tag{10}$$

where ϵ_o and μ_o are the permittivity and the permeability of free space (i.e., vacuum). If a plane wave were to be incident obliquely on a Cholesteric Elastomer slab of thickness L, we can describe the fields in terms of the column vector [9]

$$\bar{\psi}(x,y,z) = \exp[i\kappa(x\cos u_1 + y\sin u_1)] \begin{pmatrix} e_x(z) \\ e_y(z) \\ h_x(z) \\ h_y(z) \end{pmatrix}. \tag{11}$$

where u_1 the angle formed by the propagation vector and the y-axis in the $x - y$ plane. Thus, the essential part of Maxwell curl equations can then be written as [10].

$$\frac{d}{dz}\bar{\psi}(x,y,z) = i\bar{A}'(z) \cdot \bar{\psi}(x,y,z), \qquad 0 < z < L, \tag{12}$$

In this equation, the 4×4 matrix can be factorized as $\bar{A}' = B(z) \cdot \bar{A} \cdot B(-z)$ where the 4×4 matrix

$$\bar{A}(z) = \begin{pmatrix} 0 & 0 & 0 & \omega\mu_o \\ 0 & 0 & -\omega\mu_o & 0 \\ 0 & -\omega\epsilon_o\epsilon_2^{(0)} & 0 & 0 \\ \omega\epsilon_o\epsilon_d & 0 & 0 & 0 \end{pmatrix}$$

$$+ \frac{\kappa^2}{\omega\epsilon_o\epsilon_1^{(0)}} \begin{pmatrix} 0\;0 & -\sin u\cos u & -\cos^2 u \\ 0\;0 & \sin^2 u & \sin u \cos u \\ 0\;0 & 0 & 0 \\ 0\;0 & 0 & 0 \end{pmatrix}$$

$$+ \frac{\kappa^2}{\omega\mu_o} \begin{pmatrix} 0 & 0 & 0\;0 \\ 0 & 0 & 0\;0 \\ \sin u\cos u & \cos^2 u & 0\;0 \\ -\sin^2 u & -\sin u\cos u & 0\;0 \end{pmatrix}, \tag{13}$$

with $u = \frac{h\pi z}{\Omega} - u_1$ and

$$B(z) = \begin{pmatrix} \cos\phi(z) & \sin\phi(z) & 0 & 0 \\ -\sin\phi(z) & \cos\phi(z) & 0 & 0 \\ 0 & 0 & \cos\phi(z) & \sin\phi(z) \\ 0 & 0 & -\sin\phi(z) & \cos\phi(z) \end{pmatrix}. \tag{14}$$

Equation (12) can be solved by a variety of techniques. We chose to implement the piecewise homogeneity approximation method. The final result is the computation of a transfer matrix $\bar{U}'$ such that

$$\bar{\psi}(L) = \bar{M}^T(L) \cdot \bar{U}' \cdot \bar{M}(0) \cdot \bar{\psi}(0) \equiv \bar{U} \cdot \bar{\psi}(0)\,. \tag{15}$$

4. Boundary value problem

Suppose that plane waves are obliquely incident on a cholesteric of finite thickness from both half-spaces $z \leq 0$ and $z \geq L$, which are taken to be vacuous. As a result, the total electric and magnetic fields in these half-spaces may be written as

$$\left.\begin{aligned} \mathbf{e}(z) &= (a_L\hat{\mathbf{u}}_+ + a_R\hat{\mathbf{u}}_-)\exp[ik_0 z + i\kappa(x\cos u_1 + y\sin u_1)] \\ &\quad + (r_L\hat{\mathbf{u}}_- + r_R\hat{\mathbf{u}}_+)\exp[-ik_0 z + i\kappa(x\cos u_1 + y\sin u_1)] \\ \mathbf{h}(z) &= i\,(-a_L\hat{\mathbf{u}}_+ + a_R\hat{\mathbf{u}}_-)\exp[ik_0 z + i\kappa(x\cos u_1 + y\sin u_1)] \\ &\quad + i\,(-r_L\hat{\mathbf{u}}_- + r_R\hat{\mathbf{u}}_+)\exp[-ik_0 z + i\kappa(x\cos u_1 + y\sin u_1)] \end{aligned}\right\}, \quad z \leq 0, \tag{16}$$

and

$$\left.\begin{aligned} \mathbf{e}(z) &= (t_L\hat{\mathbf{u}}_+ + t_R\hat{\mathbf{u}}_-)\exp[ik_0(z-L) + i\kappa(x\cos u_1 + y\sin u_1)] \\ &+ (b_L\hat{\mathbf{u}}_+ + b_R\hat{\mathbf{u}}_-)\exp[-ik_0(z-L) + i\kappa(x\cos u_1 + y\sin u_1)] \\ \mathbf{h}(z) &= i\,(-t_L\hat{\mathbf{u}}_+ + t_R\hat{\mathbf{u}}_-)\exp[ik_0(z-L) + i\kappa(x\cos u_1 + y\sin u_1)] \\ &+ i\,(-b_L\hat{\mathbf{u}}_+ + b_R\hat{\mathbf{u}}_-)\exp[-ik_0(z-L) + i\kappa(x\cos u_1 + y\sin u_1)] \end{aligned}\right\}, \quad z \geq L\,. \tag{17}$$

where

$$\hat{\mathbf{u}}_\pm = \frac{\hat{\mathbf{u}}_x \pm i\hat{\mathbf{u}}_y}{\sqrt{2}}, \tag{18}$$

and $(\hat{\mathbf{u}}_x, \hat{\mathbf{u}}_y, \hat{\mathbf{u}}_z)$ is the triad of Cartesian unit vectors. In these equations, the complex–valued amplitudes $a_{L,R}$ and $b_{L,R}$ of the left– and right–circularly polarized (LCP and RCP) components of incident plane waves are assumed to be known, whereas the reflection amplitudes $r_{R,L}$ and transmission amplitudes $t_{L,R}$ are not known.

Continuity of the tangential components of the electric and the magnetic fields across the planes $z = 0$ and $z = L$ leads to the prescriptions of the boundary values

$$\bar{\psi}(0) = \bar{Q} \cdot \begin{pmatrix} a_R \\ a_L \\ r_R \\ r_L \end{pmatrix}, \quad \bar{\psi}(L) = \bar{Q} \cdot \begin{pmatrix} t_R \\ t_L \\ 0 \\ 0 \end{pmatrix}, \tag{19}$$

where the matrix

$$\bar{Q} = \begin{pmatrix} 1 & 1 & 1 & 1 \\ -i & i & i & -i \\ i & -i & i & -i \\ 1 & 1 & -1 & -1 \end{pmatrix}. \tag{20}$$

By virtue of Eqs. (16), (17) and (19), we obtain the matrix algebraic equation

$$\begin{pmatrix} t_R \\ t_L \\ b_R \\ b_L \end{pmatrix} = T \cdot \begin{pmatrix} a_R \\ a_L \\ r_R \\ r_L \end{pmatrix}, \tag{21}$$

where the so called transfer matrix is given by

$$T = \frac{1}{4} \bar{Q}^{\dagger} \cdot \bar{U} \cdot \bar{Q} \tag{22}$$

and the symbol † denotes the conjugate transpose.

The result of solving Eq. (21) is best arranged as

$$\begin{pmatrix} t_R \\ t_L \\ r_R \\ r_L \end{pmatrix} = S \cdot \begin{pmatrix} a_R \\ a_L \\ b_R \\ b_L \end{pmatrix} \tag{23}$$

$$S = \begin{pmatrix} t_{RRa} & t_{RLa} & t_{RRb} & t_{RLb} \\ r_{LRa} & r_{LLa} & r_{LRb} & r_{LLb} \\ t_{RRa} & t_{RLa} & t_{RRb} & t_{RLb} \\ r_{LRa} & r_{LLa} & r_{LRb} & r_{LLb} \end{pmatrix} \tag{24}$$

where t_{LRa} and t_{LRb}, etc., are the transmission coefficients and r_{LRa} and r_{LRb}, etc., are the reflection coefficients. The co–polarized transmittances engendered by the waves coming from the left-hand side are denoted by $T_{LLa} = |t_{LLa}|^2$ and $T_{RRa} = |t_{RRa}|^2$, and the cross–polarized ones by $T_{LRa} = |t_{LRa}|^2$ and $T_{RLa} = |t_{RLa}|^2$; and similarly for the reflectances

$R_{RRa} = |r_{RRa}|^2$, etc. similar definitions are valid for the waves generated by the waves coming from the right-hand side $T_{LLb} = |t_{LLb}|^2$, $T_{RRb} = |t_{RRb}|^2$ and $T_{LRb} = |t_{LRb}|^2$. The matrix involved S in Eq.(23) is known as the scattering matrix and contains the same information of the transfer matrix but order in different way. It can be straightforwardly shown by writing explicitly the algebraic equations defined by Eq.(21) and solving them for t_R, t_L, r_R and r_L that both matrices are related by the expression

$$S = (T.P_f - P_b)^{-1}.(P_f - T.P_b), \tag{25}$$

where P_f and P_b are the forward and backward projectors given by

$$P_f = \begin{pmatrix} 1\,0\,0\,0 \\ 0\,1\,0\,0 \\ 0\,0\,0\,0 \\ 0\,0\,0\,0 \end{pmatrix}, \; P_b = \begin{pmatrix} 0\,0\,0\,0 \\ 0\,0\,0\,0 \\ 0\,0\,1\,0 \\ 0\,0\,0\,1 \end{pmatrix}.$$

In what follows we shall restrict our calculation to the case in which the incident waves are coming only from the left hand side that is we take $b_R = b_L = 0$.

5. Spectra

In our calculations we have chosen the physical amounts of a real cholesteric elastomer material ($p_0 = 218$nm) [5] [11]. Fig. 6 shows that in the absence of solvent ($\alpha = 0$) there is only a band reflection for right-circularly copolarized light R_{RR} with the bandwidth of $\simeq$ 50nm and maximum amplitude. The center of the reflection band blue-shift whereas the width and amplitude band remain constant as the incidence angle θ increases. The corresponding reflectances R_{RL} and R_{LL} do no exhibit practically any band. In contrast, when $\alpha = 0.4$ there appear bands for both left- and right-copolarized reflectances R_{RR} and R_{LL} whereas the cross-polarized reflectance R_{RL} is almost zero. In this case the R_{RR} is not as well formed as in the absence of solvent because is amplitude present various peaks and the bandwidth varies as a function of the incidence angle, reaching its smaller value for angles around 45 degrees. Differently, R_{LL} is very thin since its bandwidth is around 15nm and contained in the interval where the R_{RR} band lies.

Fig. 7 presents the corresponding trasmittances to Fig.6. The case without solvent ($\alpha = 0$) confirms consistently the complemetary behavior of an undistorted cholesteric. The co-polarized T_{RR} and T_{LL} exhibit troughs where the transmittance diminish, the one for T_{RR} is well defined whereas that of T_{LL} changes is thickness a deep. The blue-shift of the trough for large incident angles is also observed. In the Fig. 8 is shown the band reflection for left and right circularly polarized ligh R_{RR} and R_{LL} for different values of parameter, c. In both cases, the band reflection for left circularly polarized light is thinner than the band reflection for right circularly polarized light.

In the Fig. 9 is shown the reflectance of left and right circularly polarized light as a function of wavelenght and the chiral order parameter to oblique incidence. Complementary, In the Fig. 10 is shown the same of the Fig.9 as function of wavelenght and the parameter c.

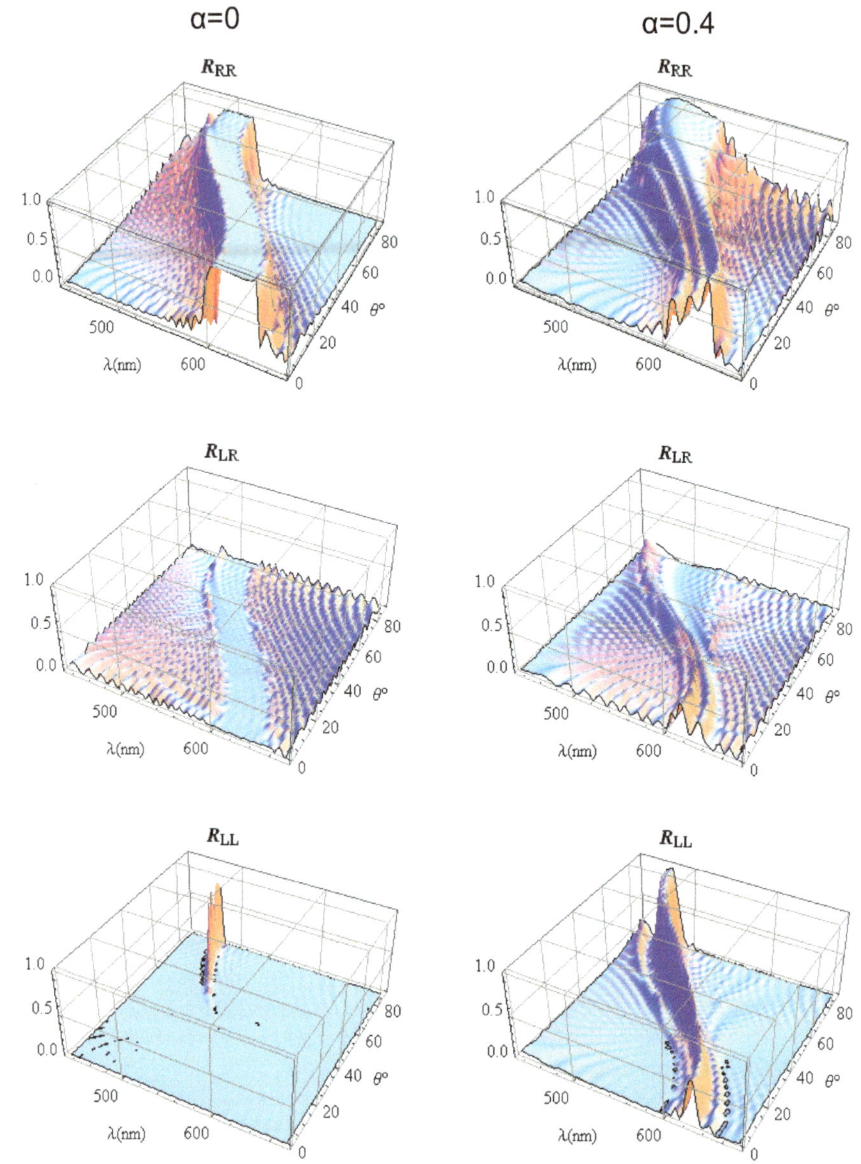

Figure 6. Circularly polarized reflectance R_{RR} and R_{LL} as function of the wavelength and angle incidence for the ideal helix case $\alpha = 0$ (left column) and $\alpha = 0.4$ (right column). Other parameters are: $\epsilon_{\perp} = 1.91$, $\epsilon_{\parallel} = 2.22$, $h = 1$ and $q_0 = 218$nm.

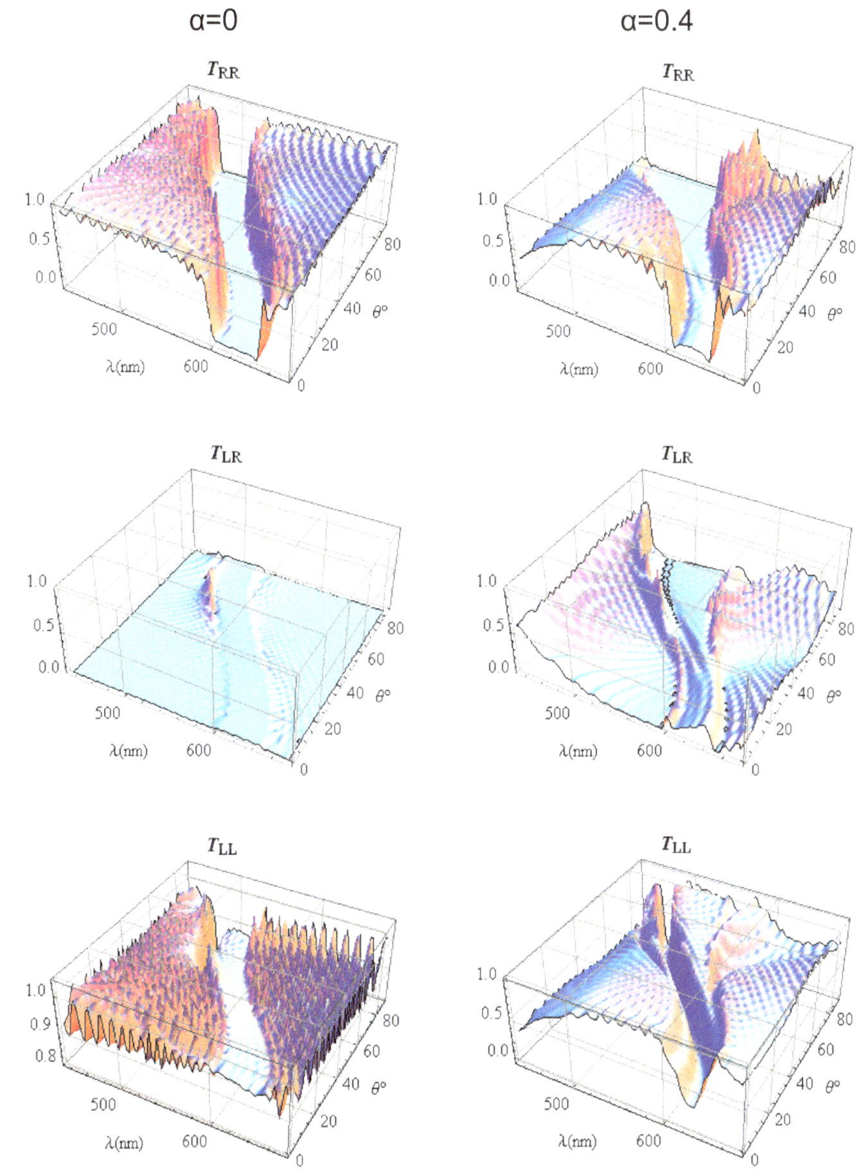

Figure 7. Circularly polarized transmittance T_{RR} and T_{LL} as function of the wavelength and angle incidence for $\alpha = 0$ (left column) and $\alpha = 0.4$ (right column).

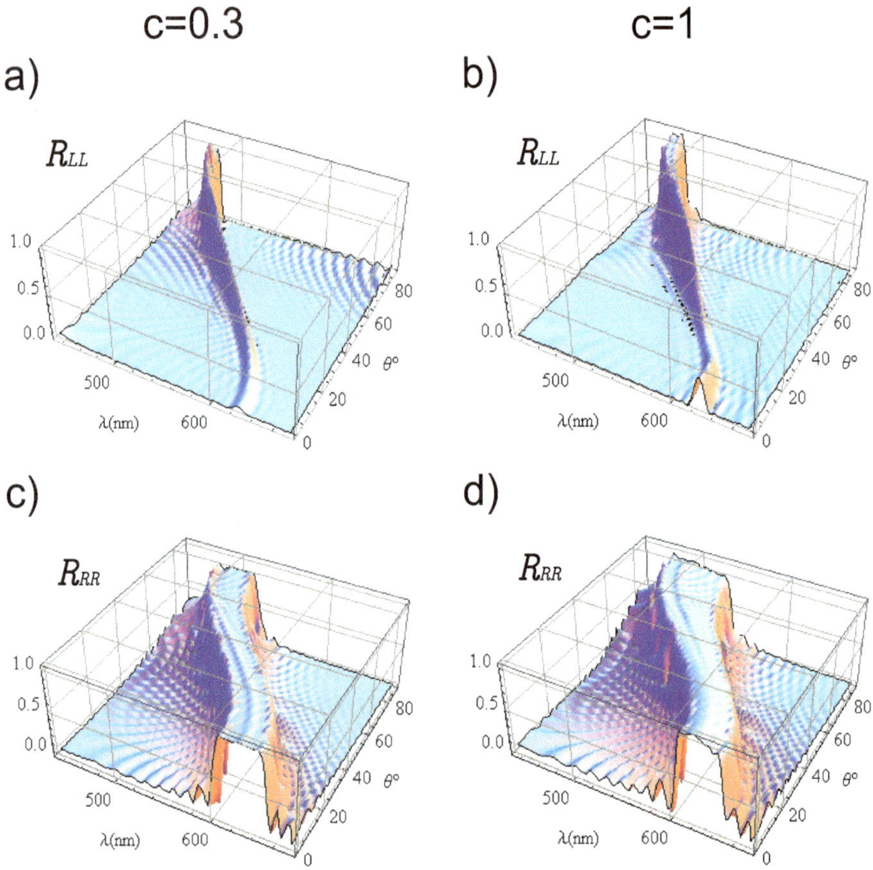

Figure 8. Circularly polarized reflectance R_{RR} and R_{LL} as function of wavelength and angle incidence for $c = 0.3$ (left column) and $c = 1$ (right column).

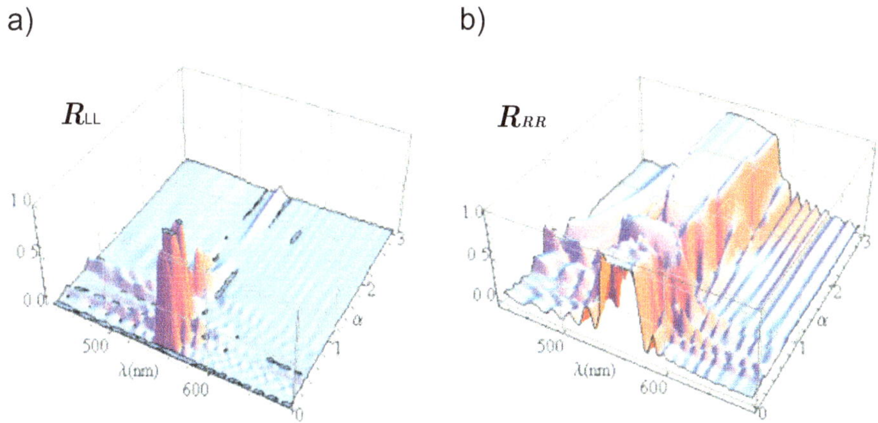

Figure 9. Circularly polarized reflectance R_{RR} and R_{LL} as function of wavelength and the chiral order parameter to oblique incidence angle of 45° and $c = 0.5$.

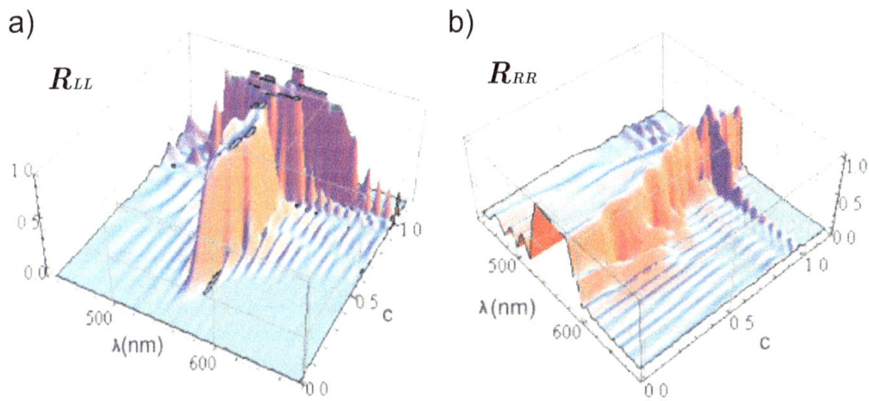

Figure 10. Circularly polarized reflectance R_{RR} and R_{LL} as function of wavelength and the c parameter to oblique incidence angle of 45° and $\alpha = 0.2$.

6. Conclusions

We have obtained the circularly polarized reflectances and transmittances as function of the chiral order parameter of a cholesteric elastomer immersed in a racemic solvent. We have found considerably changes in the bandwidth of the reflectance for left- and right-copolarized light under the presence of the solvent which are susceptible to be detected experimentally. We have obtained a R_{RR} bandwidth larger than that for an undistorted elastomer ($\alpha = 0$) for $\alpha = 0.4$ and obliquely incidence for angles larger than 60°. Also in this value we have observed a thin band reflection for both polarizations R_{RR} and R_{LL} for oblique incidence.

The modifications of the transmittance and reflectance spectra during a preferentially absorbing process suggest to utilize the optical spectra as an indirect method to determine the concentration of preferentially absorbed molecules during a segregation process.

Author details

P. Castro-Garay
Depto de Fisica, Universidad de Sonora, Apdo Post. 1626, Hermosillo, Son. Mexico 83 000, México

Jesus Manzanares-Martinez
Depto. de Investigacion en Fisica, Universidad de Sonora, Hermosillo, Son., Mexico, Apdo P. 5-088, México 83 190.

7. References

[1] A. Jakli, *et al. One and two dimensional fluids: Properties of smetics, lamellar and columnar liquid crystals (Condensed Matter Physics)*. Taylor and Francis: Florida, EU, (2006).
[2] B. Singh. *J. Phys. D: Appl. Phys.*, 40, 584 (2007).
[3] Y. Mao, M. Warner. *Phys. Rev. Lett.*, 86, 5309, (2001).
[4] M. Warner, *et al. Liquid Crystal Elastomers*. Oxford Science Publications: Oxford, EU, (2007).
[5] M. Rivera, J. A. Reyes. *Appl. Phys. Lett.*, 90, 023513, (2007).
[6] P. Cicuta, A. R. Tajbakhsh, E.M Terentjev. *Phys. Rev. E.*, 70, 011703, (2004).
[7] S. Courty, A. R.Tajbakhsh, E. M. Terentjev. *Phys. Rev. Lett.*, 91, 085503, (2003).
[8] Y. Mao,M. Warner. *Phys. Rev. Lett.*, 84, 5335, (2000).
[9] A. Lakhtakia,W.S. Weiglhofer. *Proc. R. Soc. Lond. A.*, 453, 93, (1997); correction: 454, 3275 (1998).
[10] N. Marcuvitz,J. Schwinger. *J. Appl. Phys.*, 22, 806, (1951).
[11] P. Cicuta, A. R. Tajbakhsh. E. M. Terentjev. *Phys. Rev. E.*, 65, 051704, (2002).

Chapter 13

Modelling Friction and Abrasive Wear of Elastomers

Nándor Békési

Additional information is available at the end of the chapter

1. Introduction

Tribological modelling of elastomer parts has high importance in engineering practice due to their widespread industrial use. Generally, the experimental investigation of the wear behaviour is time consuming and expensive, which led to the development of numerical techniques.

In order to properly model the friction and wear of elastomer materials, it is essential to understand some specific mechanical and tribological features. Mechanical characteristics of elastomer materials are different from those of metals and other relatively hard materials. Elastomers can bear large deformations, the stress-strain curves are non-linear, time and temperature dependent. When the elastomer is deformed, a part of the strain energy is dissipated. This feature makes the elastomer parts fulfil damping functions. The contact mechanical aspects of elastomers are also important for modelling. When an elastomer part is pressed against a counter-surface, the contact area would be larger than that estimated by the Hertzian theory, because of the effect of adhesion [1]. For the same reason, when the elastomer is contacting a rough surface, the real area of contact would be bigger than it is common for harder materials, as the elastomer tends to fill the valleys of the rough surface.

Friction of elastomers is usually considered as a joint effect of the friction on the surface by adhesion (and fluid film shearing in lubricated friction) and the internal friction generated by the hysteresis of the deformation by the counterpart [2]. Since the hysteretic contribution is caused by the visco-elastic properties of the material, it depends on the temperature and on the exciting frequency. When the elastomer is sliding on a hard, rough counter-surface, the asperities of the rough surface repeatedly deform the elastomer and so cause hysteresis loss. Thus the hysteresis contribution is dependent on the sliding speed. The hysteresis caused by the surface roughness of the counterpart is usually referred as micro-hysteresis. Hysteresis can occur on the macro level as well, where it can be a significant part of the friction.

When designing elastomer parts for tribological applications, modelling friction is inevitable in order to determine the contact conditions, which usually have an important effect on the functionality of the part. Wear, on the other hand, has less importance in short term use (an average passenger car tire can run up to 50 000 km before the need to change). However, in most cases, the reason for the elastomer parts to be replaced is the wear. In long term or when the lubrication is not ideal wear can dramatically change the geometry of the part and thus the nature of the contact. Therefore wear modelling is needed to estimate the product lifespan and to schedule maintenance operations. Beside experimental tests, which are cost and time consuming, numerical wear simulation techniques were developed in the last decades and became more and more commonly used. Although there are wear simulation methods incorporating various numerical methods, such as the boundary element method [3] or the discrete element method [4], the most successful and popular approaches deal with the finite element method, since it is a general method for mechanical stress analysis.

The most popular FE based wear simulation method was developed by Pödra and Andersson [5]. In this iterative method, firstly the contact pressure distribution is determined by FEM and the nodal wear increments are calculated from the wear equation of Archard [6]. Then the contacting nodes are moved with respect to the nodal wear values. Finally the FE contact calculation is carried out again with the modified mesh and the cycle is repeated according to the simulated wear process. This method was improved by several researchers (e.g. [7,8]) for various applications, different geometries and materials, such as metals, ceramics, polymers and composites. Kónya and Váradi in [9] improved the method in order to consider heat generation and time dependent material properties in the wear simulation.

Even though this chapter shows methods for modelling and simulation of elastomers, some experimental work is inevitable for the calculations. The parameters of the material models are usually determined based on experimental data. In my simulations I used the material model described by Pálfi et al. in [10]. It is a Mooney-Rivlin model with an attached generalized Maxwell model. This material model was constructed based on stress relaxation measurements. The tribological properties, like the *coefficient of friction* and the *specific wear rate* also require experimental tests. The basic mechanical and tribological properties of the investigated materials are summarised in [11] and in [12]. The coefficient of friction depends on many factors, yet the lubrication has one of the most important effects. Figure 1 shows the result of a *starvation* test, which was carried out at the Institute for Composite Materials, Kaiserslautern, Germany on a shaft-on-plate test rig (Dr. Tillwich GmbH, Horb-Ahldorf, Germany). In this configuration the rubber sheet was pressed against the rotating steel shaft. The normal and the tangential force were measured online. In this experiment lubricant (Hydraulan 407-1, BASF, Ludwigshafen, Germany) was applied only at the beginning of the test. Later, as the amount of lubricant decreased between the rubber and the steel shaft, the nature of the lubrication changed. In the beginning, a mostly constant coefficient of friction with a value of $\mu = 0.06$ lets us assume *soft EHD* lubrication regime (the green section in the figure) [13]. In the next two phases, the coefficient of friction is increasing, which means more and more solid body contacts as the lubricant decreases (blue and orange). In the last phase, the coefficient of friction has an approximately constant value that is higher than one, which implies technically dry friction (red).

The presented finite element (FE) models were generated with MSC.Marc Mentat software package; the calculations were done by the MSC.Marc Solver. The developed wear simulation algorithms operate the MSC.Marc model and result files, however they are independent of the FE environment and can easily be adapted to other FE systems.

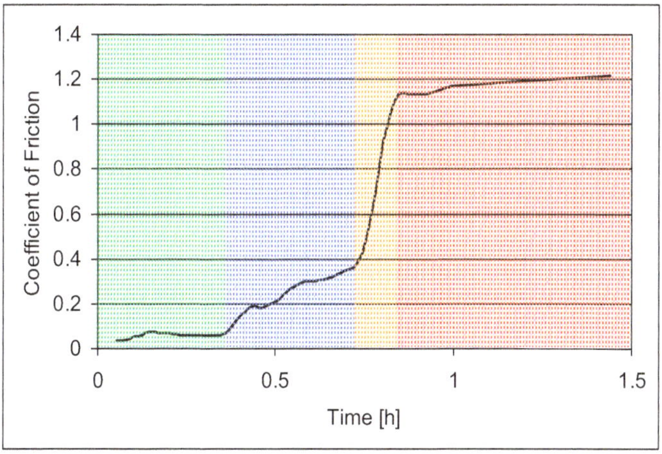

Figure 1. The coefficient of friction with starving lubrication (v = 250 mm/s F_n = 28 N)

2. Modelling internal friction of elastomers

During deformation of elastomer materials, because of the visco-elastic material behaviour, a part of the applied strain energy is transformed to heat as a result of internal friction, hysteresis [14]. The amount of this energy loss can be characterized by the loss factor (tan(δ)), which is the ratio of the loss and the storage moduli of the material. When repeated loads are present, the hysteresis contribution to the friction is more significant. In case of sliding friction between elastomer and a rough rigid counter-surface, the elastomer is subjected to repeated, cyclic deformation by the asperities of the rough surface. This kind of internal heat generation can lead to fatigue wear [15].

2.1. Modelling micro-hysteresis

In a series of numerical calculations the micro-hysteresis was investigated. First the hysteretic friction caused by a single asperity was studied. The asperity was modelled by a rigid cylinder with a radius of 250 μm (Figure 2). The asperity was pressed and rubbed against a rubber block with varying sliding speed. In this model the surface friction was neglected, so only the friction caused by the hysteresis was considered.

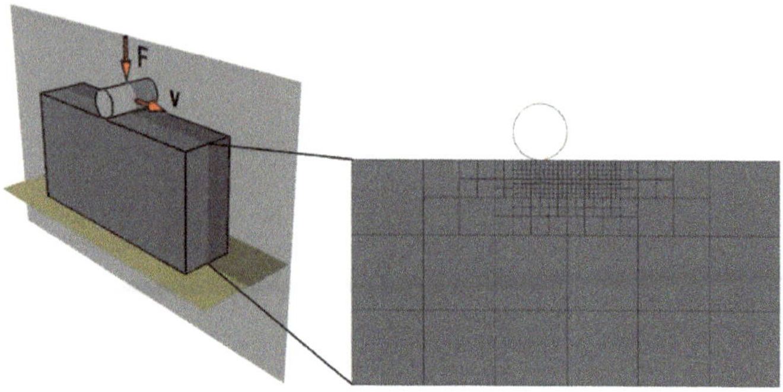

Figure 2. The model of one cylindrical asperity sliding on rubber

Figure 3 shows the results of the numerical simulations. The coefficient of friction was calculated as the ratio of the tangential and the normal reaction forces. It can be seen that the velocity dependence of the hysteretic coefficient of friction is in agreement with the measured loss factor curve.

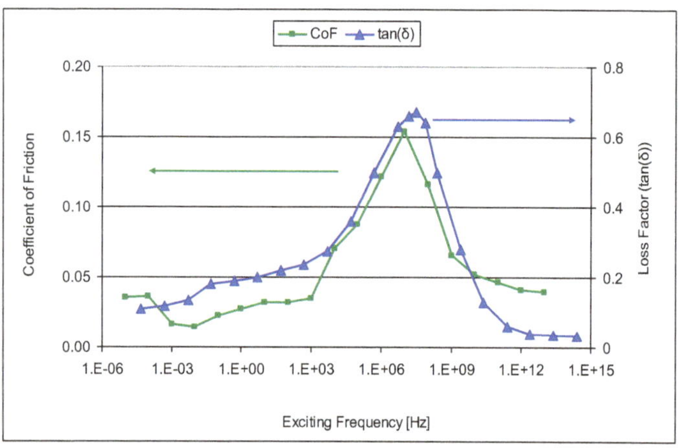

Figure 3. The hysteretic coefficient of friction vs. the exciting frequency and the corresponding loss factor curve

2.2. Modelling macro-hysteresis

A 3D finite element model was created in order to model a Pin-on-Plate test described in reference [12]. A 4 mm thick rubber sheet was modelled by 7410 3D brick elements (Figure 4). The hemispherical steel slider (r = 5 mm) was modelled as a rigid body. The rubber sheet was secured on a steel substrate, which was modelled as a rigid body too. The assumption of rigid bodies is reasonable if one considers the fact that the steel has four orders of magnitude higher elastic modulus than the rubber specimen.

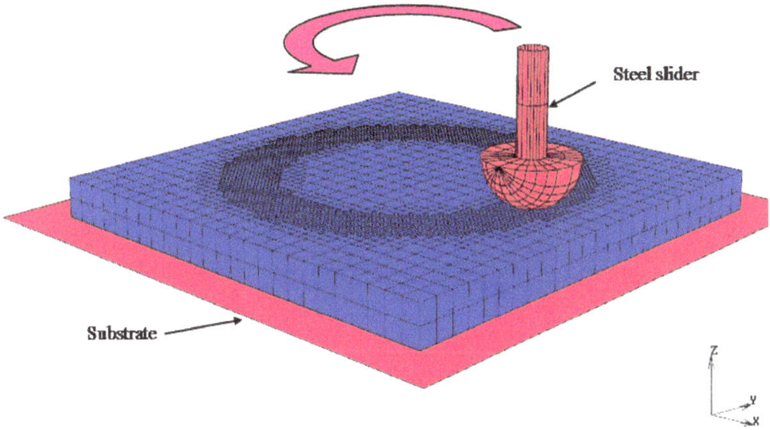

Figure 4. 3D finite element model of a Pin-on-Plate configuration (the elements are blue; the surfaces considered to be rigid are magenta). Adaptive local remeshing was applied in the contact area.

The length of the elements varied between 0.5 and 2 mm. The hemispherical counter surface was moving along a circular track of 33 mm diameter at various sliding velocities. $F_n = 100$ N normal load acted on the substrate, which pressed the rubber sheet against the slider.

No displacement was possible between the substrate and the specimen; the coefficient of friction specified between the specimen and the counter surface was $\mu = 0.06$ in order to take into consideration the friction developing on the lubricated rubbing surfaces (See Figure 1).

A series of calculations were run on this FE model with different sliding velocities (in a range between 3 and 1036 mm/s).

After the calculations, the friction corresponding to each velocity value was determined from the finite element result files. Changes in the coefficient of friction are shown in Figure 5. The figure also indicates the corresponding measured results.

A slight decrease in coefficient of friction can be observed both in the calculation and in the test results. Note that these calculations did not consider the changing viscosity of the lubricant. Any changing of the friction can be attributed to the changing of the internal friction, since the surface friction was modelled by a constant prescribed coefficient of friction. The decreasing friction can be explained by the decreasing penetration depth of the slider and the decreasing contact area. The higher the sliding velocity, the lower the penetration depth, because of the visco-elastic behaviour of the material (the rubber seems harder at higher strain rates). This causes decreasing in the excited volume, and the hysteresis part of the friction depends highly on the excited volume. Of course the hysteresis – as it was already mentioned – depends on the exciting frequency as well, however in the investigated sliding velocity range the resultant frequency (approximately from 0.3 Hz to 103.6 Hz) does not change the tan(δ) and the coefficient of friction significantly.

Figure 6 shows the contact area and the penetration depth of the counter surface for the sliding velocities examined. It can be established that the counter surface moving at a higher velocity can be pressed into the rubber surface to a lesser extent; thereby the contact area will also be smaller. There is a 0.25 mm difference between the penetration values corresponding to the lowest and the highest velocity.

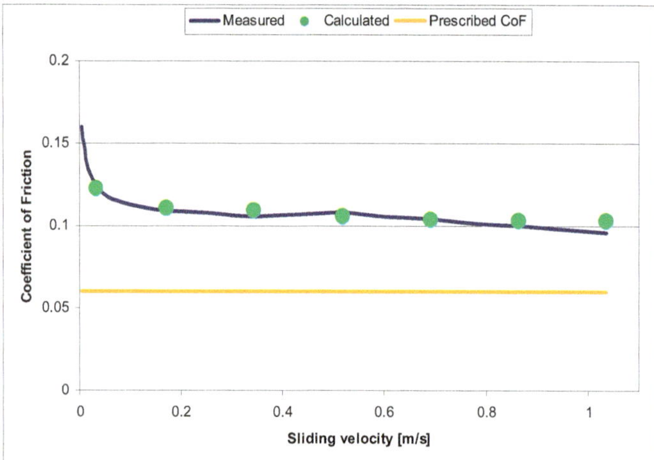

Figure 5. Coefficient of friction as a function of sliding velocity

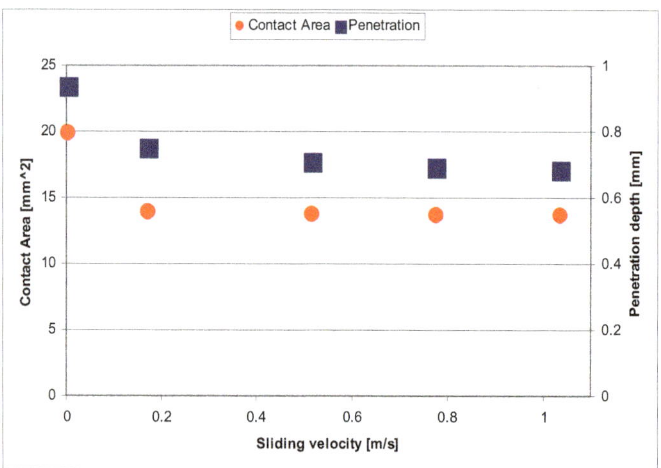

Figure 6. Contact area and penetration depth of the counter surface in function of sliding velocity

3. Modelling wear of elastomers

Three different wear simulation approaches are presented in the followings. The first one is the method of moving the contact nodes according to the nodal wear increments as proposed by Pödra and Andersson in [5]. It will be shown, that this method is suitable for modelling wear, which is much smaller than the elements in the FE model. The second and the third methods can model relatively large wear by global remeshing and by deactivation of certain elements, respectively.

3.1. Wear simulation by moving the nodes

The wear simulation method of Pödra and Andersson was adapted to model the wear process of an elastomer part considering its non-linear material properties. The simulation process is shown in Figure 7. In the modelled configuration a rubber sheet (thickness: 2 mm) was pressed against a rotating steel cylinder (diameter: 10 mm) with a normal force of 16 N. The rubber part was modelled in 2D assuming plain strain conditions (Figure 8). The rotating steel cylinder was modelled as a rigid body. A coefficient of friction with a value of 1 was prescribed between the rubber and the cylinder, according to the dry wear test results.

After the FE contact calculation, the wear increments were determined based on the stress distribution. The Δh_i nodal wear of the i-th node in the contact area by Archard [16] is defined as:

$$\Delta h_i = W_s p_i v \Delta t \tag{1}$$

where

W_s is the specific wear rate,

p_i is the nodal contact pressure at node i,

v is the sliding velocity,

Δt is the time increment.

In the next step of the simulation the contacting nodes were moved with respect to the nodal wear increments. Then the cycle was restarted with a new contact calculation using the modified geometry.

Figure 9 shows the result of the wear simulation after one minute of sliding. The profile of the worn specimen is also plotted in the figure. It can be seen that the wear is not symmetric. The simulation and the experimental results are in good agreement both by the means of the amount of wear and the worn shape. However, in longer term simulations and with finer FE mesh, this method failed to accurately estimate the experimental results.

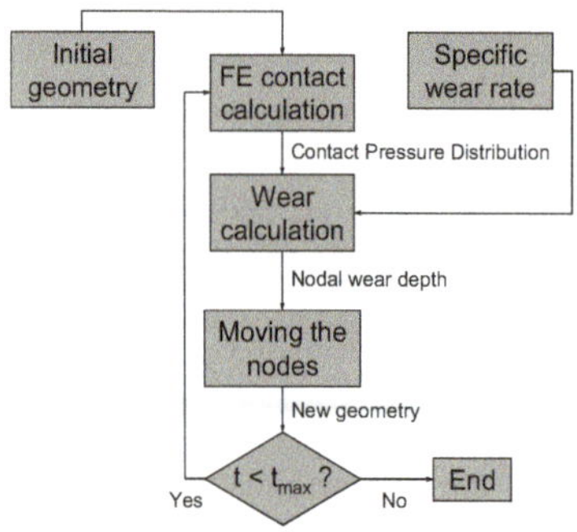

Figure 7. The flowchart of the wear simulation

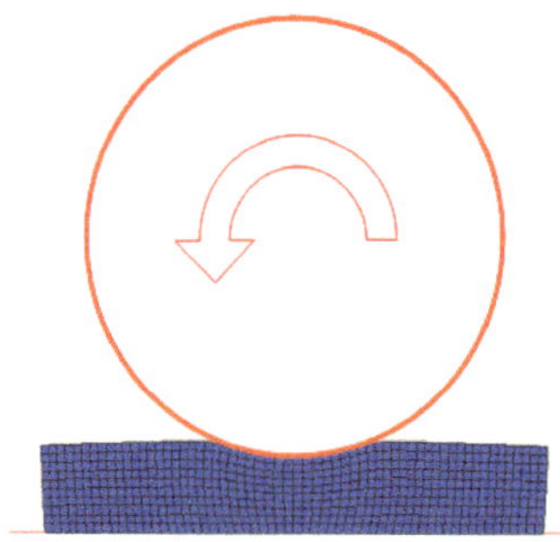

Figure 8. The finite element mesh of the shaft-on-plate configuration

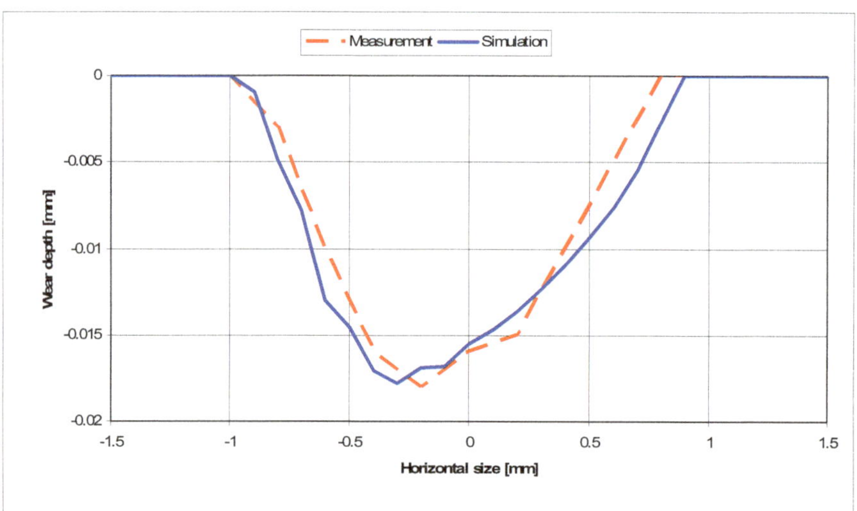

Figure 9. The calculated and measured worn profile of the rubber sheet ($t = 60$ s, $F_N = 16$ N, $\omega = 50$ rad/s)

Despite the widespread use of the Pödra-Andersson method, the wear simulation is highly limited; only the top layer of elements can be worn. This limitation does not affect the usability of the method in case of relatively hard materials such as metals or ceramics, since minimal wear can have a great effect on the contact area and pressure distribution and thus on the performance of these parts (e.g. bearings). In case of rubber parts, wear in the magnitude of some μm does not change the contact conditions significantly. The wear needed for a rubber part to malfunction is much greater than those of metal parts; therefore the wear of the top layer is insufficient in rubber applications. To increase the volume of the wear to be modelled, the size of the elements can be increased; however it is not recommended, since it decreases the accuracy of the calculation. In the followings two methods are proposed to simulate wear regardless of the element size. It will be shown that these methods can model the material loss due to wear even in a scale comparable with the size of the part.

3.2. Wear simulation by global remeshing

In order to model wear that is larger than the elements of the FE mesh, a wear simulation procedure was developed using global remeshing. To demonstrate this method a wear simulation study of a reciprocating sliding seal (Figure 10) is presented. In the investigated application a rubber seal is coupled with an aluminium rod (diameter: ø22.2 mm). Since this method integrates the FE mesh generation, the geometric model is the starting point of the wear simulation. The seal was modelled assuming axisymmetry.

The counter-surface was modelled as rigid body, since its elastic modulus is some orders of magnitude larger than the elastomer material of the seal. The seal was mounted axially by fitting it in the housing, which was considered ideally rigid. The rod was moving with alternating motion at a speed of 20 mm/s with 9 mm amplitude. The contour of the seal section was modelled by line segments (Figure 11). A fine approximation in the vicinity of the lip as well as the ridges was required; in the region of the ridges and the lip the average length of the line segments was 10 μm.

The seal section was meshed by the built-in automatic meshing procedure using three-node axisymmetric triangular finite elements (Figure 12). In the FE calculations the interference fit between the seal and the rod was also taken into consideration: the unloaded inner diameter of the seal was 1.8 mm smaller than the diameter of the rod. This caused the seal to be stretched in the first load step. In the next step the alternating motion was modelled considering the friction during the wear simulation. According to the working conditions, in the inward strokes of the rod a pressure load of $p = 2$ MPa was applied on the right side of the seal, modelling the sealed pressure. There was no pressure applied in the outward strokes. Between the seal and the rod a prescribed coefficient of friction with a value of $\mu = 0.1$ was defined in order to model the lubricated friction.

In the wear simulation process, the FE mesh was created based on the line segments describing the geometry. The automatic mesher was set to create elements of different sizes, so the elements in the vicinity of the contact zone were smaller than those deep inside the material. In the initial mesh approximately 4000 elements were distributed in a way that the regions, that are important for the wear calculations, were modelled more accurately with the small elements. Even so the total number of the elements remained in a reasonably low range, so the CPU time for one calculation was short enough to handle the iterative simulation process.

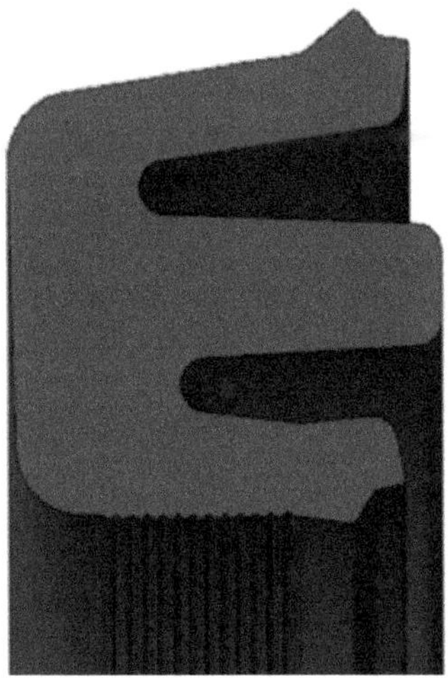

Figure 10. Section view of the investigated reciprocating seal

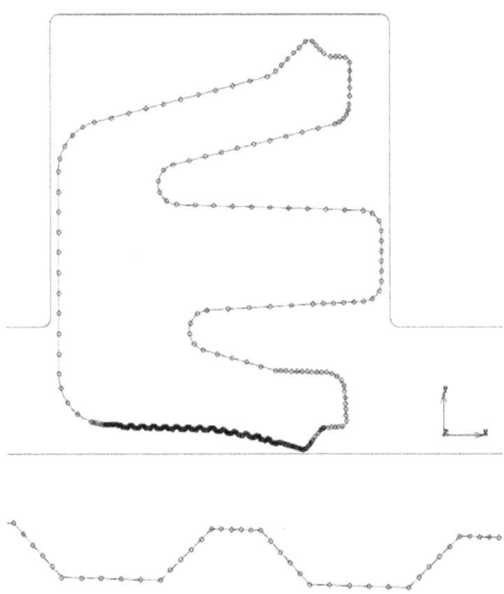

Figure 11. The geometry of the seal consisting of line segments and points

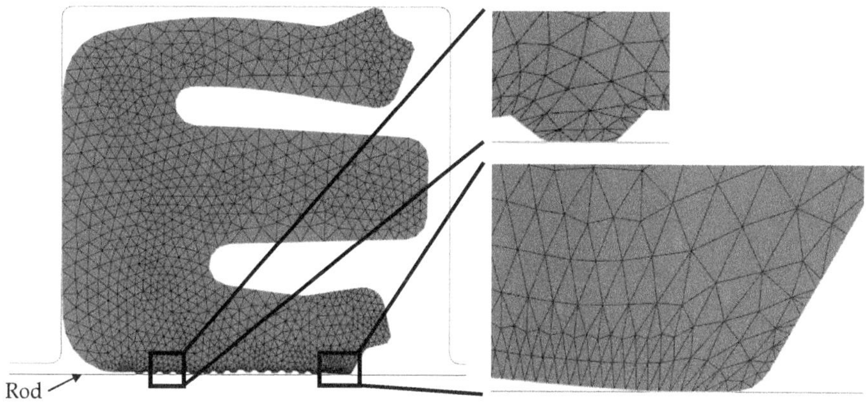

Figure 12. Deformed shape of the seal in the first outward stroke (without pressure). The curves represent the rigid bodies (the rod and the housing). The element size varies from 10 to 200 μm

The flowchart of the simulation can be seen in Figure 13. In the first step the FE mesh was created based on the initial geometry. After adjusting the simulation parameters (coefficient of friction, sliding velocity, material properties, applied pressure, time increment, etc.) the contact calculation was run. In the first stroke the rod was moving rightward. The wear calculation was performed using Equation 1. The moving of the contacting nodes according to the nodal wear increments was the next step, however in a slightly different way. The surface nodes were attached to the contour points of the seal, so if a point was moved the attached node would also move with respect to the nodal wear value of the attached node. After moving the points, the whole FE mesh was deleted. Based on the new geometry the automatic mesher created the new elements and the cycle ran over again. The direction of the rod motion was changed in each cycle, so one simulation cycle represented one outward (left) or one inward (right) stroke. The applied pressure was also changed stroke by stroke, as for the inward strokes the working pressure of 2 MPa was applied, while in the outward strokes the working pressure was turned off in the simulation process. 200 simulation cycles were calculated in the frame of this study.

Figure 14 shows the deformed shape of the seal in the first outward stroke, when the seal was not pressurized. The resulting contact pressure comes from the stretching of the seal on the rod. It can be seen that the contact area is reduced: only five ridges on the left side, one on the right side and the lip are in contact with the rod. The maximum of the contact pressure can be observed at the lip. The values of the contact pressure are generally low, since there is no working pressure applied to press the seal surface to the rod.

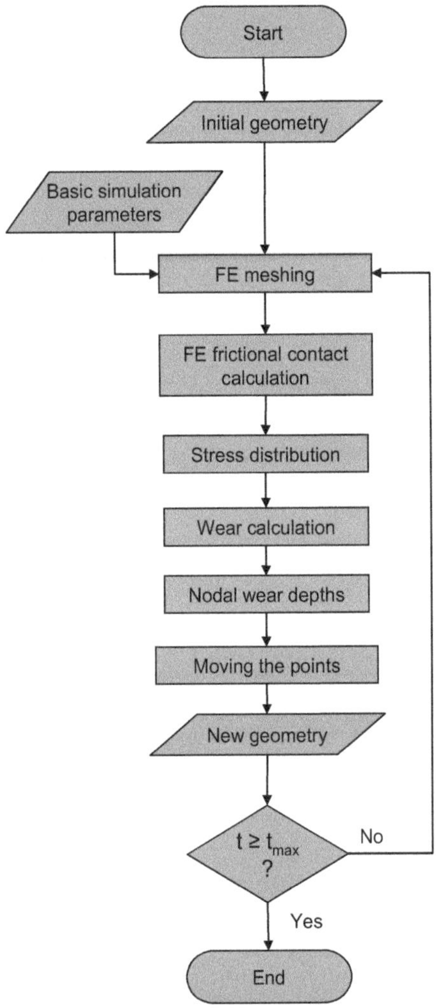

Figure 13. The flowchart of the wear simulation process

The calculated worn profiles of the seal are shown in Figure 15 after 0, 100 and 200 cycles of the simulation. It can be seen that the left side ridges of the seal and the lip edge wear first. Figure 15 also shows the contact pressure distributions in the investigated wear phases. Note that the FE mesh is different in each state, since the model was remeshed in every cycle; however the element sizes are distributed identically. It should be mentioned that the total simulated wear depth reaches 30 μm in some regions (e.g. at the lip), which is three times higher than the size of the elements in the contact area. This wear depth is so large that the Pödra-Andersson method can not be applied here.

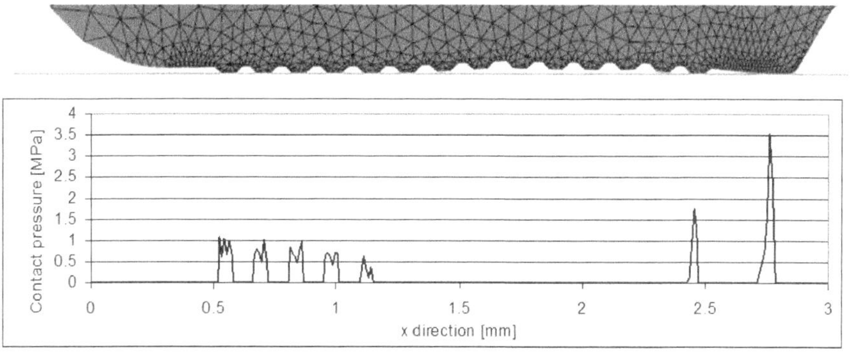

Figure 14. Deformed shape of the seal in the first outward stroke (without applied pressure) and its contact pressure distribution.

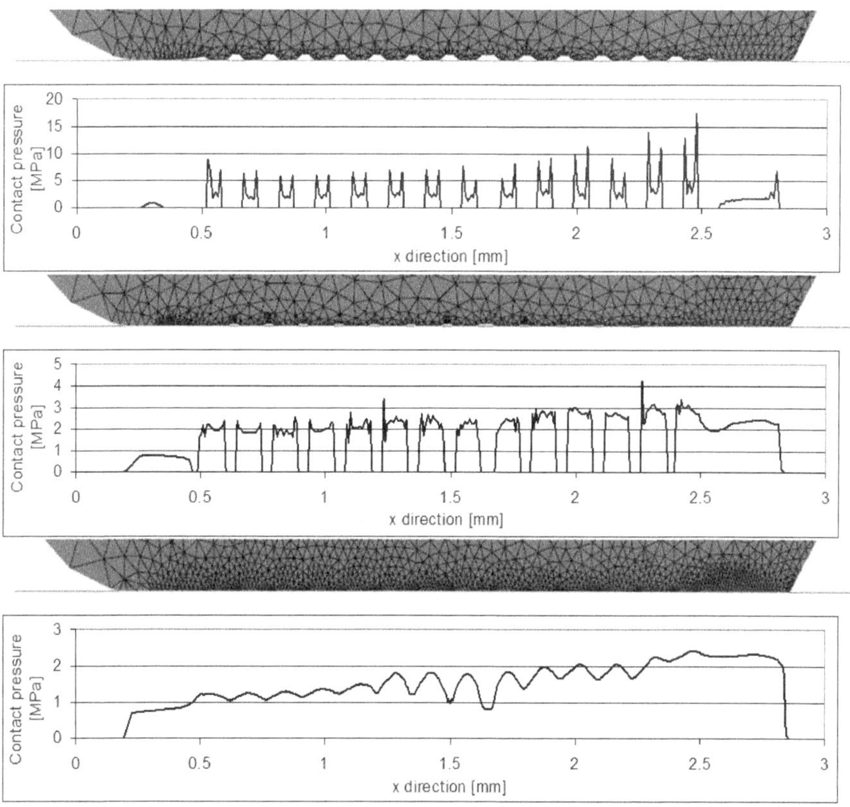

Figure 15. The calculated worn profiles after 0, 100 and 200 cycles of the simulation in inward strokes (p = 2 MPa) and the corresponding contact pressure distributions along the axis

Based on the change of contact pressure distribution the followings can be established. The area of contact increased, while the maximum values of the contact pressure decreased. This phenomenon can also be seen in Figure 16. for the contact pressure distribution of the lip over the wear process. One can see that the seal at the beginning has about three times higher contact pressure than later. The pressure reduction is caused by the wear process, which also increases the contact area. It is remarkable that at the left side of the lip the contact pressure is increasing at first, and then slightly decreasing. It can be explained by the wear of the lip. As the lip has the highest contact pressure in the beginning of the simulation, it will suffer the most severe wear in the first cycles. As the lip wears, the contact pressure distribution becomes flatter, which causes the slight increase in the lower pressure region. Later this mostly uniform contact pressure remains, only the average pressure decreases.

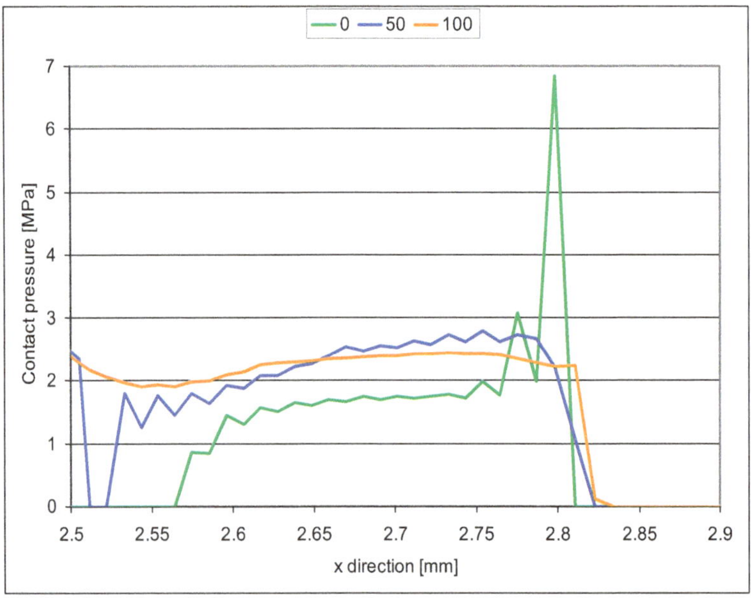

Figure 16. The change of the contact area and the contact pressure distribution at the lip over the wear process (after 0, 50 and 100 simulation cycles)

3.3. Wear simulation by deactivation of elements

In hydraulic systems the seals are well lubricated in most cases, so those parts show minimal wear and they can function for years without any sign of damage. However, in some special cases the rubber seals can work under mixed or boundary lubrication conditions. It is suspected that sometimes (typically after the machine is stopped for a longer period) the seal can squeeze the lubricant out of the space between the surfaces and some parts come to direct contact. If it happens, adhesion can occur, which significantly increases the frictional force and the wear mechanism would change completely. The different wear mechanisms of rubbers are examined by several studies e.g. in [17, 18]. In general it can be said that dry sliding wear of rubber can be traced back to the initiation and propagation of cracks in the rubber material. The modelling of this type of wear requires advanced simulation techniques.

A popular approach for modelling material detachment is the deactivation or deleting elements from the FE mesh (sometimes also referred as element death) based on the characteristic damage criterion of the given material. It is commonly used for simulation of manufacturing processes e.g. in [19, 20]. The principle of wear – from the simulation point of view – is similar to the manufacturing processes. Authors in [21] modelled the particle detachment process based on the strain history of the elements in dry sliding.

The wear process of an experimental test of the sliding seal described above was modelled. The seal was cut and a section of it was straightened and fixed in the clamping. The aluminium rod was then pressed and rubbed against the surface of the seal. The detailed description of the test can be found in [22]. In the series of wear tests, different lubrication conditions were studied, and it was found that when the amount of lubricant was decreased, the seal surface worn rather intensively. The surface of the seal was inspected with a MicroProf white light profilometry device (WLP, Fries Research & Technology GmbH., Bergisch Gladbach, Germany) before and after the test, so the wear of the seal can be evaluated and compared with the simulation results (See Figure 17). One can see that the most wear occurs in the symmetry plane of the seal, where the contact pressure was maximal; this region shows a rough surface texture, which is probably caused by the wear particles that were not removed from the contact region, since there are peaks that are higher than the original surface. In order to avoid the false results caused by the wear debris, a plane with slightly less but still significant wear was investigated, which was clear of wear debris (indicated in Figure 17).

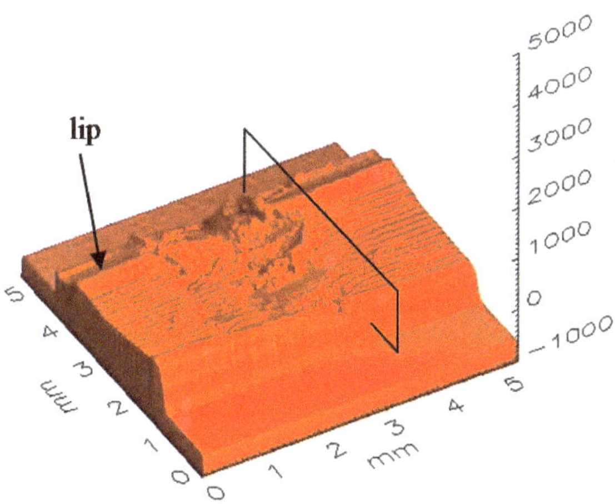

Figure 17. Worn surface of the seal

The previous wear simulation methods were not suitable to model the compound wear mechanism, which was caused by boundary lubrication. Beside the wear simulation method of moving the nodes, the deactivation of elements was integrated in the simulation procedure. The flowchart of the simulation can be seen in Figure 18.

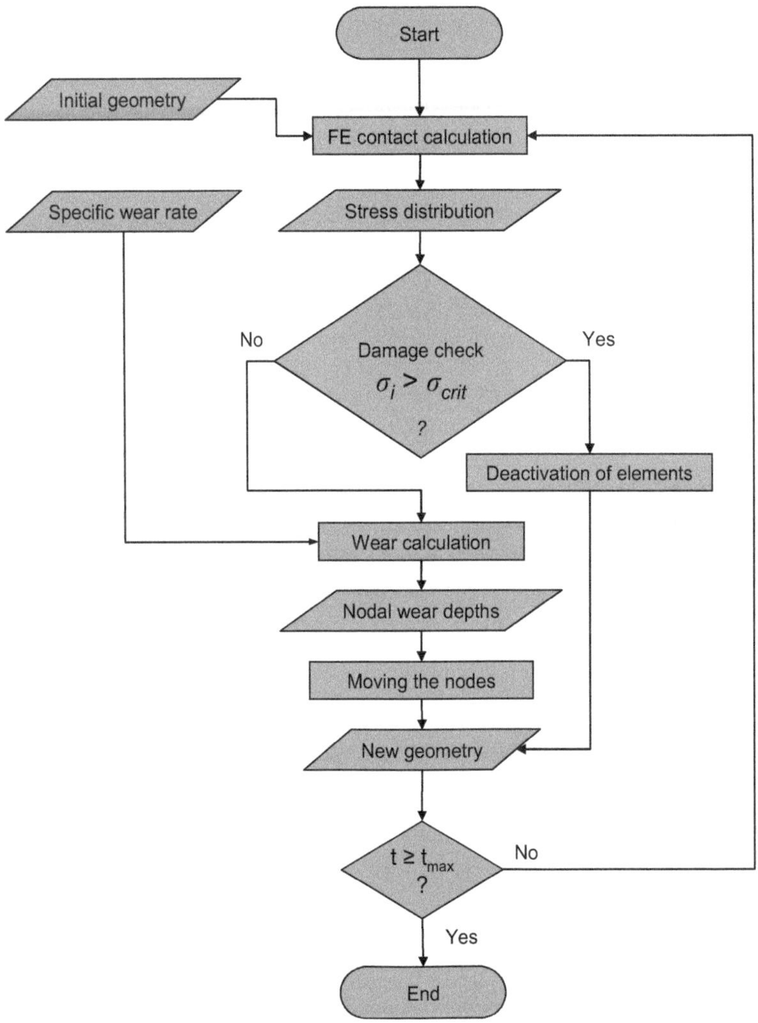

Figure 18. Flowchart of the simulation [22]

The test can be divided into repeated cycles of the inward (right) and outward (left) strokes. Since the stress distributions in the inward and outward strokes are different, the wear calculations were performed four times in every cycle (see Figure 19).

Starting with the initial geometry a frictional contact calculation is carried out assuming inward stroke. Based on the results, the normal stress of each element in the model was compared to the predefined critical stress limit, the tensile strength, which was 19.0 MPa for the material of the seal. If the tensile stress of an element was greater than the critical value, the element became deactivated. The damage check was done in every element in the following way. If $\sigma_i > \sigma_{crit}$ (where σ_i is the normal stress in the i-th element, and σ_{crit} is the ultimate tensile strength of the material), the element i became deactivated.

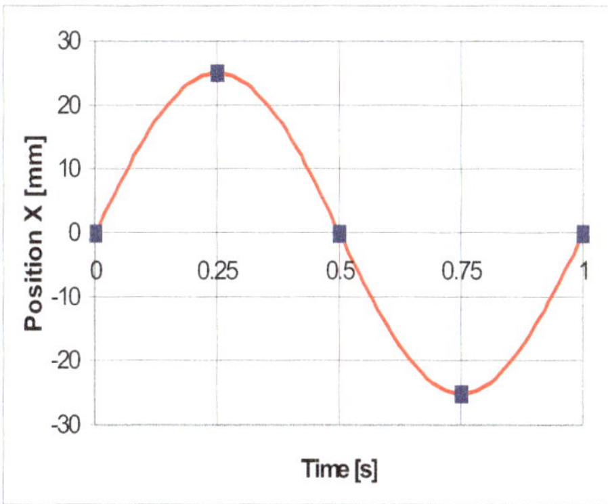

Figure 19. The motion of the rod (continuous red) and the points of the wear calculations in a cycle (blue dots)

Wear calculation was applied to determine the nodal wear of the nodes in the contact area using Equation (1). Once the nodal wear values were known, the nodes, which were in contact, were moved to their new position as in the Pödra-Andersson technique.

The seal section was modelled by four-node 2D finite elements assuming plain strain condition. The housing, the clamping, which fixed the seal to the housing, and the counterpart, was considered ideally rigid. In the first load step of the FE calculation, the seal was mounted in the holder, and the normal load was applied to the counterpart, according to the wear test. After this, the rod started its motion. There was a prescribed coefficient of friction between the seal and the rod with a value of $\mu_{lub} = 0.3$, representing the boundary lubricated friction; and another between the seal and the holder parts with a value of $\mu_{dry} = 1$, since the friction condition was dry in this case. Values of coefficient of friction in case of lubricated and dry friction were determined based on the results in [12] and [11], respectively. Note that the lubrication was taken into consideration only by the lower value of coefficient of friction. Other effects – like the lubricant separating the surfaces reducing real contact and wear – were neglected. The FE mesh with the ideally rigid parts and with the applied prescribed coefficients of friction is shown in Figure 20.

The calculated worn profiles are shown in Table 1. It can be seen that the ridges of the seal and the lip edge wear first, which is in accordance with the experimental findings. One can see that during the wear simulation the wear occurs not only in the top layer, but in several layers of elements. The maximal wear depth is about ten times bigger than the typical element edge length in the contact area.

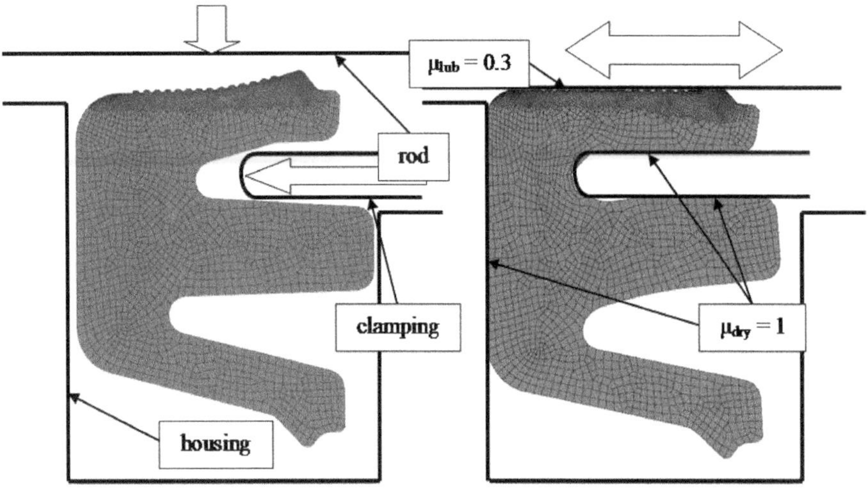

Figure 20. FE model of the seal section before (left) and after mounting in the holder (right)

The height of the elements in contact gets reduced due to wear and over time the contacting elements become more and more distorted and can even turn inside out, which can make the calculations unable to run. To avoid this and to maintain the stability of the simulation process, the following procedure was implemented. A filter was applied in the simulation algorithm, which checked the distortion of the elements before each FE calculation. Distortion in the FE software is determined at each node of the element as the cosine of the angle between the element edges at the nodes. The distortion of the element is defined as the maximum of these cosines. If the normal vector to the element changes sign from one node to the next (inside-out element), the distortion is one. If the distortion value of an element exceeded a critical value (0.99) before the load was applied, the element got deactivated. With this method the inside-out elements and the elements that were too flattened due to the wear, were became deactivated.

The wear process of one ridge can be tracked in the successive images of Figure 21. One can see the effect of the wear algorithm on the height of the elements. The dark shaded elements were deactivated by the damage check module of the simulation.

The measured and the calculated worn profiles of the seal are compared in Figure 22. The calculated worn profiles are mostly identical with the measured worn surfaces. However, in the vicinity of the lip, the simulation over predicts the wear. It is suspected that the deviation is caused by the assumption of linear wear theory in the wear calculations, while in reality wear is a non-linear function of contact pressure. This simplification nevertheless appears to be suitable for the regions with less contact pressure.

No. of cycles	Sliding time [min]	Calculated worn profile
0	0	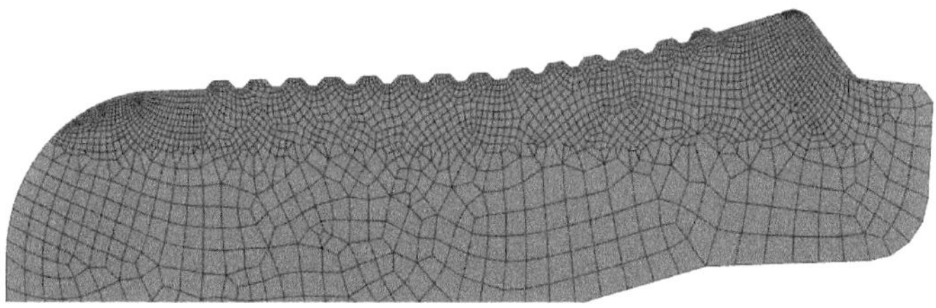
600	10	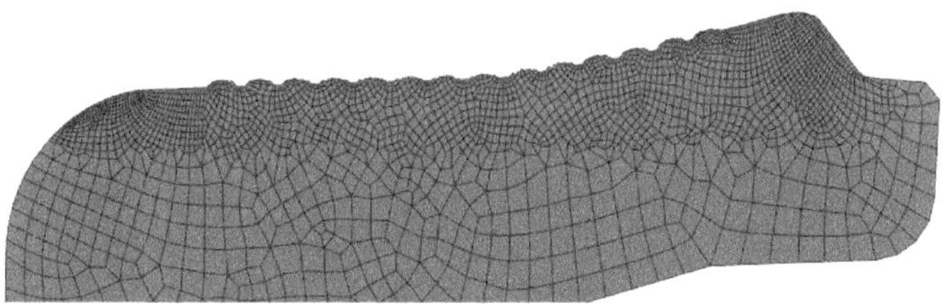
1200	20	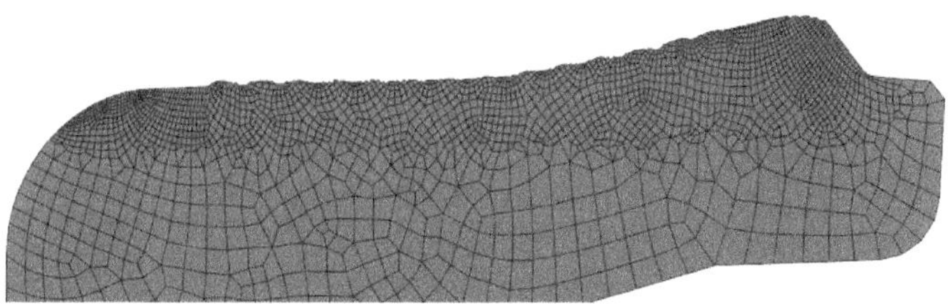
3600	60	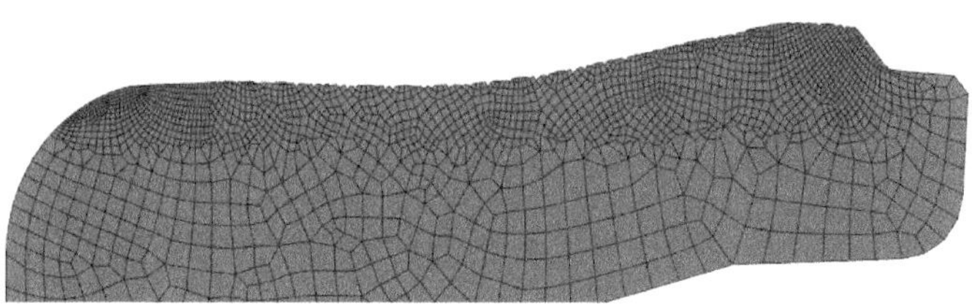

Table 1. Evolution of the calculated worn profile (The deactivated elements are not plotted.)

Figure 21. The wear process of one ridge of the seal. The dark elements are deactivated by the damage check algorithm. The images show the left ridge of the seal model after 0, 15, 30, 45 and 60 minutes of sliding, respectively.

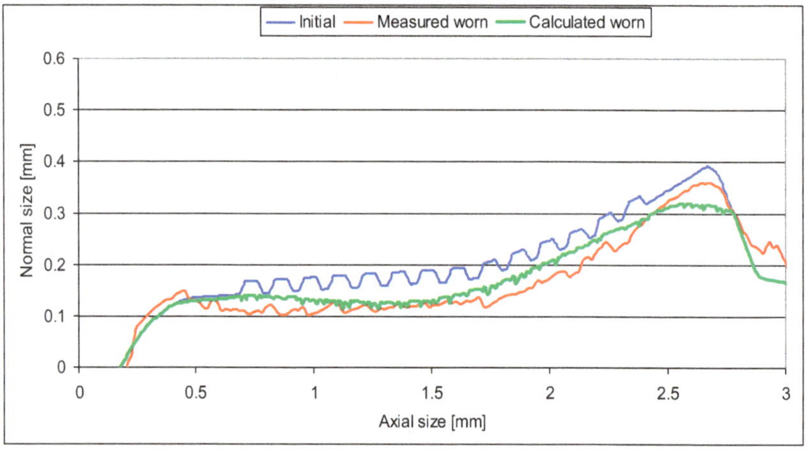

Figure 22. The measured and calculated worn profile of the reciprocating seal

The developed wear simulation algorithm was able to model the wear process even in the case, when the amount of wear was much larger than the size of the elements used in the FE calculations.

4. Conclusion

Friction of elastomers has a hysteretic part, which is caused by the energy dissipation in the visco-elastic material. The hysteretic friction depends mainly on the sliding velocity and the geometry. It was shown that the effect of sliding velocity on the coefficient of friction can be modelled numerically both in macro and in micro level. Furthermore, the penetration depth of the counter surface and the size of the contact area can be modelled as functions of sliding velocity.

Three different wear simulation techniques were presented for different purposes. The first method can be useful for modelling relatively small amount of wear. The second can model wear even bigger than the elements in the model by global remeshing. The third method deactivates the damaged elements, thus the effect of the surface rupture of the elastomer part can be modelled.

Author details

Nándor Békési
Department of Machine and Product Design,
Budapest University of Technology and Economics, Budapest, Hungary

5. References

[1] Persson, B.N.J., A.I. Volokitin (2006) Rubber friction on smooth surfaces, The European Physical Journal E 21: 69-80
[2] Moore, D.F. (1972) The friction and lubrication of elastomers. Oxford, New York: Pergamon Press
[3] Serre, I., Bonnett, M., Pradeilles-Duval, R. (2001) Modelling an abrasive experiment by the boundary element method, Comptes Rendus de l'Académie des Sciences Paris Série II b 329: 803-808
[4] Fillot, N., Iordanoff, I., Berthier, Y. (2005) Simulation of Wear Through Mass Balance in a Dry Contact, ASME Journal of Tribology 127: 230-237
[5] Pödra, P, Andersson, S. (1999) Simulating sliding wear with finite element method, Tribology International 32: 71–81, doi:10.1016/S0301-679X(99)00012-2
[6] Archard, J.F. (1953) Contact and rubbing of flat surface, J Applied Physics. 24: 981–988
[7] Söderberg, A., and Andersson, S., (2009), Simulation of Wear and Contact Pressure Distribution at the Pad-to-Rotor Interface in a Disc Brake Using General Purpose Finite Element Analysis Software, Wear, 267: 2243-2251
[8] Ashraf, M. A., Sobhi-Najafabadi, B., Göl, Ö., and Sugumar, D., (2009) Numerical Simulation of Sliding Wear for a Polymer-Polymer Sliding Contact in an Automotive Application, International Journal of Advanced Manufacturing Technology, 41: 1118-1129
[9] Kónya, L, Váradi, K. (2008) Wear simulation of a polymer–steel sliding pair considering temperature- and time-dependent material properties In: Friedrich K., Schlarb A.K. editors. Tribology of Polymeric Nanocomposites, Tribology and Interface Engineering Series, Volume 55, pp. 130-145
[10] Pálfi L., Goda T., Váradi K. (2009) Theoretical prediction of hysteretic rubber friction in ball on plate configuration by finite element method, Express Polymer Letters 3:11: 713–723 doi: 10.3144/expresspolymlett.2009.89
[11] Karger-Kocsis J, Mousa A, Major Z, Békési N (2008) Dry friction and sliding wear of EPDM rubbers against steel as a function of carbon black content, Wear 264:3-4: 359-367.
[12] Békési N., Váradi K. (2009) Experimental and numerical evaluation of lubricated friction of EPDM rubber, Periodica Polytechnica Mechanical Engineering 53:2: 1-5.
[13] Hamrock, B.J. (1994) Fundamentals of Fluid Film Lubrication, New York: McGraw-Hill
[14] Persson B.N.J. (1998) On the theory of rubber friction, Surface Science 401: 445–454
[15] Zhang S.-W. (2004) Tribology of Elastomers, Tribology and Interface Engineering Series, 47, Amsterdam: Elsevier

[16] Archard J.F. (1953) Contact and rubbing of flat surface, J Appl. Phys., 24: 981–988
[17] Schallamach, A. (1971) How does rubber slide? Wear 17: 301-312
[18] Fukahori, Y., Yamakazi, H. (1994) Mechanism of rubber abrasion. Part I: Abrasion pattern formation in natural rubber vulcanizate, Wear 171: 195-202
[19] Yan, J., and Strenkowski, J. S., (2006), A Finite Element Analysis of Orthogonal Rubber Cutting, Journal of Materials Processing Technology, 174: 102-108
[20] Ko D-C., Kim B-M., Choi J-C. (1997) Finite-element simulation of the shear process using the element-kill method, Journal of Materials Processing Technology, 72:1: 129-140, doi: http://dx.doi.org/10.1016/j.bbr.2011.03.031
[21] Eleőd A., Devecz J., Balogh T. (2000) Numerical modelling of the mechanical process of particle detachment by finite element method, Periodica Polytechnica Transportation Engineering 28:1-2: 77-90 (available:http://www.pp.bme.hu/tr/2000_1/tr2000_1_07.html)
[22] Békési N., Váradi K., Felhős D. (2011) Wear simulation of a reciprocating seal, Journal of Tribology 133:3: 031601-1-031601-6, doi: http://dx.doi.org/10.1115/1.4004301

Chapter 14

Anelastic Behavior in Filled Elastomers Under Harmonic Loading Using Distributed Rate-Dependent Elasto-Slide Elements

Wei Hu and Norman M. Wereley

Additional information is available at the end of the chapter

1. Introduction

Elastomeric materials are broadly used for stiffness and damping augmentation in various applications due to their simple design, low weight, and high reliability. However, filled elastomeric materials exhibit significant nonlinear behavior [1]. Such nonlinear characteristics include a) non-elliptical shear strain vs. stress hysteresis diagram under sinusoidal excitation, b) stiffness and damping dependent on amplitude, frequency, temperature and even preload, c) Mullins effect with reduction of stiffness at small strain following cyclic deformation at large strains, and d) low stress relaxation and creep rates. Specifically, the large reduction of damping with increasing amplitude of harmonic displacement excitation leads to excessive size and weight of dampers in order to accommodate all operating conditions. It was also found that highly damped elastomeric dampers demonstrated low loss factors at low amplitudes resulting in unacceptable limit cycle oscillations [2]. Therefore, a precise analytical model is necessary to describe the nonlinear behavior of an elastomer and to determine its dynamic characteristics.

Some prior research introduced nonlinear terms into conventional Kelvin or Zener models. The complex modulus is usually used to characterize viscoelastic materials under harmonic excitation: the storage (or in-phase) modulus is a measure of the energy stored over a cycle or period of oscillation, and the loss (or quadrature) modulus is measure of the energy dissipated over a cycle. This model can be represented as a spring and a dashpot in parallel (Kelvin chain). However, the complex modulus is a linearization method in the frequency domain that represents the nonlinear hysteresis cycle as an equivalent ellipse, and is only applicable to steady harmonic forced response analysis. Some researchers have extended the basic Kelvin chain to complicated mechanism-based modeling approaches in order to explain the nonlinear behavior of elastomers. To display basic behavioral characteristics, such as creep and relaxation, a viscoelastic solid can be represented as a spring in series with Kelvin elements (if one Kelvin element is used, then a Zener model results [3]). Gandhi and Chopra [4] developed a nonlinear viscoelastic solid model in which a nonlinear lead spring was used in series with a single linear Kelvin chain. Using this model, the variation of complex moduli with oscillation amplitude closely matched experimental data. Felker *et al.*. [5] further developed a nonlinear complex modulus model based on a single Kelvin chain, in which the spring force was a nonlinear function of the displacement, and the damping force was a nonlinear function of displacement and velocity. This model was used to describe the amplitude dependent moduli and to study dual frequency damper motions. The parameters in all of these models were identified using amplitude-dependent complex modulus data. As a result, the nonlinear hysteresis behavior of could not be captured using these models.

Elastomeric materials typically demonstrate tribo-elastic behavior [1]. Physically, the amplitude dependent behavior exhibited by the elastomer is results from the interaction of the fillers with the rubber or elastomeric matrix materials [6]. Before large deformation of a filled elastomeric damper, an intact filler structure displays a large stiffness and low loss factor for small amplitudes. As the amplitude increases, the filler structure breaks resulting in a stiffness reduction. However, the breaking of filler structures, which is similar to friction, increases the loss factor. As amplitude increases further and the frictional effect is fully manifested, both stiffness and loss factor are further reduced, which are then maintained relatively constant by the remaining polymer chains. Motivated by the above physical mechanisms, a number of researchers have combined springs and frictional slides to represent the filler and rubber compound in filled rubbers or elastomers.

In the model developed by Tarzanin *et al.* [7], the elastomeric behavior was represented by a nonlinear spring and a nonlinear Coulomb friction damper. This model was based on single frequency elastomer data and matched the value of energy dissipation per cycle. Panda *et al.* [2] replaced the Coulomb friction damping element with a variable friction damping element whose force was calculated based on the peak displacement of excitation when the velocity was zero. This model correlated well with the experimental hysteresis data, but the effectiveness of this model over a range of amplitudes and frequencies has not been demonstrated in the literature. As early as 1930, Timoshenko [8] suggested that general hysteretic systems consist of a large number of ideal elasto-plastic elements with different yield levels. Iwan [9, 10] further developed a distributed-element model to study the steady-state dynamic response of a hysteretic system. Instead of specifying a distribution function numerically to agree with experimental data, a constant band-limited statistical function was used to define yield properties of the slide elements in this model. This model proved successful in predicting steady-state frequency response of a hysteretic system using a method of linearization. However, the excitation amplitude must be known as *a priori* while this model is applied in the analysis, which is only applicable in steady-state response prediction. The theory of triboelasticity [11] also stated that the behavior of a filled elastomer can be represented by a large or infinite number of alternate springs and frictional slides in series, and each slide has a constant yield force and each spring has a constant stiffness.

Coveney *et al.* [1] developed a three-parameters standard triboelastic solid (STS) model based on the theory of triboelasticity and further developed a four-parameters rate-dependent triboelastic (RT) model. These models gave a satisfactory representation of the material behavior. However, because the yield force was fixed along different slides, these models showed less flexibility in representing the amplitude dependent behavior for different filled-level materials.

Alternative elastomeric models were developed using internal variable or nonlinear integral equations. Strganac [12] used a stress shift function to formulate a nonlinear time domain model for elastomers, but the nonlinear integral formulation in the model was difficult to implement in traditional aeromechanical analysis. Lesieutre and Bianchini [13] developed the anelastic displacement field (ADF) method to describe the frequency-dependent behavior of viscoelastic materials. It was based on the notion of scalar internal variables or augmenting thermodynamic fields (ATF) [14] that described the interaction of the displacement field with irreversible processes occurring at the material level. In the ADF approach, the effects of the thermodynamic processes were focused on the displacement field, which consists of both elastic and anelastic fields. The anelastic part may be further subdivided to consider the effects of multiple relaxation processes. Although there is no explicit physical interpretation when multi-anelastic elements are involved, one single ADF model is mechanically analogous to the Zener model. In order to capture the characteristic nonlinear hysteretic behavior of elastomeric materials, Govindswamy *et al.* [15] developed a nonlinear ADF model, in which the linear ADF parameters were replaced with nonlinear terms. The model captured the variations of the complex modulus with amplitude, and performed as well in matching the strain vs. stress hysteresis cycle. Furthermore, other functional forms for the ADF parameters were introduced in order to improve hysteresis loop predictions. Brackbill *et al.* [16] improved the nonlinear ADF model by adding rate independent nonlinearity, in which friction-damping and linear-spring elements in parallel with the baseline nonlinear ADF model were used to provide additional amplitude dependent relaxation behavior. As many as sixteen parameters were used to construct the model. Although the complex moduli were fitted well in certain amplitude and frequency ranges, the performance of the model in predicting nonlinear hysteresis behavior could still be improved. Moreover, the process to determine model parameters was complicated by the fact that some parameters were chosen by empirical observation. In a recent study, Ramrakhyani *et al.* [17] developed an ADF based model containing nonlinear fractional derivatives and frictional elements. This model used eight parameters instead of sixteen parameters to capture the amplitude-dependent and mild frequency-dependent modulus. However, the prediction of hysteresis loops was not noticeably improved, and the determination of model parameters remains complicated.

In this chapter, a hysteresis model is introduced to study the anelastic behavior of elastomers under harmonic excitation. Using this model, the behavior of the elastomer is analogous to the behavior of a nonlinear Kelvin element, in which the stiffness is a nonlinear monotonic function of displacement and the damping is a monotonic rate dependent hyperbolic tangent function. Although, the hysteresis model can capture the hysteresis behavior of the elastomer, the use of simple nonlinear spring and damping elements are not sufficient because model parameters, at the very least, are amplitude dependent. Thus, a computationally efficient and precise model for the elastomer is developed from a profound understanding of the damping mechanism within the damping material. As mentioned before, the anelastic behavior demonstrated by the elastomer is mostly based on the interaction between fillers and rubber compound inside the filled elastomeric materials. Based on this physical mechanism, a non-uniform distribution of rate-dependent elasto-slide elements is used to emulate filler structure behavior and a parallel linear spring (and linear viscous damping) is used to represent the remaining polymer stiffness (and damping). Extensive testing including single frequency and dual frequency testing is conducted for material characterization, identification of model parameters and validation of the model. It is shown that this material model can be used for damping element design and can be integrated into numerical analysis for dynamic systems. It is also shown that the non-uniform distributed rate-dependent elastomer model is applicable in complex loading conditions without any *priori* information, and the flexibility in determining model parameters provides a potential to improve the model and to apply the model for elastomers with different filler structures.

2. Characterization of elastomers

To characterize elastomeric materials under harmonic excitation, dynamic tests were conducted for two different elastomer configurations. The first elastomeric configuration was a double lap shear specimen as shown in Fig. 1a, or flat linear bearing, hereinafter called elastomeric specimen 1. The second elastomeric configuration was a linear concentric tubular bearing as shown in Fig. 1b, hereafter denoted the elastomer specimen 2. Both specimens were characterized under pure shear deformation. Testing was carried out with varying excitation amplitudes and frequencies, and all tests were conducted at room temperature: 25°C.

As shown in Fig. 1a, elastomer specimen 1 was a double lap shear speciment that is comprised of three parallel brass plates between which the elastomeric material is sandwiched symmetrically. To study the effect of preload on the behavior of the elastomer, elastomeric specimen testing was conducted with and wihtout preload. Preload was applied to the elastomeric specimen by compressing the double lap shear specimen 10% of the width of the specimen using a simple vise. The testing setup for the elastomer specimen 1 is shown in Fig. 2a. A 24.466 kN servo-hydraulic MTS test machine was used to characterize the specimen. Fixtures and grips were designed and machined to hold the specimen in place. A hydraulic power supply (HPS) unit supplied the servo fluid to the testing machine for power, and the specimen was loaded and tested on the load frame. An actuator provided the sinusoidal loading, and an LVDT sensor measured displacement while a load cell measured force. Single and dual frequency tests were conducted using this load frame. In the dual frequency test, an HP 8904A multi-function synthesizer was used to generate and sum the sinusoidal signals for both frequencies. The single frequency test was conducted with displacement control for excitation amplitude ranging from 0.25 mm to 5 mm, i.e. 2.5%

to 50% shear strain, in increments of 0.25 mm. The frequencies were chosen as 2.5 Hz, 5 Hz and 7.5 Hz. The dual frequency testing was carried out at a combination of 5 Hz and 7.5 Hz, and the amplitudes for 7.5 Hz were 0.5, 1.5, 2.5, 3.5, and 4.5 mm, respectively, while the amplitude for 5 Hz was maintained the same as for the single frequency tests.

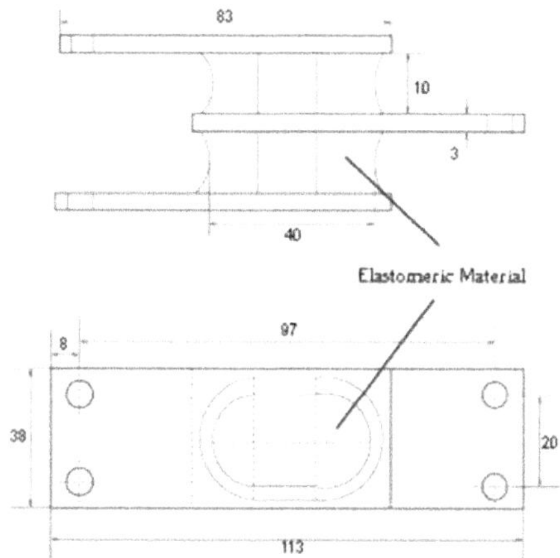

(a) Specimen 1: Double Lap Shear

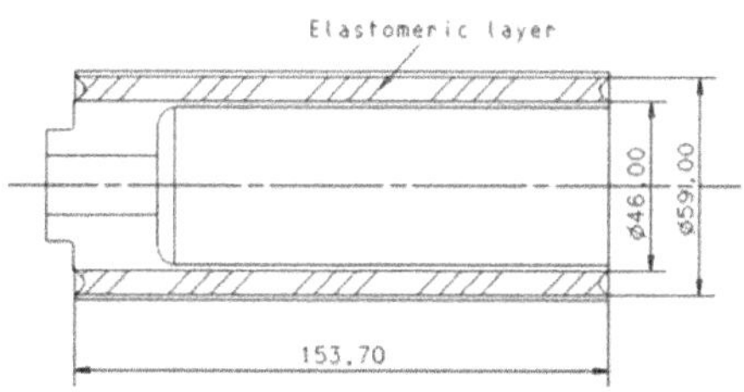

(b) Specimen 2: Cylindrical Tubular Shear

Figure 1. Elastomer Specimen (Unit: mm)

An important effect of filler materials in the filled elastomeric specimen is stress-softening. If an elastomeric sample is stretched for the first time to 100% followed by a release in the strain and then stretched again to 200%, there is a softening in the strain of up to 100% after which it continues in a manner of following the first cycle. This stress softening effect was first discovered by Mullins and is called the 'Mullins Effect' [18]. To account for this phenomenon, the test samples were first cycled and loosened before the actual tests by exciting them at 1 Hz frequency and 5 mm for 300 cycles since 5 mm is the maximum amplitude during tests. Stress relaxation is also shown in the case of dynamic loading. As the material is subjected to cycling loading, energy dissipation in the material heats up the material and results in elevated temperature softening. Usually, material self-heating and other unsteady effects require about 250 seconds to stabilize and reach a steady state. Hence, in order to ensure temperature stabilization and consistency of data, during a normal test, the elastomeric sample was typically excited at the test frequency and amplitude for 300 seconds before collecting data. For simplification, the stress softening and relaxation effects were not considered in the modeling process such that the model parameters were independent of the loading level and temperature.

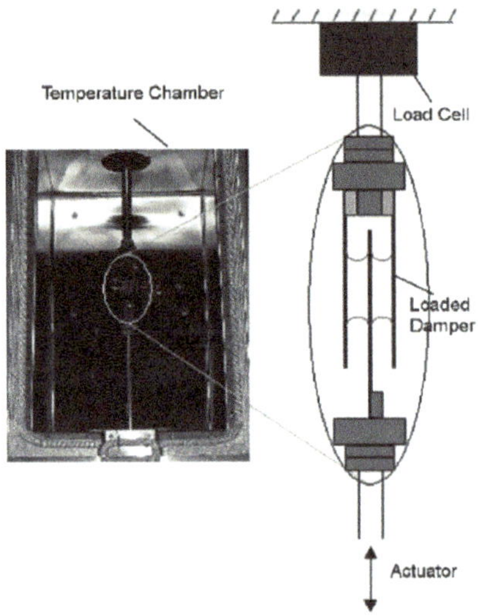

(a) Testing Setup for Specimen 1

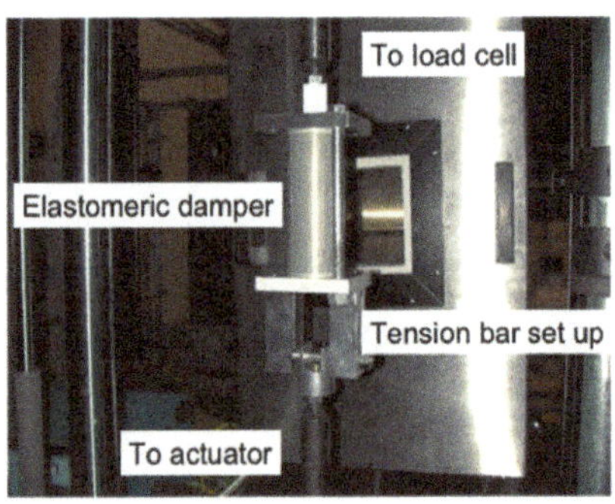

(b) Testing Setup for Specimen 2

Figure 2. Testing Setup

As shown in Fig. 1b, elastomer specimen 2 was fabricated from two concentric cylindrical metal tubes, with an elastomeric layer sandwiched between the outer and inner tubes. The volume enclosed by the inner tube forms a cylindrical inner chamber, and a threaded trapezoidal column is attached to one end of the inner tube. As shown in Fig. 2b, the specimen was installed in the MTS testing machine, the outer tube was attached to the load cell on the fixed end of the MTS machine, and the inner tube was connected to the actuator through an adapter. Thus, the axial translation of the actuator induced a relative translation between the inner tube and the outer tube which in turn led to a shear deformation of the elastomer along the tube length. The specimen was excited in displacement control by a sinusoidal signal, and the displacement and force were measured by the LVDT sensor and load cell of the MTS machine. The excitation amplitude ranged from 0.25 mm to 1 mm in increments of 0.25 mm (approximately 5% to 20% shear) at three different frequencies of 2.5, 5.0, and 7.5 Hz, respectively.

All test data were collected using a high sampling frequency (2048 Hz) such that most higher harmonic components in the measured nonlinear force were included. To reduce the noise of the sinusoidal displacement signal, a Fourier series was used to reconstruct the input displacement. The reconstructed displacement signal was then differentiated to obtain the velocity signal. The Fourier series expansion of the input displacement, $x(t)$, is

$$x(t) = \frac{x_0}{2} + \sum_{k=1}^{\infty}\left[X_{c,k}\cos(k\omega t) + X_{s,k}\sin(k\omega t)\right] \tag{1}$$

where,

$$\begin{aligned} X_{c,k} &= \frac{\omega}{\pi}\int_0^{\frac{2\pi}{\omega}} x(t)\cos(k\omega t)\,dt \\ X_{s,k} &= \frac{\omega}{\pi}\int_0^{\frac{2\pi}{\omega}} x(t)\sin(k\omega t)\,dt \end{aligned} \tag{2}$$

In Eq. (1), $x_0=X_{c,0}$. For single frequency data processing, any bias and higher harmonics were filtered, so that only the frequency of interest, ω, remained, that is 2.5, 5.0 and 7.5 Hz. For dual frequency testing, the general equation for the input dual displacement signal is written as:

$$x(t) = X_1\sin(\Omega_1 t) + X_2\cos(\Omega_2 t) \tag{3}$$

where Ω_1 and Ω_2 are correspnding frequencies of 5 and 7.5 Hz, and X_1 and X_2 are the amplitudes at each frequency. The signal is periodic with a frequency corresponding to the highest common factor of both harmonics, i.e., 2.5 Hz. The displacement signal was filtered using 2.5 Hz as the base frequency. The first three harmonics were needed to reconstruct the dual frequency displacement signal in order to capture Ω_1 and Ω_2. Due to the nonlinearity of elastomers, the higher harmonics of the measured force were not filtered.

A typical approach used for characterizing elastomer behavior is the complex stiffness. The linearized complex stiffness, K^*, is composed of an in-phase or storage stiffness, K', and a quadrature or loss stiffness, K'', as follows:

$$K^* = K' + jK'' \tag{4}$$

Therefore, the elastomer force can be written as the summation of an in-phase spring force and a quadrature damping force, and the elastomer force can be approximated by the first Fourier sine and cosine components at the charterized frequencies, i.e. 2.5, 5.0 and 7.5 Hz:

$$\begin{aligned} F(t) &= F_c \cos(\omega t) + F_s \sin(\omega t) \\ &= K'x(t) + \frac{K''}{\omega}\dot{x}(t) \end{aligned} \tag{5}$$

where F_c and F_s are the first harmonic Fourier coefficients of the measured force. The storage stiffness, K', and the loss stiffness, K'', are determined using the following equations:

$$\begin{aligned} K'(\omega) &= \frac{F_c X_c + F_s X_s}{X_c^2 + X_s^2} \\ K''(\omega) &= \frac{F_c X_s - F_s X_c}{X_c^2 + X_s^2} \end{aligned} \tag{6}$$

where, X_c and X_s are the first harmonic Fourier coefficients of $x(t)$. The loss factor, η, is also used to measure the relative levels of the loss stiffness to the storage stiffness. The ratio is written as:

$$\eta = \frac{K''}{K'} \tag{7}$$

Typical linear characterization results for elastomeric specimen 1 in the absence of preload are shown in Fig. 3. These plots indicate that the linearized storage and loss stiffness of the specimen are highly amplitude dependent at low amplitudes. For smaller amplitudes, the rate of change of the storage stiffness and the loss stiffness is much greater than one for larger amplitudes. However, the complex stiffness does not change substantially over the narrow frequency range tested. The loss factor, η, is also strongly dependent on amplitude. The maximum value of the loss factor is as high as 1.025 for this filled elastomer. This elastomeric material performs most effectively as a damping material within the amplitude range of 0.76 - 1.27 mm 7.5% to 12.5% shear strain), in which the loss factor, $\eta > 1$. Similar results were exhibited by the elastomer under preload. In general, the linearized behavior of the elastomeric specimen is highly amplitude dependent and weakly frequency dependent in this frequency range.

The measured complex modulus and loss factor of the elastomer specimen 2 are shown in Fig. 4. Both in-phase (storage) and quadrature (loss) stiffness demonstrate moderate amplitude dependence and weak frequency dependence. In contrast to the characteristics of the elastomeric specimen 1, the in-phase stiffness of specimen 2 is much greater than its quadrature stiffness, and both in-phase and quadrature stiffness vary with the displacement amplitude at a similar rate. Thus, the loss factor of the specimen is quite low (around 0.25 to 0.3) and almost constant over the range of amplitude and frequency tested.

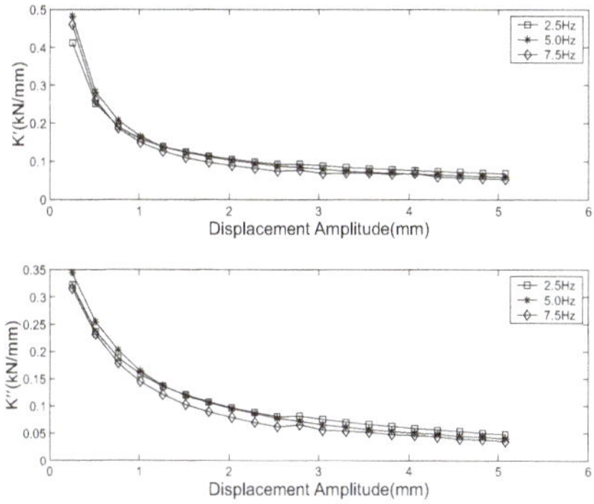

Figure 3. Linear Characterization of Specimen 1

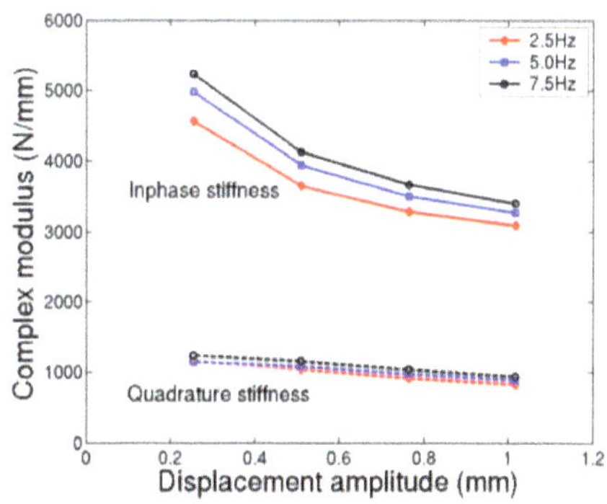

Figure 4. Linear Characterization of Specimen 2

The single frequency linear characterization can capture the general trends of the in-phase and quadrature stiffness for a filled elastomer. However, this linear analysis cannot be used to accurately reconstruct the nonlinear hysteresis behavior exhibited by the elastomer [19]. Therefore, nonlinear modeling methods are requried to accurately describe the material behavior of elastomers, and two modeling methods are decribed in the following sections

3. Nonlinear hysteresis model

3.1. Modeling approach

The mechanical properties of a linear viscoelastic material are represented by a Kelvin model, which consists of a spring and dashpot in parallel. The spring and damping coefficients are constants, and the Kelvin model is equivalent to a complex modulus approach. Although the Kelvin model cannot describe the relaxation process after a constant strain is applied to a material specimen, it can successfully characterize stiffness and damping during steady-state harmonic excitation. Therefore, it is the simplest approach to describe the nonlinear behavior of elastomeric materials based on a Kelvin model.

As mentioned before, when a linear viscous material is subjected to sinusoidal loading, the stiffness force is an in-phase force with a constant stiffness, and the damping force with a constant damping is a quadrature force. For an elastomeric specimen, both stiffness and damping are not constant. To extract the stiffness force, F_s, and damping force, F_d, from the experimental force data, F, the displacement phase angle, $\phi(t)$, at arbitrary time t should be known firstly from the reconstructed displacement signal. A single frequency sinusoidal displacement can be written as:

$$x(t)=\sqrt{X_c^2+X_s^2}\sin(\phi(t)) \tag{8}$$

where

$$\phi(t)=\omega t+\arctan\left(\frac{X_c}{X_s}\right) \tag{9}$$

From a given start time, t_0, when $\phi(t_0)=2k\pi+\pi/2$, k=0,1,.., one cycle of the force data is used to determine the stiffness force. Since the stiffness force has the same phase angle as $\phi(t)$ and the damping force is $\pi/2$ ahead of $\phi(t)$, the stiffness force, F_s, in one cycle can be obtained as:

$$\begin{aligned} &F_s[\phi(t)]=\frac{F[\phi(t)]+F[2\pi-\phi(t)]}{2} \\ &F_s[2\pi-\phi(t)]=F_s[\phi(t)] \\ &\phi(t)=\phi(t_0),\ldots,\phi(t_0)+\pi \end{aligned} \tag{10}$$

Similarly, from a different start point $\phi(t_0')=2k\pi$, k=0,1,..., the damping force, F_d, is obtained as follows:

$$\begin{aligned} &F_d[\phi(t)]=\frac{F[\phi(t)]+F[2\pi-\phi(t)]}{2} \\ &F_d[2\pi-\phi(t)]=F_d[\phi(t)] \\ &\phi(t)=\phi(t_0'),\ldots,\phi(t_0')+\pi \end{aligned} \tag{11}$$

The identified nonlinear stiffness and damping forces are shown in Fig. 5. In Figure 5a and b, the experimental force-displacement and force-velocity hysteresis cycles are plotted using dotted lines for a sinusoidal displacement excitation at 5Hz and 1.5mm amplitude. The stiffness force, F_s, and the damping force, F_d, are denoted as solid lines in the force-displacement plane and the force-velocity plane, respectively. Clearly, the stiffness force is a nonlinear monotonic function of displacement, and the damping force is a nonlinear monotonic function of velocity. Then, in order to establish a nonlinear Kelvin model, effective functional forms must be found to represent both the nonlinear stiffness and damping forces, respectively.

In Fig. 5a, it is shown that the stiffness force, F_s, is almost linear but increases with increasing displacement. By observing different stiffness lines at different displacement amplitudes, it is noted that the nonlinear behavior of stiffness always occurs while the velocity is close to zero. It implies that the nonlinearity of the stiffness is due to velocity but not displacement. Therefore, the nonlinear stiffness model is given by:

$$\hat{F}_s = \left[K + \delta K \exp\left(\lambda_k |\dot{x}|^{1.5}\right)\right] x \tag{12}$$

where, K is the linear stiffness, δK is the nonlinear stiffness increment, and λ_k is used to describe the nonlinear stiffness slope. These parameters can be determined by a nonlinear curve-fitting algorithm, in which an objective function, J_s, is minimized as follows:

$$\min_{K,\delta K,\lambda_K} J_s\left(K,\delta K,\lambda_K\right) = \sum_{i=1}^{n}\left[\hat{F}_s(t_i) - F_s(t_i)\right]^2 \tag{13}$$

where, n is the number of data points.

As shown in Fig. 5b, the shape of the nonlinear damping force, F_d, is reminiscent of a friction damping. This implies that filled elastomeric materials exhibit anelasticity instead of viscoelasticity. In some references, an inverse hyperbolic tangent [20] or exponential [21] functional approximation has been suggested to describe the tribo-elastic behavior in filled elastomers. In our model, a hyperbolic tangent function is used due to its simplicity and computational efficiency, so that the analytical friction damping force is assumed to behave as:

$$\hat{F}_d = F_y \tanh\left(\lambda \dot{x}\right) \tag{14}$$

where, F_y and λ represent the yield force and yield parameter respectively. Similarly, the nonlinear damping parameters also can be determined by minimizing the objective function, J_d, as follows:

$$\min_{F_y,\lambda} J_d\left(F_y,\lambda\right) = \sum_{i=1}^{n}\left[\hat{F}_d(t_i) - F_d(t_i)\right]^2 \tag{15}$$

The total predicted force due to the displacement excitation is the summation of the stiffness model and damping model as follows:

$$\hat{F} = \hat{F}_d + \hat{F}_s \tag{16}$$

The reconstructed force is shown as the solid line in the Fig. 5c, and the analytical force-displacement hysteresis matches the experimental data very well. Moreover, using the identified friction function, a force-displacement hysteresis due to friction damping is reconstructed. It is known that energy dissipation due to the damping is proportional to the enclosed area inside the damping force-displacement hysteresis loop. The area enclosed by the total force-displacement hysteresis loop is equal to the area enclosed by the damping hysteresis. This proves that the nonlinear damping function precisely describes the energy dissipation while the elastomer specimen is under sinusoidal loading.

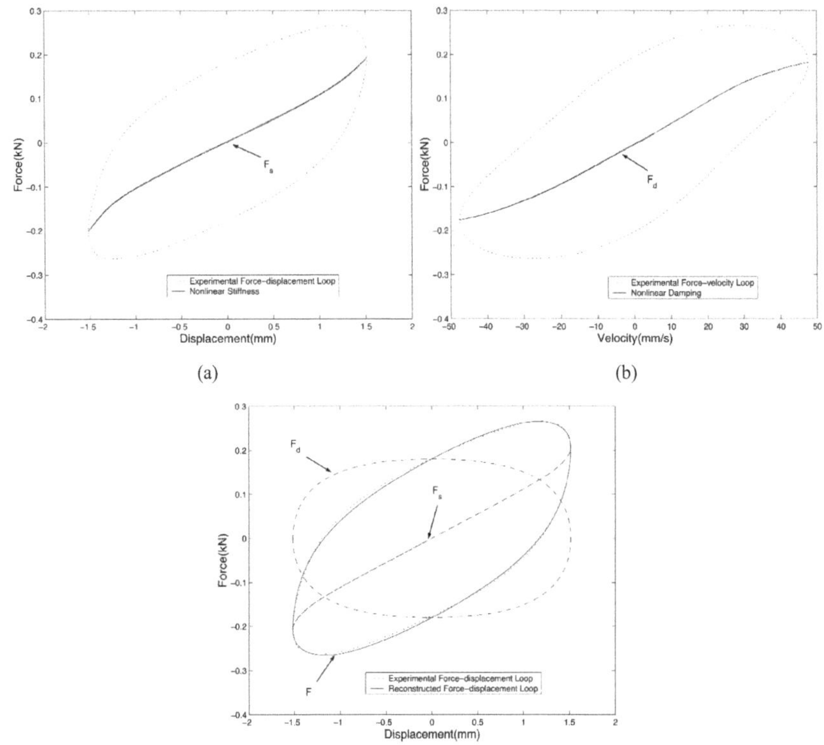

(c)

Figure 5. Hysteresis Used for Decoupling of Stiffness and Damping

Briefly, this nonlinear Kelvin model is based on a hysteresis modeling approach developed from damper modeling efforts. Since the nonlinear hysteresis loops of an elastomeric damper are described by a nonlinear monotonic stiffness and a nonlinear monotonic damping function respectively, the model parameters can be determined separately and efficiently. The modeling results can precisely capture the force-displacement time history data of the elastomer. Meanwhile, the introduction of the friction damping physically emphasizes the anelasticity of elastomeric materials. Since the hysteresis cycles for different amplitudes and frequencies are different, it is helpful to study the model parameter variations and get insight into the nonlinear behavior of elastomers by characterizing the different hysteresis cycles.

3.2. Modeling results

The model characterization was obtained based on three sets of force-displacement time history data of the elastomer specimen 1 at three different frequencies, i.e. 2.5Hz, 5Hz and 7.5Hz. For every frequency, there were twenty displacement amplitudes ranging from 0.25mm to 5mm. After the model parameters were determined, the output force was predicted using known displacement input data, and then the force-displacement hysteresis was reconstructed. In Fig. 6, the reconstructed hysteresis curves and test data are shown for different amplitudes (0.5mm, 1.5mm, 2.5mm, and 3.5mm) at 5Hz frequency with zero preloading. It can be seen that at low amplitudes, the experimental hysteresis plots are nearly elliptical in shape, while for higher amplitudes there is a deviation from this elliptical behavior. However, compared to the elastomer hysteresis models developed by Krishnan [22] and Snyder [23], the current nonlinear model more accurately captures nonlinear force-displacement time histories under sinusoidal loading.

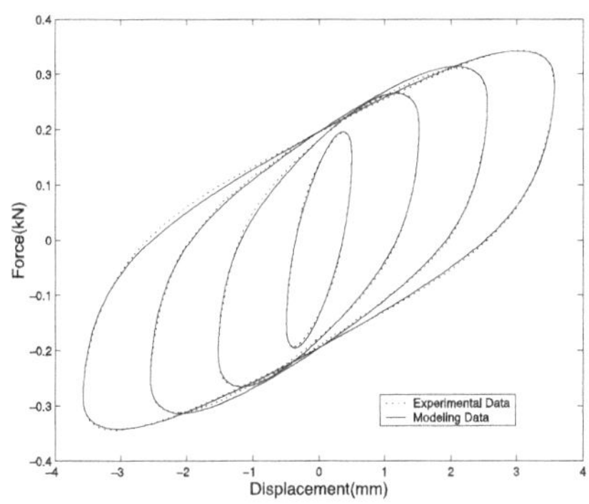

Figure 6. Modeling Results of Hysteresis Model

Furthermore, model parameter variation as a function of the amplitude at different frequencies was studied. As shown in in Fig. 7a, all three nonlinear stiffness parameters are amplitude dependent. However, the linear stiffness, K, and the nonlinear stiffness increment, δK, maintain nearly the same value at different frequencies, but the nonlinear stiffness slope, λ_k, decreases as the frequency increases. Similarly, the identified damping parameters are shown in Fig. 7b. The yield force, F_y, for amplitudes above 1.5mm is nearly constant with both amplitude and frequency. At low amplitudes (<1.5mm), the optimized yield force shows large deviations since the damping is almost linear at small amplitude excitations and the model is insensitive to F_y. The yield parameter, λ, shows strong amplitude and frequency dependent characteristics.

Some interesting characteristics are noted in Fig. 8a, in which the stiffness is shown as a function of velocity amplitude. The nonlinear stiffness slope parameters of all three frequencies follow exactly the same curve, which is inversely proportional to the velocity amplitude. This implies that the nonlinear stiffness slope, λ_k, only depends on the velocity amplitude. Similarly, as shown in Fig. 8b, the nonlinear damping yield parameter, λ, is also dependent on the acceleration amplitude. Both characteristics make this nonlinear model nearly independent of frequency and thus useful to predict dual frequency response.

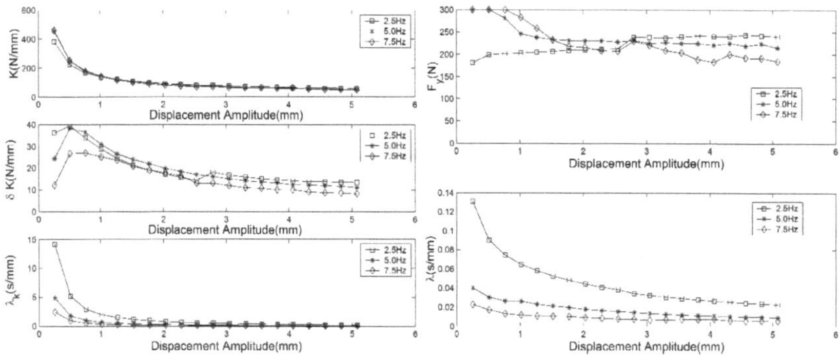

(a) Stiffness Parameters (b) Damping Parameters

Figure 7. Hysteresis Model Parameters as a Function of Amplitude and Frequency

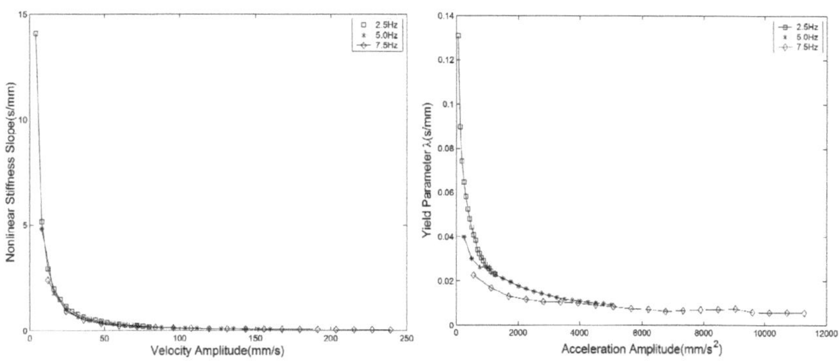

(a) Stiffness Slope VS Velocity Amplitude (b) Yield Parameter VS Acceleration Amplitude

Figure 8. Hysteresis Model Parameters as a Function of Velocity and Acceleration Amplitude

To capture the behavior of the elastomer under dual frequency excitation, the elastomer model parameters were determined without frequency information because the excitation amplitude was known a *priori*. Specifically, both the linear stiffness and stiffness increment are inversely proportional to the displacement amplitude, and the nonlinear slope can be interpolated using the known velocity amplitude. Meanwhile, for nonlinear damping, the yield force is almost a constant and the yield parameter can be determined using the acceleration amplitude. As a result, the nonlinear dual frequency behavior of the elastomer could be predicted. In Fig. 9, the dual frequency displacement excitation was a summation of two single frequency inputs, i.e., 5 Hz and 7.5 Hz, and the reconstructed hysteresis cycles using the hysteresis model are compared to experimental data. It is shown that the modeling results can accurately capture the hysteresis behavior exhibited by the elastomer.

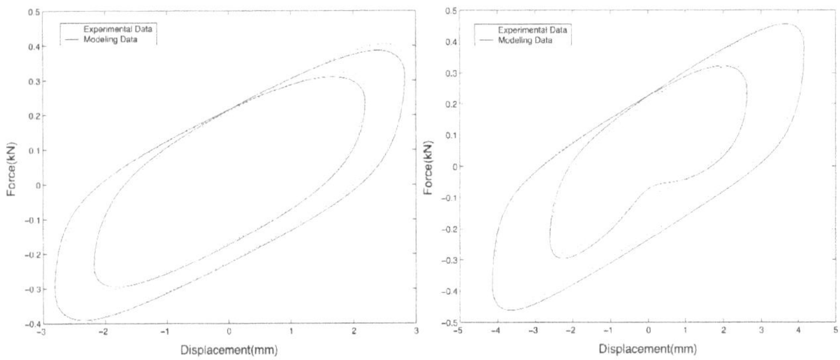

(a) 2.5 mm at 5 Hz, 0.5 mm at 7.5 Hz (b) 2.5 mm at 5 Hz, 1.5 mm at 7.5 Hz

Figure 9. Dual Frequency Modeling Results Using Hysteresis Model

As shown in the nonlinear hysteresis modeling effort, the nonlinear forced response of elastomers under harmonic excitation consists of uncoupled nonlinear stiffness force and nonlinear damping force. Thus, this model is mechanically analogous to a nonlinear Kelvin model where the stiffness is a nonlinear monotonic function of displacement and the damping is a monotonic rate dependent friction function. Since it accurately describes the characteristics of the hysteresis loops, this model can predict steady state or harmonic forced response very well. However, since the model parameters are still amplitude dependent, this model cannot be easily used to describe the transient or stress relaxation behavior of the elastomer.

4. Distributed rate-dependent elasto-slide model

4.1. Model development

The distributed rate-dependent elasto-slide model is shown in Fig. 10, in which a series of elasto-slide elements is combined in parallel with a constant linear spring. The model can be applied either in force-displacement relations or in stress-strain relations, but only the force-displacement formulation will be used in this study. Each elasto-slide element consists of a leading spring with stiffness k/N in series with a slide which has a yield force f_i^*, where N is the total number of elements. The yield force for each element is different, and the stiffness for each leading spring is assumed to be a constant. The ideal model assumes that the yield force is a Coulomb force, which is a constant as the slide moves at any speed. However, the rate-dependent elasto-slide model represents real friction behavior by representing the yield force of the slide as a function of the slide velocity. This function will be discussed later. For simplicity, the description of the model starts from using an ideal slide (which has an ideal

Coulomb force), and then the behavior of the model will be studied when the ideal slide is replaced by a rate-dependent slide.

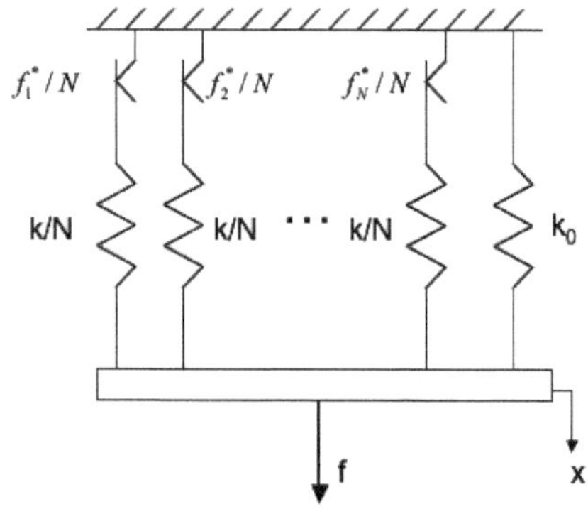

Figure 10. Distributed Elasto-Slide Model

First, we will apply a displacement, x, to the ideal elasto-slide element, which assumes a constant Coulomb force. At the beginning, the displacement is small such that the consequent leading spring force is smaller than the yield force, so that only the spring is deformed. After the spring force reaches the yield force, the slide yields and a motion is induced, and the resisting force of the element remains constant with the same value as the yield force. Since each elasto-slide element in the model is assumed to have a different yield force level, this model presents gradual stiffness reduction as amplitude increases until such a condition as all elements have yielded. At that time, only the parallel spring, k_0, remains to represent the polymer stiffness of the elastomer. Since the elastomer is a continuum, the total number of elements, N, approaches infinity, and, in the limit, the discrete yield force for a single slide is replaced with a distributed density within a certain yield force range. Alternatively, a mathematically equivalent model is shown in Fig. 11, in which f_e represents the total resisting force of the elasto-slide elements.

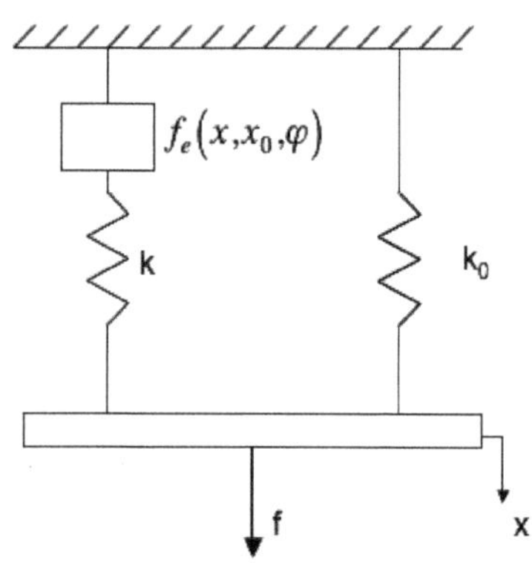

Figure 11. Mathematically Equivalent Elasto-Slide Model

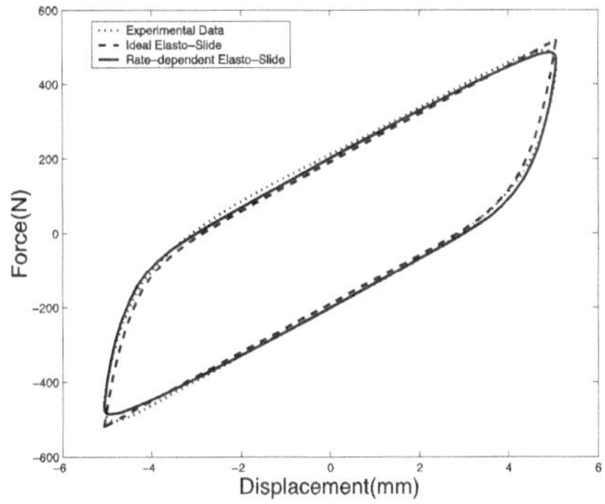

Figure 12. Effectiveness of Elasto-Slide Model

The following simulation using the model will show that the existence of the filler structures inside the elastomer can lead to hysteretic behaviors when the damper is cycled between fixed deflection limits. Analytically, a distribution function of the yield force is denoted as $\varphi(f^*)$ such that the density of the slides with yield force f^* is expressed as $\varphi(f^*)\mathrm{d}f^*$. Using the ideal slide assumption, the slide behaves as a Coulomb friction element. Thus, upon initial loading, the leading spring at a certain yield element stretches with the displacement x until the spring force reaches the maximum slide yield force, that is:

$$\begin{aligned} \mathrm{d}f &= kx\varphi\left(f^*\right)\mathrm{d}f^*, \dot{x} > 0, 0 \le x \le \frac{f^*}{k} \\ \mathrm{d}f &= f^*\varphi\left(f^*\right)\mathrm{d}f^*, \dot{x} > 0, x \ge \frac{f^*}{k} \end{aligned} \tag{17}$$

where, k is the stiffness of the leading spring. If the direction of loading is reversed, the force-deflection relation is more complicated. Including the yielded and non-yielded elements, the force-deflection relation becomes:

$$\begin{aligned} \mathrm{d}f &= \left[f^* - k\left(A - x\right)\right]\varphi\left(f^*\right)\mathrm{d}f^*, \dot{x} < 0, A - 2\frac{f^*}{k} \le x \le A, f^* \le kA \\ \mathrm{d}f &= -f^*\varphi\left(f^*\right)\mathrm{d}f^*, \dot{x} < 0, x \le A - 2\frac{f^*}{k}, f^* \le kA \\ \mathrm{d}f &= kx\varphi\left(f^*\right)\mathrm{d}f^*, \dot{x} < 0, f^* > kA \end{aligned} \tag{18}$$

where A is the maximum deflection of the elastomer. An expression similar to Eq. 18 is obtained when the loading reaches the minimum deflection and is reversed again. This process continues until the loading is terminated. Clearly, at each time, only some of the elasto-slide elements have yielded. The total resisting force can be obtained by integrating all of the elasto-slide forces along the yield distribution region and by adding the spring force due to the residual polymer stiffness. Using Eq. 17 and adding the effect of the polymer stiffness, k_0, the initial loading force is given as

$$f = \int_0^{kx} f^* \varphi(f^*) \mathrm{d}f^* + kx \int_{kx}^{\infty} \varphi(f^*) \mathrm{d}f^* + k_0 x,\ \dot{x} > 0 \tag{19}$$

Similarly, the resisting force due to the reversed loading is obtained by integrating Eq. 18:

$$f = \int_0^{\frac{k(A-x)}{2}} -f^* \varphi(f^*) \mathrm{d}f^* + \int_{\frac{k(A-x)}{2}}^{kA} \left[kx - \left(kA - f^*\right) \right] \varphi(f^*) \mathrm{d}f^* + kx \int_{kA}^{\infty} \varphi(f^*) \mathrm{d}f^* + k_0 x, \quad \dot{x} < 0,\ x \le A \tag{20}$$

Thus, while an elastomer is under a sinusoidal displacement loading, a theoretical force-deflection hysteresis cycle is shown as dashed line in Fig. 12, where A is 2.5 mm and frequency is 2.5 Hz. Compared to the experimental hysteresis shown as dotted line, the model prediction gives a good match to the experimental result.

The distributed elasto-slide model resembles the physical mechanism of an elastomer, so it can account for the nonlinear characteristics of the behavior demonstrated by the elastomer under a cyclic loading either in single frequency or multi frequencies. However, using the ideal slide with a Coulomb force, the maximum and minimum displacement of the excitation must be known for response calculation, which makes it impossible for the model to describe elastomer behavior under complex loading conditions. The ideal elasto-slide is also incapable of modeling frequency dependent properties and non-hysteretic behavior in the time domain such as stress relaxation or creep.

Actually, the Coulomb slide is only an idealized friction model. The practical friction behavior includes a preyield slip and a postyield steady resistance leading to a rate dependent damping effect [21, 24]. Thus, a rate-dependent elasto-slide model is introduced to improve the modeling performance. In the rate-dependent elasto-slide model, the Coulomb slide is replaced with a non-Coulombic friction function and the coupling between a slide and a leading spring is described by an internal displacement denoted as x_0 such that the slide force at a certain yield region is written as

$$\mathrm{d}f = f^* \varphi(f^*) \mathrm{d}f^* \left(\frac{\dot{x}_0}{v_r} \right)^{\frac{1}{p}} \tag{21}$$

where p is a positive odd integer and v_r is a characteristic reference velocity. Coupled with the lead spring $k\varphi(f^*)\mathrm{d}f^*$, the internal displacement x_0 at certain yield region is obtained using the function:

$$\frac{\dot{x}_0}{v_r} = \left[\frac{k}{f^*} \left(x - x_0 \right) \right]^p \tag{22}$$

By integration over the whole yield force region, the total force due to the elasto-slide element is obtained as:

$$f_e = \int_0^{\infty} k(x - x_0)\varphi(f^*)\mathrm{d}f^* \tag{23}$$

Adding the spring force due to the polymer stiffness, the damper force due to any deflection loading, x, is determined as:

$$f = f_e + k_0 x = \int_0^{\infty} k(x - x_0)\varphi(f^*)\mathrm{d}f^* + k_0 x \tag{24}$$

This relation can be shown in Fig. 11. Eq. 22 is a typical well-posed initial-value problem, and numerical solution for this differential equation can be obtained given an initial condition. The simplest way to guarantee a stable solution for such a stiff initial-value problems is to adopt a predictor-corrector approach with the corrector iterated to convergence (PECE) [25]. In this approach, the numerical algorithm is based on the Adams-Bashforth four-step method as the predictor step and one iteration of the Adams-Moulton three-step method as the corrector step, with the starting values obtained from a fourth-order Runge-Kutta method. In accordance with the ratio of the yield force and the stiffness, k/f^*, and an appropriate choice of p, the Adams-Bashforth-Moulton method gives relatively stable and fast convergence of a solution within a limited number time steps.

For an elastomer under a sinusoidal displacement excitation, the steady-state response predicted by the rate-dependent elasto-slide model is shown as the solid line in Fig. 12. The predicted hysteresis cycle correlates much better with the experimental data, especially at turning points of the loading deflection. It should be noted that there is no requirement for excitation amplitude information in this modeling process. Thus, the distributed rate-dependent elasto-slide model can predict time domain forced response of an elastomer under a sinusoidal displacement excitation.

In order to apply the elastomer model to a dynamic system, a numerical method using MATLAB ODE algorithm was also evaluated. For a dynamic system with a governing equation:

$$M\ddot{x} + C\dot{x} + Kx + F_d(x, \dot{x}) = F(t) \tag{25}$$

where, M, C, and K are mass matrix, equivalent damping matrix, and stiffness matrix, respectively, F_d is used to describe the force due to an elastomer, and F is a loading vector applied to the system. The size of the matrix depends on the degree of freedom of the system. For simplicity, only a 1-DOF system is considered now. In the distributed rate-dependent elasto-slide damper model, there are theoretically an infinite number of internal variables, x_0. However, the distributed yield stress usually falls within a limited range. Therefore, according to the form of a distribution function, the continuous yield force distribution area can be uniformly decomposed into n discrete elements from minimum to maximum yield force and each element has a yield force range, Δf^*. At each yield stress range, f_i^*, the corresponding distribution is equal to the area of that element, $\varphi(f_i^*)\Delta f^*$. Each element has an internal variable denoted as x_{0i}, $i=1,\ldots,n$, and each x_{0i} satisfies the Eq. 22. Thus, the elastomer force F_d can be described as:

$$F_d = k\sum_{i=1}^{n}\varphi\left(f_i^*\right)\Delta f^*\left(x - x_{0i}\right) + k_0 x \tag{26}$$

Rewriting Eq. 25 using a first order form and combining Eq. 22 and 26, the state equation of the system is expressed as

$$\begin{aligned}
\dot{x} &= \dot{x} \\
\ddot{x} &= \frac{F(t)}{M} - \frac{1}{M}\left[\left(K + k_0\right)x - k\sum_{i=1}^{n}\varphi\left(f_i^*\right)\Delta f^*\left(x - x_{0i}\right)\right] - \frac{C}{M}\dot{x} \\
\dot{x}_{01} &= \left[\frac{k}{f_1^*}\left(x - x_{01}\right)\right]^p v_r \\
\vdots \;\; &\vdots\vdots \\
\dot{x}_{0n} &= \left[\frac{k}{f_n^*}\left(x - x_{0n}\right)\right]^p v_r
\end{aligned} \tag{27}$$

This is a $(n+2)$th order state function. Using the ODE23 algorithm in MATLAB, the forced response or the transient response due to initial displacement and velocity can be solved numerically. For the system with more degrees of freedom, the state function can easily accommodate these additions by using additional number of states.

4.2. Model parameters determination

As seen in the construction of the model, the major parameters to be determined are the leading spring, k, and yield force distribution function, $\varphi(f^*)$, for the distributed elasto-slide element, and the parallel spring, k_0, for the remaining polymer stiffness. In the absence of the knowledge of the elastomer structure, the selection of these parameters is only based on experimental data in this stage. One possible selection of methods would be to make use of an experimentally determined initial loading curve.

The definition of the distribution function implies that $\varphi(f^*)$ has to obey the following three constraints:

$$\begin{aligned}
&\int_0^\infty \varphi\left(f^*\right)\mathrm{d}f^* = 1 \\
&\varphi\left(f^*\right) \geq 0 \\
&0 \leq f^* \leq \infty
\end{aligned} \tag{28}$$

From the initial loading curve Eq. 19, yields

$$\begin{aligned}
&\frac{\mathrm{d}f}{\mathrm{d}x} = k\int_{kx}^{\infty}\varphi\left(f^*\right)\mathrm{d}f^* \\
&f^* = kx
\end{aligned} \tag{29}$$

Then

$$\frac{d^2 f}{dx^2} = -k^2 \varphi\left(f^*\right) \tag{30}$$

Thus, the distribution function would be related to the curvature of the initial loading curve by the following formula

$$\varphi\left(f^*\right) = -\frac{1}{k^2}\frac{d^2 f}{dx^2} \tag{31}$$

Determination of the distribution function relies on the identification of the initial loading curve from the experimental data.

In the view of the distributed elasto-slide model using an ideal Coulomb slide, the initial loading curve is independent of loading rate such that the maximum force in the initial loading curve responds to the maximum displacement in cyclic loading as seen in Eq. 19. As a result, an initial loading curve can be obtained using a series of experimental hysteresis loops at different amplitudes. An example of the initial loading curve is shown in Fig. 13, the initial loading curve at three different frequencies are obtained from hysteresis cycles of the elastomeric specimen by identifying the force at corresponding maximum displacements. The analytical initial loading curve is determined by considering the influence of the rate-dependent slide. This curve appears elasto-plastic behavior, which is described as:

$$f = \frac{1}{\varphi}\left(1 - e^{-k\varphi_0 x}\right) + k_0 x \tag{32}$$

Notably, φ_0 is a distribution constant and an index of the post yield force level, and k and k_0 are the stiffness of the leading spring and the remaining polymer stiffness respectively. Summation of the leading spring and the polymer stiffness is just the slope of the force-displacement curve when $x \to 0$. This is conceivable since there only exists the influence of the springs while all slide elements are not yielded. Substituting Eq. 32 into Eq. 31, yields a very simple distribution function as:

$$\varphi\left(f^*\right) = \varphi_0 e^{-\varphi_0 f^*} \tag{33}$$

The distribution area for the elastomeric specimen is shown as the shaded area in Fig. 14. It is easily shown that the distribution function satisfies all the properties of Eq. 28.

As the distribution function is determined, for the distributed elasto-slide model using the Coulomb slide, the steady-state forced response of the elastomer under cyclic loading will be predicted as follows:

$$f=\frac{1}{\varphi_0}\left(1+e^{-k\varphi_0 x_{max}}-2e^{-k\varphi_0\frac{x-x_{min}}{2}}\right)+k_0 x,\ \dot{x}>0$$
$$f=-\frac{1}{\varphi_0}\left(1+e^{-k\varphi_0 x_{max}}-2e^{-k\varphi_0\frac{x_{max}-x}{2}}\right)+k_0 x,\ \dot{x}<0 \tag{34}$$

where x_{max} and x_{min} are the maximum and minimum amplitude of the cyclic loading, respectively.

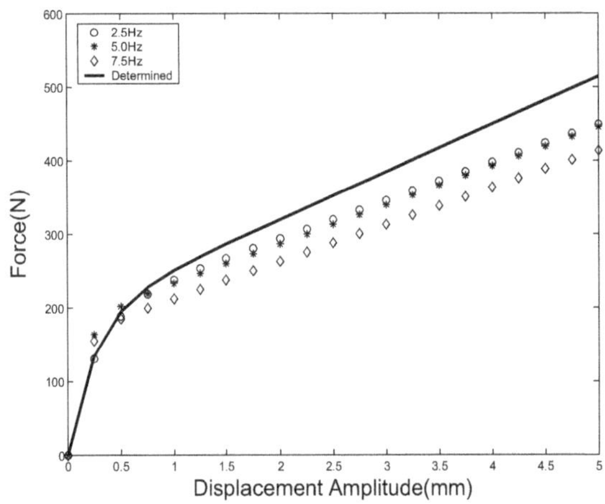

Figure 13. Initial Loading Curve

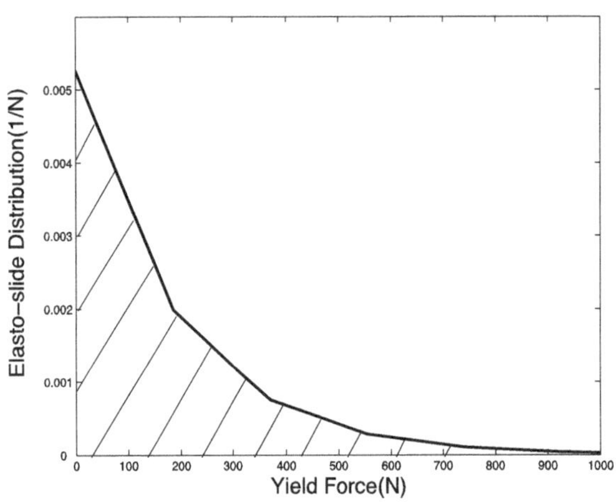

Figure 14. Yield Distribution

For the rate-dependent elasto-slide model, the reference velocity v_r and the exponent p need to be determined. The choice of v_r was based on steady state force-velocity curves in which the boundary between the pre-yield and post-yield region was approximated. Analytically, p should be as large as possible such that the post-yield force rapidly transitions to a constant value, which is similar to friction behavior. However, large values of p result in a stiffer system. Thus, p was chosen by a tradeoff between both factors. For the elastomer specimen 1, the determined model parameters are shown in Table 1, in which the elastomeric specimen has two preload conditions. Notably, the distribution constant φ_0 at 10% preload is lower than that without preload. This implies that the yield force level can be increased with the normal force in the preload condition. Similarly, the preload force also can increase the stiffness. As a result, the addition of a preload perpendicular to the loading axis tends to increase the equivalent stiffness and damping over the entire amplitude range. This effect is due to the compressive preload increasing the friction response of the filler in the elastomer, and is not reflected in the distributed elasto-slide model. Thus, the model parameters are different for different preload conditions. For the elastomer specimen 2, the determined model parameters are shown in Table 2. Notably, the elastomer specimen 2 is much stiffer than the elastomer specimen 1 since the stiffness of the leading spring and remaining spring is much higher. Thus, the loss factor of the elastomer specimen 2 appears much lower than the elastomer specimen 1 though the yield force level of the specimen 2 is higher than the specimen 1.

Parameter	No Preload	10% Preload
φ_0 (1/N)	0.0068	0.0053
k (N/mm)	505	739
k_0 (N/mm)	44.4	64.9
v_r (mm/s)	50	50
p	7	7

Table 1. Model Parameters for Elastomer Specimen 1

Parameter	No Preload
φ_0 (1/N)	0.0015
k (N/mm)	6436
k_0 (N/mm)	2915
v_r (mm/s)	15
p	7

Table 2. Model Parameters for Elastomer Specimen 2

4.3. Modeling results and validation

As stated before, the distributed rate-dependent elasto-slide model is reminiscent of the behavior of filler structures in the elastomer such that it can predict the forced harmonic response of an elastomer in the time domain. In this section, single frequency and dual frequency steady-state hysteresis data are used to validate the model. To assess the model's capability in describing elastomer behavior under complex loading conditions, the response under dual frequency loading with slowly varying amplitude is also correlated with model predictions.

For the elastomer specimen 1, three sets of single frequency hysteresis cycle data were used to assess model fidelity. Each set of data was obtained by measuring the forced response while the elastomeric specimen was under sinusoidal displacement excitation at 2.5 Hz, 5.0 Hz and 7.5 Hz, respectively. At each frequency, the displacement amplitude was chosen as 1 mm, 2 mm, 3 mm and 4 mm. In Fig. 15, the experimental data at three frequencies are shown compared to the modeling results. Generally, the modeling results correlate quite well with the experimental results while the displacement amplitude is in the moderate amplitude range, i.e. $2<x<5$ mm. In the small amplitude range, i.e. $x<2$ mm, the analytical model under-predicts the area enclosed by the hysteresis cycle. The reason for that is partly because the lower yield region for the elasto-slide element was replaced with a non-zero constant yield force for numerical consideration and the influence of this approximation was amplified at small deflection loading.

The complex modulus determined by the analytical model is also compared to the experimental result. As shown in solid lines in Fig. 16, the predicted storage and loss stiffnesses using the model have the same amplitude dependent trend as the experimental result. The experimental moduli are well matched with the analytical moduli at moderate amplitude range except that the moduli over small amplitude range are under-predicted especially for loss stiffness. Model predictions are also compared to the experimental data for the loss factor. The predicted loss factor represent common features of the elastomeric response. Clearly, at small amplitude, most of the filler structures, or corresponding elasto-slide elements, have not yielded, so that the loss factor is small. As the amplitude increases, breaking filler structures or yielding of slides leads to a rise in the loss factor. After all of the slide elements have yielded, the loss factor decreases again. In Fig. 16, it also shows that both experimental and predicted moduli are weakly dependent on frequency. This phenomenon is consistent with the tribo-elastic mechanism of elastomeric materials [11].

Similar single frequency modeling results for the elastomer specimen 2 are shown in Fig. 17 as force-displacement diagrams for different amplitudes at 2.5 Hz (Fig. 17a), 5 Hz (Fig. 17b) and 7.5 Hz (Fig. 17c), respectively. Clearly, the analytical model captures the amplitude and frequency dependent behavior of the elastomer specimen 2.

In some applications, the elastomer would experience multi-frequency excitation. Under such a circumstance, the potential loss of damping at the lower frequency due to limitation of stroke is well known [5], so it is important to predict the response of the elastomer under dual frequency excitation. Experimental dual frequency force-displacement data of the elastomer specimen 1 were used to evaluate the adaptability of the model under complex loading conditions.

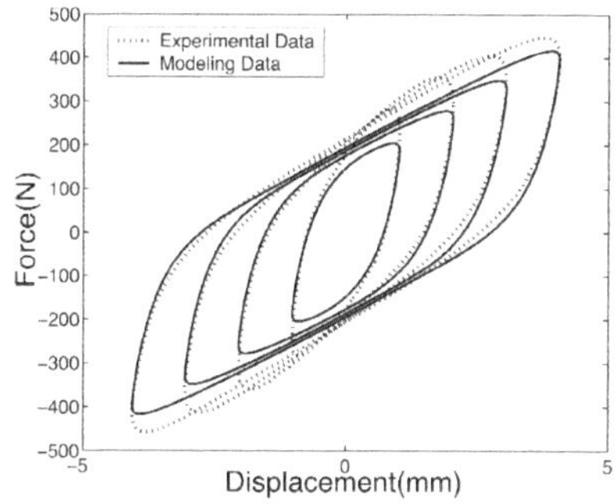

(a) 2.5 Hz

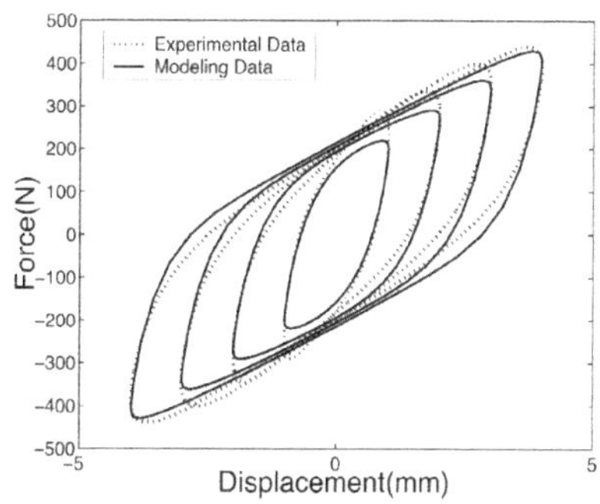

(b) 5 Hz

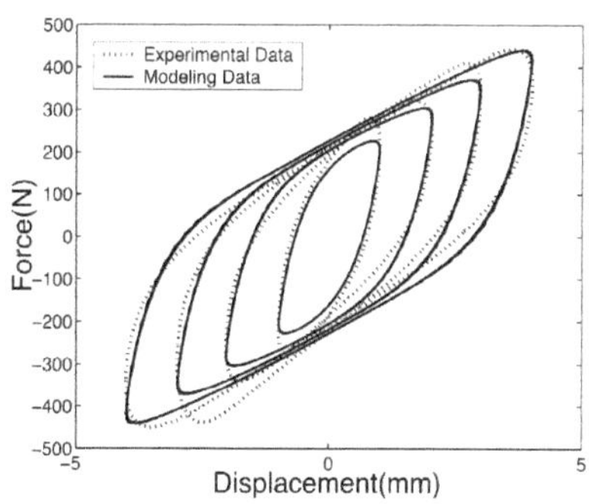

Figure 15. Single Frequency Modeling Results for Specimen 1

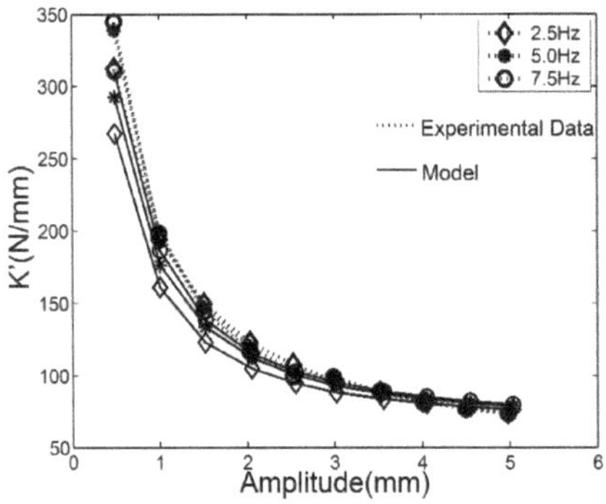

(a) In-phase Stiffness

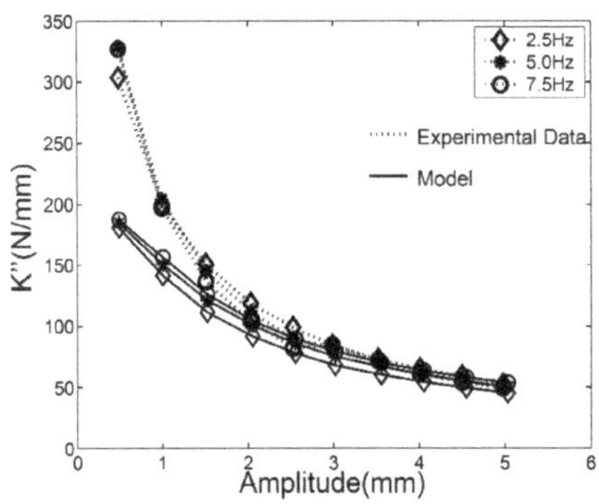

(b) Quadrature Stiffness

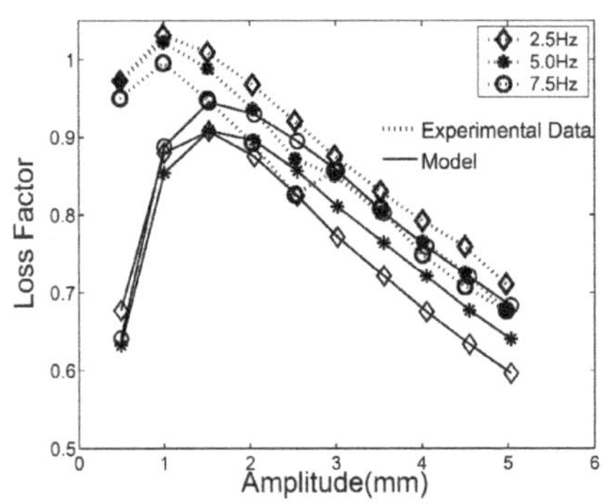

(c) Loss Factor

Figure 16. Complex Modulus for Specimen 1

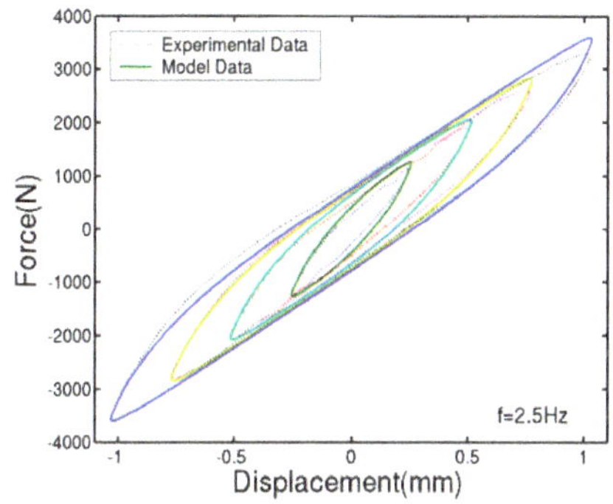

(a) 2.5 Hz

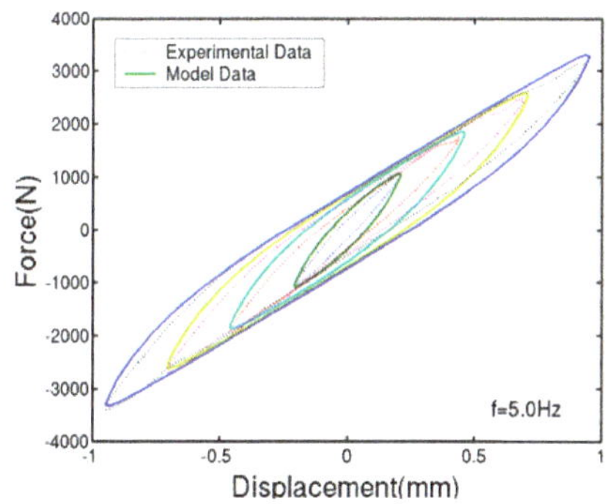

(b) 5 Hz

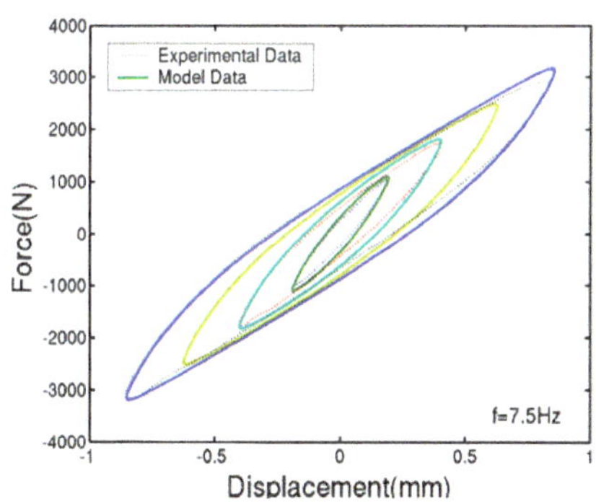

c) 7.5 Hz

Figure 17. Single Frequency Modeling Results for Specimen 2

The dual frequency test data were obtained while the amplitudes at both 5 and 7.5 Hz frequencies were held constant. For each test condition, the amplitude for 5 Hz and 7.5 Hz frequencies ranged from 0.25 mm to 5 mm, and the sum of both amplitudes must not exceed 5 mm, which corresponds to the maximum allowable strain of 50%. The modeling result at each dual frequency loading condition was correlated with the corresponding experimental result. Some of these dual frequency modeling results are presented in Fig. 18 and 19. The figures are grouped according to the 7.5 frequency amplitude. Fig. 18 shows the modeling results for 2.5 mm amplitude at 7.5 Hz and at four different amplitudes at 5 Hz. As the displacement amplitude at 7.5 Hz is 2.5mm, the experimental dual frequency behavior can be matched quite well with the modeling results. Comparatively, Fig. 19 shows the modeling results for 0.5 mm amplitude at 7.5 Hz and at four different amplitudes at 5 Hz. Notably, the model under-predicts the forced response as the total amplitude at 5 Hz and 7.5 Hz is below 2.5 mm since the high yield force region in Fig. 14 is not well described by the numerical algorithm of the model. In general, the distributed rate-dependent elasto-slide model performs well in the moderate amplitude range except it over-predicts the inner loop in some cases. The model also should be improved to predict the response over the small amplitude range.

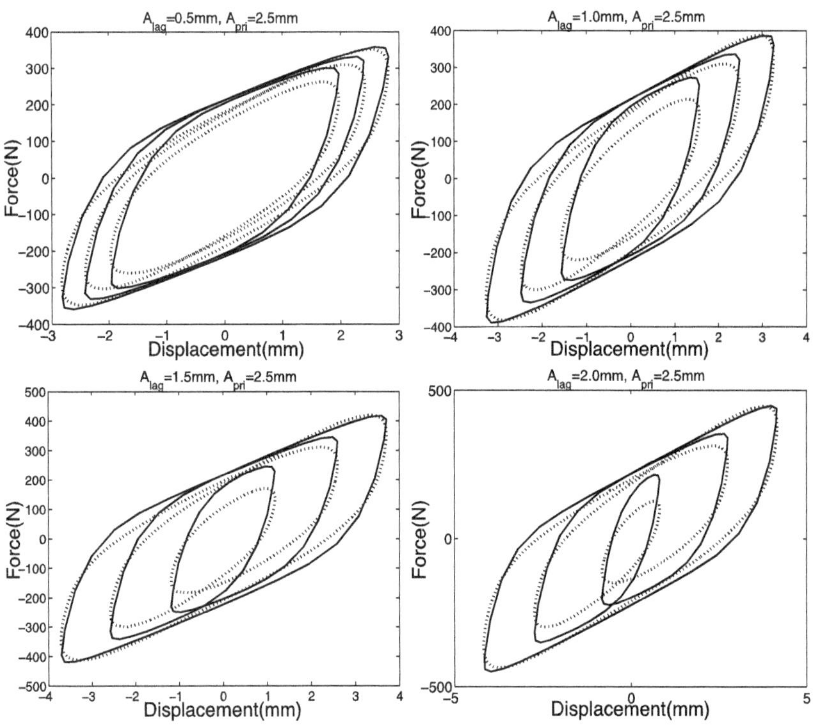

Figure 18. Dual Frequency Modeling Results for Specimen 1

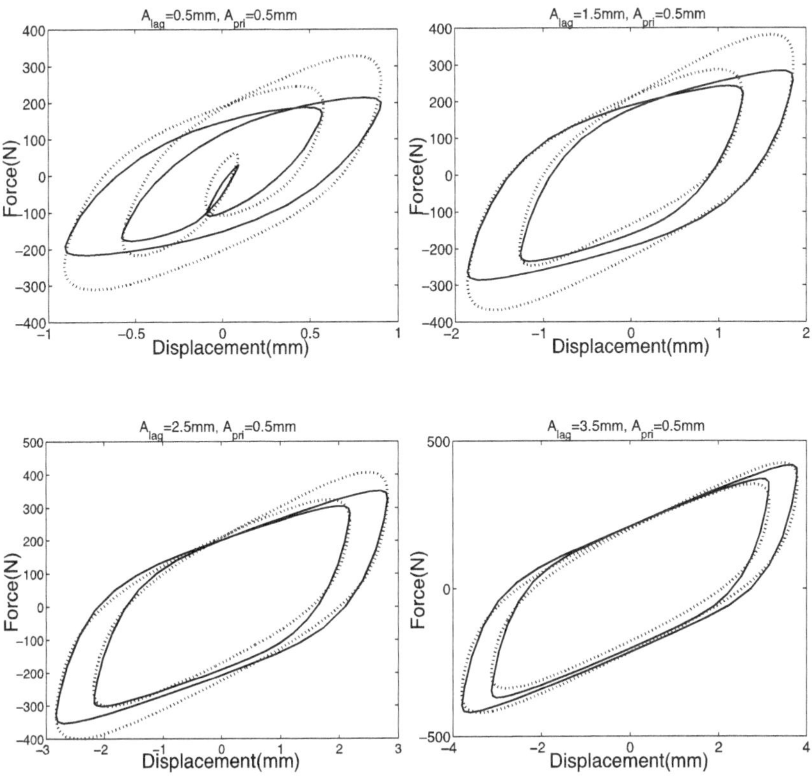

Figure 19. Dual Frequency Modeling Results for Specimen 1

The behavior of the elastomer under a dual frequency excitation with a slowly-varying amplitude modulated periodic loading can also be predicted using the analytical model. For simplicity, the analytical and experimental simulation results are only shown for one scenario, in which the amplitude for 7.5 Hz is 1.5 mm and the amplitude for 5 Hz is assumed to be as below:

$$A_{5Hz} = 1.5\left[1 + 0.2\sin\left(0.2\pi t\right)\right]\left(\mathrm{mm}\right) \tag{35}$$

The predicted forced response data shown in Fig. 20a and 20b compared well to the experimental results for two different time scales. Similarly, the modeling force-displacement hysteresis cycle is also matched well with the experimental data as shown in Fig. 20c. The predicted damper response due to the slowly varying displacement excitation exhibits the same varying trend as the observed elastomer behavior, and also the force value is tracked quite well. Clearly, the proposed elastomeric model performs fairly well in predicting dual frequency response, and especially the distributed elasto-slide model can predict the behavior of the elastomer under slowly-varying amplitude modulated periodic loadings.

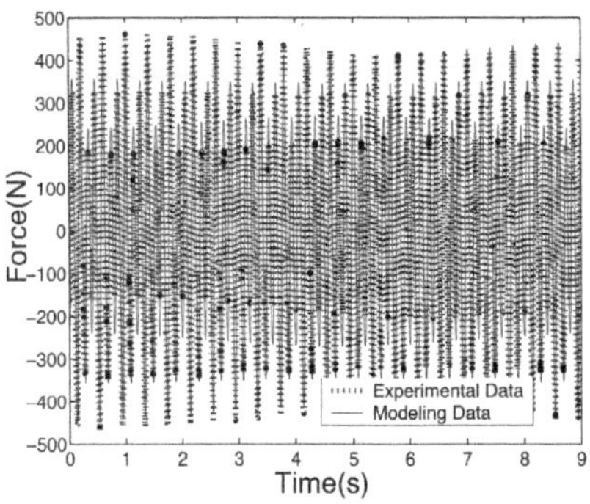

(a) Force Response (0-9 s)

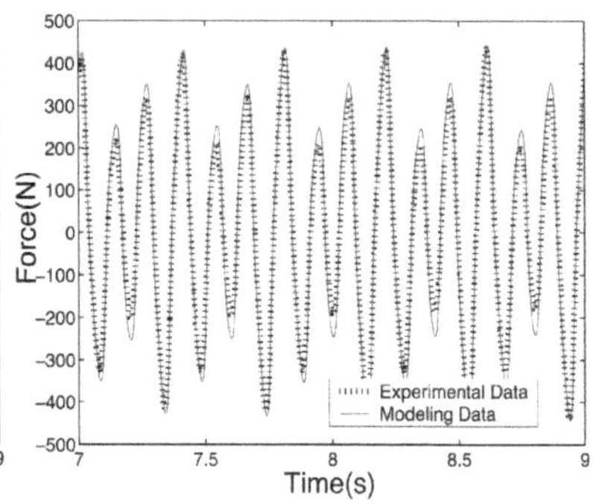

(b) Force Response (7-9 s)

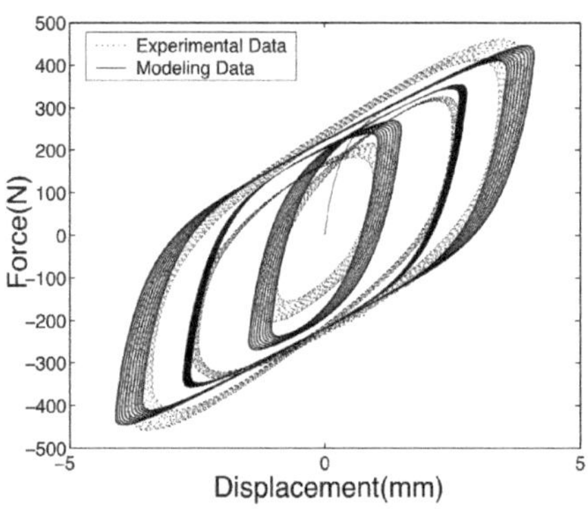

(c) Force-Displacement Hysteresis

Figure 20. Modeling Results for Dual Frequency with Slowly Varying Amplitude

5. Summary

Modeling methods for describing elastomeric material behavior were investigated. Most prior models introduced nonlinear terms into the conventional Kelvin model or Zener model. Because filled elastomers are anelastic materials, a friction mechanism damping element proves useful to model rate-independent damping. Nonlinearity in tested elastomeric materials manifested in two ways. First, the forced response of an elastomer subjected to harmonic displacements was nonlinear (non-elliptical), which meant that the response could not be predicted by linear differential or integral equation. Second, the stiffness and damping of elastomers varied as a function of amplitude and frequency. While some models capture the amplitude dependent complex moduli very well using constant parameters, such models cannot predict stress-strain or force-displacement hysteresis accurately. On the other hand, most hysteresis models can predict non-elliptical hysteresis quite well, but their parameters are usually amplitude and frequency dependent. The methods require amplitude and frequency as prior information when these models are implemented.

A nonlinear hysteresis model was developed to characterize the nonlinear behavior of the elastomer under a cyclic loading. This model is mechanically analogous to a nonlinear Kelvin model where the stiffness is a nonlinear monotonic function of displacement and the damping is a monotonic rate dependent friction function. Since it accurately describes the characteristics of the hysteresis loops, this model can predict steady state or harmonic forced response very well. However, the model parameters are still amplitude dependent. A challenge still remained to describe transient or stress relaxation behavior using this type of mechanisms-based model.

Therefore, a distributed rate-dependent elasto-slide elastomeric model was used to describe the amplitude dependent characteristics of an elastomer. This physically motivated damper model resembles the behavior of filler structures in the elastomer under cyclic loading. A method to determine the model parameters was presented. It was found that a unique exponential function could be used to describe the yield force distribution for elastomers. Numerical algorithms were developed for model applications. Dynamic test were conducted on a double lap shear elastomeric specimen and a linear concentric tubular elastomeric specimen, respectively, and the measured data were used to evaluate the modeling method. The fidelity of the model was verified by the good correlation between predicted single and dual frequency force-displacement hysteresis and the experimental results except that the damping at lower amplitude range cannot be fully predicted by the model. Since the proposed model is a time domain model, the adaptability of the model in predicting damper response under a slowly varying displacement excitation was evaluated. The predicted force response of the elastomeric specimen under this slowly varying displacement excitation correlated quite well with the corresponding experimental data.

In conclusion, the distributed rate-dependent elasto-slide elastomeric damper model is a time-domain modeling approach to capture nonlinear behavior of the elastomer. The damper model, formulated as a state space model has the advantage that it could easily be implemented into dynamic system models. Because the model is physically motivated, the flexibility in determining the distribution function provides means to improve the model performance especially over the low amplitude range. Although only a one-dimensional elastomeric model is described in this paper, the distributed elasto-slide model can also be extended into a three-dimensional form such that it can be implemented easily into a finite element analysis for a complex elastomeric damper configuration.

Author details

Wei Hu and Norman M. Wereley*
Department of Aerospace Engineering, University of Maryland, College Park, MD, USA

6. References

[1] Coveney, V.A., Johnson D.E., and Turner D.M., "A Triboelastic Model for The Cyclic Mechanical Behavior of Filled Vulcanizates," *Rubber Chemistry and Technology*, Vol. 68, No. 4, 1995, pp. 660-670.

[2] Panda, B., Mychalowycz, E. and Tarzanin, F.J., "Application of Passive Dampers to Modern Helicopters," *Smart Materials and Structures*, Vol. 5, No. 5, 1996, pp. 509-516.

[3] Zener, C.M., *Elasticity and Anelasticity of Metals*, University of Chicago Press, Chicago, IL., 1948.

[4] Gandhi, F., and Chopra, I., "A Time-domain Non-linear Viscoelastic Damper Model," *Smart Materials and Structures*, Vol. 5, No. 5, 1996, pp. 517-528.

[5] Felker, F., Lau, B., McLaughlin, S. and Johnson, W., "Nonlinear Behavior of an Elastomeric Lag Damper undergoing Dual-Frequency Motion and its Effect on Rotor Dynamics," *Journal of the American Helicopter Society*, Vol. 32, No. 4, 1987, pp. 45-53.

[6] Payne, A.R. and Whittaker, R.E., "Low Strain Dynamic Properties of Filled Rubbers," *Rubber Chemistry and Technology*, Vol. 44, No. 2, 1971, pp. 440-478.

[7] Tarzanin F.J., and Panda, B., "Development and Application of Nonlinear Elastomeric and Hydraulic Lag Damper Models," Proceedings of the 36th AIAA/ASME/ASCE/AHS/ASC Structures, Structural Dynamics, and Materials Conference, New Orleans, Louisiana, April 1995.

[8] Timoshenko, S.P., *Strength of Materials*, Part 2, D. Van Nostrand Company, New York, N.Y., 1933, pp. 679-680.

[9] Iwan, W.D., "A Distributed-Element Model for Hysteresis and Its Steady-State Dynamic Response," *Journal of Applied Mechanics*, Vol. 33, No. 4, 1966, pp. 893-900.

[10] Iwan, W.D., "On a Class of Models for the Yielding Behavior of Continuous and Composite Systems," *Journal of Applied Mechanics*, Vol. 34, September 1967, pp. 612-617.

[11] Turner, D.M., "A Triboelastic Model for the Mechanical Behavior of Rubber," *Plastics and Rubber, Processing and Applications*, Vol.9, No. 4, 1988, pp. 197-201.

* Corresponding Author

[12] Strganac, T.W., "An Experiment and Analytical Methodology to Characterize Nonlinear Elastomeric Lag Dampers," Proceeding of the 38th AIAA/ASME/ASCE/AHS/ASC Structures, Structural Dynamics, and Material Conference, Orlando, Florida, April 1997.

[13] Lesieutre, G.A., and Bianchini, E., "Time Domain Modeling of Linear Viscoelasticity Using Augmenting Thermodynamic Fields," Proceedings of the 34th AIAA/ASME/ASCE/AHS/ASC Structures, Structural Dynamics and Materials Conference, La Jolla, CA, April 1993.

[14] Lesieutre, G.A., and Mingori, D.L., "Finite Element Modeling of Frequency-Dependent Material Damping Using Augmenting Thermodynamic Fields," *Journal of Guidance, Control and Dynamics*, Vol. 13, No. 6, 1990, pp. 1040-1050.

[15] Govindswamy, K., Lesieutre, G.A., Smith, E.C., and Beale, M.R., "Characterization and Modeling of Strain-Dependent Dynamics Behavior of Viscoelastic Elastomers in Simple Shear," Proceedings of the 36th AIAA/ASME/ASCE/AHS/ASC Structures, Structural Dynamics, and Materials Conference, New Orleans, LA, April 1995.

[16] Brackbill, C.R., Lesieutre, G.A., Smith, E.C., and Govindswamy, K., "Thermomechanical Modeling of Elastomeric Materials," *Smart Materials and Structures*, Vol. 5, No. 5, 1996, pp. 529-539.

[17] Ramrakhyani, D.S., Lesieutre, G.A., and Smith, E.C., "Efficient Modeling of Elastomeric Materials Using Fractional Derivatives and Plastic Yielding," Proceedings of the 43rd AIAA/ASME/ASCE/AHS/ASC Structures, Structural Dynamics, and Materials Conference, Denver, CO, April 2002.

[18] Aklonis, J.J., and McKnight, W.J., *Introduction to Polymer Viscosity*, John Wiley and Sons, New York, 1983.

[19] Kunz, D.L., "Influence of Elastomeric Lag Damper Modeling on the Dynamic Response of Helicopter Rotors," *AIAA Journal*, Vol. 35, No. 2, 1997, pp.349-354.

[20] Constantinou, M., Mokha, A., and Reinhorn, A., "Teflon Bearings in Base Isolation, Part 2:Modeling," *Journal of Structural Engineering*, Vol. 116, No. 2, Feburary, 1990, pp. 455-474.

[21] Berger, E.J., "Friction Modeling of Dynamic System Simulation," *Applied Mechanics Review*, Vol. 55, No. 6, November 2002, pp. 535-577.

[22] Krishnan, R., "Mechanisms-based Analysis and Modeling of Elastomeric Lag Damper Behavior," *Master of Science*, University of Maryland, Aerospace Engineering, 2000.

[23] Snyder, R.A., "Mechanisms-based Modeling of Filled Elastomeric Lag Damper Behavior," *Master of Science*, University of Maryland, Aerospace Engineering, 2001.

[24] Constantinou, M., Mokha, A., and Reinhorn, A., "Teflon Bearings in Base Isolation, Part 2:Modeling," *Journal of Structural Engineering*, Vol. 116, No. 2, February, 1990, pp. 455-474.

[25] Faires, J.D. and Burden, R., *Numerical Methods,* Brooks/Cole Publishing Company, Pacific Grove, California, 1998.

Chapter 15

The Significance of Equi-Biaxial Bubble Inflation in Determining Elastomeric Fatigue Properties

Steve Jerrams, Niall Murphy and John Hanley

Additional information is available at the end of the chapter

1. Introduction

"I disapprove of certainties, said Virgil Jones. They limit one's range of vision. Doubt is one aspect of width."[1]

As yet, there are few certainties in the determination of fatigue life for rubber components. Scepticism is an essential characteristic for a researcher and the study of elastomeric fatigue is a prodigiously wide topic. Rubber components predominantly fail in fatigue and so determining elastomeric component fatigue life has acquired greater interest for materials scientists in recent years. Fatigue testing to date has generated results with specimens loaded in uniaxial tension, combined tension and torsion or in shear. While these test methods provide much useful insight into the fatigue of elastomers, they do not describe the full spectrum of elastomeric material behaviour under cyclic loading. In this light, dynamic bubble inflation offers a reliable and repeatable method for determining viscoelastic characteristics of rubber-like materials. The development of an equi-biaxial bubble inflation system is a central feature of this chapter, which also highlights the research projects that made the system development possible.

Perhaps the most comprehensive overview of elastomeric fatigue was provided by Mars and Fatemi in 2004 [1], though inevitably there has been extensive research into the topic since then. At the time, they placed factors influencing elastomer fatigue into four categories: mechanical loading history, environmental effects, rubber formulation and effects due to dissipative aspects of the constitutive response of rubber. They pointed out that fatigue failure of rubber-like materials was imperfectly understood and, despite gaining significant new knowledge in the intervening years, this remains the case. Factors influencing crack nucleation and growth were the subject of numerous previous

---

[1] A quote from the novel Grimus (ISBN: 0575018712) by Salman Rushdie, published by Gollancz (1975)

investigations [2, 3, 4]. Mechanical loading history is important in the fatigue analysis of all engineering materials and so frequencies, stress and strain ranges, strain energy and energy release rates are obvious parameters to be considered. However, the interpretation of complex loading histories is problematical and a practical approach to avoiding this difficulty is offered in this text. A further consideration for rubber is that whereas a high steady load at sub-fracture level will not induce failure in a conventional linear solid, this is not always so for elastomers, where time dependent crack growth and environmental attack can ultimately cause a component to fail. Though such failures cannot be attributed to fatigue, since no cyclic deformation takes place, non-strain crystallizing elastomers can still fail over time. However, additional consideration must be made of the role played by strain crystallization in limiting crack growth in some elastomers, notably natural rubber (NR) and also whether cyclic load applications open or close internal flaws. Many procedures for physical testing of rubber had their roots in methods that were applied to testing metals. These procedures are often far less effective for elastomers. Consider the tendency to cycle samples through constant strain limits [5, 6] which makes perfect sense for the micro-strains experienced by steel or aluminium alloy, but a rubber sample subject to set and stress relaxation will be cycled through an ever decreasing load range, so a fatigue test can be inordinately long or the component may not fail at all. Conversely, if a rubber sample is cycled between constant load limits, there is always the possibility that the strain will increase to a level where the stroke of the test machine is exceeded. Clearly, the choice of test methods and parameters in fatigue testing of rubber is far from simple. Decisions to plot fatigue data against parameters used for metals (minimum stress, maximum stress, maximum strain, stress amplitude) are arbitrary and again an alternative approach is suggested in this text. Mars and Fatemi [7] highlighted a lack of understanding of the effects of multi-axial fatigue loading of elastomers and pointed to a tendency to use thin laboratory specimens though, unlike linear elastic solids, rubbers experience significant size effects [8, 9]. They concluded that the following factors are important;

- the relationship between the applied loads and the localised load experienced by a crack,
- whether or not applied loads cause closure of initial cracks
- the dependence of material properties on the type and magnitude of the imposed deformation (e.g. the role that might be played by strain crystallisation).

Loading frequency has little effect on the fatigue properties of strain crystallising rubbers, but has a significant effect on amorphous rubbers that has been attributed to time-dependent steady crack growth [10]. At high frequencies and strains, rubber experiences a further failure mechanism termed thermal runaway, where compound temperatures increase rapidly and degradation ensues [11]. The fatigue properties of amorphous rubbers are considered to be more influenced by waveform than those of strain crystallising elastomers, but this is highly dependent on polymer and filler types [12].

Environmental conditions are particularly important in affecting fatigue processes for elastomers. Fatigue life can fall and crack growth rate increase markedly as a result of temperature increase and these outcomes are independent of chemical changes due to aging or continued vulcanisation, though temperature increases clearly accelerate chemical processes. Exposure to Ozone increases crack growth rate and reduces fatigue life [13]. In the vicinity of stress concentrations occurring at the crack tip, network chains react with ozone, causing scissions of the chains, though other chemical agents can attack rubber in the same way [14]. Oxidative aging of rubber causes the material to become more brittle and accelerates fatigue crack growth [15].

Rubber fatigue behaviour is influenced by polymer type, filler type and volume fraction. Additionally, the manufacturing process and the amount and type of antioxidants, antiozonants and curatives will have an effect. Again, whether or not the compound exhibits strain crystallisation is a key factor; strain crystallising rubbers are less prone to environmental effects [16]. Carbon black improves fatigue properties with low-structure blacks imparting superior properties to high-structure blacks [17].

In the past few years, the increased use of finite element techniques to facilitate elastomeric fatigue life prediction has become more prevalent [18], along with novel approaches to predict discontinuous crack growth [19]. Similarly, greater use of tomography has occurred, particularly in the study of heat build up during cycling [20, 21, 22] and X-ray diffraction plays an important part in investigations of fatigue mechanisms [23]. Scanning Electron Microscopy (SEM) and Digital Image Correlation (DIC) techniques have each been used more extensively to study crack morphology and growth in the last decade [24, 25]. The research of Le Cam and Toussaint is particularly illuminating in respect of crack bifurcation under non-relaxing conditions [26] and cavitation [27] and McNamara *et al* have developed Image Correlation Photogrammetry (ICP) to study stress softening in successive cycles at stress-raisers in a range of compounds [28].

Improvements in engineering materials are continually sought as functionality, reliability and resilience are demanded from the machines that we depend upon each day. As our understanding of elastomeric material behaviour has advanced, there has been an accompanying realisation that obtaining reliable multi-axial dynamic data is essential in fully characterising the hyperelastic and viscoelastic properties of rubber. When machine components fail, they should and usually do fail in fatigue. For them to fail under static loading would indicate that designs or material selections were inept. As there is an increasing threat to natural resources, ensuring that components exhibit longevity has gained heightened importance. Hence, understanding fatigue and designing to obviate premature failures is essential. This text describes interrelated fatigue research programmes spanning a fifteen year period from the inception of a static bubble inflation system, a novel approach to elastomer fatigue life prediction using Wöhler (S/N) curves, the consideration of pre-stressing and dynamic stored energy, an appraisal of the contribution made by flaw size in inducing fatigue failure in rubber samples, cyclic bubble inflation to provide reliable fatigue data and culminates in a study of swelling phenomenon in diaphragms.

2. Approaches to determining the fatigue life of elastomers

Fatigue life prediction for rubber components can be approached in two ways:

a. By using fracture mechanics methods as epitomised by the research of Busfield *et al* [29] and Thomas [30].
b. By relating the magnitude of a parameter during cycling (e.g. stress amplitude, maximum stress, maximum strain or stored energy) to the cycles to failure. These relationships are shown in Wöhler (S/N) curves derived from cycling standard test-pieces to failure (Abraham *et al* [31]).

In the first approach, Mechanical fatigue of elastomers is characterised by a reduction of physical properties as a result of crack propagation during dynamic excitation. There are levels of stress and strain below which fatigue damage will not occur, but these levels are imperfectly understood. Energy concepts of elastomeric fatigue life determination are based on energy release rates and have their origins in the theories first postulated by Griffith [32]. It is assumed that the energy of a cracked sample will be entirely converted into crack surface energy during repeated loading. However, when characterising crack propagation in rubber, two unique problems must be surmounted:

i. Fatigue properties are not merely dependent on chemical composition, but also on individual considerations such as cross linking systems and aging protection used in the compound.
ii. The combination of hyperelastic and viscoelastic properties gives rise to sensitivity to loading modes and frequencies, strain rates and temperatures.

2.1. Fracture mechanics approach

Rivlin and Thomas [33] first offered approaches to determining energies in crack growth of rubber specimens, defining tearing energy T as the decrease in elastic strain energy caused by the crack growth per unit area, giving rise to the relationship shown in Equation 1.

$$T = \left(\frac{\delta W}{\delta A} \right)_l \tag{1}$$

where W is the total elastic energy, A the crack area and l denotes that the specimen is held under constant deformation, so the application of applied external forces does no additional work.

Under tensile loading of a notched specimen, tearing energy can be expressed as

$$T = 2kW_0c \tag{2}$$

where W_0 is the strain energy per unit volume, c is the crack length and k a constant given by

$$k = \frac{\pi}{\lambda} \qquad [34, 35, 36](3)$$

Tearing energy of a notched sample under tensile loading can be calculated using Equation (2) by replacing W_0 with strain energy density W_{max} at stretch ratios of $\lambda = \lambda_{max}$. The fracture mechanics approach can be used to relate initial flaw size to fatigue life. At intermediate and high tearing energies, fatigue behaviour can be approximately described by Equation (4)

$$\frac{dc}{dN} = BT^{\beta} \qquad (4)$$

where B and β are material constants

By integrating Equation (4) and substituting Equation (2) for T, an expression for cycles to failure N is derived (Equation (5)).

$$N = \frac{1}{\beta - 1}\left[\frac{1}{\beta(2kW_0)^{\beta}}\right]\left[\frac{1}{c_0^{\beta-1}} - \frac{1}{c^{\beta-1}}\right] \qquad (5)$$

Using this expression can be problematical since material constants from short-term testing are used to extrapolate fatigue lives for un-notched specimens.

2.2. Wöhler (S/N) curves

As early as 1870, Wöhler [37] summarised his studies on fatigue in steam train axles, pointing to the importance of stress amplitude and the existence of a 'fatigue limit' for steel components. Thereafter, Wöhler (or S/N) curves became a standard method for determining fatigue life in linear elastic solids. Most commonly, alternating stress σ_a or $\log_{10} \sigma_a$ (ordinate), is plotted against cycles to failure N (abscissa) and usually the latter scale is logged ($\log_{10}$). The magnitude of mean stress σ_m, where components are pre-stressed ($\sigma_{min} \neq 0$), is a critical factor in fatigue life prediction and several approaches (e.g. Haigh and Goodman diagrams) have been used to determine its influence.

3. Uniaxial fatigue testing of elastomers

It is well documented that for strain crystallising rubber, crack growth steps during repeated loading are small if the material is not fully relaxed between each cycle [38]. A crystalline region develops at the crack tip and the crystallinity remains for repeated load cycles. Consequently crack growth is inhibited and fatigue life is prolonged. Increased fatigue (or crack growth) resistance with increased minimum strain is accompanied by an increase in hysteresis. Lake [39] showed that higher hysteresis caused lower crack growth and subsequently Lindley [40] showed that the high strain hysteresis was due to strain induced crystallisation. However, improved fatigue resistance in non-strain crystallising rubbers subjected to preloading had not been observed. When compared with strain crystallising elastomers, Gent [38] and others [41, 42] found no comparable improvement in fatigue properties for filled non-crystallising elastomers subjected to a pre-stress prior to cycling between tensile stress limits. These results were confirmed by Lindley [10]. By contrast with strain crystallising rubbers like NR, the hysteresis of non-strain crystallising rubbers results from viscoelastic behaviour and is generally of a lower order.

Possible reasons for increased fatigue life for higher mean stresses (where a pre-load is applied prior to cycling at a constant stress or strain amplitude) have been advanced and are discussed in points a) - d).

a. Compressive and tensile loading of elastomers each reduce the effective severity of natural flaws. Natural flaws are unavoidable and for many years were thought to have an effective initial size of 25 μm and were considered equivalent to a sharp edge cut of approximately 40 ± 20 μm in depth [43]. If a compressive pre-load closes a stress raiser and a tensile pre-load 'smoothes out' a stress concentration [44], it is arguable that stress cycles become less injurious throughout a fatigue test.

b. Discounting the effect of hysteresis, the strain energy or tearing energy available in each cycle to propagate a crack can be less when σ_{min} or ε_{min} is not zero. This situation is shown diagrammatically in Figure 1 where equal load amplitudes do not equate to equal values of 'dynamic stored energy'. This explanation is consistent with the tearing (energy) approaches described earlier and is considered further in the next section.
c. Lee and Donovan [45] established that carbon black increases the size of the crystallised zone at a stressed crack tip.
d. Lake *et al* [46] have shown that certain non-crystallising rubbers experienced the formation of anisotropic structures resulting from filler particle alignment or aggregates at the crack tips. In tests on samples that were not completely relaxed between cycles, this produced similar increases in fatigue strength to those observed in tests on strain crystallising rubbers.

3.1. Uniaxial fatigue testing of non-strain crystallising elastomers

Abraham [47] investigated the fatigue life and dynamic crack propagation behaviour of non-strain crystallising elastomers to determine their dependency on test parameters. The research culminated in recommendations of criteria for precise prediction of service life for components formed from these compounds. Fatigue life was investigated for EPDM and styrene-butadiene rubber (SBR) dumbbell specimens. The chemical compositions of the materials are given in Tables 1 and 2 respectively.

The specimens were cycled to failure under load control at 1 Hz. The frequency was chosen to avoid internal friction causing large increases in temperature and consequent thermal breakdown [48]. For this reason, the 1 Hz frequency was subsequently adopted by Jerrams *et al* [49] for all equi-biaxial cyclic testing. The influence on fatigue properties of varying minimum stress for a constant stress amplitude was investigated and confirmed the well-known amplitude dependence of fatigue life in filled rubbers. Raising the level of minimum stress (and consequently mean and maximum stress for a constant stress amplitude), applied to uniaxial test samples, was found to produce an increase in fatigue life. This cannot be attributed to strain crystallisation as is the case for NR. It appeared that this effect was specific to filled systems and the research showed that the fatigue behaviour of carbon black filled non-strain crystallising rubbers cannot be universally described using a maximum stress or a maximum strain criterion. As a result, an energy criterion was postulated.

EPDM	100.0 pphr
Carbon black N550	70.0 pphr
Carbon black N772	40.0 pphr
Stearin acid	1.0 pphr
Zinc-oxide	5.0 pphr
Oil Sunpar 2280	70.0 pphr
Sulphur	1.5 pphr
Vulcanising agent CZ (CBS)	1.0 pphr
Vulcanising agent Thiuram (TMTD)	0.8 pphr

Table 1. Chemical composition of the filled EPDM material.

SBR 1712	100.0 pphr
Carbon black N234	51.0 pphr
Stearin acid	1.45 pphr
Zinc-oxide	2.2 pphr
GPPD	1.45 pphr
TMQ	0.73 pphr
Antilux 500L	1.45 pphr
Sulphur	1.27 pphr
Vulcanising agent CZ (CBS)	0.73 pphr
DPG	0.29 pphr

pphr = parts per hundred rubber

Table 2. Chemical composition of the carbon black filled SBR material

Another important effect was highlighted in this research [50, 51, 52]. As load cycles were accumulated, it was observed that most mechanical properties changed, but in particular the stiffness of specimens changed throughout the full duration of the fatigue tests; an equilibrium material stiffness was never reached. This behaviour was observed in elastomers containing reinforcing fillers. For carbon black filled EPDM test specimens, it was found that failure occurred when they reached approximately 76% of their initial stiffness (or complex tensile modulus E^* was 76% of the initial tensile modulus), irrespective of the form or severity of the loads applied to the specimens during repeated cycling. Carbon black filled SBR test specimens failed in similar tests when they reached approximately 71% of their initial stiffness. The results indicated that for non-strain crystallising elastomers there is a characteristic value for the material parameter (E*) that is reached at failure.

In respect of minimum load dependence (pre-stressing), results from the dynamic crack propagation tests on carbon black filled non-strain crystallising elastomers showed similar behaviour to that observed in the fatigue to failure tests. Increasing minimum loads at constant strain (displacement) amplitude, under both pulsed and sinusoidal excitation, decreased the crack growth rate of the carbon black filled rubber material. Thus it was argued that pre-loading can lead to higher service lives for parts produced from these materials. Unfilled EPDM and SBR test specimens showed no beneficial effects from pre-loading in either the crack propagation tests or the fatigue to failure experiments. As the normal failure observed during the fatigue to failure experiments was a sudden catastrophic crack when a certain level of stiffness (or modulus) was attained, it was assumed that these tests characterised the initiation process. It was thus, an important finding of the research that crack initiation and crack growth in rubbers displayed a similar dependence on test parameters. Furthermore, initiation and growth both appeared to be energy controlled processes in rubber materials.

Abraham determined that the parameter termed 'dynamic stored energy', when plotted against cycles to failure, provided a reliable predictor of fatigue life of EPDM samples and further that this parameter worked equally well for determining fatigue life in SBR samples. The value of dynamic stored energy is shown diagrammatically for one load / displacement curve in a cycle in Figure 1, as the area of the plot beneath the hysteresis curve where the minimum load in the cycle constitutes the lower boundary. The amount of energy available in the tensile pre-loading cycle, shown in red, is smaller than that in the load / displacement cycle with zero minimum load (no pre-load applied) shown in blue.

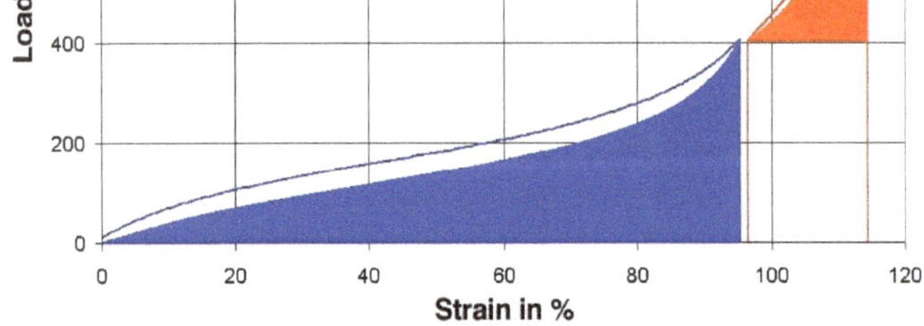

Figure 1. Diagrammatic representation of dynamic stored energy for two equal load amplitudes.

Wöhler (or S/N) curves for EPDM and SBR, where (log_{10}) dynamic stored energy was plotted as ordinate, are shown in Figures 2 and 3 respectively. Each graph includes data for tests on samples subjected to cycles with zero and non-zero minimum loads and different load amplitudes. Thus it can be seen that dynamic stored energy can provide a basis for reliable fatigue life prediction for non-strain crystallising rubbers. It was noted that for SBR, (log_{10}) of strain range plotted against (log_{10}) of cycles to failure could also be used to predict fatigue life for the material (Figure 4). However, it appears that dynamic stored energy offers a reliable parameter for predicting fatigue failure for non-strain crystallising rubbers that can be applied universally.

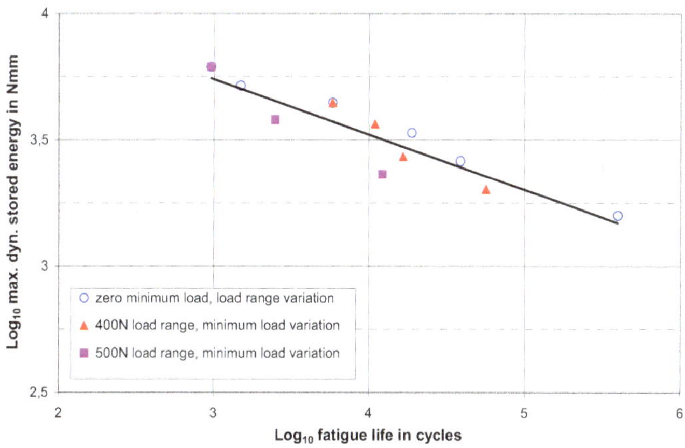

Figure 2. Log_{10} of maximum dynamic stored energy versus log_{10} of number of cycles to failure for filled EPDM for various load ranges and minimum loads.

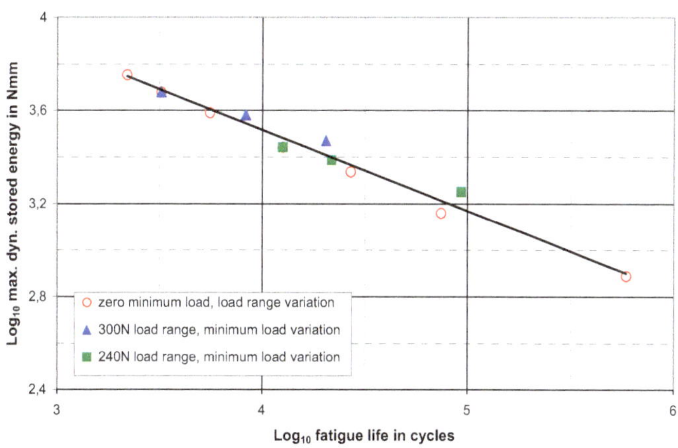

Figure 3. Log_{10} of maximum dynamic energy versus log_{10} of number of cycles to failure for filled SBR for various load ranges and minimum loads.

Figures 5 and 6 substantiate Abraham's third significant conclusion. They depict higher fatigue lives for test samples subjected to different levels of pre-stressing at different constant stress amplitudes, for EPDM and SBR respectively. This finding has prompted the hypothesis that pre-loading non-strain crystallising rubber components can enhance fatigue properties since less dynamic stored energy is available for higher stress ranges under such loading conditions.

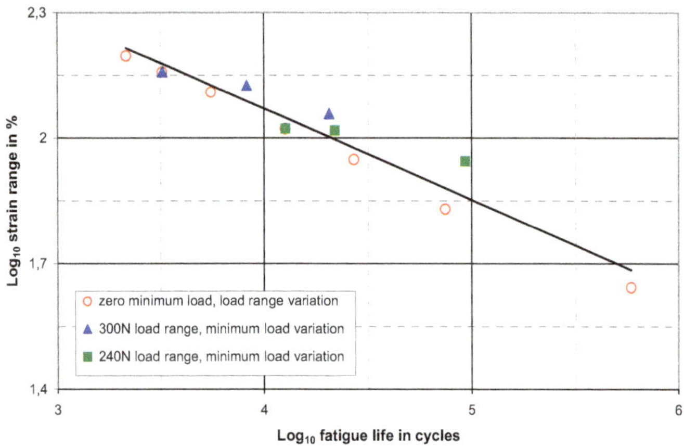

Figure 4. $\log_{10}$ of the strain range against $\log_{10}$ of number of cycles to failure for filled SBR for various load ranges and minimum loads.

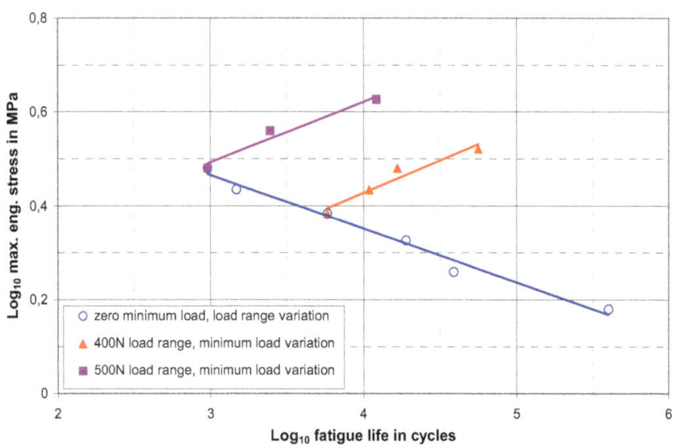

Figure 5. $\log_{10}$ of the maximum engineering (nominal) stress versus $\log_{10}$ of number of cycles to failure for filled EPDM for various load ranges and minimum loads.

The data from all the tests show that the compounds underwent changes in physical properties throughout the load controlled cycles. The dynamic displacement induced an increasing amount of set as each test progressed. Typical hysteresis curves are shown in Figure 7. It is usual for the modulus of filled rubbers to decrease significantly during the first few cycles of a physical test as a result of stress softening (the Mullins effect) [53]. However, these fatigue tests indicated that E^*, as well as the storage modulus E', loss modulus E'' and loss factor (tan δ) decreased throughout the process to final fracture. In a typical fatigue test for which four cycles are shown in Figure 7, it is worth noting that no visible cracks were observed until 7449 cycles were reached and complete rupture occurred within approximately 250 cycles (at 7709 cycles). The same phenomenon was observed under less severe test conditions when test specimens failed after a few hundred thousand cycles.

The results of the fatigue testing suggest that elastomeric materials fail after a material specific loss in complex modulus E* is reached. This predictor seems to be a characteristic of each compound because it is independent of the applied loading. It does not depend on the load amplitude or the applied minimum load and so it is potentially a very important indicator for the maintenance of machines utilising elastomeric parts. The predictor allows us to assess the condition of rubber components easily and online and to replace them just before they fail. The influence of frequency and temperature on the residual stiffness predictor was not examined in this research, though it is well documented that fatigue resistance has a high dependency on these parameters. Moreover, pauses during dynamic testing and their role in the reconstitution of the filler network were not included in this research, so there is much still to be done.

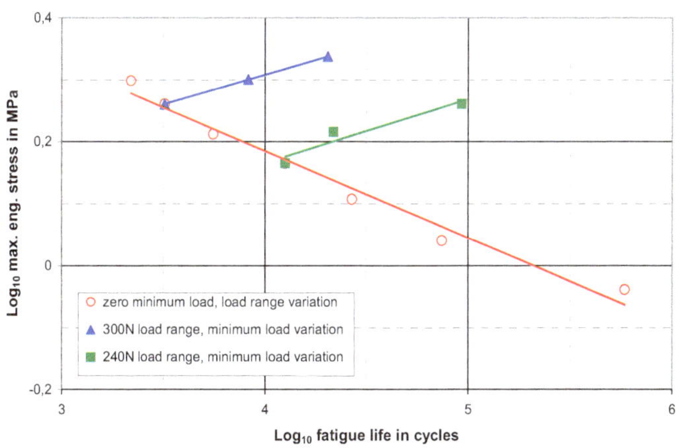

Figure 6. $\log_{10}$ of the maximum engineering (nominal) stress against $\log_{10}$ of number of cycles to failure for filled SBR for various load ranges and minimum loads.

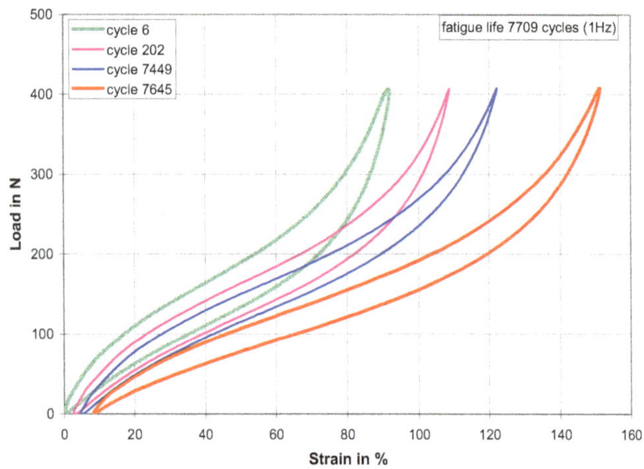

Figure 7. Hysteresis loops for a filled EPDM specimen during fatigue testing over a load range of 400N with no pre-load (1 Hz).

However, the research showed that, for the load amplitudes employed, the fatigue resistance of active carbon black filled non-strain crystallising EPDM and SBR rubber material increased when compressive or tensile preloads were applied. The increase in fatigue life resulted from the active filler and its filler-filler and filler-polymer interaction since this phenomenon did not occur in unfilled EPDM or SBR, where pre-loading reduces fatigue lives. This research showed that cyclic conditions from a minimum stress of zero are the most severe for active filled elastomers at these load amplitudes and also demonstrated that, like other solids, compression is less severe for elastomers than is tension.

The standard fatigue crack propagation testing described in the literature predominantly used displacement (strain) controlled tests. This is only applicable if service conditions under forced displacement are to be simulated. But most of the service conditions of elastomeric products are under load control and testing under displacement control would not provide the precision required. Clauss [54] compared load and displacement controlled fatigue tests. One result was that the stiffer (lower carbon black dispersion) materials failed much earlier under displacement control than softer materials. However, under load control both materials exhibited similar fatigue resistance. Clearly, for rubber the conditions of a laboratory simulation of fatigue life should be as close as possible to service conditions. If a rubber product is used under load control, then the testing should also be carried out under load control, otherwise the results can be misleading for two reasons. Firstly, stress softening occurs throughout the full life of the component. If the test was carried out under strain (displacement) control, less work would be applied in successive cycles and a pronounced modulus loss would not occur. Secondly, when materials of different stiffness are compared under displacement control, the stiffer material will always be more harshly tested than the softer material and so should fail earlier, because a higher stiffness will lead to higher stresses and strain energy densities.

Abraham's research questioned if the testing of crack growth is representative of the fatigue life of rubber components. Crack growth describes and compares approximately less than the final 5 % of the service life of a rubber product. The greater importance of deriving material or component data for the intact sample and to ascertain how long it remains intact is emphasised. Hence, the crack initiation stage represented in a Wöhler (S/N) curve is of more practical use to manufacturers than the time taken for a crack to propagate and bring about failure of a component. Thus, providing a reliable predictor of fatigue life is available, fatigue to failure tests should become the standard and relevant tests for Industry, even though they are more time consuming than crack propagation tests.

Abraham's research prompted the following questions;

i. Do non-strain crystallising elastomers exhibit a limiting value of complex modulus E* when cycled to failure under equi-biaxial loading?
ii. Do non-strain crystallising test specimens subjected to equi-biaxial loading have higher fatigue lives when they are pre-stressed ($\sigma_{min} \neq 0$)?
iii. Is there a limiting value of E* for swollen non-strain crystallising rubber samples?

Equi-biaxial bubble inflation was used to investigate these issues.

4. Equi-biaxial fatigue testing

4.1. The advantages of using the bubble inflation method

Consider two issues in respect of fatigue life prediction for elastomers and its simulation. Firstly, numerous phenomenological models for rubber used in Finite Element Analysis (FEA) are based on the James, Green and Simpson strain energy density function [55] which employs the even powered strain invariants I_1 and I_2. The formula for stress determined from the function is shown in Equation 6.

$$\sigma_1 = C_{10}\frac{\partial I_1}{\partial \lambda} + C_{01}\frac{\partial I_2}{\partial \lambda} + C_{11}\frac{\partial I_1(I_2-3)}{\partial \lambda} + C_{11}\frac{\partial I_2(I_1-3)}{\partial \lambda} + 2C_{20}\frac{\partial I_1(I_1-3)}{\partial \lambda} + 3C_{30}\frac{\partial I_1(I_1-3)^2}{\partial \lambda} \quad (6)$$

where C_{10}, C_{01}, C_{11}, C_{20} and C_{30} (MPa) are elastic constants determined from specific material test data

The significant contribution of the second strain invariant in bi-axial deformation is easily demonstrated. Consider the magnitude of the strain invariant I_2 and its derivative for biaxial deformation with respect to stretch ratio λ:

$$I_1 = 2\lambda^2 + \frac{1}{\lambda^4} \quad (7)$$

$$\frac{\delta I_1}{\delta \lambda} = 4\lambda - \frac{4}{\lambda^5} \quad (8)$$

$$I_2 = \lambda^4 + \frac{2}{\lambda^2} \tag{9}$$

$$\frac{\delta I_2}{\delta \lambda} = 4\lambda^3 - \frac{4}{\lambda^3} \tag{10}$$

So for even low values of C_{01}, C_{11} etc. terms become large for high deformations in the bi-axial load case. For example, with a biaxial stretch ratio λ of 2 and a typical value for C_{01} of say 0.04 MPa, the component of stress for the second term in the equation becomes

$$\begin{aligned} C_{01}\frac{\partial I_2}{\partial \lambda} &= C_{01}\left(4\lambda^3 - 4\lambda^{-3}\right) \\ &= 0.04\left(31.5\right) \\ &= 1.26\ MPa \end{aligned} \tag{11}$$

This value of stress alone is significant for a rubber-like material irrespective of the evaluation of the other terms.

The second issue requiring consideration is the amount of scatter present in fatigue to failure tests for rubber which is invariably large. The research carried out by Alshuth and Abraham *et al* [56, 57] on un-notched EPDM dumbbell samples generated Wöhler (S/N) curves to determine a reliable fatigue prediction criterion for this and other non-strain crystallising materials. Although, it was established that there was a relationship between the dynamic stored energy in successive stress-strain cycles and the number of cycles to failure, there were high levels of scatter in the experiments. This is typical of all fatigue tests of rubber compounds and it is believed that the scatter is a consequence of variations in flaw size at the failure cross-section. Subsequently, Abraham *et al* [58] conducted crack growth tests on EPDM single edged notched samples (SENs) and using Equation 5 correlated lives from these tests with those determined for un-notched dumbbell specimens. The flaw size in the dumbbells was controlled by evenly distributing glass spheres of constant diameter in the structure. It was found that the results from the fracture mechanics and Wöhler methods of prediction converged for samples containing glass beads of 203 μm in diameter, suggesting that cracks in the test material would emanate from naturally occurring flaws of this size. This allowed Abraham to control the amount of scatter in dumbbell fatigue tests, but at the expense of inducing unrepresentative early failures due to the severity of the flaws. Subsequent research into the prevalence of large flaws in samples of the same material was undertaken by Robin and McNamara [59] using tomography to establish the total number of large flaws in notched EPDM dumbbell specimens (shown diagrammatically in Figure 8) and the number of flaws in the critical region of a sample. Typically, it was found that 67 (N) flaws of 200 μm in diameter or greater, with an average volume of 0.394 mm^3, were present in a gauge volume of approximately 4770 mm^3. If the average volume of a large flaw is considered to be a single discrete volume, then the number of discrete volumes in the gauge volume was 12,107 (V_A) and the number of discrete volumes in the critical region (Figure 8) was 311 (V_B). The probability that critical flaws were absent from the critical region is given by:

$$\left(\frac{V_A - N}{V_A}\right)^{V_B} \tag{12}$$

Hence, for this analysis the probability equalled 0.178; therefore there was approximately an 18% chance of there being no large flaw in the critical region. The dumbbells were subjected to uniaxial cyclic loading and significantly on average one in every five to six samples exhibited an inordinately high fatigue life. This finding led to the hypothesis that notched samples had higher fatigue lives than un-notched samples due to the likelihood that flaws would often be absent from the vicinity of the stress concentration associated with the notch. Figure 9 shows an X-ray tomography image of a dumbbell sample with one large flaw (≈ 500 μm in length) close to the centre of the sample [60]. Interestingly, Kingston and Muhr [61] have recently published estimates of effective flaw sizes to initiate fatigue failure in uniaxial fatigue on parallel sided test-pieces for three materials; NR without carbon black, NR with 45 pphr N330 carbon black and SBR with 77 pphr N339 carbon black. For the latter two (rubbers containing carbon black), nominal flaw sizes at high strain rates (250%) were 209 μm and 219 μm respectively. The effective flaw size for the unfilled NR at this strain amplitude was nominally 67 μm. Effective flaw size diminished with reduction in strain amplitude for all three materials.

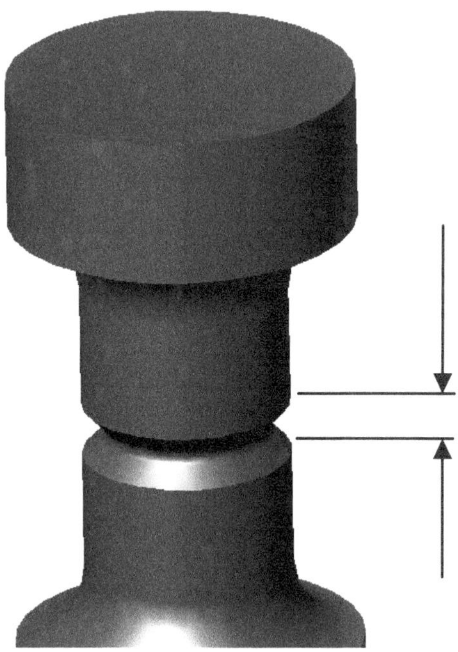

Figure 8. A diagrammatic representation of a dumbbell specimen. The volume of the critical region is indicated as that confined to the notch as shown.

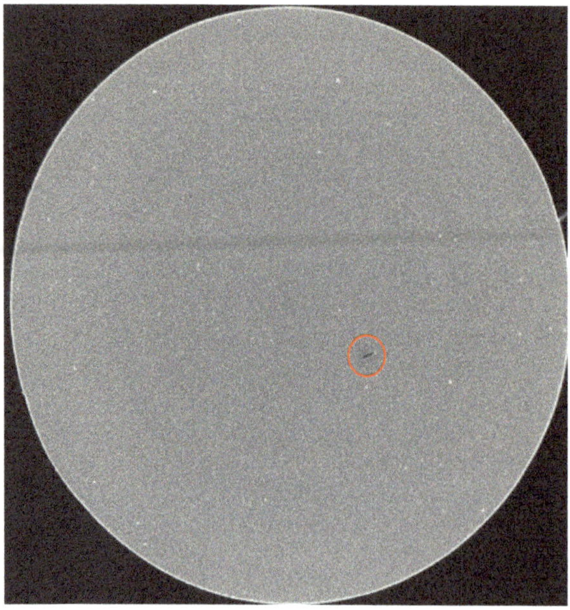

Figure 9. X-ray tomography image of a 15 mm diameter dumbbell sample showing imperfections of various sizes in one plane through the specimen. (A flaw of approximately 500 μm in length can be seen close to the sample centre. The image is reproduced from an analysis performed by S. Robin).

It is evident that two important questions are posed in relation to elastomer fatigue where dynamic equi-biaxial bubble inflation provides the solution; these questions are:-

i. How can fatigue tests be carried out on rubber samples that are representative of the multi-axial loading that components experience in service?
ii. How can fatigue to failure tests be performed on rubber specimens that do not produce results with high levels of scatter?
i. The first question is resolved using the bubble inflation method. For a circular membrane, clamped at the edge and subjected to inflation, the local deformations are a function of radius, from equi-biaxial at the pole to planar extension (approximating to pure shear) at the clamped radius. At the pole, the largest principal stresses (meridian stress σ_1 and hoop stress σ_2) are equal as shown in Figure 10, whilst the smallest principal stress (σ_3) is zero, normal to the outer surface.
ii. In a standard dog-bone equi-biaxial fatigue test-piece, failure can occur anywhere in the parallel gauge section of the specimen and so early fractures are common where a large flaw exists at some point in this section. Since most failures will occur at the point of highest principal stress (at the bubble pole) in an equi-biaxial bubble inflation test, this method has the advantage of reducing scatter in fatigue tests. If rarely, failure occurs away from the bubble pole, it is apparent that a large flaw is present at the point of failure and this test can be ignored.

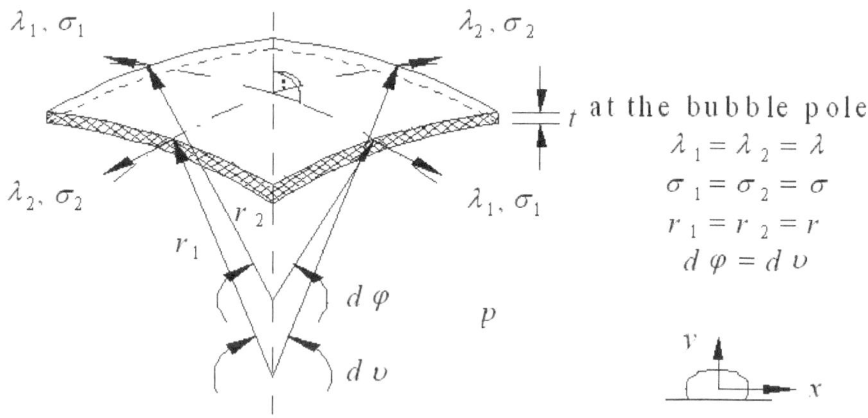

Figure 10. Membrane stresses at the pole of a bubble.

A typical stretch frame is shown in Figure 11. By comparison with the bubble inflation method, using stretch frames lead to the disadvantages of:

- High stress concentrations in the vicinity of the clamps
- Variation in stress distribution toward the specimen centre
- The small stretch ratios achieved result in stress amplitudes that either will not induce failure or alternatively make fatigue to failure tests inordinately long
- Large friction and inertia losses
- An inability to perform equi-biaxial creep tests.

4.2. Obtaining equi-biaxial stress-strain relations for rubber using bubble inflation

Bubble inflation has been used extensively in the past to study the hyperelastic behaviour of rubber [62, 63, 64, 65, 66, 67, 68, 69] and the research described here had its beginning in a project carried out in Robert Bosch GmbH. Johannknecht *et al* [70, 71, 72] pneumatically and hydraulically inflated nitrile rubber test samples to failure to determine material constants for hyperelastic FE analyses and also offered plausibility criteria for these constants [73]. Failures achieved by pneumatic and hydraulic static bubble inflation of Hydrogenated Nitrile Butadiene Rubber (HNBR) samples are shown in Figure 12. It can be seen that the fracture in the pneumatically inflated sample is far more severe than that in the hydraulically inflated sample. Bubble inflation is considered to comply with theory for thin shell structures subjected to pressure (membrane theory), in which bending stiffness is assumed to be negligible. For an ideal isotropic material and an axisymmetric set-up, the bubble contour exhibits rotational symmetry (as depicted in Figure 10). Johannknecht utilised images produced by two mutually perpendicular monochrome charge-coupled device (CCD) cameras to determine displacements at the bubble pole.

Figure 11. A stretch frame used for equi-biaxial loading of rubber specimens.

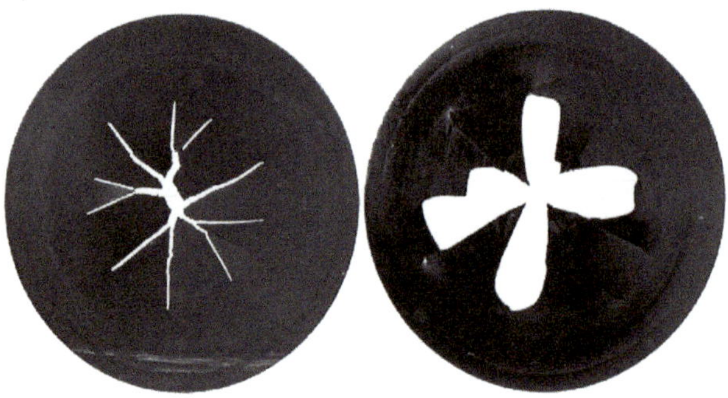

(a) (b)

Figure 12. Failure of statically inflated HNBR 50 mm samples, a) hydraulically inflated and b) pneumatically inflated

Using the simple expression for circumferential stress at the pole (σ), given in Equation 13, where P is the applied pressure, r the radius of curvature and t the specimen thickness, equi-biaxial stress-strain curves to failure were plotted that allowed material constants to be determined using the standard curve fitting procedures available in commercial FE software codes.

$$\sigma = P \cdot \frac{r}{2t} \tag{13}$$

Local stretch ratios at the pole can be determined using Equation (14).

$$\lambda = ((x_{cir} - x_{orig})/x_{orig}) + 1 \tag{14}$$

Where λ is the principal stretch ratio, x_{cir} is the circumferential point spacing at the bubble pole and x_{orig} is the original unstrained circumferential point spacing [70]. When the bubble height exceeds the radius of the inflation orifice, the bubble profile changes from spherical to elliptical which necessitated the calculation of an effective 'radius' at the bubble pole to allow engineering stresses to be computed. Murphy [74] showed that the formula for engineering stress σ_{Eng} had to be modified to include increased loading as a bubble enlarged (Equation 15). This meant that the relationship between engineering and true (Cauchy stresses) σ_{True} during dynamic bubble inflation was identical to that in uniaxial extension and equi-biaxial tension using stretch frames. Hence, the expression for Cauchy stress σ_{True} is as shown in Equation 16.

$$\sigma_{Eng} = P(r / 2t_o)\cdot\lambda \tag{15}$$

$$\sigma_{True} = P(r / 2t_o)\cdot\lambda^2 \tag{16}$$

where P and r are as previously defined and t_o is the unstrained sample thickness.

Unlike simple tensile and stretch frame testing, the total load acting on the specimen increases as the test progresses and the bubble expands under pressure.

4.3. Dynamic bubble inflation

Previously there was a dearth of research into dynamic bubble inflation [75, 76], although the method is capable of loading test-pieces in equi-biaxial tension for a large range of stress or strain amplitudes and allows the total number of cycles to failure to be recorded. Using the system advanced by Murphy [74] specimens were fatigued to failure and the resulting data when used in conjunction with other load cases, facilitated a full characterisation of fatigue properties for a particular elastomer sample subjected to complex loading. The characterisation of the equi-biaxial dynamic properties of EPDM was carried out in an initial programme to validate Abraham's findings for the uniaxial cyclic loading of non-strain crystallising rubbers [77]. However there are many difficulties associated with obtaining reliable fatigue data using the bubble inflation method. Consider the disparate requirements of determining fatigue life for specimens subjected to i) constant pressure amplitudes, ii) constant strain amplitudes iii) constant engineering stress amplitudes and iv) constant true stress amplitudes. Load case i), representative of a diaphragm in a machine working between set pressure limits, is the most easily achieved. The test-piece will grow continually throughout inflation / deflation cycles and be subject to stress softening and set. Constant strain amplitudes, load case ii), representative of the deformation at some point on a component like a belt, hose or drive shaft boot, require continual monitoring and control of specimen profile to retain an approximately equal deformation by varying the pressure limits. Load cases iii) and particularly iv) are representative of the majority of rubber components; auto exhaust hangers, tyre treads, engine mounts, etc. and require both continual monitoring of pressure and specimen profile as each parameter influences stress levels in subsequent cycles. Thus, to obtain successive cycles at a constant engineering stress amplitude in bubble inflation, continually coordinating hydraulic, vision and control systems is essential. This is made complex by the need to continually adjust inflation pressure as the load experienced by the bubble surface increases with bubble growth due to the Mullins effect [53], but it is the large reversible deformations associated with rubber that make it so useful in all spheres of technology. Thus, characterising the material using true (Cauchy) stresses is far more relevant than using engineering stresses. In a conventional linear elastic solid, this distinction is unimportant because engineering stresses are virtually equal to true stresses as the materials normally only experience micro-strains.

Accordingly, Murphy *et al* [78, 79, 80, 81, 82] have developed a bubble inflation facility to produce reliable equi-biaxial fatigue data using a system similar to that employed by Johannknecht [70]. However, they were constrained to use hydraulic inflation and deflation as the lags induced in pneumatic systems were incompatible with cyclic loading. The system can also be used to determine static failure strength, stress relaxation, creep, stress softening and set for dynamic equi-biaxial loading. As stated, the initial impetus for the research was to determine if Abraham's findings [47] in uniaxial fatigue testing of non-strain crystallising rubbers were confirmed for biaxial loading. Very little research had been carried out previously using dynamic bubble inflation of rubber, though Bhate and Kardos [75] used bubble inflation to study high frequency vibrations in elastomers and Hallett [76] studied equi-biaxial rubber fatigue without integrating the systems essential to allow full control of the tests. Hence, unlike the previous investigations, the hydraulic system was integrated with control and vision systems to allow cycling at constant amplitudes of pressure, strain and stress and offered ready access to test results. Murphy applied a system for tracking the displacement of a row of dots at the bubble pole to record circumferential displacement using two CCD cameras, but unlike Johannknecht, he achieved this objective with a camera placed on the vertical axis above the bubble pole and another inclined at 45^0 to the vertical axis. The system configuration is shown in Figure 13. An inflated test sample is shown in Figure 14.

Fatigue to failure tests were carried out on an EPDM rubber of 70 Shore A hardness, containing 110 pphr low activity carbon black with 70 pphr softener added. The bubble specimen consisted of a rubber disc of 50 mm and a thickness of 2 mm cut from an EPDM sheet, the inflating orifice was 35 mm in diameter. The first tests were carried out at constant pressure amplitudes and a typical stress-strain plot showing cycles selected at intervals in a fatigue test is given in Figure 15. Thereafter, the system was developed to facilitate repeated cycles under constant engineering stress amplitude control, using the relationship $\sigma_{Eng} = PR\lambda_1/2t_o$. Cycles selected at intervals in a typical test are shown in Figure 16.

Figure 17 depicts selected cycles for a fatigue test on an EPDM sample (zero minimum stress) under constant stress control and shows that stress softening continues for the duration of a fatigue test, as was observed by Abraham in uniaxial dynamic testing of the same material.

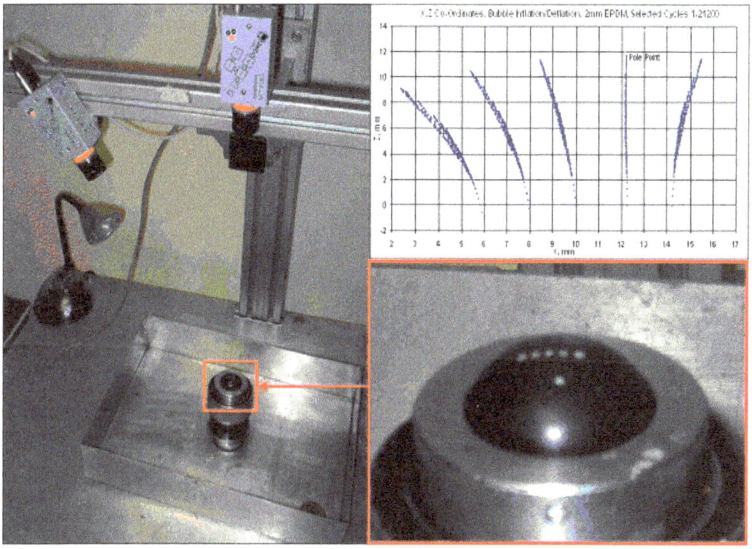

Figure 13. The final configuration of the equi-biaxial bubble inflation test system.
The graph depicts the tracking of the displacement of centres of dots at the pole of the bubble.

Figure 14. An inflated EPDM test sample

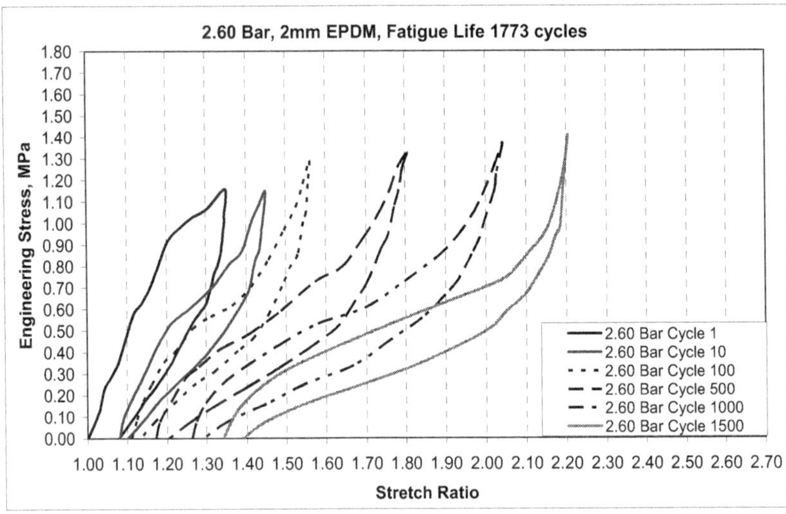

Figure 15. 2 mm thick sample cycled between constant pressure limits ~ selected cycles.

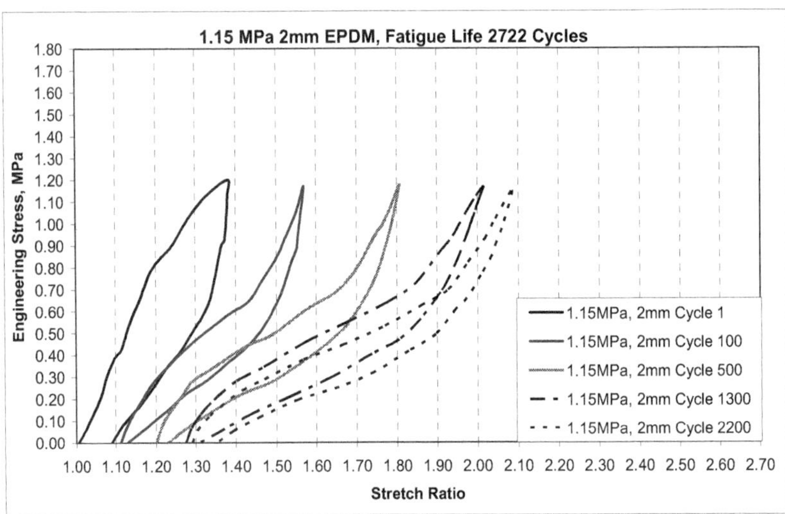

Figure 16. 2 mm thick sample cycled between constant engineering stress limits ~ selected cycles.

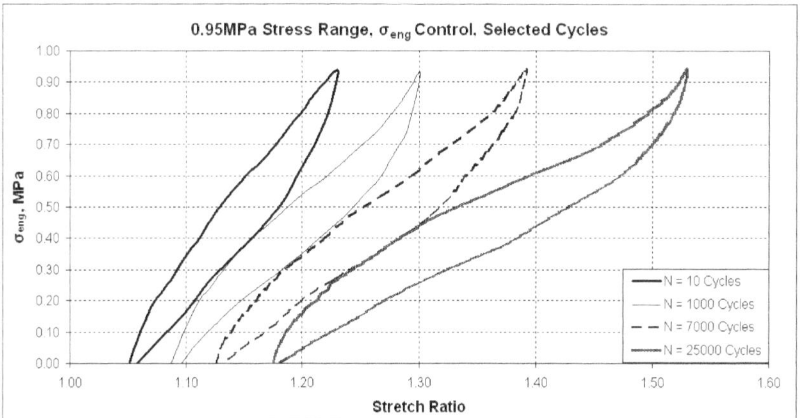

Figure 17. Hysteresis curves for selected cycles in a fatigue test on an EPDM sample.

Fatigue testing using constant engineering stress amplitudes and a non-zero positive minimum pressure did not definitively align with Abraham's observations for uniaxial pre-stressing of EPDM samples. In general, some increase in fatigue strength was observed with pre-stressing for higher pressure amplitudes, but not for smaller amplitudes. The effect of pre-stressing in dynamic equi-biaxial fatigue of EPDM shows that just below the high damage region of loading, increases in fatigue life were exhibited. It is apparent that the fatigue lives of samples cycled at higher constant equi-biaxial stress amplitudes with pre-stressing were greater than for samples cycled to the same stress amplitudes with zero pre-stressing. However, this effect is only exhibited for these high pre-stress levels as depicted in Figure 18. Figure 19 illustrates diagrammatically the variation in the effect of pre-stressing with changing engineering stress amplitudes and also highlights the relationships between the various loading regimes. Reducing dynamic stored energy due to pre-stressing in a loading cycle has less influence on the total dynamic stored energy available for lower maximum cyclic stresses and increases in stress (or pressure) ratio R for a material being tested, where:

$$R_{ratio} = \sigma_{min} / \sigma_{max} \left(\text{also } P_{min} / P_{max}\right) \tag{17}$$

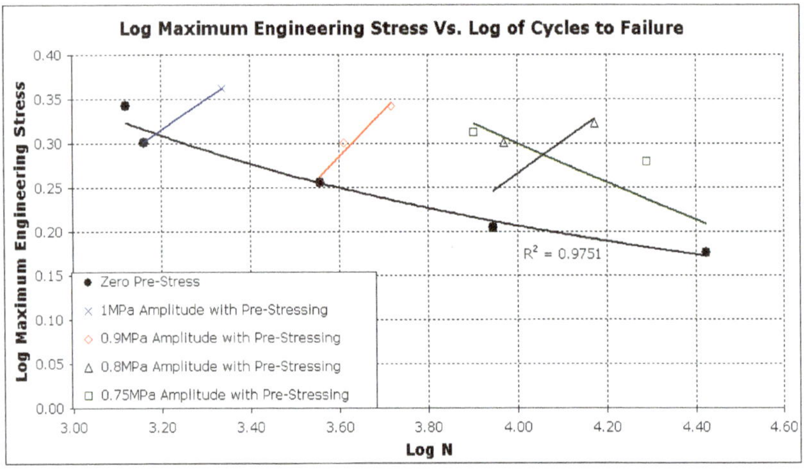

Figure 18. S-N curve for EPDM samples with and without pre-stressing, using $\sigma_{Eng} = P\,R\,\lambda_1/2\,t_o$ control.

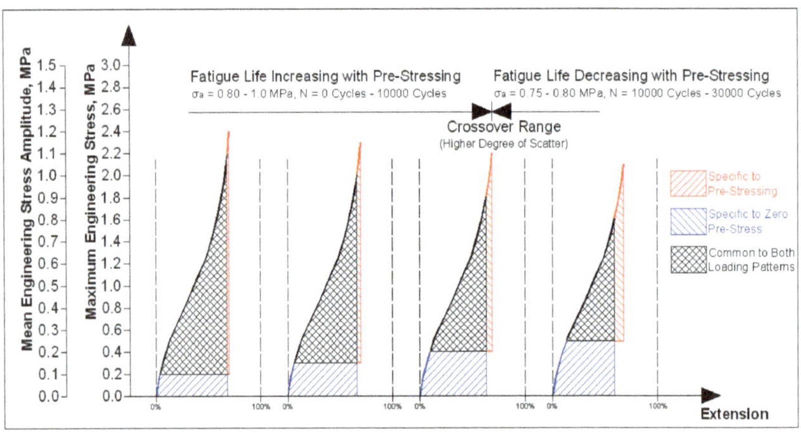

Figure 19. Effect of dynamic stored energy due to pre-stressing on fatigue life.

A point was reached below which the contribution of the stored energy solely attributable to the pre-stressing was insufficient to increase fatigue life and in fact, as the test programme continued, reduced fatigue lives were recorded for samples under pre-stressed conditions with lower maximum engineering stresses.

4.4. Equi-biaxial fatigue of elastomers subjected to swelling

There have been few investigations of the dynamic behaviour of rubber under the influence of oil swelling [83, 84, 85, 86] although swelling of elastomers has been comprehensively studied under conditions of static loading. All rubber swells to some degree in oil, but the amount of swelling can be estimated for any oil-rubber combination if the solubility parameters Δ for both components are known. If the square root of the difference between the solubility parameters of the rubber and the oil is less than 1, as in Equation (18), then the rubber will swell appreciably in that oil [87]. To control the level of oil swell in elastomers it is practical to use hydraulic fluids where the properties have been pre-determined and can be used as a reference. Difficulties arise in precisely determining the solubility parameter of a fluid when it consists of two or more fractions. However, the solubility parameter for a hydraulic fluid can be estimated in experiments from other chemical properties which are readily available (e.g. the aniline point of the oil) and an empirical relation between the solubility parameter and the aniline point of the oils under test can be determined [88].

$$(\Delta_1 - \Delta_2)^{\frac{1}{2}} < 1 \tag{18}$$

The effect of oil swelling on the fatigue life of EPDM under conditions of equi-biaxial cyclic loading using dynamic bubble inflation was studied by Jerrams *et al* [49] and the research project set out to:

1. Determine if fatigue strength reductions due to swelling in rubber specimens loaded equi-biaxially were consistent with strength reductions in samples subjected to uniaxial tests to failure.
2. Create equi-biaxial Wöhler (S-N) curves for a range of rubbers swollen in different media in tests conducted at constant stress amplitudes.
3. Establish relationships between cycles to failure and dynamic modulus and stored energy.
4. Offer predictors of fatigue life for dry/swollen specimens subjected to equi-biaxial fatigue.

An EPDM rubber of 70 Shore A hardness, cross-linked with sulphur and containing low activity carbon black was used in the experimentation. The samples were 50 mm in diameter and prior to testing, had a thickness of 2 mm. Unconditioned samples were clamped and dynamically inflated and deflated through a 35 mm diameter orifice for the bubble inflation tests. The fatigue properties were initially considered for dry EPDM samples that had been subjected to ten loads cycles to condition them.

The degree of cross-linking of the EPDM material and hence its modulus is related to swelling potential by the Flory-Rehner Equation:-

$$-\ln(1 - V_2) + V_2 + XV_2^2 = (V_1 / \upsilon M_c)(1 - 2M_c / M)(V_2^{1/3} - V_2 / 2) \quad [89](19)$$

where V_2 is the volume fraction of polymer in the swollen mass, X is the Flory-Huggins interaction parameter, V_1 is the molar volume of the solvent, M_c is the average molecular mass between crosslinks and M is the primary molecular mass.

Three different sets of EPDM samples were tested; the dry specimens and those swollen in two different ASTM reference oils,

IRM 902 and IRM 903. Specimens were subjected to varying degrees of swelling in reference mineral oils and fatigued at constant engineering stress amplitudes [90]. The reference oils used for swelling the EPDM had known aniline points, allowing the determination of rubber-oil compatibility. The silicone based oil used as the inflation fluid for fatigue testing was selected to minimise the amount of additional swelling during cycling. The samples were cycled between zero and predetermined maximum stress values. Wöhler (S-N) curves were produced for dry and swollen specimens and the changes in complex elastic modulus E* and dynamic stored energy were determined. Specimen fractures were observed using SEM. As in the equi-biaxial bubble inflation fatigue tests previously described, the specimens had a pattern of dots applied to their surface. The deformation of this pattern during inflation and deflation was recorded by the optical system previously specified [74], allowing correlation to a specific engineering stress value at the bubble pole. From ASTM standards for rubber-liquid compatibility, the reference oil IRM 903 was chosen as the most appropriate oil for high swelling of EPDM. Its aniline point was determined and compared with the solubility parameter of the EPDM to allow rubber-liquid compatibility to be gauged. Also, using reference oil, IRM 902, which imparts medium swell effects, allowed the influence of variation of oil solubility on rubber properties to be investigated [91]. Specimens were immersed in the reference oil at 100 °C for one hour. After removal from the hot oil the samples were cooled in oil at ambient temperature for a short period, before being dried and weighed. An average swelling ratio Q was calculated for the samples, where Q was expressed as,

$$Q = \frac{W_s}{W_d} \tag{20}$$

W_s is the weight of the swollen elastomer sample and W_d is the weight of the dry elastomer before swelling. A swelling ratio of 1.10 (10% increase in mass) was calculated for the EPDM swollen in IRM 903 and a ratio of 1.042 (4.2% increase in mass) for the EPDM swollen in IRM 902. EPDM specimen sets (Dry, 10% Swell, 4.2% Swell) were then cycled hydraulically between pre-set maximum and zero minimum engineering stress limits using the silicone based oil as the inflation medium. In general, fatigue tests were carried out at frequencies of 1 Hz to minimise any potential for samples to degrade due to heat build-up during cyclic loading [48]. However in bubble inflation tests there can be minor exceptions to this practice as it is essential to consider pressure or stress amplitude in combination with frequency to avoid thermal degradation. To maintain a constant maximum engineering stress in each cycle, the pressure set-point was adjusted continually throughout testing. S-N curves were produced to allow comparison of the fatigue lives of the dry and swollen rubber samples. The plot of stress amplitude (σ_a) versus cycles to failure (N) for all three specimen sets is shown in Figure 20.

As anticipated, the unswollen specimens had greater fatigue resistance than the swollen samples. The fatigue life of the material was reduced proportionally to the degree of swelling. The fatigue behaviour of the three specimen sets was subsequently analysed in respect of the complex elastic modulus, E* (as previously defined). Plots showing the decrease in modulus E* of the samples with the accumulation of cycles at a stress amplitude of σ_a = 1MPa for the three sample sets are shown in Figure 21. (The three dashed curves represent results of FE simulations for the three material sets).

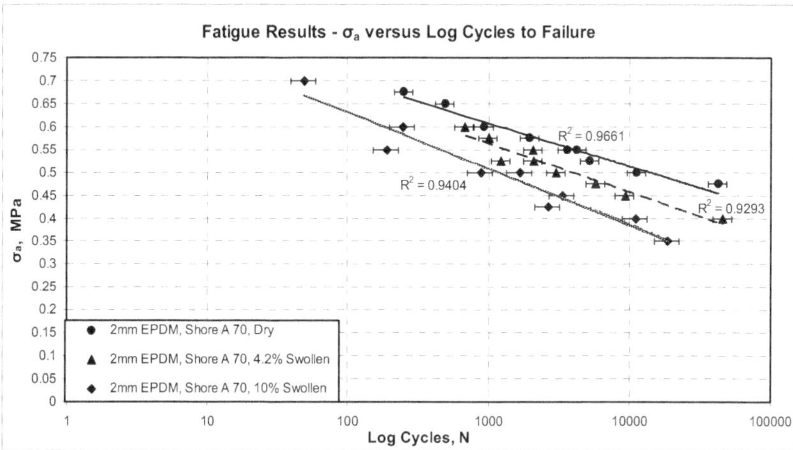

Figure 20. Plot of stress amplitude versus $\log_{10}$ cycles to failure for dry and swollen specimens.

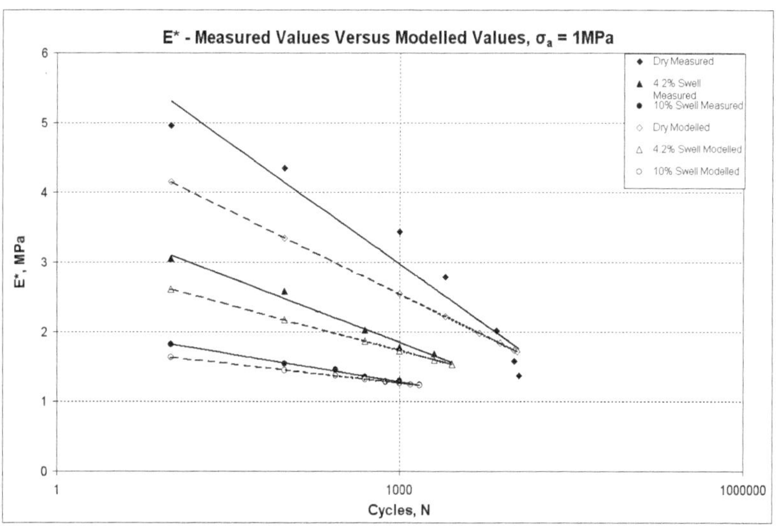

Figure 21. E* versus cycles $\log_{10}$N for dry, 4.2% and 10% swollen specimens at σ_a = 1MPa.

It was noted earlier that in tests carried out by Abraham, [47] it was found that for a given material, failure consistently occurred after a specific loss in complex modulus E*. This predictor, termed E^*_{res} can be represented as:

$$E^*_{res} = \left(E^*/E_{con}\right) \times 100\% \tag{21}$$

where E_{con} is defined as the modulus of the material after ten conditioning cycles. The dry and swollen specimens subjected to equi-biaxial cycling were analysed using this approach, where the percentage value of E^*_{res} versus cycles was plotted up to 95 % of the fatigue life of the specimen. A plot of the decrease in E^*_{res} (as a percentage of E_{con}) for the dry EPDM is shown in Figure 22. E^*_{res} was found to have an average value of 33% ±10% of its original value at 95% of the fatigue life of each dry specimen. The 4.2% and 10% swollen specimens had values of E^*_{res} of 50.5%±10% and 60.1%±5% at 95% of specimen life, respectively.

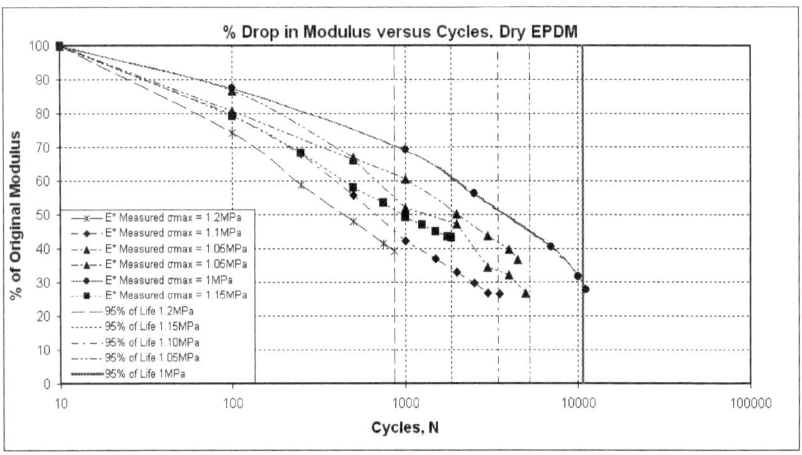

Figure 22. E^*_{res} versus $\log_{10}$ cycles N for dry EPDM for five different stress amplitudes.

The moduli (E*) for the dry and swollen samples at failure were studied by comparing the dry and the two swollen specimen sets at similar stress amplitudes. The trends for each specimen set are shown over a range of stress amplitudes in Figure23.

There is an average E^*_{res} at failure of 33% ±10% for these sets of specimens over four different stress ranges. By knowing the initial modulus E^*_{con} for the dry specimens, an approximation of the failure modulus for the swollen test-pieces can be determined. At a stress amplitude of 1MPa, comparison can be made between all three sample sets in respect of the decrease in specific modulus. Analysis of the value to which E^*_{res} decreases for each sample type at a stress amplitude of σ_a = 1MPa, with E^*_{con} taken as the modulus of the dry EPDM after ten conditioning cycles, suggested that E^*_{res} at failure fell within the 33% ±10% range for each of the three specimens. The dynamic stored energy during cycling of the three sample sets was also evaluated. Previous analyses of uniaxial test data for EPDM found good correlation between the dynamic stored energy in the specimen versus the cycles at failure and its use as a plausible fatigue life predictor was proposed [51-52]. Stress-strain measurements were made periodically throughout the equi-biaxial dynamic tests. The dynamic stored energy in a cycle was subsequently calculated for the specific cycles measured. After conditioning, the dynamic stored energy increased linearly when plotted against cycles to failure, confirming that using this parameter provides a plausible predictor of fatigue life for elastomers, irrespective of the degree of swelling present in the material. The plots of dynamic stored energy to failure versus $\log_{10}$ cycles for the dry EPDM over a range of stress amplitudes are shown in Figure24 and the graphs of dynamic stored energy versus $\log_{10}$ cycles to failure for all three material sets are shown in Figure 25. At similar stress amplitudes, the dry EPDM samples had higher dynamic stored energy at failure than the swollen specimens, with the 10% swollen rubber having the lowest dynamic stored energy for a given stress amplitude. As would be expected, this result is due to a reduction in stiffness with increasing levels of swelling.

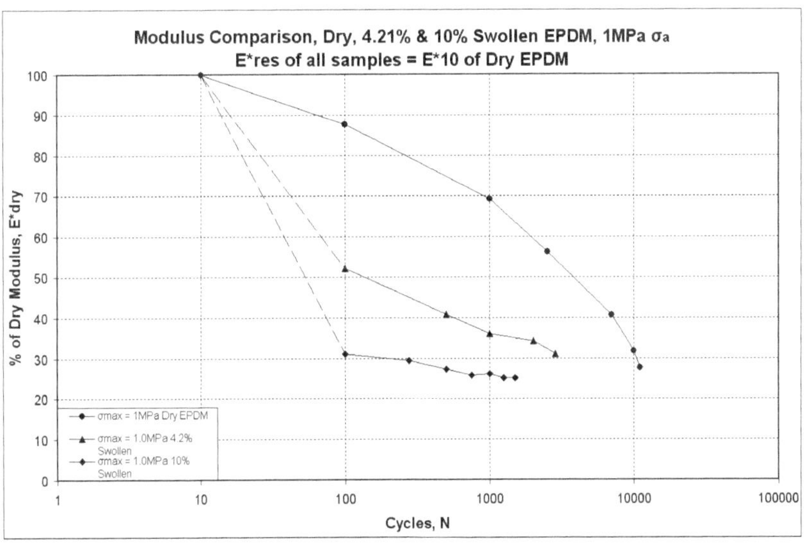

Figure 23. E^*_{res} expressed as a percentage of E^*_{con} versus $\log_{10}$ cycles N for dry and 4.2% and 10% swollen EPDM for different stress amplitudes.

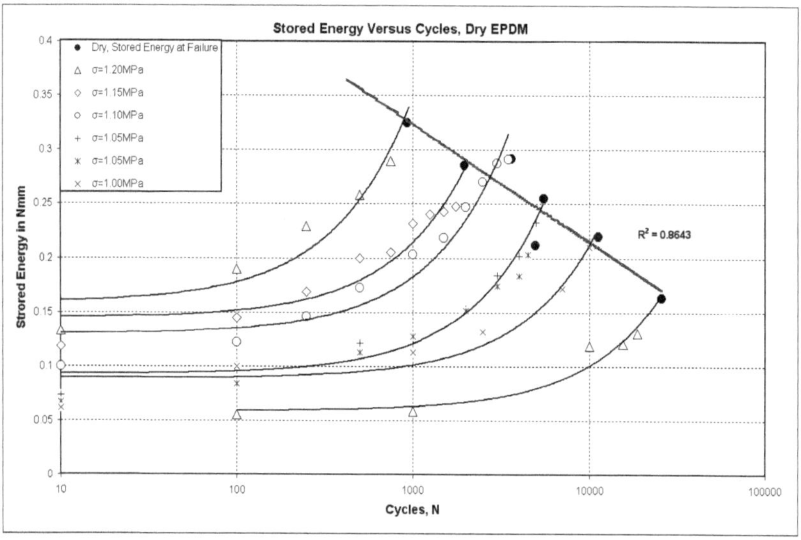

Figure 24. Dynamic stored energy versus log10 cycles to failure for dry EPDM samples

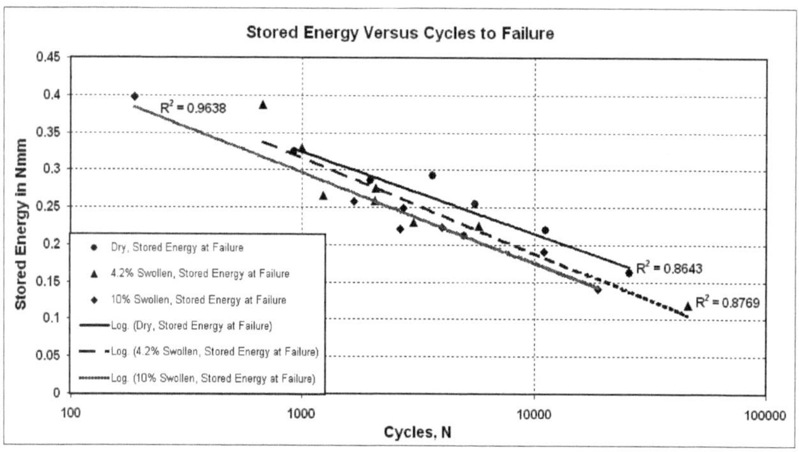

Figure 25. Plot of Dynamic Stored Energy versus log10 cycles to failure for the three specimen sets.

The surface morphology of the fractures for each sample set was observed. The fracture surfaces of the dry samples were analysed using SEM. A magnification factor of X400 at 2.0 kV was used to view failure surfaces. For low lives of less than one hundred cycles, the failure mode was similar to that observed for the fracture surfaces of the material from a static inflation test to destruction. The surface morphology was fibrous in nature and in some instances showed delamination at the failure surface. Failures that occurred beyond 100 cycles showed clear evidence of crack propagation and subsequent rupture. Cracks propagated predominately in the bubble pole region. This behaviour was common to both the dry and swollen specimens. SEM images of the surface morphologies of the specimens at stress amplitude of 1 MPa are shown in Figure 26. In the dry samples there is a coarser failure surface at lower cycles than for failures at higher cycles. Previous research indicated that the surface of the dry samples exhibited blunt tearing, while the surfaces of swollen test-pieces indicated sharp tearing [75]. SEM images from this study confirmed these observations, where the blunter surfaces of the dry specimens were in stark contrast to the smoother surfaces of the swollen test-pieces. This blunt tearing was attributed to greater polymer filler interaction in the dry material, as the material showed higher toughness for equal values of applied stress. The SEM imaging suggests that the swollen failure surfaces 'flow' more readily over one another than those of the drier specimens.

Dry **4.2% Swollen** **10% Swollen**

Figure 26. SEM Imaging of specimen fracture surfaces, σ_{eng} = 1 MPa (ROI = Region of interest).

Flaws were more abundant in the swollen material. As the flaws were potential stress raisers, the probability of crack growth being initiated at them would be high and the areas around the flaws would tend to have a lower complex modulus than other points in the network.

5. Conclusions

Using load controlled uniaxial cyclic deformation of dumbbell specimens, Abraham *et al* [47] correlated fracture mechanics and SN curve approaches for determining fatigue life in non-strain crystallising rubbers. In subsequent analyses they were able to show that failure in fatigue would ensue from flaws of 200 μm or more in length if such flaws were in the critical region of the loaded specimen. By applying tomographic techniques, McNamara *et al* [60] were later able to confirm the presence of flaws of this magnitude in standard rubber samples and account for the large scatter in elastomeric fatigue tests by highlighting the possibility of these flaws occurring on critical sections in a test-piece or component. Abraham's research resulted in three other key findings in respect of fatigue resilience under uniaxial loading in amorphous elastomers. These can be summarised as:

i. Fatigue life can be improved by pre-loading these materials.
ii. 'Dynamic stored energy' can be used as a basis for predicting fatigue life
iii. There is a limiting value of complex elastic modulus E* below which fatigue failure can be anticipated.

These findings prompted numerous questions:-

a. Will the influence of pre-loading be significant under more complex loading regimes?
b. Can 'dynamic stored energy' be used as a predictor of fatigue life for multi-axial loading of rubber samples?
c. Is there a limiting value of E* that is applicable for non-strain crystallising rubbers loaded biaxially?
d. Can the dynamic stored energy criterion and limiting value of complex modulus be applied to dynamic equi-biaxial cyclic loading of swollen amorphous rubbers?

In consequence a programme of equi-biaxial dynamic testing ensued. Johannknecht *et al* [72] had already demonstrated that controlled bubble inflation was a feasible method of analysing equi-biaxial stress-strain relations for rubber-like materials. Thereafter, Murphy *et al* [77] created an equi-biaxial dynamic test facility for rubber using the bubble inflation method. Cyclic testing of EPDM samples appeared to show that Abraham's findings for pre-stressing were only applicable for high load ranges, where the material's failure could be anticipated in relatively few cycles.

Hanley [90] developed the cyclic bubble inflation method further and applied it to swollen equi-biaxial disc samples. It was found that the fatigue life results produced in the equi-biaxial tests agreed with those from dynamic uniaxial swelling experiments; that is, the fatigue lives of specimens were reduced in proportion to the amount of swelling. Unsurprisingly, the fatigue life of an EPDM sample under dynamic equi-biaxial loading was greatly reduced in the presence of oil in the rubber network, even for relatively small amounts of swelling. Though only in contact with the oil for an hour at 100 °C, the dynamic properties of complex modulus E* and dynamic stored energy were fundamentally altered from those of untreated EPDM. The changes in these properties and the lower fatigue lives of the swollen specimens were attributed to a number of physical and chemical factors. Physical factors included the presence of larger voids in the network and a lower initial complex modulus due to swelling. Chemical factors included changes in the network structure due to oil swelling, where there may have been a reduction in the number of cross-links resisting the tensile force or where the swelling led to differences in the equilibrium length of individual chains. Also, possibly the reformation of polysulphidic linkages during loading cycles was inhibited in the presence of oil [92].

The relationship between dynamic stored energy and cycles to failure was influenced by the degree of swelling, with the energy at failure reducing as swelling levels increased. It was proposed that a practical approach to predicting realistic fatigue lives in rubber compounds was to use the limiting value of the residual modulus E^*_{res} if the residual modulus for swollen specimens is based on a calculation using the initial conditioned modulus of dry rubber samples at similar stress amplitudes. E^*_{res} values exhibited reasonable correlation with a limiting value of 33%±10% for both the dry and swollen specimens, so by knowing the limiting value of E^*_{res}, (in this case 43% of E^*_{con}), the effect of swelling on fatigue life of non-strain crystallising rubber components can be determined. Clearly, this approach could have practical significance in the design and maintenance of elastomeric components subjected to fatigue loading conditions. The practical significance of this study was focused on the effect of contamination on the mechanical performance of non-strain crystallizing elastomers subjected to realistic loading conditions. EPDM components are predominantly used in automotive applications, where they provide the advantage of having improved functionality at higher temperatures than SBR or NR. However, some of these advantages are diminished by the material's poor resistance to numerous fluids used in vehicles. Hanley's research illustrated that if a risk of oil contamination is sufficiently high and the automotive design engineer has failed to appreciate the potential detrimental influence of swelling, safety factors may be too small and life expectations for components may be unrealistically high.

5.1. The rationale for equi-biaxial bubble inflation fatigue testing of rubber

The broad outcome of the research described in this chapter is that dynamic equi-biaxial bubble inflation provides rubber technologists with a robust and adaptable method of determining fatigue resilience in elastomeric materials. Physical testing is possible under a wide range of amplitude controls, pre-stressing is easily incorporated into testing, results scatter is reduced and the study of numerous other viscoelastic phenomena in rubber is simply achieved. The test facility is under continuous development to provide viscoelastic material data for loading regimes experienced in 'real world' applications. Most recently, the equi-biaxial bubble inflation system has undergone design changes to deliver dynamic characterisation of the physical properties of magnetorheological elastomers (MREs) [93].

Author details

Steve Jerrams and Niall Murphy
Centre for Elastomer Research (CER), Dublin Institute of Technology, Ireland

John Hanley
FDT Consulting Engineers and Project Managers, Dublin, Ireland

Acknowledgement

The authors wish to express their gratitude to all of the researchers who have participated in the various elastomer fatigue programmes described in this chapter, but in particular to Thomas Alshuth, Raphaël Johannknecht, Georg Clauss, Frank Abraham, Stefan Robin and John McNamara. They also place on record their sincere thanks to the Deutsches Institut für Kautschuktechnologie e.V for an enduring research partnership and to Enterprise Ireland who supported the design and development of the Dynamet equi-biaxial dynamic bubble inflation system under its 'Proof of Concept' programme.

6. References

[1] Mars W, Fatemi A (2004) Factors that Affect the Fatigue Life of Rubber: A literature Survey, Rubber Chemistry and Technology, Vol 77, No 3, 391-312.
[2] Lake G (1995) Rubber Chemistry and Technology. 68, 435.
[3] Thomas A (1994) Rubber Chemistry and Technology. 67, G50.
[4] Ellul M, (1991) 'Mechanical Fatigue', Engineering with Rubber – How to design rubber components, Editor Gent A, Hanser Verlag, Berlin, Germany, Chapter 6, 129.
[5] Young D (1991) Fatigue and fracture of elastomeric materials, Rubber World.
[6] Bauman T (2008) Selecting Strain Amplitude, Fatigue Stress and Strain of Rubber Components: Guide for Design Engineers, Hanser, ISBN: 978-3-446-41681-9, 8.2.2, 118.
[7] Mars W, Fatemi A (2004) Factors that Affect the Fatigue Life of Rubber: A literature Survey, Rubber Chemistry and Technology, Vol 77, No 3, 398.
[8] Kadir A, Thomas A (1981) Tear behavior of rubbers over a wide range of rates. Rubber Chemistry and Technology 54:15–23.
[9] Mazich K, Morman K, Oblinger F, Pan T, Killgoar P, (1989) The effect of specimen thickness on the tearing energy of a gum vulcanizate, Rubber Chemistry and Technology, 62:850–62.
[10] Lindley P (1974) Non-relaxing crack growth and fatigue in a non-crystallizing rubber, Rubber Chemistry and Technology 47:1253–1264.
[11] Beatty J (1964) Fatigue of rubber, Rubber Chemistry and Technology, 37, 1341–1364.
[12] Hardy D, Moneypenny H, Holderied M, Harris J, Campion R, Morgan G (1999) Influence of low surface area carbon blacks on air permeation and fracture mechanical behaviour of tyre innerliner compounds, Rubber Chem 99, Antwerp, Belgium.
[13] Gent A, Hirakawa H (1968) Rubber Chemistry and Technology, Vol. 41, 1294-1299.
[14] Gent A, Hirakawa H (1968) Journal of Polymer Science A-2 6, 1481.
[15] Kim S, Lee S (1994) Rubber Chemistry and Technology, Vol. 67, 649.
[16] Gent A, Hindi M (1990) Effect of oxygen on the tear strength of elastomers. Rubber Chemistry and Technology Vol 63:123–34.
[17] Zhao J, Ghebremeskel G (2000) Rubber Plast. News, 27, 14.
[18] Busfield J, Jha V, Liang H, Papadopoulos I, Thomas A (2005) Prediction of fatigue crack growth using finite element analysis techniques applied to three dimensional elastomeric components, Plastics , Rubbers, Composites, Vol. 34 349-356.
[19] Sakulkaew K, Thomas A, Busfield J (2012) A new approach to categorize the onset of tearing in rubber, Constitutive Models for Rubber VII, Editors Jerrams S and Murphy N, CRC Press, ISBN 978-0-415-68389-0, 185-189.
[20] Marco Y, Le Saux V, Calloch S, Charrier, P (2012) Heat-build up and micro-tomography used to describe the fatigue mechanisms and to evaluate the fatigue lifetime of elastomers, Constitutive Models for Rubber VII, Editors Jerrams S and Murphy N, CRC Press, ISBN 978-0-415-68389-0, 347-352.
[21] Le Chenadec Y, Raoult I, Stolz C, Nguyen-Tajan M, (2009) Cyclic approximation of the heat equation in finite strains for the heat build-up problem of rubber, Journal of Mechanics of Materials and Structures 4 (2) 309-318.

[22] Le Saux V, Marco Y, Calloch S, Doudard C, Charrier P (2010) Fast evaluation of the fatigue lifetime of rubber-like materials based on a heat build-up protocol and microtomography measurements, International Journal of Fatigue 32, 1582-1590.
[23] Beurrot S, Huneau B, Verron E, Rublon P, Thiaudière D, Mocuta C, Zozulya A (2012) In-situ synchtron X-ray diffraction study of strain-induced crystallisation of natural rubber during fatigue tests, Constitutive Models for Rubber VII, Editors Jerrams S and Murphy N, CRC Press, ISBN 978-0-415-68389-0, 353-358.
[24] Le Cam J-B, Toussaint E, (2012) The mechanism of fatigue crack growth in rubbers under severe loading: The effect of stress induced crystallisation, Constitutive Models for Rubber VII, Editors Jerrams S and Murphy N, CRC Press, ISBN 978-0-415-68389-0, 69-75.
[25] Hainsworth S (2007) An environmental scanning electron microscopy investigation of fatigue crack initiation and propagation in elastomers, Polymer Testing 26, 60-70.
[26] Le Cam J-B, Toussaint E (2012) The mechanism of fatigue crack growth in rubbers under severe loading: The effect of stress induced crystallization, Constitutive Models for Rubber VII, Editors Jerrams S and Murphy N, CRC Press, ISBN 978-0-415-68389-0, 69-75.
[27] Le Cam J-B, Toussaint E (2008) Volume variation in stretched natural rubber: competition between cavitation and stress induced crystallisation, Macromolecules 41, 7579- 7583.
[28] Alshuth T, McNamara J, Jerrams S (2007) Improving the Prediction of Stress Softening in Rubber Components, Kautschuk, Gummi, Kunststoffe (KGK) Iss 12.
[29] Busfield J, Thomas a, Ngah M (1999) Application of fracture mechanics for the fatigue life prediction of carbon black filled elastomers, Constitutive Models for Rubber, Editors Dorfman A, Muhr A, 249-256, Balkema, ISBN 9058091139.
[30] Thomas A (1994) The development of fracture mechanics for elastomers, Rubber Chemistry and Technology, Charles Goodyear Medal Address.
[31] Alshuth T, Abraham F, Jerrams S (2002) Parameter Dependence and Prediction of Fatigue Properties of Elastomer Products. Rubber Chemistry and Technology, 75, 365.
[32] Griffith A, Phil. Trans. Royal Society, London,Ser. A 221 11929 163.
[33] Rivlin R, Thomas A (1965) Rupture of Rubber, 1, Characteristic Energy for Tearing. Journal of Polymer Science. Vol 10, No. 3, 291.
[34] Ellul M, (1991) Engineering with Rubber – how to design rubber components, Ch. 6 Mechanical Fatigue, Editor A.N Gent, Hanser Verlag, Berlin, Germany, 129.
[35] Lake G, Lindley P (1965) The Mechanical Fatigue Limit for Rubber', Journal of Applied Polymer Science. Vol 9, 1233.
[36] Seldén R (1995) Fracture Mechanics Analysis of Fatigue of Rubber – A Review, Progress in Rubber and Plastics Technology. 11, 56 – 83.
[37] Wöhler A (1870) Über Festigkeitsversuche mit Eisen und Stahl, Zeitschrift Bauwesen 20, Spalten 73-106.
[38] Gent A (1994) Strength of Elastomers, Science and Technology of Rubber, Editors Mark J, Erman B, Eirich F, 2nd ed., Academic Press, Ch No. 10, 471 - 512.

[39] Lake G, (1972) Mechanical Fatigue of Rubber, Rubber Chemistry and Technology, 45, 309.
[40] Lindley P, (1974) Non-Relaxing Crack Growth and Fatigue in a Non-Crystallizing Rubber, Rubber Chemistry and Technology, 47, 1253.
[41] Cadwell S, Merrill R, Sloman C, Yost F (1940) Ind. Eng. Chem., Anal. Ed., 12, 19.
[42] Fielding J (1943) Ind. Eng. Chem. 35, 1259.
[43] Berry J (1972) Fracture of Non-Metals and Composites, Fracture: An Advanced Treatise, Vol 7, Ch. 2, Editor Liebowitz H, Academic Press, New York.
[44] Jerrams S, Tabakovic A (2001) The Influence of Mean Stress on the Fatigue Properties of Natural Rubber, IRC 2001, Birmingham, UK, 327.
[45] Lee D, Donovan J (1987) Microstructural Changes in the Crack Tip Region of Carbon Black Filled Natural Rubber, Rubber Chemistry and Technology, 60 910.
[46] Lake G, Samura A, Teo S, Vaja J (1991) Polymer, 32, (16), (1991) 2963.
[47] Abraham F (2002) The Influence of Minimum Stress on the Fatigue Properties of Non Strain-Crystallising Elastomers, PhD thesis, Coventry University, Coventry, UK.
[48] Seldén R (1995) Fracture mechanics analysis of fatigue of rubber – a review, Progress in Rubber and Plastics Technology, Vol 11, 1, The Institute of Materials/RAPRA.
[49] Jerrams S, Hanley J, Murphy N, Ali H (2008) Equi-biaxial Fatigue of Elastomers – The Effect of Oil Swelling in Specimen Fatigue Life, Rubber Chemistry and Technology, Vol 81, Issue 4, 638-649.
[50] Abraham F,. Alshuth T, Jerrams S (2005) The effect of minimum stress and stress amplitude on the fatigue life of non strain crystallising elastomers, Journal of Materials and Design, Vol 26, Issue 3, 239-245 (Elsevier).
[51] Alshuth T, Abraham F, Jerrams S (2002) Parameter Dependence and Prediction of Fatigue Properties of Elastomer Products, Rubber Chemistry and Technology, Vol 75, Issue 4, 365.
[52] Abraham F, Alshuth T, Jerrams S (2001) Dependence on mean stress and stress amplitude of fatigue life of EPDM elastomers, Institute of Materials, Vol 30 No. 9 421-425, ISSN 1465-8011.
[53] Mullins L, (1969) Rubber Chemistry and Technology, 42, 339.
[54] Clauss G (1999) Das Hysteresis-Messverfahren zur dynamischen Prüfung von Kunststoffen - ein Verfahren zur Lebensdauervorhersage auch für Elastomer-Bauteile, Lebensdauer und Ermüdung von Elastomeren, DIK Fortbildungsseminar, Hannover, Germany.
[55] James A, Green A, Simpson G (1975) Strain energy functions of rubber, 1: Characterisation of gum vulcanisates, Journal of Applied Polymer Science, 19.
[56] Alshuth T, Abraham F, Jerrams S (2002) Parameter Dependence and Prediction of Fatigue Properties of Elastomer Products. Rubber Chemistry and Technology, 75, 365.
[57] Abraham F (2002) The Influence of Minimum Stress on the Fatigue Properties of Non Strain-Crystallising Elastomers, PhD thesis, Coventry University, Coventry, United Kingdom, 52-57.

[58] Abraham F, Alshuth T, Clauss G (2005) Testing and simulation of the influence of glass spheres on fatigue life and dynamic crack propagation of elastomer. Constitutive Models for Rubber IV, Editors Austrell P, Kari, L, Balkema, ISBN: 0415383463.

[59] Mc Namara J (2011) PhD thesis, Dublin Institute of Technology, Ireland, Novel approaches to the analysis of localised stress concentrations in deformed elastomers, 126-130.

[60] Mc Namara J, Jerrams S, Alshuth T, Robin S (2010) New insights into the distribution and size of flaws in rubber compounds produced by commercial processes, Materials Ireland Conference.

[61] Kingston J, Muhr A (2012) Determination of effective flaw size for fatigue life prediction, Constitutive models for Rubber VII, Editors Jerrams S and Murphy N, ISBN 978-0-415-68389-0, 337-342.

[62] Flint, C, Naunton W (1937) Physical Testing of Latex Films, Transactions of The Institution of the Rubber Industry, Volume 12, P.367-406.

[63] Treloar L (1944) Strains in an Inflated Rubber Sheet, and the Mechanism of Bursting, Rubber Chemistry and Technology, 17, 957-967.

[64] Adkins J, Rivlin R (1952) Large elastic deformations of isotropic materials. XI The deformation of thin shells, F.R.S., Vol. 244, A. 888.

[65] Bhate P, Kardos J (1984) A Novel Technique for the Determination of High Frequency Equi-biaxial Stress-Deformation Behaviour of Viscoelastic Elastomers, Polymer Engineering and Science, Vol. 24, No. 11.

[66] Kong D, White J (1986) Inflation Characteristics of Unvulcanised Gum and Compounded Rubber Sheets, Rubber Chemistry and Technology, 59, 315-327.

[67] Song W, Mirza F, Vlachopoulos J (1991) Finite element analysis of inflation of an axisymmetric sheet of finite thickness, The Society of Rheology, Inc.

[68] Khayat R, Derdouri A (1995) Stretch and Inflation of Hyperelastic Membrane as Applied to Blow Molding, Polymer Engineering and Science, Vol. 35, No. 23.

[69] Mott P, Roland C, Hassan S (2003) Strains in an Inflated Rubber Sheet, Rubber Chemistry and Technology, 76, 326-333.

[70] Johannknecht R (1999) The Physical Testing and Modelling of Hyperelastic Materials for Finite Element Analysis, PhD Thesis, (Coventry University in collaboration with Robert Bosch GmbH).

[71] Johannknecht R, Jerrams S, (1999) The need for equi-biaxial testing to determine elastomeric material properties, Constitutive Models for Rubber, Editors Dorfman A, Muhr A, 73-76, Balkema, ISBN 9058091139.

[72] Johannknecht R, Jerrams S, Clauss G (2002) Determination of non-linear, large equal biaxial stresses and strains in thin elastomeric sheets by bubble inflation, Proceedings of the Institute of Mechanical Engineers, Vol 216 Part L, No L4, 233-243, (ISSN 1464-4207) Journal of Materials, Design and Applications.

[73] Johannknecht R, Clauss G, Jerrams S (1997) The Uncertainty of Implemented Curve Fitting Procedures in Finite Element Software, Finite Element Analysis of Elastomers. IMechE Seminar, Finite Element Analysis of Elastomers, 141-151.

[74] Murphy N (2010) Providing Stress Controlled Equi-Biaxial Fatigue Test Data for Elastomers using the Bubble Inflation Method, PhD thesis, Dublin Institute of Technology.
[75] Bhate P, Kardos J (1984) A Novel Technique for the Determination of High Frequency Equi-biaxial Stress-Deformation Behaviour of Viscoelastic Elastomers, Polymer Engineering and Science, Vol. 24, No. 11.
[76] Hallett, J (1997) Multi-axial Strength and Fatigue of Rubber Compounds, PhD Dissertation, Loughborough University, England.
[77] Murphy N, Jerrams S,Hanley J, McCartin J, Lanigan B, McLoughlin S, Clauss G, Johannknecht R (2005) Determining Multiaxial Fatigue in Elastomers using Bubble Inflation, Constitutive Models for Rubber IV', Editors Rustrell P, Kari L, 65-70, Balkema, ISBN 0 415 38346 3.
[78] Murphy N, Hanley J, Jerrams S (2009) The Effect of Pre-Stressing on the Equi-Biaxial Fatigue Life of EPDM, Constitutive Models for Rubber VI. Editors Heinrich G, Kaliske M, Lion A, Reese S, 269-273, CRC Press, ISBN: 978-3-00-025427-7
[79] Murphy N, Hanley J, Ali H, Jerrams S (2007) The Effect of Specimen Geometry on the Multi-axial Deformation of Elastomers, Constitutive Models for Rubber V, Editors Boukamel A, Laiarinandrasana L, Meo S and Verron E, 61-65, Taylor and Francis, ISBN 0415454425.
[80] McLoughlin S, Murphy N, Hanley J, Lanigan B, Markham C (2005) Measuring biaxial strain in rubber using stereo vision, Proceedings of the Irish Machine Vision and Image Processing Conference, IMVIP2005, Belfast.
[81] Murphy N, Jerrams S, Spratt C, Ronan S (2003) A test rig to provide multi-axial fatigue data in elastomers, Matrib, Velaluka, Croatia.
[82] Murphy N, Jerrams S, Spratt C, Ronan S, Johannknecht R (2003) Method for determining equi-biaxial fatigue in elastomers, Constitutive Models for Rubber III, Editors Busfield J, Muhr A, 21-26, Balkema, ISBN 9058095665.
[83] Cho K, Wook J, Daeho L, Hyunaee C, Young-Wook C, (2000) Polymer. 41, 179.
[84] Gul V, Fedyukin D, Dogadkin B, (1959) Rubber Chemistry and Technology 32, 454.
[85] Dogadkin A, Gul V, (1951) Rubber Chemistry and Technology 24, 344.
[86] Neogi C, Bhattacharya A, Bhowmick A, (1990) Rubber Chemistry and Technology 63, 651.
[87] Beerbower A, Pattison D, Staffin G, (1964) Rubber Chemistry and Technology. 37, 246.
[88] Abhimanyu P, Coolbaugh T, (2005) Rubber Chemistry and Technology 78, 516.
[89] Flory P, Rehner J, (1943) Statistical Mechanics of Cross-Linked Polymer Networks II. Swelling, J. Chem. Phys. 11, 521.
[90] Hanley J (2008) Swelling Effects in Dynamic Equi-Biaxial Testing of EPDM Elastomers by the Bubble Inflation Method, PhD thesis, Dublin Institute of Technology (DIT).
[91] ASTM Standard D471-95, (1995) Standard Test Method for Rubber Property-Effect of Liquids, Standard 09.01, 86.
[92] Cho K, Wook J, Daeho L, Hyunaee C, Young-Wook C (2000) Polymer 41, 179.
[93] Chen L, Jerrams S (2011) A rheological model of the dynamic behaviour of magnetorheological elastomers, Journal of Applied Physics 110, 013513.

Permissions

The contributors of this book come from diverse backgrounds, making this book a truly international effort. This book will bring forth new frontiers with its revolutionizing research information and detailed analysis of the nascent developments around the world.

We would like to thank Anna Boczkowska, for lending her expertise to make the book truly unique. She has played a crucial role in the development of this book. Without her invaluable contribution this book wouldn't have been possible. She has made vital efforts to compile up to date information on the varied aspects of this subject to make this book a valuable addition to the collection of many professionals and students.

This book was conceptualized with the vision of imparting up-to-date information and advanced data in this field. To ensure the same, a matchless editorial board was set up. Every individual on the board went through rigorous rounds of assessment to prove their worth. After which they invested a large part of their time researching and compiling the most relevant data for our readers. Conferences and sessions were held from time to time between the editorial board and the contributing authors to present the data in the most comprehensible form. The editorial team has worked tirelessly to provide valuable and valid information to help people across the globe.

Every chapter published in this book has been scrutinized by our experts. Their significance has been extensively debated. The topics covered herein carry significant findings which will fuel the growth of the discipline. They may even be implemented as practical applications or may be referred to as a beginning point for another development.

The editorial board has been involved in producing this book since its inception. They have spent rigorous hours researching and exploring the diverse topics which have resulted in the successful publishing of this book. They have passed on their knowledge of decades through this book. To expedite this challenging task, the publisher supported the team at every step. A small team of assistant editors was also appointed to further simplify the editing procedure and attain best results for the readers.

Our editorial team has been hand-picked from every corner of the world. Their multi-ethnicity adds dynamic inputs to the discussions which result in innovative outcomes. These outcomes are then further discussed with the researchers and contributors who give their valuable feedback and opinion regarding the same. The feedback is then collaborated with the researches and they are edited in a comprehensive manner to aid the understanding of the subject.

Apart from the editorial board, the designing team has also invested a significant amount of their time in understanding the subject and creating the most relevant covers. They scrutinized every image to scout for the most suitable representation of the subject and create an appropriate cover for the book.

The publishing team has been involved in this book since its early stages. They were actively engaged in every process, be it collecting the data, connecting with the contributors or procuring relevant information. The team has been an ardent support to the editorial, designing and production team. Their endless efforts to recruit the best for this project, has resulted in the accomplishment of this book. They are a veteran in the field of academics and their pool of knowledge is as vast as their experience in printing. Their expertise and guidance has proved useful at every step. Their uncompromising quality standards have made this book an exceptional effort. Their encouragement from time to time has been an inspiration for everyone.

The publisher and the editorial board hope that this book will prove to be a valuable piece of knowledge for researchers, students, practitioners and scholars across the globe.

List of Contributors

Elena Manaila and Gabriela Craciun
National Research and Development Institute for Laser, Plasma and Radiation Physics, Magurele, Bucharest, Romania

Maria Daniela Stelescu
National Research and Development Institute for Textiles and Leather - Leather and Footwear Research Institute, Bucharest, Romania
National Research and Development Institute for Laser, Plasma and Radiation Physics, Magurele, Bucharest, Romania

Jaume Garcia-Amorós and Dolores Velasco
Grup de Materials Orgànics, Institut de Nanociència i Nanotecnologia (IN2UB), Departament de Química Orgànica, Universitat de Barcelona, Barcelona, Spain

Kokou D. (Honorat) Dorkenoo, Hervé Bulou, Grégory Taupier, and Alex Boeglin
IPCMS, UMR 7504, 23 rue du Loess, BP 43, F-67034 Strasbourg Cedex 2, France

Emel Sungur
Bilkent University, Department of Physics, Advanced Research Laboratories, 06800, Ankara, Turkey

Anna Marzec and Marian Zaborski
Institute of Polymer and Dye Technology, Technical University of Lodz, Poland

Maurizio Galimberti
Politecnico di Milano, Dipartimento di Chimica, Materiali e Ingegneria Chimica G. Natta, Milano, Italia

Anna Boczkowska and Stefan Awietjan
Warsaw University of Technology, Faculty of Materials Science and Engineering, Warsaw, Poland

Magdalena Maciejewska and Alicja Krzywania-Kaliszewska
Technical University of Lodz, Institute of Polymer and Dye Technology, Poland

Shi Cheng
Ericsson AB, Stockholm, Sweden

Shuang (Jake) Yang
Department of Pathology, Johns Hopkins University, Baltimore, USA
Department of Mechanical Engineering, University of Maryland, USA
Department of Chemistry and Biochemistry, University of Maryland, USA

Kunqiang Jiang
Department of Mechanical Engineering, University of Maryland, USA
Department of Chemistry and Biochemistry, University of Maryland, USA

Koji Matsuura
Research Core for Interdisciplinary Sciences, Okayama University, Okayama, Japan

Keiji Naruse
Cardiovascular Physiology, Graduate School of Medicine, Dentistry and Pharmaceutical Sciences, Okayama University, Okayama, Japan

Bala Muhammad
Bayero University Kano, Nigeria

P. Castro-Garay
Depto de Fisica, Universidad de Sonora, Apdo Post. 1626, Hermosillo, Son. Mexico 83 000, México

Jesus Manzanares-Martinez
Depto.de Investigacion en Fisica, Universidad de Sonora, Hermosillo, Son., Mexico, Apdo P. 5-088, México 83 190

Nándor Békési
Department of Machine and Product Design, Budapest University of Technology and Economics, Budapest, Hungary

Wei Hu and Norman M. Wereley
Department of Aerospace Engineering, University of Maryland, College Park, MD, USA

Steve Jerrams and Niall Murphy
Centre for Elastomer Research (CER), Dublin Institute of Technology, Ireland

John Hanley
FDT Consulting Engineers and Project Managers, Dublin, Ireland